CASE STUDIES ON HUMAN RIGHTS AND FUNDAMENTAL FREEDOMS

CASE STUDIES ON HUMAN RIGHTS AND FUNDAMENTAL FREEDOMS

A World Survey

VOLUME ONE

WILLEM A. VEENHOVEN

editor-in-chief

WINIFRED CRUM EWING

assistant to the editor-in-chief

CLEMENS AMELUNXEN   JAN PRINS
KURT GLASER   NIC RHOODIE
STEFAN POSSONY   JIRO SUZUKI
L.P. VIDYARTHI

associate editors

PUBLISHED FOR THE FOUNDATION FOR THE STUDY OF
PLURAL SOCIETIES

BY

MARTINUS NIJHOFF / THE HAGUE
1975

ISBN 90 247 1779 5
90 247 1780 9 *(vol. I)*

PRINTED IN THE NETHERLANDS

Table of Contents

Colin Tatz

Paul V. Hyer

Introduction

ATTEMPTS by the world community of nations to find an operational international norm for identifying and measuring violations of fundamental human rights are of fairly recent vintage. The First World War played an important role in focusing world attention on the "rights" of the so-called "national minorities" in Europe, culminating in the rights of self-determination to which theorists as far apart as Wilson and Lenin gave attention. The Second World War, and the founding of the United Nations in particular, led to an unprecedented pre-occupation with virtually the full spectrum of human rights. The formulation of the human rights provisions of the United Nations Charter (1945) and of the Universal Declaration on Human Rights (1948) brought clear proof that a considerable measure of common ground existed within the world community of nations for recognizing at least certain human rights. However reluctantly in some individual cases, the nations, by adhering to the Charter and participating in the human rights activities of the United Nations, accepted the premise that every individual has the same minimum fundamental rights, irrespective of the cultural and political system to which he belongs. It was in the context of these rights that the concept of "discrimination" assumed its current international meaning.

As a result of the more sophisticated nations' acuter sensitivity for the human rights cause, and particularly the U.N.'s growing involvement in alleged violations of these rights, interest in so-called "discriminatory" situations increased universally. Special attention was given to a considerable number of new discriminatory situations to which decolonization in the Third World gave birth. Because the boundaries of the former colonial territories were mostly kept unchanged, new minority problems arose in practically all the newly independent African and Asian states with heterogeneous racial and/or cultural composition. A few examples that immediately come to mind concern the Indians in Burma, the Biharis in East Pakistan (now Bangladesh), the Chinese in Malaysia and Indonesia, the Ibos in Nigeria, the Blacks in the Southern Sudan, the Eritereans in Ethiopia, the Hutu in Burundi, and so forth. In countries of West European culture and mainly or wholly Caucasian governing strata such as Canada, the U.S.A., Brazil, South Africa, Australia, New Zealand and the member states of the European Economic Community (E.E.C.), the position of subordinate "Black" communities has shifted increasingly into the spotlight. A similar concern is expressed for the fate of certain ethnic and religious groups in the two major Communist blocks, such as the Volga Germans, Crimean Tartars, Jewish and Baltic peoples in Russia; the Tibetans and Mongolians in Communist China; and religious believers in both blocks.

Modern mass media, with their sophisticated technology, have seen to it that human rights issues in general, and discriminatory practices in particular, are continually brought within the ambit of public debate. In this manner an extensive international detection apparatus has come into being whereby news or rumours of discriminatory situations almost anywhere in the world are investigated and the facts made known to the world community.

Few developments relating to our knowledge and understanding of discrimination in the context of fundamental human rights have contributed more to our insight in this matter than the proliferation of social science studies in the field of intergroup relations over the past two decades. It befell a new generation of sociologists, social psychologists, cultural and social anthropologists, and political scientists to expose the confusion, ignorance and distorted perspective in this field. They have, in particular, revealed the mind-boggling complexity of intergroup problems in plural societies. While social scientists, like the alchemists of yore, are sometimes bemused by their own incantations and have a human weakness for finding complex words for simple things, there is no doubt that their contribution to our understanding of the ways human groups and their members interact has, on balance, been positive. Case studies in many countries, especially comparative analyses based on cross-national research, have brought into view the universality of discrimination and the many disguises and forms in which it manifests itself in social, economic and political life.

It is in this perspective that the present compendium of case studies should be viewed. The Editors therefore hope that it will provide the reader with relevant knowledge and some guidance in developing his own understanding of the complex problems of intergroup relations in various parts of the world – problems that may affect him personally since they are the Dragon's Teeth of wars and turmoil. The case studies will show the many guises and shapes in which violations of basic human rights reveal themselves. They will also show that these violations emanate from differential human needs within societies in which the resources that sustain power, privilege, wealth and prestige are both finite and scarce. They will certainly underscore that no society is immune from problems centering upon the protection and extension of human rights. The compendium will in particular help to support a viewpoint on which definite concensus exists among objective observers namely that no *one* society or country can be singled out to serve as the international model for assessing the scope, intensity and censurability of discrimination.

Today social scientists widely agree that no racially and/or culturally plural society can escape at least a modicum of discrimination in those spheres of man's existence which together constitute his social and cultural life. Discrimination relating to fundamental human rights is the result of group differentiation based mainly on sex; culture (including religion, language, life-style and value system); race (in the sense of identifiable biological descent, especially as registered socially by somatic traits, chiefly skin colour); and nationality (which normally involves racial and/or cultural differentials). The *differentiation* leads to *discrimination* because the groups or categories involved are *ranked*. The case studies in the present compendium deal mainly with major socially defined stratification or ranking systems involving four main clusters of socially defined groups known generally as (a) *sex groups*, (b) *classes*, (c) *castes*, and (d) *minorities*. The term "minority", as used here, is operational, not statistical: it refers to inferior position in the power structure and the distribution of social benefits. Minorities can be subdivided into two subcategories, namely *racial* and *cultural* minorities. Racial minorities relate to groups socially defined mainly in terms of biological descent and involving visible physical characteristics. Cultural minorities relate to groups socially defined mainly in terms of ethnicity, language, religion, historicity, life-style and/or national origin. The case studies selected for the compendium will also support a generalization that has become a commonplace in social science. This is that as long as members of a Society have different talents, dispositions, needs, group affiliations and backgrounds, and occupy different positions in the means of production and distribution, some form of ranking will be an inevitable structural characteristic of that society. In each society, various groups compete for a limited and finite supply of "scarce goods", which includes not only material goods but such "intangibles" as power and prestige. Each society consequently forms a pool of resources, but because of differences in temperament, ambition, aptitude, ideal, values, group identification, power, wealth and occupational and educational attainment, people will inevitably find themselves unequally positioned in competing for scarce goods. This inequality crystallizes into a system of ranking; that is, the stratification (hierarchical structuring) of people in different group contexts, such as groups differentiated on the basis of sex, race or language. In real life the situational realities of ranking with regard to power status, rewards and opportunities ensure that social, cultural and political "rights" and "liberties" within a national society can never be "equal" in an absolute sense. As long as a society is differentiated in groups on the basis of any of the characteristics mentioned, some people will always enjoy their "equal" rights more fully than others.

No society has ever attempted complete equality in the allocation of scarce goods: to do so would be to negate any social structure and to make impossible the system of rewards and punishments that is the motor of social behaviour. But it is equally important not to let inequality reach the point where anyone is denied his or her fundamental rights. Enlightened social policy is therefore concerned with overcoming *gross* discrimination, and many of our case studies will not only recite facts of discrimination but will also describe measures or proposals to assure fundamental rights.

The greatest attention is currently focussed on *racial* discrimination, and this kind is usually held up as the most abhorrent and reprehensible. One reason for this is that, except for sex (the fixity of which is now less than absolute), race is the category affording the least mobility. A person can change his life style or his religion: he cannot shed the colour of his skin. Yet the contention that discrimination relating to basic human rights causes more suffering when motivated by *racial* reasons cannot be substantiated by an in-depth comparative analysis of the world's current intergroup problems. Such an analysis will provide ample proof that group differentiation conceived in terms of religion, language, life-style and other cultural criteria to-day rates far higher as a cause of extreme human misery than group conflict defined in purely racial terms (that is, in terms of differences in biological history). In fact, the number of people who since the close of World War II have lost their lives in so-called racial conflicts are only a small fraction of the mortality caused by physical strife involving disparate ethnic, linguistic, religious and national groups. And if racial conflicts are limited to those between Whites and non-Whites, this fraction would become miniscule.

A detailed inventory of current and recent human disasters involving high mortality will show that racial confrontation as such (that is, where the adversaries define their conflict of interests primarily in terms of divergent biological histories) does not figure nearly as prominently as the man in the street might have been led to believe. And when this type of confrontation is further scaled down to Black-White relations, it will be seen that conflict in this sphere pales into insignificance when compared with the slaughter, hunger and poverty caused by conflicts centering around ideological and political disagreements and national, linguistic, religious and other disparities. Furthermore, racialism and colour prejudice can also be made less repulsive by redefinition as *class* differentiation – a tactic followed by several Latin American countries trying to obviate accusations of racial discrimination. It is therefore imperative to draw a clear distinction between the main structural characteristics of social ranking and the criteria in terms of which the ranking occurs, on the one hand, and the *results* of the ranking process in the form of prejudice and discrimination, on the other. In the case of our four basic stratification systems (involving sex, class, caste and cultural and racial minority groups) the results – in the form of restricted life-chances and socio-economic and political deprivation – can be precisely the same. Whether an individual's depressed status can be ascribed to class, caste, sex, cultural or racial (somatic) criteria, the result for the individual can be exactly the same. Different criteria may result in precisely the same disadvantageous position in society's reward and opportunity structure. Consequently, the above ranking systems do not each develop separate and exclusive sets of discriminatory practices. In fact, most concrete acts and policies of discrimination may flourish in all four ranking systems.

In subscribing to the Charter of the United Nations, all Member States pledged themselves to take joint and separate action for the achievement of universal respect for, and observance of, human rights and fundamental freedoms for all without distinction as to race, sex, language or religion. In other words, the Charter identified four major areas in which rights and freedoms could be denied or violated. The Human Rights Commission of the UN has been in existence since 1946, and in 1948 the Central Assembly adopted a Universal Declaration on Human Rights, which embodied a wide range of freedoms and basic human rights.

Subsequently, however, a vast gap developed between theory and practice, with the UN deciding to concentrate on one of the four major areas, viz. race, and, furthermore, mainly to restrict its concern to the Southern portion of Africa.

At its twenty-sixth session (1971) the United Nations General Assembly invited the Economic and Social Council to request the Commission on Human Rights to submit suggestions concerning a "Decade for vigorous and continued mobilization against racism and racial discrimination in all its forms". The Commission accordingly drafted a programme of action to be followed during the envisaged decade. In November 1973, the UN General Assembly formally launched the Decade for Action to Combat Racism and Racial Discrimination and decided to inaugurate the activities proposed in the programme of action drawn up by the Commission on Human Rights, on 10 December 1973, the twenty-fifth anniversary of the Universal Declaration of Human Rights.

Was the UN's concern for the respect of human rights and basic freedoms being implemented on the broad front envisaged by the words of its Charter? The answer to this question must unfortunately be in the negative.

On the first of March, 1973, United States delegate Philip E. Hoffman told the Human Rights Commission in Geneva that racial discrimination was to be found in many and varied forms, some blatant, others more subtle. He emphasized that racial discrimination

... exists everywhere in one form or another – and this applies to North and South America, Europe (East and West), all of Africa and Asia. No continent, country or people is free of some form of racial discrimination. Violations of human rights can be found to-day in discrimination not only of Whites against Blacks but Blacks against Whites, Whites against Whites and Blacks against Blacks.

Mr. Hoffman expressed concern that the UN decade for action to Combat Racism and Racial Discrimination would fall short of its objectives if countries maintain that discrimination exists "only somewhere else" and make Southern Africa the sole preoccupation of the decade.

On March 6, 1975, Ambassador John Scali, United States Representative to the UN, addressed the Twelfth North American Invitational Model United Nations, Washington, D.C., on the subject of human rights. He did so, he said, because in no other area was the contrast between what nations say and what they do so stark and so vivid on no other issue was the gap between the ideally desirable and the practically attainable so frustatingly wide.

During its thirty-year history, Scali claimed, the United Nations perhaps had made less progress in the area of human rights than in any other major field of activity. Why had progress been so slow? The Ambassador answered this question as follows:

Not only have the United Nations members remained divided over how far the organization is permitted to go in promoting human rights, but there is also no consensus on the very definition of the term – human rights. Thus, from the earliest days of the United Nations representatives of East and West found themselves using the same words – freedom, justice, rights and dignity – but sharply disagreeing on their meaning.

Ambassador Scali pointed out that while the Western tradition of human rights centered on the individual, it was sometimes forgotten that the vast majority of the world's peoples had not been raised in this Western tradition, that they did not automatically share Western values and that they, therefore, viewed human rights issues from a different perspective. Moreover, even Western society had produced social thinkers, such as Karl Marx, who had taken a completely different approach to human rights. Marxism almost by definition adopted the collective view of human rights. Many Third World leaders felt that the extension of civil liberties had to take second place to economic development. Many nations of the Third World also tended to share the Soviet block's very restrictive interpretation of the United Nations human rights mandate, at least as it applied to them.

Ambassador Scali summed up the position as follows: "Because the United Nations remains divided about what it is trying to achieve in the field of human rights, its pronouncements on this subject are often so vague and generalized as to be devoid of meaning. Because all nations are reluctant to embarrass their friends, the United Nations tends to confine its critical spotlight to states which for one reason or another find themselves relatively friendless in the international community". He warned that politically motivated, partisan criticism, in the guise of objective international study, only weakened the moral authority of the

United Nations and thus diminished its ability to protect human rights everywhere.

Just some weeks prior to Mr. Scali's address in Washington, the UN Commission on Human Rights was holding its 31st (1975) session in Geneva and illustrating the same point. The representative of Amnesty International referred to "mass violations of human rights which were continuing in Brazil and Indonesia" and said that his organization "hoped that the Commission would henceforth deal unambiguously with all flagrant and systematic violations of human rights, regardless of the ideologies involved to justify them."

The representative of the World Conference on Religion and Peace declared that it was no secret that many observers for non-governmental organizations were deeply disillusioned at the continued lack of progress in elaborating instruments against religious intolerance, whereas a historic declaration and a useful convention had been adopted against racial intolerance. Religious intolerance and discrimination gave rise to patterns of gross violation of human rights involving many States, continents and, indeed, religions.

He referred to countries such as the Philippines ("The Moslems in the Philippines continued to be subjected to political and religious discrimination, despite the demise of colonialism"), Pakistan ("In 1974 some dozens of Ahmadiyyas had been killed in riots and thousands had become refugees. The National Assembly of Pakistan had, in September 1974, amended the Constitution of the Islamic Republic in such a way as to reduce the Ahmadiyyas to the status of a non-Moslem minority in that theocratic Islamic State"), Syria (where 4 000 Jews – the last representatives of a community of 40 000 – were discriminated against in many ways), Cyprus (desecration of churches and ill-treatment of priests), Egypt (where members of the Coptic Church were exposed to systematic persecution: That large minority representing 10 to 20 percent of the population (between 3.3 and 6 million persons) was refused equality of opportunity with the rest of the population in respect of higher education and employment in the Government service and private business"), and Czechoslovakia (where priests continue to need authorization to perform their pastoral duties and one-fifth have been refused such permission). He also pointed out that a form of religious intolerance was prevalent in the Third World, directed against Christianity as a foreign and western religion "or even a religion of the White race".

The representative came under bitter attack for his remarks. Subsequently the Commission passed a resolution (in the form of a draft re-commended for adoption by the Economic and Social Council) clearly aimed at discouraging similar happenings. The resolution claimed that "some non-governmental organizations have occasionally failed to ob-serve the requirements of confidentiality" and that "the oral interven-tions of some non-governmental organizations on matters affecting member States have often shown disregard for proper discretion", and then laid down that any such organization "failing to show proper dis-cretion in oral or written statements may render itself subject to sus-pension of its Consultative status". To the history of inter-group rela-tions, such disposal of unwelcome complaints calls to mind the proce-dure of the Minorities Commission of the old League of Nations. The members of that Commission were in large part delegates of the states *against* which minorities raised complaints, and behaved accordingly.

The compendium will consist of three volumes of which the first two will be published simultaneously. The third will be published early in 1976. All articles are original contributions that were specially solicited for this source book. The Foundation for the Study of Plural Societies would like to express a special word of thanks to the respective authors who contributed towards this compendium. We also wish to record our gratitude towards the Editors without whose enthusiasm and unstinting efforts these volumes could not have materialized. In conclusion, the ideas, assumptions and viewpoints expressed by individual authors are not necessarily those of the Foundation.

WILLEM A. VEENHOVEN
editor-in-chief

Chairman, Foundation for the Study of Plural Societies

The Hague
15 May 1975

From Gulag to Guitk: Political Prisons in the USSR Today

STEFAN T. POSSONY

STEFAN T. POSSONY was born in Vienna, Austria, and completed his Ph. D. at the University of Vienna in 1935. An outspoken opponent of National Socialism, Dr. Possony left Austria for France, from where he proceeded to the United States and joined the Institute for Advanced Study in Princeton, New Jersey. During World War II he served as a psychological warfare officer in the U.S. Office of Naval Intelligence. From 1946 to 1961 he was a Professor of International Politics at Georgetown University in Washington, D.C., and in 1961 he joined the Hoover Institution on War, Revolution and Peace at Stanford University, where he is now Senior Fellow. He has served on the editorial boards of eleven publications and is the Strategic Affairs Editor of the American Security Council. His books include *A Century of Conflict; Lenin, the Compulsive Revolutionary; Aggression and Self-Defense; The Geography of Intellect* (with Nathaniel Weyl); and *International Relations* (with Robert Strausz-Hupé). His titles written in German include *Strategie des Friedens* (Strategy of Peace) and *Zur Bewältigung der Kriegsschuldfrage* (Mastering the War-Guilt Question). His most recent book, *Waking Up the Giant*, a comprehensive analysis of the problems faced by current American foreign policy, was published by Arlington House in 1974.

From Gulag to Guitk: Political Prisons in the USSR Today

STEFAN T. POSSONY

We need not look for hell in the netherworld. We can prove the existence of hell far more conclusively on the surface of the earth – a hell for the still-living, a hell organized and maintained since 1918 by the Kremlin in the USSR. *Terribilis est locus iste.*

Alexander Solzhenitsyn's *Gulag Archipelago* shocked the world in 1974 as though the Western public, which has not yet forgotten the nazi concentration camps, never heard about the concentration camps run by the communists. Yet a bibliography of 1958 which was far from complete listed more than 150 titles which were devoted to the subject, including books published as early as 1930.[1]

Subsequently, the subject was investigated repeatedly, e.g. by the AFL in 1949, by the International Confederation of Free Trade Unions in 1951, the German Social Democratic Party during the same period, and the Kampfgruppe gegen Unmenschlichkeit, as well as the International Commission against Concentration Camp Practices during the early and mid-1950's. This Commission, in 1959, defined a concentration camp regime as consisting of three features:

> Arbitrary deprivation of freedom
> Forced labour for the benefit of the state
> Inhumane conditions of detention

The facts about the USSR were reported by former prisoners of soviet camps, of numerous nationalities; by former officials of the camp administration and the USSR's political police; by those opposed to the regime and exiles; and by communist victims. They were the object of diplomatic representations and of petitions to the U.N.

I am heavily indebted to Rosemarie Kantner and Regina Possony without whose help this analysis could not have been written.

In 1950, two leading French communists asserted that the stories about concentration camps in the USSR were mendacious, and that those stories really replicate conditions in *nazi* camps and were falsely attributed to the USSR which runs the most humane and advanced penal system in the world.[2] These communists were hauled before French courts for defamation. The trial, in which former communists of international rank testified, like General El Campesino who, during the Spanish Civil War had been the foremost CP hero, brought out the fact that the Belgian communist party had published a French translation of the Russian penal code which contained a decree of 7 August 1932 and an appendix to article 107, which provided for 5 to 10 year sentences in "concentration camps": this was *precisely* the expression which was used.[3] On 12 January 1951, the two communists and the party organ, *Les Lettres Françaises*,[4] were convicted. They had to pay heavy fines and were compelled to publish the judgment which stated, *inter alia*, that the plaintiffs had proved unable to back up their allegation, namely that there were no camps in the USSR. The communists appealed but on 6 July 1953 lost again before the Eleventh Chamber of the Paris Cour d'Appel. Meanwhile, Stalin had died; and in 1956 Khrushchev was to make his famous speech which, for a while, headed off further communist endeavors to deny the reality of the Gulag Archipelago.

Since that time, information from and about the USSR has been flowing more easily and the tribulations of the political opposition to the Kremlin are being reported frequently in the press. Thus, by the end of October 1974, it was widely known that Valentin Moroz, Ukrainian historian, had been on hunger strike for more than 100 days and that he was confined in the harsh prison of Vladimir. Russian writers were warning the Free World that torment and hunger had been stepped up once again and that the Gulag prisoners are faced with destruction.

To be technically accurate, Gulag exists no more. The word "lager" was tabooed because of its association with the nazis. "Lager" was transformed into "colony" and Gulag became Guitk.[5] The term has the advantage of being unpronounceable (as well as unspeakable). But the phenomenon described by Gulag and Guitk continues to exist. As evidenced by the issuance of pertinent legal texts in 1972, the interest of the communist authorities in the peculiar institution is becoming stronger once again.

What do these legal texts tell us?

I. Penal Policy in the USSR

The USSR has an elaborate system of punishments revolving mainly around the concept of "deprivation of freedom" *(lisheniye svobody)*. The various forms of punishment are defined in article 21 of the criminal code which was redone in 1962 and remains in force in the 1962 version.[6] Elaborate, multiple and interlocking institutions *(uchrezhdeniya)* have been created for the purpose of punishment and deterrence; as well as to instill an honest attitude toward labour, exact compliance with the laws, and observe the rules of socialist community life. Article 20 claims that "punishment is not aimed at the infliction of physical suffering, nor at the degradation of human dignity". Implementation of article 20 is utterly ineffective.

How long has this system been functioning? David Dallin, who was a foremost authority on the subject thought as late as 1949 that "the soviet government resorted to the system of labour camps about 1930". Actually, the camps were started on 5 September 1918 when the Cheka was given unlimited powers to imprison certain categories of persons and suspects even if there was no evidence against them. On 15 April 1919, a decree establishing camps for forced labour and a Central Administration of Camps was signed by Mikhail Kalinin. On 17 May 1919, *Izvestiya* disclosed details on the camp construction programme. By that time the NKVD's department *(otdel)* of forced labour had become operative. Some 25 years later, Roosevelt and Churchill apparently were uninformed about the institution, for in 1944 they consented to "the use of German labour" in the USSR.[8] This plan was shelved but the USSR carried it out by itself on a more limited scale. Did Moscow have a secret agreement with Washington and London?

A. Social Punishments

The concept of "corrective labour" implies that the person should be not only punished but also be transformed into a useful member of the soviet society. The idea is clear enough, but ideas on implementation have remained vague. Except for rudimentary provisions concerning "education" within the institutions, there are no statutes which provide for measures which could lead to human improvements. On the contrary, the whole body of relevant legislation can only lead to the deterioration of the "human material".[9]

Some punishments call for "corrective labour without deprivation of freedom."[10] Certain of these penalties also are used as additions to punishments through which freedom is restricted or abolished. Under those provisions, a person is corrected without being sent to jail or camp. Instead, he is subjected to economic hardship. Only a portion of his earned income will be paid out to him, yet he must not leave his assigned place of employment and he must continue with his normal duties; alternatively he may have to discharge less satisfactory job assignments, sometimes with hard labour. During the punishment period no social security benefits *(stazh)* accrue and he earns no vacation. The penalty is registered in the labour book.

Social punishments which are regarded as additions apparently are rarely meted out singly. They include fines and firings, compensatory payments for damage the individual is supposed to have caused, and confiscations of property. However, since reprimands are in the nature of basic punishment, they are often handed out by themselves, and seem to have a deleterious impact on careers.

Banishment from the place of the person's home takes the form of *vysylka* or *ssylka*. In case of *vysylka* the punished individual is – according to the legal provisions – entitled to choose the rayon in which he wants to live and within that rayon to pick his job; except that certain towns and jobs are out of bounds. In case of *ssylka*, the individual, for a period of two to five years, is assigned both residence and work. He can take along his family to the *vysylka* as well as to the *ssylka*.[11]

Persons subjected to economic penalties and banished from their homes usually are not accused of "counter-revolutionary" acts or serious crimes. However, since they are tied to particular jobs at particular places, none of which are of their own choosing, they are in the legal position of temporary serfs. To the extent that industrial workers, and especially *kolkhoz* and *sovkhoz* peasants, are legally and materially prevented from changing jobs, those "free" and "unpunished" persons are subjected to rules of serfdom. The laws which forbid the *kolkhoznik* to leave the farm without permission no longer seem to be enforced but they can be invoked against select defendants at the discretion of the "competent organs".

B. Corrective Labour Linked to Deprivation of Freedom

The so-called settlement colonies *(kolonii poseleniya)*[12] are a mixture between freedom (soviet style) and non-freedom. The individual is not
constantly watched by guards, instead he is under surveillance. He is
not in jail, nor in a camp barrack, but he lives within a camp territory.
His movements are not upon command, instead he is allowed to move
freely within the camp during daytime, and he can obtain occasional
authorization to go outside the area. Furthermore, he lives in private
quarters and can have his family with him. Usually, this particular
penalty is imposed upon persons who are liberated from labour camps
as an additional punishment and as a method of keeping control. The
settlement colonies or areas cluster around closed camps and are indispensable to sustain the logistics and the economic life of the camps.

Soviet legislation considers prison *(tyuremnoye zaklyucheniye)* as the
most severe form of deprivation of freedom. Prison terms are given to
felons, political criminals, repeaters or "recidivists", and to prisoners
who behaved unlawfully in camp. The prisons are divided into those
with general (obshchi) and strict or severe (strogi) regimes; in the latter
the prisoners are isolated in cells *(kartser)*. Since soviet prisons usually
are over-crowded, isolation often is carried out in punishment cells
(strafny isolator).[13] Otherwise the regimes are differentiated in terms of
"privileges" or lack of those, e.g. under the general regime the prisoner
is allowed a daily walk of one hour, while under the strict regime he is
allowed only half-an-hour daily.[14] Similar discriminations are applied
to food, money, correspondence, packages, etc.

Soviet practice has been to keep the majority of prisoners in corrective labour colonies where they are expected to correct themselves
through labour *(ispravitelny trud)*. A previous rule[15] was to serve all
sentences *above* three years in camps. To what extent this rule still holds
is not known, but apparently more emphasis is now placed on prisons
and on corrective labour indoors in workshops.[16]

II. The Camp Regimes

The colonies are divided into four main groups: the general, hard,
strict, and special regime.[17]

A first offender is assigned to the general regime, and a first offender

felon to the hard regime. Recidivists and *all* convicted politicals are assigned to the strict *(strogi)* regime, and dangerous criminals and political recidivists, plus those whose death penalty has been commuted, are
subjected to the special *(osoby)* regime.[18] The rule of the old penal code,
that delicts against the foundation of the soviet regime and counter-
revolutionary activity are the worst crimes that can be committed in
the USSR still stands. The old practice of giving top penalties to about
60% of the politicals[19] may also still be in force.

The differences between the various types of camps are not systematically spelled out;[20] only a few points are mentioned in the legal
sources. For example, with respect to visiting rights, the prisoner in the
general regime can receive three short and two long visits, in the hard
regime two short and two long visits, in the strict regime two short and
one long visit, in the special regime one short and one long visit; and in
prison he is only entitled to two or one short visit.[21] Cases have been
known when all visits were denied. (A short visit is defined as a four-
hour meeting in the presence of guards, whereas a long visit is supposed
to last three days.)

Similar gradations presumably extend to most factors which are
important in the life of prisoners, such as money (earnings, remittances,
and the right to spend), packages, and clothing, as well as living quarters and the choice of the camp itself.[22] The statutes make it clear that
the special *(osoby)* regime is, as a rule, one of hard labour. This stipu-
lation, however, does not seem to exclude hard labour from the *strogi* or
even the "lighter" regimes.[23]

There are several types of special purpose prisons *(tyurma osobogo
naznacheniya)*, notably prisons for preliminary detention (i.e. interroga-
tion), transit and distribution prisons. Those do not *per se* serve the
purposes of punishment and correction but vitally affect the fate of the
prisoners. The investigative prisons may house the prisoner for a long
time. They are the place where he is most likely to be confronted by
torture. If he is sentenced to death, that's where he dies. After sentencing he may stay in the investigative prison *(sledstvenny isolator)* for
two to four months, or longer.[24]

As to the transit and distribution prisons, the prisoners who are being
moved about frequently spend much time there. From these prisons the
individual may be sent to a better or worse camp. In those camps
where mortality seems to be the highest he may become infected and
die. Health hazards, often aggravated by lack of water, food, sleep,
sanitation, etc., threaten the prisoner during transfer which is by rail

and ship, and may last weeks or even months. The dangers of transit and transfer are proportionate to the number of prisoners transported.[25]

Women prisoners are assigned to special camps and supposedly are completely segregated from male prisoners. In practice, this separation does not seem to be effective. Prostitution and related abuses appear to be widespread in the camps.[26] In general, women are assigned to the first regime. However, they can be subjected to the stricter regimes, including the special regime. There are special provisions for women with children. Mothers who are arrested and who have children below two years of age, take the children along into the "institutions".[27] Mothers are allowed to keep their children to the age of two, at which point the children are sent away. The mothers are forced to work, and since baby-sitting arrangements are deficient, the mortality of those children is high. However, quite a few children have been born in camps and were brought up by the MVD.

So-called educational labour colonies *(vospitatelno-trudovye kolonii)* exist for convicted or administratively confined juveniles between the ages of 12 and 18. The legislation on children dates from an *ukaz* of 7 April, 1935,[28] and was preceded in the 1920's by genocidal operations against the *bezprisornyi*, i.e. free roaming bands of orphaned children. During the purges most of the children of purge victims were placed into "educational" camps. The fact that such camps still are being operated suggests that there remains a high rate of imprisonment for adults and that, furthermore, there is substantial criminality. The imprisoned youngsters are feared because of their especially vicious behaviour. Many of them grow up and form the criminal elements of the Archipelago *(blatnye)*. The stipulations about the education which juvenile prisoners are to be given are extremely rudimentary.

Military offenders are assigned to punishment battalions which in time of war are mainly used for particularly dangerous missions frequently resulting in the death of the entire unit.

According to article 58 of the 1962 code, individuals can be subjected to compulsory medical measures.[29] Persons selected for medical treatment may be assigned to a general or special psychiatric hospital. The latter type of establishment specializes in the handling of "socially dangerous persons" requiring special security.[30]

Until 1937, the communists were content with sentences up to 10 years. After 2 October 1937, 10 to 25 year sentences were authorized, and during Stalin's later years, 25-year sentences to corrective labour camps were entirely commonplace. Usually they were handed out administratively: no need to bother the courts. At present, according to paragraph 24 of the 1962 criminal code, the court decides the punishment. It would seem that the maximum sentence is 15 years.

Under Stalin, the politicals usually got maximal punishment, while the criminals got off with lower comparative penalties. Whether, and to what extent, this dichotomy still prevails, is not known. Furthermore the matter of "maximal penalty" is not classified by the codes. A person can be punished for each of several crimes, and transgressions in prison or camp may be added to the original sentence. The legislative language is uninformative about consecutive terms but it would seem that the law can be bent easily to keep prisoners locked up indefinitely. Andrei Amalrik got a three-year sentence but before he completed it, he got still another three year sentence – in Anno Domini of 1973. The other practice exists that a basic delict is coupled with attempts and preparations for other delicts, and that conspiracy also may be used as a factor of aggravation.[31] In any any event, once a person is released from camp, he frequently is assigned to a settlement colony or to a *ssylka-vysylka* area and remains subjected to administrative surveillance by the police.[32] Such surveillance lasts for a minimum of one year and involves regular reporting to the militia, travel and residence restrictions, curfews, etc. Violations entail administrative or criminal penalties and can return the person to the camp as a "recidivist".

A. The Death Penalty and Camp Mortality

Bolshevik propaganda was very critical of the death penalty as it was practiced by the tsarist regime and promised great reforms in penal law, such as the abolition of the "supreme punishment". Under the tsarist regime,[33] between 1876 and 1904, there were 17 executions per annum. During the revolutionary period of 1905–1908, 2,200 persons or 540 yearly were executed The communists were indeed very progressive in that between June 1918 and October 1919 they attained an annual rate of 12,000 executions. The estimated number of those executed during 1937–1938, the peak period of the purges, ranges between 500,000 and 1.7 million plus 480,000 criminals *(blatnye)*.[34] This writer knows of no estimates concerning the number of executions under Brezhnev.[35] It is clear that he has not matched Stalin's record by any means, but it is most unlikely that the number of executions is below the rate under the tsar. The USSR does not publish statistics on its current use of the death sentence.

The death penalty was abolished in 1917, restored in 1918, enshrined in article 21 of the 1919 criminal code for serious crimes threatening soviet power,[36] abolished and restored again in 1920, and once again abolished in 1927, except for crimes under article 58, dealing with political delicts, and except also for military crimes and "banditry" (meaning rebellion). In 1932, in the course of the collectivization of agriculture, the so-called "wrecking" crimes were added. The new delicts were formulated in such a way that "theft" of an apple from the tree was potentially punishable by death. The death penalty was in force during 1937 and was made more broadly applicable in 1943. In 1947, after the military victory, the death penalty was replaced by 25 years of detention in special camps, but in 1950 the death penalty was reintroduced.[37] Applicability was extended in 1954, i.e. *after* Stalin's death. Despite repeated "abolitions" the death penalty was always applied. In 1954, 1961,[38] and 1962 further delicts were added which could be punished by death unless committed by juveniles below the age of 18 or by pregnant women. The death penalty is now stipulated in article 23 of the RSFSR code.[39]

The new capital crimes falling under those revisions of the statute include rape, bribery, murder, theft of state property, counterfeiting, violations of currency laws, threatening the lives of policemen and of *druzhinniki* (communist activists), and terrorism at places of imprisonment (i.e. in the camps). The last provision has an obvious bearing on the question of "maximum penalty" discussed above.[40]

It may be mentioned that, according to information obtained from defectors, Colonel Oleg Penkovsky was executed in a crematorium where he was burned alive. Yuri Galanskov died after he was denied the proper diet to control his ulcers and after the camp doctors had operated on him.

Formally pronounced death penalties are usually executed in prisons, including the investigative prisons. In previous times, executions, including mass executions, were often performed in camps. This practice may have become rare or may have been stopped altogether.[41] However, camp personnel are authorized to use their weapons in case of *soprotivleniye* (resistance) by inmates.[42]

The mortality of the soviet penal system is not chiefly derived from executions. Most of the fatalities occur in the corrective labour camps. Some of those deaths have been due to the general conditions of camp

life. It is believed that after Stalin's demise orders were given to keep
mortality below a certain undisclosed rate.[43]

In the absence of deliberate liquidations, camp mortality is significantly affected by over-crowding. To the extent that, presumably, over-
crowding is avoided at present, mortality probably went down. But,
according to Solzhenitsyn, "the main killer was camp labour." This is
still true today, and it is necessarily true whenever the camp authorities
and guards do refrain from suppressive actions of violence.

Solzhenitsyn dismissed the notion that there were good camps as
"simply a delusion". However, he added that "the newcomer was not
mistaken in thinking that there were whole camps which were death
camps".[44] This being the situation, and considering also that politicals
are assigned to the strict regime anyway, there really is no need to pro-
nounce death sentences and risk legal and political complications. If
the Kremlin wants a person to perish, he can easily be made to die a
"natural" death in camp.

B. Extermination Camps

It is easy to indicate the mechanism of extermination. First and foremost, assignment to a camp in an arctic climate is by itself enough to
reduce the chances of survival *drastically*. In Lenin's time, the prisoners
the bolsheviks wanted to eliminate inobtrusively were sent to the Solovietsky Islands which are located in the large bay west of Archangel
known as the White Sea. This area is about two degrees south of the
Arctic Circle. The climate of Solovietsky is a great deal better than that
of the islands north of the Arctic Circle and in the Arctic Ocean, e.g.
Novaya Zemlya, as well as of the mainland areas defined by the January isotherm of −30C which runs roughly from the Yenisei estuary
through Bodaibo to 200 miles west from Komsomolsk and from there
to a point southeast of Wrangel Island. About 500,000 square miles of
Yakutia lie between the −40C and −50C January isotherms.

In 1958, *Kampfgruppe gegen Unmenschlichkeit* published a list of camps
which was based upon information supplied by German military and
civilian prisoners who had been repatriated from the Archipelago.[45]
This list identified the 125 most important "forced labour camps" as of
1948 and added 50 smaller local camps. Yakovlev listed 165 camp establishments. The two lists show overlaps and some differences.

It is important to recognize that many of the well-known camps were in reality *clusters* of camps. The Archipelago is hierarchically organized: "colonies" (i.e. individual camps) are grouped into divisions, divisions into regions, and regions into systems.[46] It is rarely known which "rank" a particular "camp" occupies in this structure.

Finn listed ten camps under the identification of "Vorkuta" and six under "Sakhalin". His list really identified 141 camps. He lacked information about the other clusters, but identified altogether 43 *groups* of camps. The true figure hidden behind this list may be anything between 300 and 500 camps. It seems most prudent to talk about 125–165 *camp districts*, as of 1948 and 1953. In 1941, there were only 80 such administrative units or districts.[47] Thus, the camp population grew massively during and after World War II.[48]

Although individual camps were moved, abolished and added, the overall geographic configuration of the Archipelago changed but little, to judge from recent defector information. The geographic distribution has been roughly as follows:

Area	*% of Camp Districts*
Eastern Siberia and Far East	34
Northwest Russia	16
Northeast Russia	9
Western Siberia	7
Urals	7
Central Asia	6
North Central Siberia	6
Central Russia	5
South Russia	5
East Russia	5
	100

Some 54% of those camp districts are located in what could be called acceptable climates; 46% are in zones of harsh climate. Of the total 6% are on islands in the Arctic Ocean, e.g. Franz Joseph Land, Novosibirsky Islands, Wrangel Island, Askold Island, and Vaigach Island in Kara Sea; and 12% in some of the world's coldest areas, including Chukotsk, Kamchatka, the North Siberian littoral,[49] Kolyma, and Yakutia, where the "cold pole" is located near Verkhoyansk. Nearly one-fifth of the camp districts are situated in a zone of killing climate.

It is to be observed that zones with "killing climate" are part of the USSR and require settlers for diverse economic and strategic reasons.

As such, no earth climate is bad enough to preclude settlement. However, if persons sent to such areas are not properly fed, equipped, and protected, they will inevitably die a tortured and early death. The USSR never yet set up its arctic camps in the way they could be organized and equipped to enable the inmates to survive and perform valuable work. Yet if the Kremlin really intended to reform its opponents, it would give them a chance – and it might develop its arctic possessions more effectively.

C. Hard Labour

During the Stalinist period, two-thirds of the prisoners were in camp, and the rest in jail. The current proportion is probably different, but a large portion is still assigned to camps.

According to Solzhenitsyn, about 80 percent of the fatalities in Stalinist camps occurred in connection with "general assignment work". More specifically, Solzhenitsyn asserted that 80 percent of the prisoners are given this type of work, and that all of them die of it.[50] General assignment work can be very hard in terms of physical exertion. It may be imposed for many more hours than the eight hours which are the legal norm, outside and inside the camps. There seems to be a six day week with only two or three holidays per year. This regime has been in existence, together with prolonged overtime as a prominent feature, since 1918.[51]

The physical effort may take place under conditions of extended exposures to cold or wetness. And the work may be dangerous, such as in poorly constructed mines where accidents occur and where dust, poisonous fumes, radiation, and lack of oxygen may affect the miners' health and result in early death.[52] Work detachments usually are limited to the eight-hour day, but there also are night shifts, and not infrequently the sleeping period is held down to 6:30 hours.

Hard labour is defined as work underground or involving heavy loads. Semi-hard labour can take the form of road building, for example. Normal industrial and agricultural labour is classified as "light". Sick people perform light work for four hours daily. Invalids are, theoretically, freed from work obligations. No limits are set on overtime.

Furthermore, every worker and every labour detachment must reach "norms" which during the Stalinist period were set so high that only

10% of the prisoners were able to attain them. In case of non-fulfilment, the workers in Stalinist camps either were forced to work as long as needed to complete their task or they were punished by reductions in their food ration. If groups were given a norm, they had to do the complete work regardless of whether they were at full strength or not. The norms exceeded those of tsarist institutions by a factor of 5 to 10.

The rule is that the prisoners must work to be useful to society and that they should do the type of work for which they are best equipped and most gifted.[53] Little attention is paid to this idea. The real rule is that everybody puts in the heaviest stultifying labour he is capable of doing.

Whether this insane system which may be described as a spiral toward death still exists, is not known. Indications are that only alleviations were effected.

Forced labour is done in transportation (canal building and maintenance, railway, road, and airfield construction, loading and unloading), mining, timber cutting and moving, brick-making, fishing, and oil production (especially the preparation of oil fields). During the Stalin period, 85% of the prisoners were engaged in earth moving and forestry work, and 7% worked in mines.[54] At present, indications are that much work is performed on the Tyumen oilfields, as well as on the Baikal-Amur-Sea (BAM) railway.[55] BAMLAG camps have existed since before 1935, but at present BAM has become a high priority project. Unless elaborate precautions were taken, work on this railway which must be performed in the tundra – in swampy and permafrost areas and in mountains, and which involves the bridging of numerous large rivers – could entail large fatality rates.

According to Articles 39 and 40 of the corrective labour code, the work performed in the camps carries a salary corresponding to the type of job that is being done. From this income costs for food and board are deducted, and so are taxes. The rest, which must be more than 10% of the gross income, is paid into a personal account. (The minimum deposit for invalids is 25%, for minors 45%, and for individuals in settlement camps 50%.)[56] However, if the prisoner was given disciplinary punishment, his net earnings are cut. In brief, there is horrendous exploitation but the exploiter obtains little economic advantage from the arrangement.

The camps were formally established on 5 September 1918 when "red terror" was proclaimed after an attempt on Lenin's life. On 15 April 1919, *Izvestiya* published a decree legalizing forced labour camps *(lageri prinudityelnykh rabot)*. The concept of the corrective labour was first codified on 16 October 1924. The corrective labour code of 1 August 1933 dropped the expression "forced labour", henceforth only the term "corrective labour" was employed. Finally, on 8 May 1934, the old concept of "social defence" was replaced by "punishment" *(nakazaniye)*. These changes in nomenclature are merely a semantic game. The reality was conceived, and has remained, that of forced labour.[57]

The soviets are entirely cynical about "correction" – all they care about is a signed "confession". The Chinese, by contrast, are most serious about co-ordinating punishment and supervision with reformulation of thought, and of coupling work output with political education. This is the point where "brain washing" or "remoulding of thought" comes in as a unique feature of the Maoist penal system.[58] The objective is reached if and when forced labour gradually becomes "identical with voluntary labour so that the criminal may become a new man through re-education".[59] In comparison even the soviet system has an occasional point.

D. Food

Food is still another key factor of survival. Prisoners are supposed to receive rations which "assure normal vital activity of the organism".[60] It is remarkable that soviet authorities never did manage to formulate a more satisfactory norm.

Through the history of the camps caloric intake was notoriously low, with some camps being worse off than others. Experience with soviet camp management in Germany showed that the soviets calculate too high a caloric content of the food they provided. The food is often spoiled. If so, it is never replaced. It is virtually always of poor quality.

Food supplies often break down so that meals are being missed. Quality foods with prophylactic effects are supplied only in minuscule quantities. Unless prisoners are allowed to purchase and receive food packages, their diet cannot assure the normal activity of the organism and results in avitaminoses and undernourishment.[61] The normal fare is bread, dried fish, gruel, grits, soup, a little sugar, and barely *any* meat, fat, fruit, or even vegetables. The food rations which tsarist prisoners were receiving were substantially more healthful. According to information obtained by D. L. W. Ashton from inside the USSR a prisoner who is to be punished "can be put on the 'severe' intake of 1300 calories".

Prisoners pay for their food through deductions from their earnings. The packages they receive are limited in numbers and weight; and their purchases are limited to a few rubles. Both packages and purchases may be restricted as punishment. There is some clandestine storing of food and limited black market trading but there also is extensive theft within the camps.

The major abomination is this: prisoners who cannot fulfil the work norm receive lower rations. As they work overtime to reach the norm, they only get the normal ration. As they grow weaker, their rations go down further. Prisoners who are allowed to do minimal work and invalids who don't work at all receive the lowest rations, on the grounds that he who does not work, shall not eat. The logic of this system is that those who are weak and sick should not recover but die.

Food supply appears to have been improved somewhat but the basic system still operates. A 1973 report reflecting considerable research stated:

In the strict and special regimes the worst factor is the constant hunger: prisoners are left on a starvation diet while being required to do hard physical labour ... As hunger is ... used as a deliberate instrument to destroy the physical and psychological morale of the prisoner, the diet may be considered to be a form of torture.[62]

E. Additional Destructive Factors

It is believed that torture, in the technical sense of the word, is not regularly used in the camps, but physical coercion cannot be totally lacking. For example, the use of strait-jackets is permitted except on women and juveniles.[63] Punishments in dark and small isolators, deprivation of sleep, and excessive overtime also have deleterious effects. Self-mutilations and hunger strikes are by no means rare. (Nor are suicides.) In addition, the prisoners are allowed only 2 square meters of space and thus are subjected to crowding.[64] They are not allowed to sleep with the lights being turned off.

The barracks in which the prisoners are housed are defective and provide only inadequate protection. Heating is minimal, at best. Washing and laundry facilities are rudimentary. Toilet facilities are scandalous – a situation which is often utilized for torments of a particularly degrading type. Together with inadequate clothing, these conditions result in dirt, unclean bodies, vermin, and infection.

In this unsanitary situation, disease is bound to be rampant. The most prominent diseases are frost bites, scabies, tuberculosis, respiratory diseases including those induced by underground mining, furunculosis, C-avitaminoses like scorbut and pellagra, scurvy, dysentry, typhus, and syphilis; depressions are frequent. Vaccinations are not administered regularly, persons with contagious diseases are not promptly segregated, hospitals are often lacking or ill-equipped, hospital and dispensary beds are too few, and special ambulance and evacuation services do not exist. Emergency surgery can rarely be performed competently. The inadequacy of laboratory facilities, medical supplies, and nursing is striking for a modern state. Above all, the lack of doctors is appalling. There simply is no evidence that the medical situation in the camps is improving – all indications show the contrary. However, the prisoners have the legal right to obtain hygienic and medical care, and adequate food.[65]

In the special regime as practiced *at present* in the special camp No. 10 of the Mordovian complex, people can be locked up for years in windowless cells, without proper heating and plumbing, on starvation rations, and under conditions of overcrowding. As reported by Mr Ashton, the Ukrainian prisoner Karavansky described "confinement in the special camps as a complete transformation of a human being into an animal, the destruction of an individual".

A person who is being worked extremely hard in an extremely cold climate and under extremely unsanitary conditions, with a minimal intake of food and minimal medical help, has only very low survival chances. If in spite of all those hazards a prisoner manages to survive, yet his demise is still desired, he can be softened up by more punishment and by ensuring that he "accidentally" freezes to death. He might even be finished off through terrorism by the criminal element within the camp, which is a frequent happening and usually goes unpunished. However, there is some consolation. If after release from the camp, the prisoner has incurred health damage and his work ability is impaired, he is entitled to restitution for the damage and he also has earned pension rights.[66]

III. Informal Types of "Correction"

A person falls into the clutches of the police because of suspicion, denunciation, entrapment, or accident; or because he is directly or indirectly linked to a suspect or is related to one; or belongs to a category of people who at a given time are the target of persecution; or he runs afoul of a "mass arrest" operation. In the course of the arrest procedure and the subsequent interrogations, many things may happen which offset his situation for better or worse. He may be subjected to physical pressures which include deprivation of sleep, subjection to bright lights or utter darkness, starvation, drugging, and actual torture, and to psychological pressure, including threats to close relatives. Agreement with the *kompetentnye organy* to serve as an informer may help him out of his predicament, at least temporarily. Since these *organy* are interested in such recruitment, they may arrest people for the sole purpose of keeping the police work going.

The "suspect" may or may not be deprived of the services of a lawyer, and the lawyer may help him or add to his misfortune.[67] In the end, his fate is decided by the courts which are pretty much under the influence of the KGB and the Ministry of the Interior. The MVD is in charge of the camps[68] and also regulates their influx, with the help of the procurators (who, of course, belong to the Ministry of Justice). The Justice Ministry is supposedly in charge of "overseeing" conditions in the "colonies", but this arrangement has little practical effect. The choice before which court a case is to be tried appears to be flexible. The practice of adding "public accusers" to the prosecution – supposedly experts about the defendant's alleged deeds – also merits attention.[69]

Still another piece of legal fiction should be noted, namely that the camps are under the administrative responsibility of local soviets. Can one really assume that the soviet of the Koryak national oblast is able to administer the camps within "its" territory??

Within the camp, there are MVD representatives who acting alone or with a board, may determine whether a prisoner has not yet been properly corrected and therefore needs a more severe treatment or conversely, they might give him a few breaks. According to article 16 of the corrective labour code of April 1930, signed by A. I. Rykov, the prisoners, after proving themselves in an initial or basic regime may be transferred into an alleviated and finally into a privileged regime. Most of the relevant decisions are made on the basis of reports by stool-pigeons, which are often malicious and mendacious, and sometimes are submitted by request. Prisoners who are capable of working effectively may be able to get along with the guards, but those who are too weak or too sick to fulfil their work norm get into trouble. Nevertheless, personal relations, connections, and bribery, as well as politics and antipathies play a role.

The personnel assigned to labour camps rarely includes gifted types capable of first class careers. Virtually all of the guards are transferred to camp duties because they are under a cloud or can't be used effectively elsewhere. Some of the jailers are sour at the entire world and regard the prisoners as their enemies. Others realizing that they soon may find themselves in the position of a prisoner behave more leniently. Personnel has been known to support the prisoners in their petitions, provided no criticism was voiced against the camp administration as such. The camp personnel also includes former inmates who were promoted to guard and office positions because they were willing to collaborate and proved to be useful to the MVD. Naturally, the entire personnel is under steady observation. Anyone who falls down on the job of hounding the prisoners may be demoted or punished. Many guards become tormentors in the service of their own self-interest and for the sake of self-preservation.

IV. The Hidden Power Structure

The guards are quite unable to run the camps by themselves. As a result *de facto* power structures have come into existence through which the prisoners are being administered. The bureaucratic work is largely done by prisoner-clerks who thereby acquire power and ability to affect the future of their fellow-inmates. In a different mode, female prisoners also are able to gain substantial influence.

The major hidden power structure arises in connection with the mixing of politicals and criminals, whenever such mixing occurs. The regime was, and is essentially allied with the criminal element. This is evidenced by the fact that the sentences of *political* prisoners are rarely commuted and that the amnesties which were declared largely benefitted the *criminals*.

The criminals have been traditionally divided into two groups, the *suki* (bitches) and *blatnye* (thieves, but the word is used to describe criminals of all types). They are groups in the sense that the criminal inmate is associated with either one or the other faction, that both *suki* and *blatnye* have their own command structures within the camps, and that both follow specific albeit unwritten rules.

The authorities used the *suki* repeatedly to destroy the *blatnye*. They were employing them as informers, non-office camp personnel, enforcers, and executioners. Some of the *suki* joined the MVD or militia on a regular basis. The *suki* can be described as experienced criminals without scruples who know exactly what they want and do not care how they get it.

The *blatnye* also are criminals with their own code but they are more of the wild type. Few *blatnye* live very long. They have been quite willing to kill one another, but as pointed out, whenever they became unruly the authorities relied on the *suki* to get rid of them. As soon as they have a chance, the *blatnye* terrorize the politicals and steal from them food and clothing. They pose the particular hazard of transit camps. The guards usually avoid fighting with the *blatnye*, instead they like to trade in goods "expropriated" from the political prisoners. This trade is used to ease the living conditions of the criminals as well as of the camp personnel. The result is not only an added mental oppression of the political prisoners but, more importantly, a partial or complete nullification of remittances and food packages as prerequisites of survival.

On principle, the *blatnye* have been hostile to the so-called *frayera*. This term refers to non-criminal prisoners but in the postwar camp world it often was substituted for "fascists", i.e. former collaborators with the nazis, and Balts and Ukrainians. It was also applied to those whose death penalty had been commuted. The *frayera* are frightened people who want to work their passage home but are prevented from doing so by the *blatnye*.[70]

In the past the *suki* and *blatnye* often fought for control, and the criminal who got into a camp controlled by the other fraternity rarely survived. The prisoners are supposedly under a regime of self-administration. The self-administrators usually are drawn from the ranks of the criminals, yet the *suki* tend to dominate both the *blatnye* and the politicals. Such fights affect the chances of the politicals adversely. In general, the torment to which the politicals are subjected comes in large part from the criminals. To what extent the *suki-blatnye* situation has been cleared up is not known but friction is unavoidable whenever large numbers of criminals and politicals are thrown together.

Stronger ethnic consciousness has been modifying the constellation inasmuch as, for example, Ukrainians repeatedly managed to combine their forces and act under a common command. In such cases, the criminals who as a group dislike to take risks were beaten back. As a result of better organization among the politicals, the networks of the criminals were disrupted during the 1950's.[71] It is unlikely that this success was permanent; but it is probable that whenever the politicals organize effectively, they can protect themselves against the criminals more effectively than during Stalin's time.

In the psychiatric hospital-prisons, which are treated in a later section, the situation is, by contrast, out of control.

V. Victims of the Concentration-Camp Regime

No group or person accessible to the KGB-MVD is protected against "deprivation of freedom". The following main groups may be distinguished:

1. *Relatives* of "political prisoners" whom the Stalin regime wanted to punish severely may be sent to camps, e.g. wives, children, and parents. Travellers abroad are often requested to leave their families behind as hostages who in case of defection may be sent to camp. To what extent this practice of *Sippenhaft* continues is not known. During the purges many wives and children of the victims went to the lagers, and not all of them returned.[72]

2. *Social Groups.* Under Stalin, persons who stole food,[73] or acquired it through black market deals, or who left a kolkhoz, or another place of work, or three times reported late to work,[74] or persons committing "crimes" of similar severity, were accused of theft of socialist property, of wrecking and of economic sabotage. Those types were sent to the camps in very large numbers. This practice seems to have been discontinued, but legislation is available to reactivate it at any time. Whenever in the past the communist regime became infatuated with a particular slogan, which it hoped would allow it to win an economic "battle", or whenever there was revulsion against a particular type of crime, such as rape, there were waves of arrest.[75] If in future the soviet government were to crack down on traffic accidents, a wave of automobile drivers might be expected to appear in the camps. Group arrests even from the streets, have occurred, but the liquidations of the *kulaks*,[76] the capitalists, and the aristocrats were the main phenomenon

under this heading. At the time of writing, there is no evidence of genocidal measures of this type.

3. *Ethnic Groups.* At one time or another, ethnic groups of the "soviet people" were the targets of persecutions and of more or less extensive genocide, through incarceration in labour camps, "losses" in transport, shootings, and famine. In alphabetical order: Buryat (Mongols), Chechen, Chinese, Cossacks, Crimean Tatars, Czechs, Estonians, Finns, Germans, Greeks, Hungarians, Ingush, Yakuts, Japanese, Jews, Kalmyks, Kazakhs, Koreans, Kozaks, Latvians, Lithuanians, Moldavians (Rumanians), Moslems, Poles, Tatars, Ukrainians, and Uzbeks, etc. This list comprises only those groups who were settled within the USSR, e.g. the Volga Germans and the Greeks of the Black Sea region. Some of those groups were virtually exterminated such as the Chinese and the Koreans who had settled in the Soviet Far East.[77] In addition, Austrian militant socialists of the Schutzbund who after their revolution of 1934 had sought refuge in the USSR; Greek communists who had fought an unsuccessful guerrilla war in the 1940's; the Latvian riflemen who saved Lenin's regime back in 1918; and Polish communists who had fled the Pilsudski regime were decimated. German and Rumanian Jews who had fled from Hitler landed in the camp district of Karaganda and were given camp treatment; only few of them survived. Communist and socialist veterans of the Spanish civil war as well as Spanish pilots who had been trained in the USSR, Spanish seamen, and Spanish refugee children and their guides ended in the camps:[78] of 6000 Spanish refugees who thought they had found a safe haven only 1200 survived.[79]

A substantial number of German communists and party leaders also lost their lives. The number of foreign refugees who passed through the camps and suffered substantial losses was never determined.[80]

4. *Russians.* The camp system served to decimate all political parties, including the CPSU itself which, during Stalin's repeated purges, suffered enormous losses. Officers who escaped death during the purges went to the camps in large numbers. Leading NKVD officers who were in charge of the purges were purged in turn; those who escaped the bullet in the neck ended in the camps. Soviet soldiers who had become prisoners of war, and civilians who had been impressed for forced labour by the nazis, after the war returned to their country, spent a painful time in investigation lagers, and ended up in labour camps. Those who did not wish to return to the communist *rodina* and were coerced to return, through the shame of the United States and Britain, were either

executed or condemned to 25-year terms in special camps. Former Russian citizens who had escaped abroad at various times after the revolution and who had settled in countries overrun by the Red Army, were returned and finished off in the camps. After the war, the USSR invited emigrants back and most of those who were naive enough to go "home", ended up in the camp purgatory. Those particular persecutions extended to *all* ethnic groups within the USSR. Some of them, notably the Ukrainians and the Turks, were hit worse than the Great Russians. But the Russian people itself suffered enormous losses.

5. *Other Groups.* The principle that entire groups as collectives must be punished and liquidated entered soviet jurisprudence in 1919 when the registration of former landowners, capitalists, and officials, as well as the imprisonment of categories of persons were decreed.[81] Furthermore, adherents of the main religions, such as Catholics, Baptists, Buddhists, Jews, Russian Orthodox, and also Tolstoyans suffered repression at various times. Priests and nuns were persecuted harshly. Historians, archeologists, geneticists, plant-breeders, economists, statisticians, experts in linguistics, philosophers, and for a while even physicists and aeronautical engineers were marched off to the camps, although some of the scientists were put to work in factories and behind barbed wire. Persecutions were directed, unbelievably, at esperantists, at prostitutes, and at the mentally sick while dissenters were alleged to be schizophrenics. Writers and actors were among the favourite targets. It should be stressed that many of those people were deprived of freedom not because they were convicted of any particular criminal act but because they belonged to a collective which sometimes for serious, but more often for trivial reasons ran afoul of communist despotism. Furthermore there were those who rightly or wrongly were accused of having engaged in acts hostile to the regime, such as espionage, agitation, propaganda, the spreading of news which they received via foreign broadcasts, the unauthorized disclosure of secrets, etc., in brief, the victims of article 58. This article is at present replicated in articles 64–88, and delicts against the foundations of the Soviet state – and all other workers' states[82] – still are regarded as the most heinous crimes possible (articles 64–73).[83]

6. *Foreigners.* During the war the camps grew in number and size. The so-called "enemies of the people" were not released, although many went to punishment battalions. It also appears that many prisoners of war, including most of those arrested after the war, were put into corrective labour camps. Civilians from countries occupied by

soviet troops were subjected to mass and group arrests and deported to lagers in the USSR – a procedure which violated international law. Abductions also took place. All this shows why the existence of soviet concentration and forced labour camps cannot possibly be regarded as a purely internal soviet matter. In East Germany, approximately one-half of the prisoners had been arrested without legal justification – they were just "interned". The soviets continued to operate eleven nazi concentration camps until 1950, including the infamous Buchenwald lager, and they did *not* improve the mortality figures. Indeed, mortality in select East German camps was particularly high:

Mortality in Soviet Camps in East Germany [84]
(1945–1946)

Bautzen	26%
Weesow	28%
Torgau	30%
Landsberg a.d. W.	50%
Buchenwald	72%

The literature contains only infrequent references to the fact that the communists kept Buchenwald going for five years.

After the camps were "abolished" by the soviets, the East Germans continued to operate a few for their own purposes. Some 30,000 to 40,000 German politicals were deported to the USSR, including all those sentenced to 25 years hard labour. The current practice still seems to be that the satellites, though probably not all of them, deport their long-sentence politicals to camps in the USSR. There is little doubt that "counter-revolutionaries" who were involved in the Berlin and Poznan risings of 1953, the Hungarian revolution of 1956, and the Czech "revisionist" regime of 1968 were dispatched into soviet lagers.

During Solzhenitsyn's time, about half of the inmates in the camps were "political". The implication is that the other half were criminals. The truth, however, seems to be that, as pointed out by Thomas P. Whitney, the meritorious translator of *Gulag Archipelago*, "the enormous mass of the prisoners" were made up of *bytoviki* who were neither political nor criminal.[85] Those hapless persons were imprisoned because they belonged to a particular category of people, such as kapitalist, kulak, and kin, or simply because they were arrested falsely or accidentally, and they were never released.

VI. The Economics of Communist Camp-Colonialism

The "profitability" of the camps must be discussed under three headings:

1. Is the purpose of corrective labour achieved? No data are available on criminals and criminality rates in the USSR. It is most unlikely that the camps rehabilitate many criminals, however, Free World penal systems also fail to rehabilitate. As to the politicals, most don't need a "correction" in the first place. The probability that sojourns in the camps predispose anyone favourably toward the soviet regime is surely minimal. If the camps do not fashion the better "soviet man", do the terror and the fear which emanate from them lessen the chance of the overthrow of the soviet system? In the last analysis, the fate of the soviet system depends upon its performance in circumstances the despots cannot anticipate and control; and upon the support of the population. In terms of internal security, therefore, the camps are largely irrelevant and they probably are counter-productive.

2. Has forced labour enriched the USSR? Several large canals were built by camp inmates, notably the White Sea Canal, and probably a few additional big constructions, including railways and refineries. At present, there is remarkably little propaganda on the subject, and some of those constructions are attributed to the *Komsomol*. Current plans apparently envisage the construction of oilfields in Western Siberia and of the BAM railway. Granting past "accomplishments", the man-years which were available in the camps during 45 years may be estimated at about 200 million. This is the equivalent of the energy represented by the current entire soviet labour force during 2 years. Accordingly the labour force should have created something like 700 billion dollars of value. Nothing of the sort has happened.[86] The chances are the camps delayed the development of technologies and techniques through which the arctic and sub-arctic regions of the USSR could have been opened up for productive utilization.[87]

3. Does the camp administration break even by "leasing" its labour force at or above market wages? For example, in one transaction during Stalin's time the sustenance of one prisoner per month was calculated at 140 rubles and the prisoner was rented out at 360 rubles.[88] Other figures show that the actual cost of a prisoner in camp was 366 rubles a month,[89] if the real costs of the entire Gulag establishment were properly taken into account. It is also to be observed that only 36–40% of

the entire camp force were ever actually working, and that not all of the fit prisoners can always be rented out. In a camp, where during 1939–1941 some activities such as furniture making produced a profit, the actual sustenance costs per prisoner run to 8.60 rubles, and if transports were involved to 13 rubles per day.[90] Thus, it is unlikely that the camps are profitable. However, David Dallin believed that the MVD made huge profits which enabled it to finance its "immense functions" and its military force.[91] Perhaps it depends on the bookkeeping. But if indeed the camp administration were ever to live up to article 56 of the corrective labour camp code and feed the prisoners so as "to provide normal vital activity of the organism", and also give them proper shelter, clothing, sanitation, and hospital care, as the law demands, the camps must operate at a loss.

VII. Psychiatry as a System of Internal Security

Some soviet authorities, including elements of the KGB and MVD, can't fail to realize that unending institutionalized repression of many millions of citizens must have harmful effects on the entire country and its development. They understand that security efforts should be concentrated, not on innocents for whom the interrogators themselves have to produce an utterly fictitious *legenda*, but on persons who are real and effective opponents of the regime. They also must have learned that a repressive system which is lawless, capricious and murderous poses enormous dangers to the executioners themselves and that in the interest of the survival of KGB-MVD personnel "socialist legitimacy" must be observed in most of the cases and for most of the time. Actually, few true believers remain within the political police.[92]

Consequently, the Kremlin has been willing to reduce the magnitude of the Archipelago establishment. Possibly they debated whether the whole system was to be changed or abolished. The decision was clearly to upgrade the use of "psychiatric" repression and otherwise continue the camp system, albeit with some alleviations.

A. The "Legal" Basis of Political Psychiatry

Accordingly, a new set of legal texts dealing with the labour camps was issued in 1972. Through these provisions none of the relevant laws was abrogated. The system, as it was first ratified on 16 October 1924,[93] is, fifty years later, still substantially intact. So the legal evidence shows. The current practice is amply documented by the disclosures and writings of the victims of the system.[94]

When the "mass rehabilitations"[95] began in 1956, only a few hundred politicals were confined in psychiatric prison-hospitals, and most of those appear to have been released. In 1961, the law was changed to the effect that psychiatrists no longer were criminally responsible for placing persons in psychiatric institutions without adequate grounds.[96] On 15 May, 1969, Decree No. 345–209 was issued on "measures for preventing dangerous behaviour (acts) on the part of mentally ill persons". The practice of having undesirables hauled into confinement by psychiatrists was thereby ratified. Under this routine the psychiatrists are told whom they should examine, and they may entrap this person to come to the hospital or fetch him with the assistance of the police.[97] The doctors double as arresting officers and as interrogators. The psychiatrists fabricate a diagnosis requiring confinement, and no court judgment is required for jailing the person indefinitely. Indeed, in line with decree No. 04–14 (32) of 1961,[98] a diagnosis not merely of "incorrect behaviour" such as writing what the party disproves, but of *potentially* incorrect behaviour is quite enough to take a person out of circulation.

Within this framework, the psychiatric prison-hospitals provide special examinations for politicals. Moreover, it seems that these "hospitals" have different departments which parallel the four different camp regimes. In any event, there is a strict regime for supposedly psychiatric patients.[99] After release, the ex-patients must present themselves for periodic examination. Certain diagnoses, whether correct or not, lead to disbarment from key professions (e.g. teaching) and to extensive discrimination in the offering of jobs on psychiatric pretexts.

The experiences of Zhores Medvedev show that this routine is accomplished through wholesale violations of legal provisions and medical regulations, and through extraordinary mendacity by physicians, bureaucrats, and party officials. They also seem to suggest that entrapment is widely practiced and that such abuse often originates from political intrigue.[100] It also may be due to the pressure by local and self-appointed troublemakers.[101]

B. Torture-Psychiatry

"The political prisoners are left in crowded wards with insane inmates, where they are constantly exposed to violent and aggressive patients... beatings and humiliations are frequent; the staff are often recruited from the police and patient population."[102] This exposure to the violently insane, instead of the violent criminals, may be the basic motive why mentally healthy persons who oppose the regime are nowadays often confined in psychiatric wards.

This prolonged torture is truly diabolic and sadistic. In fact, the special psychiatric hospital is one *locus terribilis* in the USSR where torture is probably still practiced. Evidence on the subject has been presented by Alexander Yesenin-Volpin before the U.S. Senate Internal Security Subcommittee on 26th September, 1972; and a report by Vladimir Bukovsky was submitted to the same forum.

Bukovsky reported on the special mental hospital in Leningrad where political prisoners and "insane murderers" were held together, for a total of 1000 persons. The guards must be bribed lest they beat the political prisoners and recommend "medical punishment".[103] With this invention the soviets clearly scored a "first".

Within those institutions, trustees and guards rule supreme; and most of them are common criminals serving prison terms. Physicians are accomplices, torturers, and police agents, and often act as despots satisfying their own sadistic needs.

The medical punishment consists in applying drugs which produce intense pain, stomach cramps, and fever up to 104 degrees F. Another drug induces sleep and dulls the brain. It is given for 10 days. The victims "wake up as human vegetables. Some regained their senses after two months, others did not." The third punishment is the canvas bandage, in which the victim would be swathed. As the wet canvas dries, it shrinks. According to Amnesty International, patients are "strapped to their beds for several days without any provision for sanitation". To top it all, "medical treatment for those in need has... been reported by former inmates... to be quite inadequate".[104]

It would seem that this sort of treatment is designed to "influence the mind of dissenters". In addition, the use of hospitals instead of prisons prevents the defendant from obtaining legal help before the courts; it allows indefinite confinement; and it discredits the prisoners and their ideas. Furthermore, in this fashion open trials are avoided

whenever those are undesirable. Legally valid judgments against dissenters are often hard to obtain because untruth and evil intent of
"propaganda" must be proved. "Article 190/1" stipulates that "dissatisfaction or disagreement themselves cannot be regarded as criminal
since such attitudes come within the sphere of a person's inner convictions and his view of particular social and political measures only."
But precisely the criticism of "individual measures or actions of the
party, the government, or their agencies" is most unacceptable. Hence,
soviet authorities gleefully adopted the advice of psychiatrist Georgy
V. Morozov: "Why bother with political trials when we have psychiatric clinics?"[105]

Morozov is a member of the USSR Academy of Medical Sciences
and director of the Serbsky Institute of Forensic Psychiatry which is
the centre of psychiatric repression.

A further troublesome aspect is that soviet physicians, under the im-
pulsion of A. V. Snezhnevsky, concocted a "Pavlovian theory of schi-
zophrenia" on the strength of which they diagnose this disease in political oppositionists. The scientists attributed their theory to Pavlov
quite illegitimately; at the same time the medical authorities used the
dispute to purge Jewish psychiatrists.[106] This scientific scandal is a
worthy parallel to the Lysenko affair.

Amnesty International concluded that at present "physical torture
as an administrative practice does not appear to occur in the prisons",
while in the camps "the diet may be considered to be a form of torture". It is worth quoting what else they had to say: "Torture does
represent a component of the treatment of political prisoners detained
in prison psychiatric hospitals for indeterminate periods. The psychological and physical treatment they receive in these institutions appears to constitute tortures as an administrative practice."[107] It merely
needs to be added that the courts do not provide protection against
this abuse.

So far eleven special-type psychiatric hospitals have been identified,
including seven which were opened between 1965 and 1972. In addition, 15 general psychiatric hospitals have "special departments for
compulsory treatment under surveillance".[108] This would suggest that
there are at least 20,000 political psychiatric prisoners in the USSR.
Most probably additional "hospitals" are under construction.

VIII. Continuing Violation of Human Rights

During Stalin's time, according to Solzhenitsyn's estimate, there were not, at any one time, "more than twelve million in the camps... and not more than half of them were politicals".[109] This estimate is below the 15–20 million figure estimated by analysts outside the USSR.[110] No one assumes this still is the magnitude of "deprivation of freedom". Current estimates about the camp population vary between 1 and 2 million. The Brussels Committee for the defence of Human Rights in the USSR appears to opt for the one-million estimate.

Die Welt published a map of 209 current lagers, the origin of which was not explained.[111] A U.S. Senate hearing produced an outline map showing about 120 "single camps or small groups of camps" and 19 "major camp complexes, the largest of which may contain hundreds of camps and many hundreds of thousands of prisoners."[112] However, this map which is based on defector information may reflect data which are out of date.

One difficulty with low estimates should be readily apparent: the six million criminals who were in camp during the early 1950's were about 3% of the population. Only very few of those criminals were victims of Stalins' purge mania, most were guilty of crimes. Whether criminality in the USSR has increased or declined, the number of convicted criminals is hardly less than 2%. If only half of those got sentences longer than three years, they should be imprisoned in camps. Hence there must be a criminal camp population of 2.5 million. This riddle cannot be solved here. Suffice it to state that large numbers of camps still are in existence; that, according to the testimony of defectors the conditions therein were not – in 1973 – any better than they had been 10–15 years earlier; and that the build-up of prisons suggests the soviets are anticipating that the number of those who are to be deprived of freedom will be hovering around 5 to 6 million.

There is evidence that as late as 1962 German nazis, Italian fascists, officers of the Vlassov army, a few Spaniards, Dr. Trushnovich, the leader of NTS, and Raoul Wallenberg, the long lost Swedish diplomat, were surviving on Wrangel Island. The camp on this island "was an experimental camp where experiments were conducted on living people". These experiments served to determine "the effects of radiation on the human body," others involved "the effects of prolonged submersion at great depths".[113]

During the current period, there was organized in the Mordovian ASSR the Dubrovlag district for "special political prisoners", of which a map was published showing 12 prisons and 15 camps.[114] This area is climatically quite good, except that some of the area is swampy. Much of the labour is devoted to furniture making.

The evidence shows that camps *still* can be rated in four basic categories: bad, worse, repressive, and extremely cruel. There is no evidence that the logistics problems of camps in remote areas in the Arctic and Sub-Arctic have been solved. There is no evidence that due legal process is ensured, and that suspects and defendants are able to obtain effective legal defence. Nor is there any evidence that penalties have become fair and moderate. Valentin Moroz was sentenced in 1970 to six years in prison, four years in a labour camp, and four years in *ssylka*. Ukrainian history must really frighten the Kremlin.

Whatever the archipelago population might be, by definition it would include the juveniles in the educational labour camps. But it would *not* encompass the number of those in the prisons *(tyurmy)*, including the prisons for investigations and the special purpose prisons *(tyurmy osobovo naznacheniya)*. Nor would it encompass the number of "free" persons under the regimes of *ssylka* and *vysilka*, and in settlement colonies. It does not include the persons who have been assigned far-away-from-home residence by purely administrative action. It does not include the number of those who are punished to receive only fractions of their earnings and who are obligated to participate in public works within their home area. We may estimate the number of the slaves and of the doomed, but we cannot even guess the number of the serfs and the jailed.

Between 1920 and 1953, some 40 to 50 million victims passed through the slave labour system. Yet this institution still fails to accomplish the objective which the communist government strives to achieve, namely to eliminate opposition to its existence and policies. Dimitry Panin, a co-sufferer of Solzenitsyn's, estimated that the communists killed 60 million persons since 1917. Ivan A. Kurganov[115] made a careful statistical analysis and calculated a loss of 66 million during fifty years (1917–1967). Whatever the figure, the Kremlin never ceased killing. Murder generates murder: that is the causal nexus which moves despotism. Communist policies – which are implemented by lawyers, doctors, scientists, writers, soldiers, and administrators, as well as by the USSR's criminals – are necessarily based on the utter and unceasing violation of human rights.

NOTES

1. *Est & Ouest*, 16–30 September 1974, p. 29, in an excellent issue devoted to the Gulag book, lists 10 books published during the early 1920's, including works by Karl Kautsky and Victor Chernov, virtually all written by socialists and socialist organizations.
2. *Loc. cit.*, pp. 33f. for other instances of Communist mendacity on the subject.
3. Théo Bernard et Gérard Rosenthal, *Le Procès de la Déportation sans Jugement devant la Cour d'Appel de Paris*, Paris: Le Pavois, 1954, p. 19.
4. According to Vladimir Maximov, *Les Lettres Françaises* called Gulag a "grandiose enterprise" for which mankind must be grateful to the "first socialist country". (*Die Welt*, 19 October 1974, p. II).
5. *Gosudarstvennoye upravleniye ispravitelno-trudovykh kolonii* (State administration of corrective labour colonies).
6. RSFSR Ministry of Justice, *Ugolovnyi Kodeks RSFSR*, Moscow 1962. The equivalent to article 21 of *the RSFSR code* is article 23 in the *Ukrainian* code. All articles of the RSFSR code dealing with punishments and deprivations of freedom have their equivalents in the codes of the union republics.
7. David J. Dallin, *The Economics of Slave Labor*, Chicago: Regnery, 1949, p. 8.
8. *Ibid.*, pp. 9f.
9. *Kommentarii k osnovam ispravitelno-trudovogo zakonodatelstva soyuza SSR i soyuznikh respublik*, Moscow, Yuridicheskaya Literatura, 1972, articles 42–44.
10. I. L. Averbakh, *Ot prestupleniya k trudu*, Moscow, OGIZ, 1936, wrote: "The work in the camps is hard work. The discipline is the hardest that can be demanded." (Quoted from David Rousset.) Averbakh was a friend of Vyshinsky's and he was soon given an opportunity to taste his own recipe.
11. *Kodeks RSFSR*, articles 25–26.
12. *Kommentarii*, article 20.
13. For some of these data, see *Kommentarii*, article 11 and related commentary.
14. *Kommentarii*, article 21.
15. Article 28 of the 1947 penal code.
16. Since 1956/57 considerable numbers of large prisons have been built. See Paul Barton, *L'Institution concentrationnaire en Russie (1930–1957)*, with preface by David Rousset, Paris: Plon, 1959, p. 384. Article 24 of the 1962 criminal code stipulates that camp prisoners who become guilty of offences may be, by court decision, transferred to jail (for added punishment). Those jail sentences are limited to three years.
17. *Kommentarii*, article 14. The technical Russian terms are: *obshchi usilenny, strogi* and *osoby* regimes. (See also RSFSR Ministry of Justice, *Ispravitelno-trudovoi Kodeks RSFSR*, Moscow 1972, article 61).
18. *Kommentarii*, p. 61.
19. "Rapport de la Commission d'Instruction", *Livre Blanc sur les Camps de Concentration Soviétiques*, Brussels: Commission Internationale contre le Régime concentrationnaire, 1951, p. 185.
20. Chinese camps do not seem to be classified according to the soviet pattern. Instead they are divided into permanent camps, mobile camps, agricultural camps, and forced labor shops within prisons. (Commission Internationale contre le Régime concentrationnaire, *White Book on Forced Labor and Concentration Camps in the People's Republic of China*, vol. I, Paris, 1957, p. 269.
21. *Kommentarii*, article 24.
22. Some of those factors are mentioned in *Kommentarii*, article 19, except that

nothing is said there about the differences between work assignments and between camps. Also consult *Isprav.-trud. Kodeks*, articles 62–66.

23. *Isprav.-trud. Kodeks*, article 37. (It is interesting to note that the 1927 code on the camps carried the slogan, "Proletarians, of all countries, Unite!" The current version of this code features no such admonition.)

24. *Ibid.*, art. 15.

25. Alexander Solzhenitsyn (*The Gulag Archipelago 1918–1956*, New York: Harper and Row, 1974, pp. 539f.) tells of a transit prison, now dissolved, which during the winter of 1944 to 1945 had a death rate of almost one percent per day so that under those conditions "a person might manage to last five months."

26. David Rousset, el al., *Le Procès concentrationnaire pour la vérité sur les camps*, Paris: Editions du Pavois, 1951, p. 182. (Testimony by Margarete Buber-Neumann).

27. *Sobraniye Zakonov i Rasporyazhenii*, issue No. 22, item 248, 20 April, 1930, article 11, quoted from Statute on the Corrective Labor Camps, USSR, published by MIT Center for International Studies, Cambridge, Mass., 7 July, 1955, p. 7. The children over two years of age may be left with persons designated by the parents or assigned to an institution for children (*dyetskii dom*).

28. *Kommentarii*, article 11. See also 13. B. Yakovlev, *Kontsentratsionnye lageri SSSR*, Munich: Institute for the Study of the USSR, 1955, p. 233.

29. They also can be subjected to compulsory measures of an educational character. Article 7 of the 1919 criminal code stipulates that "measures of social defense court correction, as well as medical and medical-pedagogical measures" can be imposed. (See Yakovlev, p. 219).

30. Article 59 of the 1962 criminal code.

31. Socially dangerous activity may also consist of "inaction" (*bezdeistviye*), according to the stipulations cited by Yakovlev, p. 219.

32. *Kommentarii*, article 49.

33. Solzhenitsyn, p. 434.

34. *Ibid.*, p. 438. The progression from 1905–1908 to 1937–1938 was by a factor of 1852 within thirty years.

35. Mao invented the death penalty which is *suspended* for two years. (Article 13 of Regulations of 7 September 1954. See *White Book*, vol. II, p. 49).

36. Article 37 of the same code provided that an attempt to escape from camp was to be punished by increasing the original penalty up to ten times. Furthermore, if an individual was indicted before a revolutionary tribunal for the second time, he could incur the death penalty.

37. *Ukaz* of 12 January 1950 (Yakovlev, p. 239f). It should be observed that the Korean war began in June of the same year.

38. In 1961, the death penalty was extended to "economic crimes". This is extremely harsh legislation which seems to have no parallel outside the communist countries.

39. The equivalent in the Ukrainian code is article 24.

40. Note that according to article 58/1, as promulgated on 6 June 1927, any crime committed against a workers state, even when not committed in the USSR, was to be punished in line with soviet law. (See Yakovlev, p. 224). This law may still exist on a *secret* basis.

41. *Livre Blanc*, p. 75. A *hanging* of a priest in camp was reported during this writing.

42. *Kommentarii*, article 39, and *Isprav.-trud. Kodeks*, article 31.

43. Barton, p. 303.

44. Solzhenitsyn, p. 559.

45. Gerhard Finn, *Die politischen Häftlinge der Sowjetzone 1945–1958*, Berlin, 1958, pp. 56–61.

46. *Livre Blanc*, p. 250.
47. *Livre Blanc*, p. 103. A map of 297 Chinese camps during the early 1950's can be found in the *White Book*, vol. I, p. 270.
48. In 1929, the camp population was still small (*Livre Blanc*, p. 20). The dramatic growth period was between 1936 and 1953.
49. The industrial complex of Norilsk and the port Dudinka are within the −30C isotherm zone; so is the entire course of the Kolyma. The Lena north of Yakutsk till about 150 miles south of Tiksi is within the −40C area.
50. Solzhenitsyn, p. 564.
51. It was codified on 17 May 1919 (Yakovlev, p. 215).
52. See Sopade Informationsdienst, Denkschrift 27, *Der Uranbergbau in der Sowjetzone*, Hannover: SPD, 1950.
53. *Kommentarii*, art. 27.
54. *Livre Blanc*, p. 188.
55. The Baikal-Amur road was also built by forced labour.
56. *Kommentarii*, article 29. The decree of 17 May 1919 stipulated (article 34) that deductions must not exceed 75% of earnings (Yakovlev, p. 216).
57. *Est & Ouest*, loc. cit., p. 17; Yakovlev, pp. 19, 211f, 221.
58. *White Book*, vol. I, p. 270; vol. II, p. 56.
59. Regulation of 7 September 1954, article 25. Quoted from *White Book*, vol. I, p. 280.
60. *Isprav.-trud. Kodeks*, article 56. The phrasing of the 17 May 1919 decree was: "according to the measure of the norm of food for people employed in physical work".
61. *Livre Blanc*, p. 186.
62. Amnesty International, *Report on Torture*, London; Duckworth, 1973, p. 175.
63. *Isprav.-trud. Kodeks*, article 31.
64. *Isprav.-trud. Kodeks*, article 56. Space allocation in prisons and educational camps is 2.5 square meters. The punishment isolators probably are not larger than 3 cubic meters, i.e. $1.45 \times 1.45 \times 1.45$m.
65. *Kommentarii*, art. 36.
66. *Isprav.-trud. Kodeks*, art. 42.
67. According to *Isprav.-trud. Kodeks*, article 27, the sentenced person (*osuzhdenny*) may "apply in writing" for legal help. Since the text talks about the "sentenced person", this would seem to indicate that legal help is a matter for the post-trial period. Interpretation: the convict can get legal help while he is serving his sentence. This interpretation is probably correct, but the text is so vague that this provision can be circumvented. As to participation by defense lawyers in trials, this is regulated by Article 22 of the USSR's procedure code (*Zakonodatelstvo ob ugolovnom sudoproizvodstve* soyuza SSR). The lawyer is allowed to participate in the proceedings *after* the suspect has been fully interrogated and has been indicted. This procedure virtually precludes effective legal defense.
68. *Isprav.-trud. Kodeks*, articles 5, 14, 18, 19, 31, 57, 80, 82 and 89.
69. For example, two members of the writers union were admitted to the trial of Andrei Sinyavsky and Yuli Daniel in 1966. (See Max Hayward (ed.), *On Trial, the Soviet State versus "Abram Tertz" and "Nikolai Arzhak"*, New York: Harper and Row, 1966, pp. 41 and 50). They were hostile to the defendants.
70. This complicated subject is discussed by M. De Santerre, *Sovyetskiye poslevoyennye kontslageri i ikh obitateli*, Munich: Institute for the Study of the USSR, 1960, pp. 68–97.
71. Barton, p. 312ff.
72. The practice of arresting and imposing camp sentences on relatives (who also

were subjected to *ssylka* in distant areas of Siberia and confiscations) was based, *inter alia*, on articles 58/11 and 58/12 dealing with "organization" (or conspiracy) and failure to make a denunciation. The co-responsibility of kin was made most explicit in article 58/16 dealing with the military but under the heading of treason is also applicable to civilians. (See Solzhenitsyn, pp. 6of., 66f., Yakovlev, pp. 255, 228). In the 1962 code, the co-responsibility of relatives is taken care of by article 88/1 dealing with "non-reporting" (*nedoneseniye*) of crimes against the state. Article 88/2 punishes non-conspirational concealment.

73. Minor theft of food was punished by fines. *Ukaz* of 4 June 1947 imposed camp terms of 4–7 years (Yakovlev p. 238).
74. *Ukaz* of 27 June 1940 (Yakovlev, p. 237).
75. Solzhenitsyn, p. 87.
76. Collectivization through the liquidation of the Kulaks was specifically based on articles 60 and 61 of 30 March 1930 and 15 February 1931 which became largely inoperational after 1933 (Yakovlev, p. 229f).
77. Solzhenitsyn, p. 247.
78. *Livre Blanc*, p. 125.
79. *Vérité sur les Camps concentrationnaires*, p. 173.
80. The imprisonment of foreigners is in an incidental manner envisaged in Article 18 of *Isprav.-trud. Kodeks*.
81. Decree of 15 April 1919 and regulation of 23 September 1919 (Yakovleb, p. 211). The decree on registration by categories was published in *Izvestiya*, 26 September 1919.
82. Article 73.
83. The major category is that of "especially dangerous crimes against the state" (osobo opasnye gosudarstvennye pristupleniya). National and racial discrimination, i.e. limitation of rights or granting of privileges on a racial or national basis is punished as "another crime against the state" by 6 months to 3 years of deprivation of freedom or 2 to 5 years of *ssylka*, according to article 74 of the 1962 code.
84. Data based on Sopade Informationsdienst, Denkschrift 28, *Das System des Kommunistischen Terrors in der Sowjetzone*, Hannover: SPD-Vorstand, 1950, pp. 35–52. This analysis was prepared for submission to the International Socialist Conference at Copenhagen in June 1950.
85. Solzhenitsyn, p. 619.
86. Naum Yasny calculated that in 1941 the NKVD obtained 17% of the national investment and produced 1.2% of the national output (*Livre Blanc*, p. 207f). In that year their budget was 2 billion rubles, just about as much as the budget of the entire non-ferrous metals production.
87. The idea that forced labour should participate in the "construction of socialism" was sold to Stalin in 1929 by Naftali Aronovich Frenkel, a former "capitalist". This worthy was sent to the Solovietsky camp during 1927 for illicit trading in gold and currency. Stalin put him in charge of building the 140-mile White Sea Canal which task Frenkel accomplished within 20 months (1931–1933), at the cost of 100,000 fatalities, including children and women. (The death rate was 1% per day.) After his feat, Frenkel was once again imprisoned but was released in 1939 to build railways in Carelia, preparatory to the Soviet attack on Finland. He became a lieutenant-general in the NKVD and died peacefully in Moscow during the 1950's. Stalin's antisemitism did not extend to Frenkel, a man who hurt more people than even Heinrich Himmler. The story of this most successful camp criminal is treated

in Solzhenitsyn's forthcoming second volume (See *Est & Ouest*, loc. cit., p. 17f.)

88. *Vérité sur les Camps concentrationnaires*, p. 148.
89. *Livre Blanc*, p. 91.
90. *Livre Blanc*, p. 108.
91. Dallin, p. 16.
92. The history of the Okhrana which lost its faith in the tsarist system may be repeating itself.
93. RSFSR People's Commissariat of Justice, *Ispravitelno-trudovoi Kodeks RSFSR*, Moscow: Yuridicheskoye Isdatelstvo, 1927, p. 3.
94. Much of this evidence is collected in the *Arkhiv Samisdata*. Its enlarged and revised *Register of Documents* of September 1973 lists 1478 items. The data can be obtained from the Hoover Institution and additionally from three American and four European libraries. The Archive itself is located in the Research Department of Radio Liberty, München, FRG. The New York daily *Novoye Russkoye Slovo*, almost on a daily basis, publishes testimonies, memoirs, accounts and documents on repression and concentration camps in the USSR.
95. This term does not imply that the camps were emptied.
96. Medvedev, p. 149f. Article 126 of the 1962 code punishes illegal deprivation of freedom *(nezakonnoye lisheniye svobody)* with up to six months jail or camp. If force is used and damage caused, the penalty may go up to 3 years.
97. This method is described in detail by the Medvedevs. It is helped by the fact that all soviet doctors are within a hierarchical and political chain of command.
98. The two decrees are quoted by Zhores Medvedev in a report to the Committee on Human Rights, Brussels, reprinted in U.S. Senate, *Abuse of Psychiatry*, pp. 23–27. (See Note 103.)
99. Medvedev, p. 100f. Daniel R. Lunts who heads the special examination section at the Serbsky Institute, where he evidently examines the most important "patients", and develops the proper methods of extracting confessions and imposing medical punishment, is a KGB Colonel (*Abuse of Psychiatry*, p. 240).
100. Medvedev's troubles may have been aggravated or even caused by the Lysenko clique.
101. Alexandra Dubrova reported that her son was arrested by a psychiatrist and a policeman four days before he was to emigrate. The psychiatrist asserted that the youngster was deranged because "he said he does believe in God . . . he intends to go abroad." The attempt aimed at enrolling the person in the army, despite the fact that 500 rubles had been paid for the exit visa and he no longer was a soviet citizen. He was put into a hospital but declared unfit for military service and released by a second psychiatrist. "Kak delayut 'sumasshedshikh' v SSSR", *Novoye Russkoye Slovo*, August 26, 1973, pp. 2 and 5.
102. *Report on Torture*, p. 177.
103. U.S. Senate, 92nd Congress, Internal Security Subcommittee, *Abuse of Psychiatry for Political Repression in the Soviet Union*, Washington, D.C. GPO, 1972, p. 34.
104. *Report on Torture*, p. 177f. This source *(ibid.)* also describes drugs, one of which causes "depression and rigors". For further information on drugs, see U.S. Senate, *Abuse of Psychiatry*, p. 241. According to Zhores Medvedev insulin shocks are also given.
105. Quotes from Zhores A. Medvedev and Roy A. Medvedev, *A Question of Madness*, New York, Knopf, 1971, pp. 218f, 215, and 67. The Medvedevs failed to provide an exact reference for article 190/1 and I was unable to verify the quote which, given Soviet reality, is of a rather unexpected nature.

106. Medvedev, p. 68f.
107. *Report on Torture*, p. 175. This "psychiatric" practice was downgraded after Stalin's death but has been re-emerging since 1965, i.e. it must be associated with the Brezhnev regime.
108. U.S. Senate, *Abuse of Psychiatry*, p. 238f.
109. Solzhenitsyn, p. 595. The second volume of *Gulag* reportedly envisages a figure of 15 million.
110. One well informed prisoner on office duty found that there were 6 million prisoners in camp during September 1938, not counting inmates of jails, and settlement colonies, and deportees. (Dallin, p. 25).
111. *Die Welt*, 2 November 1974, p. 1.
112. U.S. Senate, 93rd Congress, Internal Security Subcommittee, *USSR Labor Camps*, Washington, GPO, 1973, p. 2.
113. U.S. Congress, *USSR Labor Camps*, pp. 31f, 49 and 51.
114. *Ibid.*, p. 43. One of the camps is for women.
115. Kurganov's book, *The family in Russia*, is discussed in *Est & Ouest, loc. cit.*, p. 15.

Discrimination Against American Blacks

J. MILTON YINGER

J. MILTON YINGER is Professor of Sociology and Anthropology at Oberlin College, Oberlin, Ohio. He obtained his B. A. degree at DePauw University, and his Ph. D. at the University of Wisconsin in 1942. From 1941 to 1947 he was on the faculty of Ohio Wesleyan University. In the latter year he joined the faculty of Oberlin, where he has been a full Professor since 1952 and Chairman of the Department of Sociology and Anthropology since 1969. He has held visiting professorships at the Universities of Michigan, Washington, and Hawaii and at Wayne State University in Detroit. Professor Yinger has served on the editorial boards of *Social Problems*, the *American Sociological Review*, and the *Journal of Conflict Resolution*. He received (with G. E. Simpson) the Anisfield-Wolf Award in Race Relations (1958), and served as a Senior Specialist at the East-West Center (1968–69). Among his numerous publications, several books pertinent to the present study may be mentioned: *Racial and Cultural Minorities: An Analysis of Prejudice and Discrimination* (with George E. Simpson), New York, 1953 (4th edition, 1972); *Religion, Society, and the Individual*, New York, 1957 (also translated into Italian, French, and Spanish); and *A Minority Group in American Society*, New York, 1965. Pertinent articles include: "Contraculture and Subculture," *American Sociological Review* 25:625–35 (1960); "On Anomie,", *Journal for the Scientific Study of Religion* 3:158–73 (1964); and "Anomie, Alienation and Political Behavior," in Jeanne Knutson, ed., *Handbook of Political Psychology*, San Francisco, 1973.

Discrimination Against American Blacks

J. MILTON YINGER

I. THE DEMOGRAPHIC PICTURE

When the United States took its first census, in 1790, persons of African descent constituted 19.3 per cent of the 3.2 million population. Although that percentage fell steadily for the next 140 years, due to the heavy immigration from Europe, the number of Blacks increased rapidly, to 3.6 million (15.7%) in 1850, to 8.8 million in 1900 (11.6%), to 11.9 million (9.7%) in 1930. Since 1930, the number of Blacks has continued to increase and the rate of growth has been slightly higher than the rate among the rest of the population. By 1974, there were 24 million persons of African or part-African descent in the United States, constituting 11.4% of the 210 million population. By now, the great majority of Negroes are of mixed ancestry; but almost all regard themselves and are regarded as Black. (The terms Black, Negro, and person of African descent will be used as synonyms, following various current usages. Each term has its ambiguities, and preferences vary widely, but in order to keep confusion to the minimum and to avoid ideological discussion as to their meaning, the terms should be read as interchangeable.)

Until recently, most of America's Black population lived in the South, with a high proportion employed on farms and plantations. In 1900, only ten per cent lived in the East, Middle West, and West, although these regions contained over three-quarters of the total population. The South was the least developed economically; Negroes therefore shared in the general disadvantages of that region, in addition to suffering the burdens of slavery (which was the circumstance of the great majority until the Emancipation Proclamation in 1863)[1] and the continuing deprivations that followed their release from legal bondage.

This situation began to change during the First World War, during which time many Negroes migrated to northern cities to take industrial jobs. There has been extensive migration ever since, supported by continued industrial expansion, restrictions on foreign immigration after 1924, depressed conditions in southern agriculture, and by the persistent hope among Negroes that life would be better in the North and West. A steady shift to southern cities has also occurred. As a result of this migration, only half of the Blacks now live in the South and only one-quarter live in rural areas. A half century ago, the "typical" Negro was a southern farm labourer or a "share cropper" growing cotton on a small, rented farm. Today he is an industrial worker living in a segregated area in a large city. There is, however, a great deal of variation. In particular, a substantial middle class has developed.

This transition has had enormous implications for American Negroes and for the society as a whole. In this brief essay, it is not possible to explore the background factors.[2] But the following questions can be examined: What is the current status of this large sub-nation? Are there signs of discrimination against its members? What are the trends?

II. Negroes in the American Social Structure

To assess the extent of discrimination against Negroes in the United States, attention may be directed primarily to three sets of measures: those dealing with income and occupation, with education, and with politics. These are, of course, highly interdependent. Indeed, they are part of a larger system of which Negro actions and responses are also a part. It is necessary to look at these effects of discrimination, to see whether they enter into the cycle of causes that sustain discrimination, or whether they are breaking into that cycle to reduce discrimination.

A. Income and Occupations Amongst American Blacks

In 1930, the median family income amongst American Blacks was 30 percent of the median White family income. This large differential reflected the facts that a large proportion of Negroes lived in the South, worked in agriculture, had significantly lower levels of training, and faced persistent discrimination. All these facts have changed, some of them drastically, with the result that the income disparity between White and Black families has been reduced, although it still remains large. By 1952, the median Black family income was 57 per cent of the median White family income. It fell back to 51 per cent in 1958, climbed to 61 per cent in 1971, and fell back again to 59 per cent in 1972.[3] There is little doubt about the long run gains; neither is there doubt that the gains have been slow, measured against national aspirations. They are highly vulnerable to changes in the economic situation, hence periods of gain have been followed by periods of loss.

It is important to differentiate among Black families, because some have experienced income gains, while others have remained deep in poverty. Among young (under 35 years of age) Black families in the North where both husband and wife are present in the household, the annual income has, for several years, been more than 90 per cent of similar White families. For most of the families with only a mother present, however, incomes have remained at or below the poverty level. This is the situation of more than one-fourth of all Black families.

B. Employment Patterns

Since income is mainly dependent upon occupation, some discussion of employment patterns amongst American Blacks is essential. Two things need to be said: They are substantially under-represented in the higher status and higher income jobs; but this disadvantage is being reduced. Table 1 shows the trends.

Although one cannot depict the trends precisely from broad occupational categories, these observations may be useful: The proportion of Non-White workers in white collar jobs nearly doubled during the twelve-year period (from 16.1 to 29.8 per cent); the disadvantage relative to White workers was reduced from 30.5% to 20.2% in white collar occupations; near equivalence in blue collar jobs was attained, but Whites continued to occupy many more of the higher paying jobs (craftsmen in particular); there was some reduction in the number of Non-White workers in the relatively poorly paid service occupations.

On a purely statistical basis, if the trends of these 12 years were to continue for another 25 years, America would have attained substantial occupational equality between White and Non-White workers. It would be hazardous to predict that this will occur, however, because the easier gains have come first. It should also be noted that Negroes

TABLE 1: *Employment Patterns for Whites and Non-Whites in the United States, 1960 and 1972* [4] *(in per cent)*

| | 1960 | | 1972 | |
	White	Non-White	White	Non-White
Professional and Technical Workers	12.1	4.8	14.6	9.5
Managers and Administrators	11.7	2.6	10.6	3.7
Sales Workers	7.0	1.5	7.1	2.2
Clerical Workers	15.7	7.3	17.8	14.4
Total, White Collar	*46.6*	*16.1*	*50.0*	*29.8*
Craftsmen	13.8	6.0	13.8	8.7
Operatives and Transport Workers	17.9	20.4	16.0	21.3
Non-Farm Labourers	4.4	13.7	4.6	9.9
Total, Blue Collar	*36.2*	*40.1*	*34.4*	*39.9*
Service Workers	9.9	31.7	11.8	27.2
Farm Workers	7.4	12.1	3.8	3.0

are among the first ones hurt by any slackening of economic activity – a fact made clear by a persistent unemployment rate that is twice as high among Black workers as among White workers. Only if there are substantial changes in other sectors of the society – in residential and educational patterns, in personal attitudes, in political processes – will the favourable occupational trends among Negroes continue with the strength shown in the last several years.

Economic gains for American Negroes during the last twenty-five years have been, to an important degree, dependent on general prosperity and on the active support of the federal government. Although there have been several recessions, through most of this period the economy has been strong, and Black workers have profited from the strength. It is remarkable that their improvements in occupational status has occurred despite the elimination of a substantial number of jobs through automation. Many of these jobs were in agriculture, but many were the kind of semi-skilled jobs that earlier migrants to the cities had found available. Beginning in 1941, when President Franklin Roosevelt issued an executive order requiring non-discriminatory employment practices, through a succession of Civil Rights bills passed by the United States Congress (of which the 1964 Act is most important for employment) and other governmental actions on the local, state, and national levels, Black workers have had increasing support from public officials. This has not meant, of course, the sudden elimination of occupational differentials. Governmental actions are only one set of forces in a complex pattern; administration has often been lax or inconsistent; earlier patterns persist in many industries despite the law. On balance, however, governmental actions have contributed significantly to the improvement of job opportunities for America's Black population.[5]

C. Progress in Overcoming Employment Discrimination

The present occupational situation in the United States, with regard to racial employment patterns, is highly dynamic. There are strong forces at work to reduce discrimination; but there is also strong resistance. One needs only to follow the daily news to see these points documented, as the following illustrations will show:

> Seven big trucking companies, threatened by a major federal lawsuit, agreed today to adopt "hiring goals" under which 33 to 50 per cent of their new workers would be blacks or persons with Spanish surnames.
> The Department of Justice predicted that a consent decree, which was filed in Federal District Court here, would be expanded into an industry-wide agreement within a few weeks. It would be the first industry-wide accord ever negotiated by Government civil rights officials. (*The New York Times*, March 21, 1974, p. 19. Persons of Spanish surname constitute five per cent of population of the United States.)

> Nine major steel companies and the steel workers union accepted today what was hailed as the most sweeping program to end job discrimination in the history of American industry. The program, announced at a joint news conference by the Departments of Justice and Labor and the Equal Employment Opportunity Commission, is designed to give back pay aggregating 30.9 million dollars to victims of job discrimination on the basis of race or sex. ... In addition, the agreement sets up "goals and timetables" to increase the number of minority workers and women in jobs where they have been under represented. (*The New York Times*, April 16, 1974, p. 64).

It should be noted that the National Association for the Advancement of Coloured People opposed this settlement, contending that it was insufficient. It is also important to note that a labour union was part of this agreement. Labour unions have, to a substantial degree, been part of the discriminatory pattern of employment, although some have contributed to the absorbtion of Black workers into industry.[6]

Another quote from *The New York Times*, April 1, 1974, p. 15:

In two important actions that could speed the elimination of job discrimination in the South, the city of Jackson has agreed after long negotiation to grant hiring preferences to blacks. ... At the same time, Jackson has agreed to award up to 1,000 dollars each to black employees, such as janitors, who have never been offered opportunities to work in better-paying jobs. ... The Jackson agreement is thought by Justice Department lawyers, who helped negotiate it, to be the most comprehensive ever reached within an American city, Northern or Southern. ... Within five years or so, it will result in a city work force that is 40 per cent Negro, the same percentage as the city's over-all population.

All three of the agreements just cited involve the question of quotas or, in the terms being used today in the United States, "hiring goals" and "affirmative action." Two principles are in contradiction in such matters. According to one, each person should be hired on his or her individual merits; the other principle declares that strong action is needed to overcome patterns of discrimination and to eliminate the handicap of lack of experience and seniority that many Black workers carry. The outcome of the current policy debate in the United States on this question will be an important factor in the employment situation during the next several years. This has its importance also for education.

One can document fairly precisely the differences between Blacks and Whites in the United States in occupations and income. It is much more difficult to explain the differences. There are some who would argue that the variation can be accounted for by noting differences in training, skill, and motivation. In a free job market, according to this view, individuals are rewarded according to their contribution. Others, pointing to earlier legal and customary barriers imposed on Negroes, reaching back to slavery, explain present differences in status as the result of past discrimination. Even with perfectly equal opportunity today, inequality is bound to be the result of past disprivilege that blocked adequate training, lowered morale, kept Blacks low on ladders of seniority, and prevented entrance into many high status occupations. A third explanation emphasises present discrimination. Inferior opportunities for education, barriers that block off apprenticeship programmes for training in skilled trades, informal hiring practices in businesses and labour unions that restrict opportunities to persons of background and culture similar to those making the decision, application of irrelevant criteria, and the like, are among the forms of discrimination frequently cited.

Past discrimination and present discrimination combine to produce the income and occupational differences between Blacks and Whites in the United States. Insofar as lack of skill and training are involved, they should be seen as effects, not causes, of the differences; although they enter into the cycle of influences that shape attitudes of the dominant group and determine its policies.

One way to measure the level of present discrimination is to note the income differentials between Whites and Negroes even when they have attained the same educational level. (It should be noted that "the same" means, in this context, that they have completed the same number of years of education. Possible qualitative differences are not taken into account; hence the measurement of discrimination is not entirely convincing.) To reduce the influence of past discrimination, only young workers are compared in Table 2.

TABLE 2: *Comparison of Median Earnings of Negro and White Workers (Ages 25 to 34) in the United States, 1970, by Educational Level* [7]

Educational Attainment	Male			Female		
	Negro	White	Ratio of Negro to White	Negro	White	Ratio of Negro to White
8 years or less	$4,743	$6,618	0.72	$2,935	$3,980	0.74
High school graduate	6,789	8,613	0.79	4,592	5,037	0.91
College graduate	8,715	11,212	0.78	6,971	7,206	0.97

It is particularly among males that discrimination is shown in Table 2. A Negro male high school graduate has about the same income as a White male with four fewer years of education. A Negro male college graduate has about the same income as a White male high school graduate. The disadvantage is smaller for Negro females (partly because both White and Black females have lower earnings than the males).

To measure the extent of occupational and income discrimination against Negroes, one must take account of several factors that influence their place in the economy: How recent is their migration from rural areas to cities, their disproportionate entrance into war-related industries (and thus their vulnerability to rapid changes in demand), and the impact of automation are compounded with discrimination. It is difficult to separate the influence of these various factors.[8] Many social scientists would agree with Killingsworth conclusion that

...racial discrimination, as a present source of economic disadvantage is probably less important than is commonly assumed. This conclusion does not deny that discrimination persists; it does not deny that discrimination makes important indirect contributions to Negro disadvantage through segregated housing and segregated education; nor does it deny the necessity for continued educational and legislative efforts to combat discrimination. What the conclusion does imply is that an anti-discrimination campaign by itself, no matter how effective, is not likely to improve the Negro's economic status significantly, at least in the short run.[9]

D. Discrimination in Housing

The economic status of a group is affected, not only by its occupations and income, but also by patterns of expenditure. America's Black population suffers severe disadvantages as consumers. This is partly due to their relatively lower incomes, which means that as purchasers, they operate under several constraints: they are more likely to buy on credit (with heavy interest charges), to buy goods of poorer quality because they cannot afford the initially higher costs of goods of superior quality, and to buy at inopportune times.[10] Discrimination also plays a part, however. This is particularly true in their search for housing, whether for rental or for purchase. Despite laws designed to prevent racial discrimination in housing and despite avowed policies of housing contractors and real estate agents, Negroes are severely limited in their choices and pay more than Whites for equivalent housing. Segregated housing has several concomitants in American cities: the neighbourhoods in which Negroes live are more deteriorated; they have fewer amenities; mortgages are more difficult to obtain; schools, hospitals, and recreational facilities are inferior. There is good evidence that racial segregation is not simply the result of income differences, but is substantially the consequence of discrimination.[11]

Segregation also has indirect effects that increase economic discrimination: it reduces job opportunities and it leads to poorer education. Since residential segregation has been only slightly reduced in the United States during the last 25 years, it stands as one of the critical supports for discrimination against Negroes.[12]

Segregation persists despite a sharp reduction in opposition to interracial neighbourhoods, as measured by national public opinion polls.

The following percentages of Whites have indicated that "it would make no difference if a Negro with the same income and education moved into my block":[13]

1942	35%
1956	51%
1963	64%
1972	84%

At the same time, the great majority of Negroes continue to prefer integrated, racially-mixed neighbourhoods. In a 15 city survey in 1968, only 13 per cent said that they preferred to live in an all-Black or mostly-Black section of the city.[14]

III. THE EDUCATION OF AMERICAN BLACKS

In the modern world, no factor is more important in the cycle of causes that sustain racial discrimination than differences in education. A group without access to education is trapped in low status jobs and reduced in political power. For many decades America has been dedicated, by official policy and by custom, to extensive education for all. Schooling is compulsory until age 16; and an elaborate system of colleges and universities – more than 2,000 – make higher education accessible. Racial discrimination is legally forbidden and receives virtually no open support in attitudes that are publicly expressed.

The facts, however, do not entirely correspond with the ideology and the law. Educational discrimination against American Negroes is indirectly produced by segregated residential patterns and by income differentials. And it is directly supported by official acts and public pressures that set the boundaries of school-attendance areas, influence curricula, and affect the attitudes of some school personnel toward children of different races.

The educational situation in the United States has changed quite extensively in the last twenty years, mainly to the effect that racial discrimination has been reduced. There is strong resistance to these trends, however, and many obstacles to equality in education are produced by discrimination in other sectors of the society.[15] Since the situation is complex, only a few major aspects and the dominant trends can be discussed in the space available.

A. The 1954 Anti-Segregation Ruling

The year 1954 was a turning point in the racial aspect of American education. Until that time, not only was racial segregation in schools widespread, but it was legally supported in the states of the South, where more than half the Negroes lived. Almost universally, segregation meant inequality, "an excuse for discrimination," as Gunnar Myrdal put it, with Black children attending inferior schools, taught by teachers with poorer training, and using inadequate facilities. There had been some reduction in inequality after 1940,[16] and the Supreme Court had begun to challenge segregation, particularly on the university level. It was the 1954 decision of the Supreme Court, however, which decisively changed the terms of national policy by declaring that segregated education was unconstitutional. The Justices said, in part:

Does segregation of children in public schools solely on the basis of race, even though the physical facilities and other "tangible" factors may be equal, deprive the children of the minority group of equal educational opportunities? We believe that it does. ... To separate them from others of similar age and qualifications solely because of their race generates a feeling of inferiority as to their status in the community that may affect their hearts and minds in a way unlikely ever to be undone. ...

We conclude that in the field of public education the doctrine of "separate but equal" has no place. Separate educational facilities are inherently unequal.[17]

In the twenty years since this Supreme Court decision, which forbade any form of school segregation based on official policy or action, extensive changes have occurred in American education. There has been substantial integration of schools in the South; differences in expenditure between Black and White students have been reduced; and the gap between them in the amount of education they receive has been narrowed significantly. But the gains have been slow in coming and leave the country a long way from the goal of equality of education. There has been some increase in segregation in northern cities; and thousands of "private schools" have quickly been created in the South by White parents to avoid racial integration. The United States Office of Education estimates that 700,000 White children – nearly ten per cent of the White children in the South – were attending such schools by 1972.

B. *Measures of Continuing Segregation*

In the evaluation of several of these points, the same difficulty arises as in the discussion of economic issues, namely that of separating the effects of discrimination from the effects of other factors. The long-run trend toward suburbanization, the desire of parents to ensure a good education for their children, economic differences, variation in preparation for school, and other factors are entwined with discrimination – and, indeed, with efforts to reduce discrimination – to produce the contemporary educational situation in the United States.

It is paradoxical that in the South, where school segregation was universal in 1954, there are now many integrated schools, while in the North many school systems have become more segregated. One way of indicating these facts is to record the percentage of Negro pupils attending schools in which more than 50 per cent of the pupils are Black. In the country as a whole, about 14 per cent of the pupils are Black. In 1972, 54 per cent of the Black pupils in the South were attending schools that were more than 50 per cent Black. In the border states, the proportion was 68; in the North it was 72. (*New York Times*, May 13, 1974, p. 24).

Segregation is even more severe in the schools of the largest cities. "More than four-fifths of New York's black pupils attend schools with black enrolments of 50 percent or more. And nearly half attend schools that are virtually all-black." *(loc. cit.)* Half of Chicago's elementary schools have Black enrolments of 90 percent or more. "The number of all-black elementary schools increased to 144 from 128 in 1972." *(loc. cit.)*

These indications of continuing school segregation need to be seen alongside other measures of education in the United States. The proportion of Black children attending school has increased steadily and quite rapidly. Between 1967 and 1972, the enrolment of Black students increased 23 per cent in high schools and 97 per cent in colleges and universities, compared with 8 and 26 per cent increases for White students. Put another way, during this six year period, the proportion of Blacks aged 18 to 24 who were attending colleges and universities increased from 13% to 18%. During the same period, the proportion of Whites decreased from 27% to 26%.[18] It should be noted, however, that this trend toward equalization has been at least temporarily stopped. In 1972, 8.7 percent of the freshmen entering American colleges and universities were Black. In 1973, this percentage fell to 7.8 (*The New York Times*, February 3, 1974, p. 43). Had Blacks been represented in proportion to their place in the population eighteen years of age, they would have constituted about 12.5 percent of the freshmen. This one-year decline, however, should be seen against a longer time perspective. In the last generation, the differences between Blacks and Whites in educational level attained, at least as measured quantitatively, have been sharply reduced. The median number of years of schooling obtained by Whites born between 1895 and 1904 is 8.9. For those born between 1940 and 1944, the median is 12.6. Amongst Blacks, using the same age cohorts, the increase was from a median of 5.1 years to one of 12.2.[19]

C. Problems of Educational Policy

Both public and private efforts to remove the effects of discrimination from American education have raised difficult questions regarding priorities. These are illustrated particularly clearly in a current debate over policies for admission of students to graduate training in law, medicine, science, and other fields. Some are saying: remove all bias in admissions; make certain that every applicant has an opportunity equal to that of all others and then judge each individual on his merits. Others are saying: such a "neutral" policy is insufficient, because the effects of past discrimination, even if no prejudice is found today, is to put some applicants at a disadvantage. At least for a time, according to this point of view, special efforts are needed to insure that Black students, and those from other disadvantaged groups, are adequately represented.

There is no doubt that special efforts have been made. "Ten years ago there were about 700 black law school students in the United States; today, following intensive efforts by previously all-white schools to increase their number, there are about 4,800. Similar multiple increases in minority enrolments have occurred in nearly all graduate and undergraduate schools." (*The New York Times*, April 7, 1974, p. 48.) Even this number is less than 5 percent of the total law-school enrolment, it should be noted.

The question now being debated is whether or not this rapid increase is simply the result of the removal of discrimination, or whether it represents "reverse discrimination." And if it is the latter whether it is a wise policy of "affirmative action," or an unwise quota system – no more desirable than previous discriminatory quotas that have been used to deprive the members of disadvantaged groups in many societies of equal opportunity.

This issue has recently received a great deal of public attention as a result of a suit brought against the University of Washington. A student with an excellent undergraduate record, Marco DeFunis, Jr., was denied admission to its law school, despite the fact that 37 Black and other ethnic-group students with poorer records were admitted. He won his suit in a lower court, but lost it in the highest court of the state of Washington. By a judge's order, he was allowed to attend classes, although he was not formally admitted, pending an appeal to the United States Supreme Court. Because this process took three years, he had virtually completed his law school training before the Supreme Court acted. The justices took advantage of this situation to avoid making a definitive decision. They ruled the case moot, because De-Funis was going to get his law degree without ever having been formally admitted. As everyone recognized, however, the issue was not thus resolved. It will be a matter of sharp public debate for many years.

IV. The Role of Blacks in American Politics

When the American Civil War ended in 1865, there was a promise, for a few years, of political equality for the Black population. Within a decade, however, this promise had begun to fade; and by 1900, the political activity of Negroes in the South – still the home of the vast majority – was almost eliminated. A number of procedures were used to exclude Negroes from politics: educational tests, unfairly applied; property requirements and poll taxes which bore most heavily on the poor, and therefore were particularly burdensome to Negroes; the fiction that political parties were "private" groups able to choose their own members – a device that maintained the "white primary" elections in states where, in effect, only one party was active. Where such restraints as these were insufficient, intimidation and violence were employed.

The situation in the North and West was more favourable to Negro political activity during the first half of the twentieth century. But the Black population was small and, like other low income groups, relatively less active in politics. In the large cities, Negro political organi-

zations did begin to develop, after World War I, often allied with the
dominant white organizations. They were able to secure a few favours
and win a few elections. By 1950, for example, there were three Black
congressmen in the House of Representatives in Washington, out of a
total of 435 congressmen. Despite such exceptions, it would be only a
slight exaggeration, to say that Negroes were almost completely excluded from politics in the United States before World War II.

In recent decades there have been significant political changes,
leading to an enlarged role for Blacks. These changes are partly the
result of indirect influences from economic, educational, and demographic forces. As Negroes have become concentrated in the cities, with
higher incomes and more education, they have, like other groups, become more active politically. The increased political participation,
however, is partly the result of recent court decisions, new laws, and
organizational work that deal directly with politics. The following
brief description of recent developments refers only to these.

A. Broadening the Black Franchise

Perhaps 1944 was the political turning point for America's Black population. In that year, the Supreme Court declared that "white primaries" were unconstitutional. Henceforth no person could legally be
prevented from participating in primary elections because of race.[20]
There followed a series of other judicial and legal actions that reduced
racial discrimination in politics. To mention only two: in 1965 a Voting
Rights Act was passed by the United States Congress that greatly increased the power of federal examiners to insure equal rights. The Act
was designed especially to guarantee the right to vote to Negro citizens
living in southern counties where discrimination had been most severe
and where they constituted a high proportion of the population.[21] In
1966, the Supreme Court declared that poll taxes were unconstitutional. (Harper vs. Virginia Board of Elections, 383 U.S. 663 [1966]). This
extended the coverage of the 24th amendment to the constitution, 1964,
which forbids poll taxes in connection with federal elections.

Such legal and judicial acts were accompanied by private actions
that also reduced political discrimination significantly. Local, state,
and national organisations, most of whose members and leaders are
Black, have been effective in getting Blacks to register, entitling them
to vote, and also to express their political views and to run for office.

These developments have changed the political situation in the United States in many ways. They have reduced, but by no means eliminated, discrimination against Negroes in politics. There has been a large increase in the number of Negroes serving on juries, thus increasing the likelihood of a fair trial.[22] The use of racist appeals in elections has been curtailed, since candidates must take account of more Negro voters. Candidates who a few years ago were openly appealing to the prejudices of White voters, Governor George Wallace of Alabama, for example, now try to win the support of Black voters as well, both by their words and their actions.

Perhaps the most important political development has been the rapid increase in the number of registered Negro voters, particularly in the South. Precise registration figures by race are not available; but Table 3 indicates a trend.

TABLE 3: *Voter Registration of Negroes in Southern United States* [23]

	Number	*Per cent of voting age population*
1944	250,000	5
1952	1,009,000	20
1960	1,414,000	28
1966	2,306,000	46
1968	3,690,000	62
1972	4,450,000	64

By 1972, the percentage of eligible Black voters registered was almost as high in the South as in the North and West (64% vs. 67%); and the national figure for Blacks was not much lower than the proportion of eligible White voters registered (65% vs. 73%). Actual voting, however, is still somewhat lower among Blacks, as shown in Table 4.

TABLE 4: *Voter Participation of Registered Voters by Race and Region* [24]

	1968	*1972*
South		
Negro	84	75
White	87	82
North and West		
Negro	90	85
White	93	90

The combination of lower rates of registration and lower rates of voting continues to deprive Negroes of their full political influence in the United States; but the differences between Whites and Blacks in this regard have been sharply reduced in the last ten years. The difference that remains is only partly due to discrimination. It is due also to class and educational variation and to the residual effects of past discrimination, which continue to affect political attitudes and behaviour.

B. The Problem of Black Representation

Another way to measure the extent of political discrimination and its trends is to note the number of elected officials who are Black. Two things emerge clearly from the evidence: Blacks continue to be seriously under-represented among elected officials; but there has been a sharp increase in the number of Blacks elected to office in the last few years. There are over half a million elected officials in the United States, when local, state, and national levels are combined. These include, among others, city councilmen, school board members, and mayors, state legislators, judges, and governors, United States congressmen and the nation's chief executive officers.

Scarcely more than one-half of one per cent of these officials were Black – 2,991, as of April 1, 1974, according to the Joint Centre for Political Studies. Yet by comparison with the recent past, this low proportion (11.4%, rather than 0.5% would correspond to the population ratio of Blacks) does not seem so low. In 1964, 72 Blacks held elective office in the South; ten years later there were 1307. In the country as a whole, the number increased in five years, 1969–1974, from 1,185 to 2,991. (*The New York Times*, April 23, 1974, p. 20.)

Since 1967, when Cleveland elected its first Black mayor, a number of major cities have elected Black chief executives. There are now at least 108 Negro mayors, including those in Detroit, Los Angeles, Atlanta, and Newark. In some instances, this trend indicates not simply a reduction in discrimination, but a delicate balance of political power. Many Whites having moved to the suburbs, the population of the central cities has become more than 50% Black (as in Newark and Atlanta), or more than one-third Black (as in Cleveland and Detroit). Nevertheless, in nearly every instance, a sizeable minority of White voters have preferred the Black candidate; and in some cases, Los Angeles for example, the Black won only with substantial White support.[25]

On the national level, there has been a significant increase in Black congressmen in recent years, from 3 in 1954 to 5 in 1964 to 16 (one Senator and 15 members of the House of Representatives) in 1974. Despite these gains, Negroes are still seriously under-represented in the national legislature (as are other groups of similar economic, occupational, and educational status). Were they represented in proportion to their population, there would be eleven senators, rather than one, and fifty members of the House of Representatives, rather than fifteen. This is only to suggest what the approximate number of Black congressmen would be were race a matter of indifference in the process of selecting candidates and in voting. The statement is not intended to support a formal proportional representation procedure in preference to the existing territorial procedure. This writer's preference is for the latter if it can be free of systematic bias, such as that based on racial discrimination.

In national elections, as in other aspects of American society, the trend is favourable, but major differences between the races remain, at least partly the result of continuing discrimination.

V. Black Responses and the Cycle of Discrimination

Discrimination against American Negroes, as against other minority groups, has many causes. The causes combine to create a situation in which Negroes are seriously disadvantaged. But they are not passive recipients. Targets of discrimination respond in many ways, some of which may intensify their disadvantages while other responses reduce them. Any study of discrimination must be aware of this complex system of interacting causes and seek to discover which parts of the system are changing, with what consequences.

This is not the place to develop a theory of discrimination;[26] but it is generally agreed that the three major sources are economic and political conflict, individual prejudice, and cultural stereotypes and traditions. These can be seen as the causes, in various combinations, of particular discriminatory acts. These acts, in turn, lead to institutionalized forms of discrimination and to reactions from the oppressed group that may re-enforce the cycle. If these several factors are thought of as a closed system, they might be diagrammed in the way suggested in Figure 1.

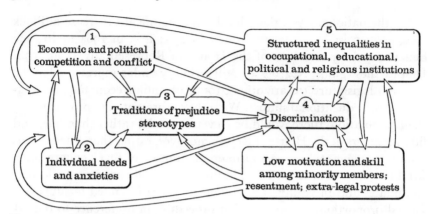

FIGURE 1 : *The System of Discrimination* [27]

It is not, in fact, a closed system. The cycle of discrimination against Blacks in the United States, as in other settings, is part of a larger system. Changes in the society, in international relations, in economic processes, in the demographic situation, perhaps even in knowledge of the causes and consequences of discrimination influence discriminatory patterns. Some of these influences may strengthen the cycle; but in the writer's judgment, most of the forces in the United States today are weakening the cycle. Slowly, but fairly certainly, discrimination against Blacks is being reduced. Several of the comments in the discussion up to this point may document the fact that discrimination against Blacks is less well supported now than formerly by economic and political competition. Thus items 1 and 5 in Figure 1, are weaker in their contribution to the cycle of discrimination. This does not mean the collapse of the system, because it can be supported, often for lengthy periods, by the other factors. These other factors may, indeed, deflect or eliminate the effect of changes in one part of the system by various "homeostatic" processes. Perhaps this rather technical point can be illustrated by a situation in which potentially improved opportunities, due to legal changes, are not in fact improved. The persistence of traditional discriminatory practices, despite the law, and the perpetuation of defeatist attitudes among the minority can cancel out the influence of a favourable change.

This is not the major process in the United States, however, with reference to discrimination against Blacks, for there is also evidence that individual prejudices and cultural stereotypes less strongly support discrimination than was true even a few years ago. This can be illustrated by reference to public-opinion poll data drawn from national samples. The Harris Survey asked a cross-section of American Whites to agree or disagree with a number of statements. The numbers given in Table 5 indicate the percentage of respondents who agree with the various statements.

TABLE 5: *White Stereotypes of Blacks in the United States* [28]

Percentage of a sample of Whites who agree that	1963	1971
Blacks have less ambition than Whites	66	52
Blacks have lower morals than Whites	55	40
Blacks have less native intelligence	39	37
Blacks breed crime	35	27
Blacks are inferior to White people	31	22

These data indicate that stereotypes continue to support the cycle of discrimination, but show also that there has been some reduction. Something of the same picture is found in a survey which demonstrates that Whites are becoming more aware of discrimination against Blacks, while Blacks are perceiving their situation as less discriminatory. There is still a "perception gap," as shown in Table 6 (next page), but it has been reduced even in a three year period.

On these five questions, the average percentage gap between White and Black perceptions was 46 in 1969 and 27 in 1972. It is reasonable to assume that an increase in White recognition of the presence of discrimination and a decrease in the Black perception of discrimination are good indices of the reduction of individual and cultural supports for discrimination.

The cycle of discrimination against Blacks in the United States is also being affected in powerful ways by the actions of the Blacks. There is still some support, as a result of their responses, for the "closed system," as suggested in item 6 of Figure 1: Lack of opportunity, feelings of resentment, and low motivation can help to create facts that seem to justify the discriminations of dominant group members. There are other forces at work, however: A smaller proportion of Negroes are "rural immigrants"; they have climbed significantly up the educational ladder from the position reached by the previous generation; they have been influenced by the rise of independent African states; powerful leaders have emerged. These changes make it more difficult to discrim-

TABLE 6: *Perceived Discrimination among Blacks and Whites in the United States* [22]

Percentage of samples who believe that Blacks are discriminated against in:	1969	1972
Getting decent housing		
Whites	46	51
Blacks	83	66
Getting white collar jobs		
Whites	38	40
Blacks	82	68
Getting skilled labour jobs		
Whites	35	40
Blacks	83	66
Way treated as human beings		
Whites	35	38
Blacks	77	64
Way treated by police		
Whites	19	25
Blacks	76	66

inate at the same time that they reduce the inducements to discriminate that were based on lower levels of skill and other signs of alleged "inferiority" among the minority group.

The reduction of discrimination has meant, not the lessening of Black protest against their disprivileges, but increased protest. As is often the case, it is relative deprivation, not absolute deprivation, that is most strongly resented. As opportunities expand, hopes and aspirations expand even more rapidly, creating a "revolutionary gap." The decade 1954–1964 was a period of reduction in discrimination against racial minorities in the United States; but the promises were more extensive than the performance. In the few years that followed, there were a number of riots in American cities, during which hundreds of persons were killed, most of them Black, and hundreds of millions of dollars worth of property destroyed. Although a relatively small proportion of Negroes participated in these riots, they expressed a widely held belief that America was not fulfilling its promises of equal treatment.[30]

Judgments differ widely on the effects of these riots on the level of discrimination. They called dramatic attention to difficult problems that had been neglected; they may have reduced the sense of powerlessness felt by many Blacks; they made other drastic, but peaceful, programmes calling for "Black Power" seem more moderate and acceptable. At the same time, riots intensify racial hatreds and fears in some people; they break lines of communication between races; they are not a programme but a "cry of pain" that deflects attention away from needed planning.

Laws designed to reduce discrimination were sometimes passed after a period of rioting; some Black officials were elected. Such facts have led some observers to the belief that riots reduced discrimination. But the civil rights movement that had produced significant changes in the twenty years before 1968 was sharply reduced in influence after the rioting; and a conservative national administration, which has done little to reduce discrimination, was elected. These too may be, in part, consequences of rioting. Although it is only possible to speak tentatively on this question it seems that on balance the violent civil disorders served to re-enforce the cycle of discrimination. Other factors were undoubtedly involved, however, in the slowing up of the civil rights movement. The death of Martin Luther King, Jr. in 1968 deprived the movement of one of its most vital leaders. The Vietnam War deflected public attention and divided American society. The prosperity of the 1960s was followed by several years of economic difficulty, making the costs of the programme to reduce discrimination more visible.

It is not clear how persistent the effects of such influences will be. In the writer's judgment, the reduction of educational disadvantages for Blacks, the virtual elimination of any legal foundation for discrimination, and the decline of prejudice among the dominant group will support the resurgence, although in a quieter form, of the civil rights movement by the end of the 1970s.

VI. Conclusion

In this brief examination of discrimination against American Negroes, only a few of the relevant topics have been discussed. Perhaps enough has been said, however, to justify the following conclusions:

1. Extensive discrimination against Black citizens of the United States persists more than a century after they were freed from bondage.

2. The disadvantages under which they suffer are even more extensive than those produced by present discrimination, because they are reenforced by the effects of past discrimination.

3. Beginning slowly in the 1930s, gaining some strength in the 1940s, and developing very strongly during the 1950s and 1960s, a civil rights

movement of major proportions has reduced racial discrimination significantly.

4. Only in the field of political action has the civil rights movement continued, so far, into the 1970s with the strength of the previous two decades.

5. The legal basis for discrimination has now been nearly stripped away as a result of a long series of Supreme Court decisions, beginning in the late 1930s, and a series of ever-stronger civil rights statutes passed by the United States Congress, beginning in 1957.

6. These laws are only partly enforced; and private forms of discrimination beyond the reach of law persist. Even private forms of discrimination, however, appear to be fading.

7. If discrimination is reduced as much in the next generation as it has been in the preceding one, by the end of the twentieth century, the United States will have accomplished a major redefinition of its race relations. Although American Negroes will still suffer from some of the effects of past discrimination, something approaching equality of treatment will have been attained.

Notes

1. There is a vast literature dealing with slavery in the United States. For two valuable recent works, see Carl N. Degler, *Neither Black nor White*, New York: Macmillan Co., 1971; Robert W. Fogel and Stanley L. Engerman *Time on the Cross: The Economics of American Negro Slavery*, Boston: Little Brown and Co., 1974.
2. For valuable discussions of the period from 1865 to 1945, see Gunnar Myrdal, *An American Dilemma*, New York: Harper and Brothers, 1944: and C. Vann Woodward, *The Strange Career of Jim Crow*, New York: Oxford Univ. Press, 1955.
3. See the periodic reports of the Bureau of the Census, United States Department of Commerce, entitled *The Social and Economic Status of the Black Population in the United States*.
4. Adapted from the *Social and Economic Status of the Black Population in the United States*, U.S. Department of Commerce, Bureau of the Census, Washington: Government Printing Office, 1973, p. 49. Note that "Non-White" in this table is not strictly comparable with Black. However, Blacks constitute over 90% of the Non-Whites.
5. See George E. Simpson and J. Milton Yinger, *Racial and Cultural Minorities: An Analysis of Prejudice and Discrimination*, 4th edition, New York, Harper and Row, 1972, chapters 11 and 12.
6. See for example, William B. Gould, "Black Workers Inside the House of Labor," *Annals of the American Academy of Political and Social Science*, Vol. 407, May 1973, pp. 78-90.
7. Adapted from *The Social and Economic Status of the Black Population in the United States, op. cit.*, p. 25.

8. For valuable studies of economic factors in discrimination, see Gary S. Becker, *The Economics of Discrimination*, Chicago: University of Chicago Press, 1957; Lester C. Thurow, *Poverty and Discrimination*, Washington: The Brookings Institution, 1969; Arthur M. Ross and Herbert Hill, *Employment, Race and Poverty*, New York: Harcourt, Brace and World, 1967; Harry J. Gilman, "Economic Discrimination and Unemployment," *American Economic Review*, Vol. 55, December 1965, pp. 107–96.

9. Charles C. Killingsworth, *"Jobs and Income for Negroes,"* in *Race and the Social Sciences* (Irwin Katz and Patricia Gurin, editors), New York: Basic Books, 1969, pp. 231–232. See also Stanley Lieberson and Glenn V. Fuguitt, "Negro-White Occupational Differences in the Absence of Discrimination," *American Journal of Sociology*, Vol. 32, Sept., 1967, pp. 188–200.

10. See David Caplovitz, *The Poor Pay More: Consumer Practices of Low-Income Families*, New York: The Free Press, 1963.

11. See Karl E. Taeuber and Alma F. Taeuber, *Negroes in Cities: Residential Segregation and Neighbourhood Change*, New York: Atheneum, 1969; and Lester C. Thurow, *op. cit.*, pp. 125–129.

12. For further discussions on discrimination and housing, see Charles Abrams, *The City is the Frontier*, New York: Harper and Row, 1965; George E. Simpson and J. Milton Yinger, *op. cit.*, pp. 434–449; and David McEntire, *Residence and Race*, Berkeley: Univ. of California Press, 1960.

13. Albert I. Hermalin and Reynolds Farley, "The Potential for Residential Integration in Cities and Suburbs: Implications for the Busing Controversy," *American Sociological Review*, Vol. 38, October, 1973, p. 598.

14. See Angus Campbell and Howard Schuman, *Racial Attitudes in Fifteen Cities*, Ann Arbor: Institute for Social Research, 1968.

15. For discussions of racial factors in American education, see James S. Coleman et al., *Equality of Educational Opportunity*, Washington: United States Office of Education, 1966; Frederick Mosteller and Daniel P. Moynihan, editors, *On Equality of Educational Opportunity*, New York: Random House, 1972; Thomas F. Pettigrew, "The Negro and Education: Problems and Proposals," in Irwin Katz and Patricia Gurin, *op. cit.*, pp. 49–112; Henry M. Levin, editor, *Community Control of Schools*, Washington: The Brookings Institution, 1970; Simpson and Yinger, *op. cit.*, pp. 546–611; Marian W. Edelman, "Southern School Desegration, 1954–1973; A Judicial-Political Overview," *Annals of the American Academy of Political and Social Science*, Vol. 407, May, 1973, pp. 32–43; Robert L. Herbst, "The Legal Struggle to Integrate Schools in the North," *Ibid.*, pp. 43–62.

16. For example, in seven states of the "deep South" per capita expenditure for Negro schools was 43 per cent of per capita expenditure for White schools in 1940. By 1952 it was 75 per cent. See Harry S. Ashmore, *The Negro and the Schools*, Chapel Hill: University of North Carolina Press, 1954, p. 155.

17. From the decision of the United States Supreme Court, *Brown v. Board of Education*, May 17, 1954.

18. *The Social and Economic Status of the Black Population in the United States*, 1972, *op. cit.*, pp. 60 and 62.

19. Christopher Jencks, *Inequality: A Reassessment of the Effects of Family and Schooling in America*, New York: Basic Books, 1972, p. 21.

20. *Smith v Allwright*, 321 U.S. 649 (1944). Since political parties have no constitutional or legal basis in the United States, the fiction had developed that they were purely private associations, able to select their own members. Southern states had used this belief to exclude Black citizens from participation in primary elections. Various state laws and Supreme Court decisions had supported this

view. But in 1941 (*United States* v. *Classic*. 313 U.S. 299) the Court had affirmed the right of Congress to regulate primary elections; and in the 1944 decision explicitly outlawed "white primaries."

21. For a full statement of the provisions of this Act, see United States Commission on Civil Rights, special publication No. 4, *The Voting Rights Act of 1965*. Washington, Government Printing Office, 1965.

22. See, for example, John Nordheimer, "Black Juryman: New Figure in the South," *The New York Times* November 26, 1970, p. 44. The extent of discrimination in the administration of justice is too complex an issue to explore here. Its importance is shown in "Race, Judicial Discretion and the Death Penalty," by Marvin E. Wolfgang and Marc Riedel, *Annals of the American Academy of Political and Social Science*, Vol. 407, May, 1973, pp. 119–133. The authors conclude: "Strong statistically significant differences in the proportions of blacks sentenced to death, compared to whites, when a variety of nonracial aggravating circumstances are considered, permit the conclusion that the sentencing differentials are the product of racial discrimination." p. 119.

23. Adapted from U.S. Census data estimates in Gunnar Myrdal, *op. cit.*, pp. 486–488; "The Negro Voter in the South – 1958," *Special Report*, Southern Regional Council, 1958; U.S. Commission on Civil Rights, *Voting* 1961; Donald R. Matthews and James W. Prothro, *Negroes and the New Southern Politics*, New York: Harcourt Brace Jovanovich, 1966, p. 18; and *The Social and Economic Status of the Black Population in the United States*, 1972, *op. cit.*, p. 97.

24. Adapted from *The Social and Economic Status of the Black Population in the United States*, 1972, *op. cit*, p. 98.

25. See Roger M. Williams, "America's Black Mayors: Are They Saving The Cities," *Saturday Review: World*, May 4, 1974, pp. 10–13, 66.

26. See Simpson and Yinger, *op. cit.*, chapters 3–7.

27. From *Ibid.*, p. 157.

28. Adapted from the Harris Survey, reported in the *Cleveland Plain Dealer*, October 4, 1971, p. 5–b.

29. Adapted from the Harris Survey, reported in the *Cleveland Plain Dealer*, January 15, 1973, p. 9–A.

30. For a variety of interpretations of recent civil disorders and riots in the United States, see United States Advisory Commission on Civil Disorder, *Report*, New York: Bantam Books, 1968; Robert H. Connery, editor, *Urban Riots: Violence and Social Change*, New York: Academy of Political Science, 1968; Seymour Spilerman, "The Causes of Racial Disturbances: A Comparison of Alternative Explanations," *American Sociological Review*, Vol. 35, August, 1970, pp. 627–649.

Anti-Jewish Discrimination Since the End of World War II

JOSEPH DUNNER

JOSEPH DUNNER, Professor Emeritus of Political Science at Yeshiva University in New York City, was born on May 10, 1908, at Fuerth, Germany. He studied at the Universities of Berlin, Frankfurt am Main and Basle. Having written various articles and brochures on the menace of Hitlerism to the cause of human freedom, he had to leave Germany in the Spring of 1933. He received his doctorate at the University of Basle in February, 1934, accepted an invitation to the United States from the Brookings Institution of Washington, D.C., and has since held lectureships at New York University and Harvard and a full professorship at Grinnell College. He served many years as a consultant to the American Jewish Joint Distribution Committee and was amongst the founders of the United Jewish Appeal. In 1963 he was offered the Petegorsky Chair of Political Science and International Relations at Yeshiva University. During some thirty-five years of teaching, he has held guest professorships at the Universities of Freiburg and Cologne and the Hebrew University in Jerusalem. During the Second World War he served as Chief of Intelligence for the U.S. Office of War Information in London. After hostilities ended in Europe he headed the Military Government Section of the United States Third Army, and helped restore a democratic press in his native Bavaria. Amongst his publications are: *The Republic of Israel*, the first textbook about that state; *Baruch Spinoza and Western Democracy*; *A Dictionary of Political Science*; and a *Handbook of World History*. In 1971, the Desch Verlag in Munich published his German-written memoirs under the title *Zu Protokoll gegeben*. He is now writing a book on American foreign policy since World War II.

Anti-Jewish Discrimination Since The End of World War II

JOSEPH DUNNER

If discrimination implies the distinction between individuals on the basis of the group to which they belong rather than their individual characteristics, anti-Jewish discrimination must be defined as some form of hostility to those persons who identify themselves or are identified by others as Jewish. Hostility to Jews as individuals identified as Jewish, to Judaism as a religious faith or to the Jewish people as an ethnic, cultural or racial group has during the last one hundred years been called "anti-Semitism," a term coined by Wilhelm Marr,[1] who wished to express thereby his dislike of Jews but not of others who on account of ethnic background or language could be classified as Semites (e.g. the Arabs). In this respect the post World War II disclosure of the details of the Hitlerian Holocaust has generated, at least for the time being, such general revulsion that overt anti-Semitism (e.g. the legal forms of anti-Semitism) became taboo. Since the mass murder of Jews in the concentration camps of the Third Reich appeared to be the ultimate consequence of legalized anti-Jewish discrimination, rather few non-Jews (whatever their sentiments may be) wish to call themselves or to be called "anti-Semitic."

As is well known, the leadership of the National Socialist German Workers' Party, abbreviated as Nazi Party, had seen in the Jews chiefly the members of a race characterized by particular biological phenomena as determined by the religion of the grandparents. Assuming, or at least pretending that the biological differences which exist or are presumed to exist between different races form the most significant criteria for the relations between human groups, believing in a hierarchy of races and the "right" if not the "duty," of superior races, however defined, to isolate, subjugate or even exterminate the inferior races, the Nazis upon entrenching their political power in Germany prepared a programme for the genocide, i.e. the systematic killing of the Jews of Europe.[2] As is also well known by now, it was the Nazi-Soviet Non-Aggression Pact, signed in Moscow by Molotov, the then Soviet Commissar of Foreign Affairs, and Ribbentrop, the Nazi Foreign Minister, which permitted the realization of what became known as the "final solution." On September 1, 1939, one day after the ratification of the Nazi-Soviet pact, the German armies, with the full backing of the Soviets, invaded Poland, a country which at that time harboured 3,350,000 Jews, ten percent of the Polish population. It was on Polish soil, in the concentration camps of Belzec, Chelmno, Majdanek, Oswiescim (Auschwitz), Treblinka, that the majority of the millions of Jewish victims of racist anti-Semitism were starved to death, beaten to death, shot or gassed.

Although the defeat of the Third Reich cannot be said to have ended all appeals to a Nazi-type racialism with its genocidal potentials, the suggestion that Jews should be exterminated is today but rarely heard and publicly rejected in most parts of the world. On the other hand, despite the overthrow of the Nazi regime and the drastic diminution of the Jews in Europe, anti-Jewish discrimination continues, albeit under somewhat new disguises and rationales. This analysis discusses three different situations. Firstly, anti-Jewish discrimination as it is practiced since 1945 in countries whose political regimes can be held responsible for generating anti-Jewish policies, such as the Soviet Union, Communist Eastern Europe and the Arab states. Secondly, countries in which government cannot be held responsible for anti-Jewish discrimination but in which religious tradition, economic conditions and political instability render the Jewish position somewhat precarious will be considered next. In this category must be mentioned some of the states of Latin America, Spain and Portugal. Finally, anti-Jewish discrimination in countries stressing constitutionalism (i.e. the Rule of Law, majority decision and minority protection). In these countries the Jewish position appears to be secure. However, since among the Jews of these countries memories of past anti-Jewish discrimination have engendered a variety of emotionally toned ideas centered on past expressions of anti-Semitism and either real or imaginary handicaps in the present time, the way these Jews perceive themselves has to be considered since it conditions their responses to their non-Jewish contemporaries.

I. Anti-Jewish Discrimination in the USSR

Foremost today is the problem of Soviet anti-Jewish discrimination. In the nationalities statistics of the 1970 Soviet census, the Jewish population of the USSR is shown to be 2,150,000 out of a total population of 242,000,000. Compared with the 1959 census, which indicated a Soviet Jewish population of 2,267,814, the new official data show a decline of 117,000 or 5.2%. This drop would be the more striking since according to the 1970 census the general Soviet population increased during the period 1959–1970 by sixteen percent. If in 1959 persons identified as Jews constituted 1.1% of the total Soviet population, they constituted but 0.9% in 1970. Recent Soviet Jewish immigrants in Israel have suggested that the Soviet Jewish population is between 3 and 4 million. If this figure were too high, the reduction of the Jewish population needs explanation. However, it is probable that a substantial number of Jews, when given the opportunity to hide their identity (particularly in a mixed marriage situation), may have preferred to identify themselves as Russians rather than as Jews.

A. The Legal Status of Soviet Jews

Under the Soviet system, Jews are classified not as a religious community of persons who freely adhere to the Jewish faith but as a nationality. The legal factor that determines a person's Jewish nationality is the biological fact that he or she is born of Jewish parents. If both the mother and the father are classified as Jewish, the children are Jewish. Only when one of the parents, father or mother, has a non-Jewish nationality, can the children, when they reach the age of sixteen, choose between the Jewish and the non-Jewish identification.[3] Unlike most other Soviet nationalities which have a distinct geographical base, the Jews are dispersed throughout all of the fifteen Union Republics and twenty Autonomous Republics. While the other nationalities are characterized by a linguistic attribute, most of the Soviet Jews, due to their dispersion as well as their high degree of urbanization, have experienced a fairly rapid linguistic assimilation. In the 1926 census over seventy percent of the Jews in the USSR still specified Yiddish as their native language. According to the 1959 census, but 405,000 Jews, about eighteen percent of Soviet Jewry, designated Yiddish as their mother tongue. In the 1970 census a further decrease was revealed in the number of those Jews who considered Yiddish or some other "Jewish language" (e.g. the Judaeo-Georgian spoken by Georgian Jews) as their mother tongue. To all other nationalities the Soviet regulations apply the principle of nationality by birth in one of the Soviet republics. The Jews are the only inhabitants of the USSR whose nationality is derived from their religious origin, a practice stemming from the Czarist system which was thereby able to distinguish between Jews and non-Jews, restricting Jews to residence in certain areas of the country (the Pale of settlement). Anti-Jewish discrimination under the Czarist regime was abolished by the Kerensky provisional government which proclaimed equal rights for Jews. Lenin, as Chairman of the Soviet of People's Commissars, published on July 27, 1918 a decree stating, "The Soviet of People's Commissars orders all soviets in the country to take decisive measures for the rooting out of the anti-Semitic movement. It orders that pogromists and those carrying on a pogromist agitation be outlawed."[4] However, this decree, frequently cited by Soviet officials as proof that the USSR opposes anti-Semitism, quite apart from its vague formulation, was never embodied in the official "collection of laws and government regulations" of the USSR and thus had no real influence on the Soviet administration and courts.[5] While in 1927 some Soviet leaders and the then existing Jewish section of the Communist Party of the USSR (Yevsekzia) suggested that the Biro-Bijan area in Siberia north of Manchuria should be designated as a "Jewish Autonomous Region" to give the Jews a territorial Jewish environment, Stalin and his successors pursued policies which, on the one hand, denied the Jews any genuine national identity and, on the other hand, emulated the Czarist regime by subjecting persons identified as Jews to discriminatory and humiliating treatment.

As a religious community, the Jews were attacked (as were the adherents of other religious faiths) since October 1917 when the Bolsheviks annulled all religious freedoms instituted by the provisional governments of Prince Lvov and Alexander Kerensky during the March-October 1917 period. After Trotsky's banishment from the USSR in 1927, Jewish clergymen were accused of supporting Trotsky (who rejected any Jewish identification) and arrested and deported as "Trotskyites", although their governmental accusers knew these charges were unfounded.[6] Judaism was stigmatized by Communist propaganda and administrative measures as "the means by which the exploiters and their agents, the clericals, transform the workers into a passive, frightened and defeated mass, and keep them from fighting for a better life."[7] The observance of the Sabbath or of other Jewish holidays was made impossible for all but the very old and very young since these particular days were decreed ordinary working days and since abstention from work on those days came under special scrutiny by the Communist Party and the governmental administration. Under the decree of the Soviet of People's Commissars of January 23, 1918, religious instruction in schools was prohibited. In accordance with this decree "the local authorities, with the help of the *Yevsekzia*, embarked on the forceful liquidation of the primary and secondary Jewish schools,"[8] preventing thereby Jewish parents from giving their children a religious education and any systematic knowledge of Jewish history. Even religious ceremonies conducted in the few synagogues which were not confiscated by the regime, and in homes, are not exempt from official harassment. Persons known to have participated in the Passover *Seder* (reminding the Jews of the Exodus of the Hebrews from the Pharaonic slave system in Egypt and linking them with the land of Israel through the prayer, "Next year in Jerusalem") are listed as "untrustworthy elements." The Hebrew language is outlawed which makes it legally impossible for children to study Hebrew, a blow not only to "Zionism" but also to the Jewish religion with its Hebrew prayers.

B. Israel and Anti-Jewish Propaganda

The discriminatory policy of the Soviet government toward Soviet Jews as a nationality and the foreign policy of the USSR toward the State of Israel have undoubtedly influenced Soviet policy toward Judaism. During the 1959–1962 period, the largest synagogues of Lvov and Chernovtsi were closed with the excuse that the residents of these two cities demanded their confiscation to terminate the "pro-Israel and anti-Soviet propaganda" in these synagogues.[9] Characteristic of Soviet anti-Jewish propaganda is Trofim Korneyevich Kichko's *Judaism Without Embellishment (Judaizm bez prikras)*, published in 1963 by the Ukrainian Academy of Sciences in Kiev, the first volume of its kind to appear under these auspices, which were also responsible for the Foreword to the book asserting: "Judaism, one of the oldest religions of the world, has incorporated and condensed everything that is most reactionary and most anti-humane in the writings of contemporary religions." In his 190-page book Kichko writes that "the Talmud is imbued with contempt for toil and for the toilers" (p. 40); that according to Judaism "Jehovah has handed over to the Jews all riches of the non-Jews" (p. 92); that "the speculation in matzoths, theft, fraud and depravity are the real face of the leaders of the synagogue" (p. 96). Conscious of Soviet-Jewish sympathies for Israel, Kichko portrays Judaism as the promoter of a reactionary, imperialistic conspiracy involving Zionism, Israel and Western capitalism. "It is impossible to understand the history of the mutual relationship between American capital, Zionism and Judaism," he writes, "without taking into account the affairs of the Rockefeller billionaires who for decades have been trying to seize part of Palestine – the Negev with its oil deposits... Israel, Zionism and Judaism are regarded by the American imperialists as the front line, as the reserve weapon for an attack on the Arab world." (pp. 171–172). The book, actually a compilation of three articles published before, included thirty caricatures which would have done justice to the Nazi organ, *Der Stürmer*. The cover of the book depicts a long-nosed rabbi, his hands dipped in blood, dressed for prayer in the traditional Jewish prayer shawl. After the American Jewish Committee, one of the leading Jewish organizations of American Jewry, distributed a number of copies of Kichko's *Judaism Without Embellishment* among the members of the U.N. Commission on Human Rights, the public airing in the Commission had the effect that even the Communist parties of France and Italy demanded that it be withdrawn from circulation.

This, however, did not prevent another Soviet writer, F.S. Mayatsky, from asserting in a widespread 96-page pamphlet entitled, *Contemporary Judaism and Zionism (Sovremenny Judaizm i Zionizm)*, published by the State Publishing House in Kiev in 1964, "Judaism is the worst of all religions, pessimistic, nationalistic, anti-feminine and anti-popular... Born from such principles the State of Israel could only become the worst of all states."[10] Nor did it prevent Kichko from adding further anti-Semitic books such as *Judaism and Zionism* (published by the Znanie Society in Kiev in 1968) and *Zionism – The Enemy of Youth* (published by the Central Committee of the Ukrainian Leninist Young Communist League in Kiev in 1972) in which under the euphemism of "anti-Zionism" is found the following: "The killing of the young, not only of *Goyim* but also of the Jewish young, is even preached in the Torah and was practiced by the believers of Judaism, the forerunners of the Zionists, for generations. The idea of sacrificing the young is now used by the Israeli rabbinate with the aim of inculcating indif-

ference not only toward people of other faiths but also toward orthodox
Jews. The militant followers of Judaism bless not only the bloody violence against the Arabs but also the atomic weapons, nurturing plans
to destroy the entire world."[11] In 1968 Kichko was awarded the "Certificate of Honour" by the Supreme Soviet of the Ukrainian Soviet
Socialist Republic and hailed as a "foremost Soviet sociologist of the
Jews" – a clear demonstration that even extreme statements about Judaism, Zionism, and Israel are not considered "anti-Semitism" in the
Soviet Union.

C. Anti-Religious Measures Against Judaism

Having ostracized Jewish believers by holding them up as reactiona-
ries and agents of imperialism to their non-Jewish neighbours, the
Soviets no doubt succeeded in reducing the number of Jews willing to
defy the authorities by joining a synagogue. Since under the Law on
Religious Associations of April 8, 1929,[12] any place of worship in the
USSR can be closed if it fails to be maintained by a group of at least
twenty citizens, the number of synagogues has declined from 1,103 in
1926 (already less than one-third of the number of synagogues existing
under the Czarist regime) to 60 in 1969.[13] Of the sixty synagogues, 30
are located in the non-European parts of the USSR, an area inhabited
by less than ten percent of the total Jewish population of the country,
since it is in that area (Kazakhstan, Uzbekistan, Kirgistan, Tadzhi-
kistan and Georgia) that the entire local population has resisted secularization measures more doggedly than the population in the European parts of the USSR. But even if due allowance is made for the unwillingness of many Jews to suffer the consequences of synagogue membership (such as loss of jobs, loss of priority in acquiring an apartment,
expulsion from the unions and other social organizations), the ratio of
synagogues to the Jewish population (one synagogue for 35,000) is
much lower than the ratio of churches and mosques to their believers.
To serve these synagogues there are only three ordained rabbis in the
whole of the USSR with the consequence that services in most synagogues must be conducted by laymen. In contrast to all other religious
groups, Jews have been prevented from training clergymen. In 1957
there was one *yeshiva* (rabbinical school) attached to the Central Synagogue in Moscow. It had thirty students. As in 1962 students from outside Moscow were refused residence permits, their number dwindled to

five in 1963; and since 1969 that *yeshiva*, the only one in the USSR, is no longer in operation.[14] In contrast to all other cults, Jews have been prevented from publishing Jewish literature of a religious nature. The exception was a photostat edition of a Hebrew prayer book originally published by the late Rabbi Yehuda Leib Levin. It was republished by the State Publishing House for the Moscow Jewish community in 1957 with an edition of three thousand copies. The Committee of State Security (KGB), which appoints the Jewish administrators of the synagogues, realizing that the three thousand copies were sold within a few days, prohibited any further republication. On the other hand, the same KGB convoked in March 1971 the first official gathering of the representatives of all registered "Jewish religious congregations" since the Leninist *coup d'état* in 1917, by asking the heads of the synagogues of the USSR to assemble for the sole purpose of denouncing Zionism and the State of Israel – an effort to counteract Jewish protests from abroad, voiced particularly by the World Conference of Jewish Communities on Soviet Jewry in Brussels on February 25, 1971.[15]

Although official Communist atheism affects all religious groups, the anti-religious attacks of the regime on the religions of the Russian Orthodox, the Georgian Orthodox or the Moslems (to cite but these few examples) do not impugn the patriotism and loyalty of non-believing Russians, Georgians or Uzbeks. The Jewish case, in this respect, is *sui generis*. The continuous attacks on the Jewish religion as a doctrine encouraging money grubbing and loyalty for Israel rather than the Soviet state; the Soviet indictment of the Jewish rite of circumcision as a "barbarous ritual" in which the "priests of the synagogue offer their regular sacrifices to their god, Jehovah"; the Soviet portrayal of the Jewish image in the stereotypes of "Trotskyite", "cosmopolite", "speculator", "agent of American imperialism"; the obvious efforts to exert blackmail against religious Jews by publishing the names and positions of those who have attended Jewish religious functions in either synagogues or homes, involve the non-religious Jew virtually as much as the religious Jew. For even if he succeeds in Russianizing his name, the moment he has to show his "internal passport", the entry "Yevrey" (Jew) is bound to evoke anti-Jewish sentiments and reactions amongst those non-Jews who fail to appreciate the subtle difference between the Jew merely born to Jewish parents and the Jew who consciously adheres to Judaism as a religious faith. Moreover, Jewish group existence is probably unique in that its secular culture is infused with religious values. This is true even of those cultural practices that had their origin in the non-Jewish cultures of the Diaspora (e.g. Yiddish, "Jewish" folk dances, "Jewish" foods). Any attack on Judaism is, therefore, a menace to the survival of the Jew *qua* Jew, irrespective of the way in which he sees his Jewish identification.

D. Discrimination Against Jews as State Policy

As a result of the Soviet government's suspicion of the Jewish nationality, Jews have been virtually eliminated from all security-sensitive areas of Soviet public life. In contrast to the years from 1917 to 1939 (the year of the Nazi-Soviet Pact), Jews have disappeared from the higher echelons of the diplomatic service and, with but few exceptions, the officers' corps of the armed forces. Also since 1939 the proportion of Jews in higher education, science, the professions has been steadily declining. In higher education (Soviet statistics lump together universities, teacher training schools, schools of journalism and music conservatories), the Jews today represent not quite three percent of all students, as contrasted with thirteen percent in 1935. Considering the fact that the Jews are the most urbanized of all Soviet groupings, this decline within a framework of an expanding economy can only be explained by anti-Jewish discrimination. Although membership in the Communist Party, the chief vehicle for the attainment of public office, is not analyzed in terms of nationality, it is well known that Benjamin Dimshitz is presently the only Jew (from the nationality point of view) who holds membership in the Central Committee of the Communist Party (with its over 250 members). Since the removal of Lazar Kaganovich in 1957, there has been no Politburo member of Jewish origin in the CPSU. The same pattern emerges with regard to the Supreme Soviet, "the highest organ of state power in the USSR." Of its 1,500 deputies but five are Jews, far below the proportion of Jews in the total population. If the selection and election of these deputies were a matter in which the members of the Party or the total population had a say, the elimination of Jewish deputies might be regarded as a sign of popular dislike of Jews. Since, however, the selection and election of these deputies depend exclusively on the choices made by the top levels of the Party machine and since there has never as yet been a contested election, the conclusion is self-evident. Interestingly enough, one of the five deputies "represents" to this day the Jews of Biro Bijan which according to the 1959 census numbered but 14,269 Jews out of a total population of 162,856. Even Khrushchev, in an interview with the correspondent of *Figaro*,[16] admitted that the plan to create a Jewish autonomous region of Biro Bijan had failed since, as he put it, Jews were too individualistic to do collective work. Yet the fiction of a "Jewish Biro Bijan" is officially maintained by the Supreme Soviet.

The key to the decrease of Jews in the political, economic and scientific positions of responsibility is the Soviet system of nationality quotas in university admissions. Although officially denied, there exists a *numerus clausus* for Jews in most universities and a *numerus nullus* in some. On May 12, 1957 Khrushchev in an interview conducted by a parliamentary delegation of the French Socialist Party stated, "At the outset of the Revolution we had many Jews in the leadership of the Party and the state. They were more educated, maybe more revolutionary than the average Russian. In due course we have created new cadres... our own intelligentsia. Should the Jews want to occupy the foremost positions in our Republic now, it would be taken amiss by the indigenous inhabitants. The latter would ill receive these pretentions, especially, as they do not consider themselves less intelligent or less capable than the Jews."[17]

Between June 22, 1941 (when the Nazis invaded the USSR) and the spring of 1945, Jews in the USSR had been encouraged to display (rather than suppress) their Jewish identity for the purpose of creating a favourable Jewish opinion in the United Kingdom and in the USA. After the Second World War, when the help of the Western Allies for the fight against Hitler was no longer imperative, all forms of Jewish culture (e.g. the Yiddish theatre, publications in Yiddish) were severely repressed. In 1952 hundreds of Jewish writers, actors, singers were shot, and it has been stated that Stalin at that time contemplated the deportation of all Jews to concentration camps. In early 1953, three months after the nineteenth Party Congress which ominously glorified the Stalin purges of the 1930's, the Soviet press announced that nine of the most distinguished medical specialists in the country had confessed to the murder of Zhdanov, the highest member of the Secretariat of the CPSU next to Stalin and Malenkov, and a plan to poison the leading political figures of the Soviet Union. Seven of the nine doctors were identified as "Jewish agents of the American Intelligence Service" which they were said to have contacted through an American sponsored Jewish philanthropic association (presumably the American Jewish Joint Distribution Committee). Only Stalin's death and the subsequent exposure of the plot as a fraud perpetrated by the KGB ended the public incitement against Jews. This, however, did not prevent Khrushchev from writing to Bertrand Russell in February 1963, "There never has been and there is not any policy of anti-Semitism in the Soviet Union."[18] Khrushchev failed to see the anti-Semitic character of the "doctors' plot". The ideological commission of the Central Committee of the CPSU first condoned Kichko's *Judaism Without Embellishment* as anti-religious propaganda, protected by the Soviet Constitution, and only later (in response to the criticism from abroad) observed that the book "might be interpreted in a spirit of anti-Semitism".[19] The Soviet authorities in the years from 1961 to 1964 engaged in a campaign against "economic crimes" in which the leading defendants were clearly stigmatized as persons of the Jewish nationality. In thirty-six trials prosecutors and press invariably portrayed the Jewish culprits as the masterminds while the non-Jews involved in these "economic crimes" were depicted as "naive, trusting officials" who had succumbed to "bribery." In Vilna, where food shortages had provoked one of the major economic trials, all eight accused were Jews. Four of them received the death penalty, four got lengthy prison terms. "Of the reported cases between 1960 and 1963 in which the death penalty was imposed, 55% involved Jews. In the Ukraine where Jews formed 2% of the population, 90% of those sentenced to death were Jews, and in the RSFSR where Jews form 0.7% of the population the percentage of Jews executed was 64."[20]

Since the State of Israel was proclaimed in May, 1948, anti-Semitic appeals amongst the apparatchiks and the continuance of state-sanctioned anti-Jewish policies were markedly accentuated by the Communist ideology of anti-Zionism. In 1947 and 1948 Stalin supported the founding of the Israeli state as a diversionist tactic against the Western powers. But the Jewish people hardly owed a debt of gratitude to the Soviet regime, which had persecuted Zionism and Zionists since the 1917 October Revolution. The enthusiasm with which Moscow's Jews welcomed the first Israeli delegation headed by Ambassadress Golda Meir; the fact that on October 16, 1948 some 100,000 Jews crowded around Moscow's Central Synagogue to join in the Jewish New Year's service, attended by the Israeli legation; the further fact that soon afterwards Jews began to apply for visas to emigrate to Israel – all these evoked swift and ruthless reprisals by the Soviet authorities. As Israel showed no inclination to become a Soviet client, anti-Israel, anti-Zionist and anti-Jewish charges were from now on interconnected. Typically, on the day *Pravda* revealed the "doctors' plot", January 13,

1953, *Izvestia* called this affair "the dirty face of Zionist espionage." Typical is also the Soviet reaction to the passionate applause given by huge Jewish crowds to the Israeli delegation which, in July 1957, participated in the International Youth Festival in Moscow. Shortly after the foreign delegations had left, hundreds of Soviet Jews, guilty of the offence of "fraternization with foreigners," were arrested. Some (the exact number is unknown) were sent to the Vorkuta hard labour camp in Siberia.

Since 1954 when Soviet foreign policy began to concentrate on the Middle East, Soviet hostility to Zionism as "Jewish bourgeois nationalism" and to Israel as an "agency of American imperialism" was used with increasing fervour to court the Arabs and gain Arab clients for the USSR. After the June 5–11, 1967 war in which Israel's defence forces inflicted a smashing defeat on the armies of three Arab states, destroying hundreds of Soviet-made MIG's and tanks, the Soviet Union (also Poland, Czechoslovakia, Hungary, Bulgaria and Yugoslavia) broke off diplomatic relations with Israel. In an intensive radio and press campaign Zionism was described as a sort of international Mafia organization which is active behind the international scene through the wide network of Jewish organizations.

In *Beware Zionism*, a 173-page book written by Yuri Ivanov and published in 1969 by the Political Literature Publishing House in Moscow (which issued a first edition of 75,000 copies at twenty-seven kopeks each), the reader learns that "modern Zionism is the ideology, the ramified system of organizations and the political practice of the big Jewish bourgeoisie, fused with the monopolistic circles of the U.S.A. and of other imperialistic powers." In a rephrasing of the notorious Czarist forgery "Protocols of the Elders of Zion," Ivanov asserts "that the Zionist leaders had always seen the creation of a Jewish state, not as a goal in itself, but as the means of reaching other, much wider aims: the restoration of control over the Jewish masses and all possible enrichment for the sake of power, parasitical prosperity and the strengthening of imperialism." To achieve these aims Zionists "concluded a tacit alliance between Zionism and Fascism" and maintained "close mutual relations with Eichmann."[21] In his novel, *The Promised Land*, Yuri Kolesnikov depicts the Nazi official, Adolf Eichmann, as having been a Zionist agent who arranged to send young and healthy Jews to Palestine while committing others to the gas chambers. Appearing in two installments in the September 1972 and October 1972 issues of the Soviet literary monthly, *Oktyabr* (with a circulation of 169,000), this novel obviously sought to release the Nazis from their responsibility for their massacre of Jews and indict the Zionist movement for collaborating with Hitler. To underline this accusation a Zionist leader is quoted to have said, "If there had been no Adolf Hitler, we Zionists would have had to invent him."[22]

In *The Bible and Modern Times* by Nikolai A. Reshetnikov (published by Mysl Publishing House in Moscow in 1968), Israel is called "the synagogue of Satan" and Biblical texts are interpreted as blueprints for a Jewish drive to world hegemony. "With the aid of the Bible, which Jews are taught to take literally, they are told that 'every place whereon the soles of your feet shall tread shall be yours.' "[23] In *Fascism Under the Blue Star* by Yevgeny Yevseyev, an official publication of Komsomol, the Communist Youth Organization, (75,000 copies issued in 1971), Franz Mehring's statement in his biography of Karl Marx that in the struggle for religious and social emancipation in the Enlightenment era of the eighteenth century "the Jewish community as such had taken no part" is twisted into the general verdict that "in the great intellectual achievements of our greatest thinkers and poets, Jews have no share."[24] As far as Zionists are concerned, only pejorative references are permitted, such as the libel that they participated in the killing of their fellow Jews at Babi Yar and the characterization of Ahad Haam and Martin Buber as "racists and chauvinists." Even Stalin's "Hofjude," Ilya Ehrenburg, is held up as an untrustworthy Jew for praising "the orthodox Zionist Marc Chagall." In the light of the events in Czechoslovakia in 1968 "is it really necessary to have more convincing proof of the existence of the Jewish world power?" Yevseyev asks, adding that the roots of "Zionist subversive activities are to be found in Judaism" which, he insists, is permeated by hatred of all other peoples and races.[25]

If one remembers that there is not a single book or article on the Soviet market which has not gone through the censorship office of the CPSU, the overt anti-Semitism of both the Communist Party and the Soviet administration cannot possibly be denied. While the Soviets, much more sophisticated than the Czarist authorities and the Nazis, employ their totalitarian control of the media of information to withhold news of pogroms from becoming known abroad, pogroms have taken place in various parts of the Soviet Union. On February 15, 1968, twenty-six Soviet citizens of Jewish origin addressed a letter to A. Snieckus, Chairman of the Central Committee of the Lithuanian Communist Party. It contains this passage: "We do not wish to overstate our case. We know that the situation of Jews is considerably better in Lithuania than in the other parts of the USSR, especially in the Ukraine where the persecution of our people is particularly terrible. During the entire post war period in Lithuania, there was only one bloody pogrom, in Plunge in 1958, whereas in the same period at least twenty pogroms occurred in the Ukraine."[26] Similar examples could be quoted *ad nauseam*. There would be little point in stressing them were it not for the fact that the Soviet attack on Judaism, Zionism and Israel has not only heightened domestic anti-Jewish policy but also caused a noticeable Jewish alienation *vis-a-vis* the Soviet regime, coupled with a growing Soviet Jewish pride in Israel's achievements. In other words, a vicious circle has been created by the Soviet system: anti-Jewish discrimination leads to Jewish estrangement from the regime and an emotional attachment to Israel as the Jewish homeland. This attachment in turn serves as a rationalization for the stepped-up anti-Jewish policies of the regime.

E. Soviet Jews and Israel

The most striking manifestation of the sense of alienation among Soviet Jews are the petitions which Jews as individuals or in groups have forwarded to the U.N., the President of the U.S.A., the government of Israel and Soviet officials (with copies deliberately leaked abroad) asking for the right to emigrate to Israel.[27] As could be expected, the initial answer of the authorities consisted in a series of well publicized arrests and trials in which the Jewish activists were denounced as "Nazi collaborators" and "allies of the Zionist storm troopers in the U.S.A." (an allusion to the Jewish Defence League, a small group whose attacks on Soviet officials in the USA have been disowned by most American Jewish organizations). The best known trial charging ten Jewish men and one Jewish woman with the crime of plotting to hijack a plane to leave the Soviet Union illegally for Israel took place in Leningrad. It began on December 15, 1970. On December 24, the court imposed death sentences on the pilot and one man accused of conceiving the escape plan; all the others received sentences ranging from four to 15 years at hard labour.[28] Since this show trial raised a storm of protest abroad, several other projected trials of Jews were postponed or held behind closed doors.

If the Soviet regime expected that Jews in the USSR would eventually assimilate and abandon their Jewish identification, its discriminatory policies clearly had the opposite effect. There can be little doubt that a number of Jews consciously try to evade Soviet attacks by stressing their loyalty to the Soviet system and by professing their indifference, if not hostility, to all aspects of the Jewish tradition. But there can also be no doubt that the Soviet policy of distrust and suspicion of the Jew *qua* Jew has produced a unique revival of Jewish national and religious consciousness among Soviet Jews in the past decade. Embarrassed and uncertain as to the methods of dealing with this unprecedented Jewish renaissance, Soviet efforts to eradicate every trace of an identifiable Jewish group by intimidation were in the end matched by a comparatively substantial grant of exit permits and a fairly steady flow of actual emigration of Soviet Jews to Israel.

While the Soviet authorities might have hoped that the grant of approximately 60,000 exit visas during the period 1968–72 would relieve them of the most persistent activists and extinguish thereby a flame that could easily spread to other nationalities, the fact that the desire to emigrate did not seem to dry up among the Soviet Jews caused the Supreme Soviet to decree on August 3, 1972 the so-called "diploma tax." Under this tax Jews requesting emigration visas to Israel were asked to reimburse the Soviet state for all funds spent on their education. Depending on training, academic degrees and length of service in their professional work, the rates varied from 4,500 rubles (for the equivalent of an American B.A. degree) to 25,000 rubles (for the equivalent of an American full professor).[29] This virtually prohibitive tax, obviously designed to put an end to Jewish emigration, created a stir throughout the Western world. It engendered i.a. the Jackson-Mills-Vanik Bill in both houses of the US Congress barring non-Common Market countries (read the USSR) from gaining US most-favoured-nation status if they denied their citizens the right or opportunity to emigrate or imposed more than a nominal tax on emigration. Toward the end of 1972 some 2,500 Soviet Jewish arrivals in Israel reported that they were allowed to leave without paying the education tax (a gesture interpreted as a Soviet attempt to curry favour with the U.S. Congress). Similar to Western public opinion, the non-Jewish democratic movement of Soviet dissidents, led by Andrei Sakharov and other leading Soviet scientists and writers, appears to sympathize with those Soviet Jews who insist on the right to emigrate. On May 20, 1971 the various groups of dissidents addressed to the Presidium of the USSR Supreme Soviet a document stating, "Our press portrays Zionism as a reactionary, practically Fascist political trend. Yet Zionism is no more than the idea of Jewish statehood, and one can only admire the persistence of an ancient and persecuted people who in very difficult conditions have resurrected a long vanished state... We call for an end to all persecution of Jews seeking repatriation to Israel and urge that there be no more violations of the obvious right of man to leave any country."[30]

Paradoxically enough, representatives of the USSR in the U.N. have publicly acknowledged that the USSR considers itself an adherent to the Universal Declaration of Human Rights. In February 1963 representatives of the USSR specified in written documentation, deposited with the U.N. Secretariat, only three grounds on which an application for an exit visa would be rejected: if a person has been charged with an offence and judgment was yet pending; if a person had been convicted and was serving a court imposed sentence; if a person had yet to discharge his obligation of service in the army or navy. Aware of these exceptions, Soviet authorities have in recent years frequently answered petitions for an exit visa by sacking the petitioner from his job and then charging him with "parasitism." In some cases Soviet doctors, cooperating with the KGB, have labelled petitioners for exit visas persons "inclined to malicious hooliganism." As a consequence, hundreds of such cases are pending in the courts, permitting the Soviet authorities to prevent "legally" the emigration of these Jewish petitioners.

II. Anti-Jewish Discrimination in Communist Eastern Europe

A. Poland

At the end of World War II the Communist Party, calling itself the Polish United Workers' Party (PUWP), came to power in Poland on the bayonets of the Soviet Red Army, which broke up and liquidated all other political parties through legal chicanery, intimidation and terror. Led by Boleslaw Bierut, a Moscow trained Pole who purged the Secretary General of the PUWP, Wladyslaw Gomulka, for "nationalist deviation," the PUWP had in its top echelon a number of "Jewish" Communists, persons who refused to identify themselves with the Jewish people and Judaism, thought of themselves as Marxist-Leninist-Internationalists but could be identified as biological Jews, i.e. the children of Jewish parents. As is well known, Poland is the country in

which most of the European Jews, among them over three million
Polish Jews, were killed by the Nazis who apparently enjoyed the active assistance of Polish collaborationists.[31] After Pilsudski's death in
1935, four years before the division and occupation of Poland by the
Nazis and the Soviets, Endeks, Naras and other avowed anti-Semitic
groups had introduced "Aryan" paragraphs in professional and economic associations, "ghetto benches" in the universities and legislation
for the forced emigration of Jews to Madagascar or other places.[32]
With noticeable exceptions disapproving of the Nazi "final solution,"
they created the political and cultural environment which enabled the
German authorities to round up the Jews of Poland and to transport
Jews from other Nazi occupied countries to the death camps and gas
chambers on Polish soil. Of some 250,000 Polish Jews who had escaped
into the interior of the USSR, circa 170,000 were later repatriated to
"People's Poland" as it emerged from the Yalta Agreement in February 1945. Since most of the Jews who had sought refuge in the
USSR (and this includes those who were repatriated to Poland) subsequently emigrated, primarily to Israel, the Jewish population of
People's Poland including children of mixed marriages and even so-
called quarter Jews was estimated in 1967 to number 25,000 in a general population of over 32 million.

Although General Mieczyslaw Moczar, head of the Polish secret police from 1956 to 1968, is usually mentioned as the mastermind behind
the official anti-Semitism and introduction of racist criteria in the definition of "who is a Jew," Gomulka, First Secretary of the PUWP
since 1956, was as willing as Moczar to exploit anti-Semitic tendencies
when it suited his convenience. After the outbreak of student riots due
to the suppression of public performances of "Dziady," the master-
piece of Mickiewicz, Gomulka, in a widely quoted speech to the Trade
Union Congress on March 19, 1968, denounced "students of Jewish
origin or nationality" as the instigators of the demonstrations.[33] Moreover, the fact that a special "Jewish Department" of the Ministry of
Internal Affairs (MIA) was working for years on the compilation of
Nazi-type "ancestral charts" of Jews, and persons suspected of Jewish
origin, could not possibly have been unknown to Gomulka.[34] Some of
the Jews were no doubt old, destitute survivors of the Nazi massacre
who considered Poland their homeland and shunned the ardours of
emigration. Instead of finding sympathy among their countrymen,
they more often than not incurred open hostility on the part of those
non-Jewish Poles who as "custodians" of Jewish properties had be-

come the beneficiaries of the Nazi occupation of Poland and opposed the claims of Jewish survivors for restitution. The others were either members or sympathizers of the PUWP who cherished the idea that the Communist movement with its Marxist-Leninist theology encouraged their de-Judaization. By participating in the forced collectivization of peasants, the suppression of the dominant Catholic Church as well as the suppression of all religious groups (Judaism included), by staffing the leading positions of the governing Party, the military and the police in a country in which their fathers and relatives had been at best tolerated aliens, they could only intensify whatever latent anti-Semitic sentiments existed in the general population. When in October 1956 the Communist government, imposed on the Poles by Stalin, was replaced by a new Communist government which stressed its national Polish character, a number of Jewish "Muscovites," as all of the Stalinists in the PUWP were called, shared the occupational hazard of leadership in a totalitarian system.

One might be tempted to write off the Polish "Aryanization" campaign in 1967–1968 as an internal factional struggle within the Polish Communist Party had it not become evident that this campaign went far beyond the membership of the Party and the various branches of the government administration. While the "partisans," as the followers of General Moczar within the PUWP were called, used the Jewish origin of the "Moscovites" for their inner Party struggle, the basis for the anti-Semitic explosion in 1967 must be seen in a grandiose effort to blame the Jews, all Jews, even some non-Jews who were arbitrarily labelled Jews (because they fitted the popular image of Jews as people with dark hair and long noses) as the cause for the most notorious crimes committed by the secret police during the Stalinist period as well as the organization of recent demonstrations of university students demanding freedom of expression and an end to censorship.

As in the USSR, the Communist leaders of Poland at first camouflaged their anti-Jewish campaign as a necessary defence of Poland against a "Zionist conspiracy." "The protagonists of the Zionists were primarily discredited people who in 1945 returned to the country in army uniform from the Soviet Union, eliminated Gomulka and other native Polish leaders, placed their Jewish friends in important positions, and in October 1956 rapidly changed sides. These petty-bourgeois elements, always alien to Communist ideology and the Polish nation, later adopted revisionist and Zionist positions. Frustrated and embittered, the leading officials of Jewish origin, once primarily responsible for the horrors of Stalinism and now creating a revisionist climate, turned against the Party and the state. Whether in or out of power, they have always been well known for their national nihilism and cosmopolitanism."[35] Having come to the conclusion that Jewish blood distorts thinking and that neither a Jew nor a half-Jew nor a quarter-Jew can be trusted, the PUWP decided that there was no room for 25,000 persons of Jewish origin in Poland. It removed those persons from the Party, the faculties and the student bodies of the universities, the press, the orchestras, the courts and in some cities (e.g. Lodz) even from manual jobs. To make sure that no male Jew would escape this purge, police squads forced suspects who did not voluntarily admit their Jewish origin to show whether or not they were circumcised. As in Nazi Germany, the entire population was encouraged to write denunciations of persons of Jewish origin and to bring to the attention of the Ministry of Internal Affairs (MIA) any person, man, woman or child, who could not produce authentic certificates of baptism or proof of non-Jewish parents. As in Nazi Germany, bonuses were offered by the MIA to those who would assist the authorities in making Poland "Judenrein."

At a briefing of the members of the security police in 1968, General Slachcic, known as Moczar's righthand man, stated, "About 7,000 of the Jews have left. We shall have another 12,000 to 15,000 of them leave." Eight thousand "Jews" are known to have left during the following four years, chiefly for the Scandinavian countries. Also known is that some eighty persons of Jewish origin committed suicide and that a few hundred persons of "mixed Jewish blood" appeared to have emulated the Marranos of fifteenth century Spain and Portugal (Jews who converted to Catholicism to escape the persecution of the Inquisition) by official conversion to the Catholic Church. Provided with forged documents, they apparently hoped to avoid persecution and emigration.

B. Czechoslovakia

The Czechoslovak census of 1930 showed 356,830 Jews, distributed as follows: 117,551 in Bohemia-Moravia; 136,737 in Slovakia; and 102,542 in Ruthenia. About 40,000 Jews, mainly from Bohemia-Moravia, emigrated after the Munich Agreement in the summer of 1938. The Jewish death roll for all of Czechoslovakia within the pre-Munich frontiers is estimated from 233,000 to 243,000.[36] When the Communists came to power in 1948, over ¾ of the Jewish survivors of the Nazi occupation left for Israel, leaving behind them not quite 19,000 Jews, $1/15$ of 1% of the general population. Having been culturally assimilated, having little or no interest in the Jewish tradition, most of these Jews were Jewish only by virtue of the fact that they had been born of Jewish parents. As in Poland, a number of these biological Jews had been Moscow-trained Communists who, during the initial period of Communist rule, occupied leading Party and governmental positions. When the USSR began to befriend the Arabs, these biological Jews who had been lifelong anti-Zionists were accused of being members of a "Zionist conspiracy," an accusation which, if advisable, could easily be changed to "Jewish conspiracy."

This was the fate of Rudolf Slansky who had been tortured by his fellow Communists in the Czechoslovakian secret police into the admission that he belonged to a "Trotskyite-Titoist-Zionist conspiratorial centre" and that he had aspired to become a "Czech Tito." Together with ten other Communists, described as "Jewish bourgeois nationalists, cosmopolitans and Zionists," he was hanged on December 3, 1952. As during Stalin's Great Purges in the 1930's, the confessions, false though they were, served an ulterior purpose. This time the purpose was to demonstrate to the masses of the USSR and the satellite countries of Eastern Europe that no Jew, whatever his political record, could be trusted; that all Jews were potential traitors in the service of "American imperialism" and "its Zionist agency in the Middle East."

Artur London, Deputy Foreign Minister of Czechoslovakia in the years 1948–51 who escaped the hangman only because his brother-in-law, a member of the Politburo of the French Communist Party, was able to mobilize the French government on behalf of the accused, in his memoirs reports that after his arrest the Communist prison commander greeted him with the words, "You and your dirty race, we shall eliminate you. Not everything Hitler did was bad. He killed the Jews and that was a good thing." Having been a "Spaniard" (i.e. a person who fought with the International Brigade in the Spanish Civil War), he was continuously pressured by the Communist security officials in Prague to indict other arrested veterans of the Spanish Civil War as "Jewish traitors." When he remarked that apart from himself and one older man, the other accused were not of Jewish origin, the investigator replied, "You forget their wives. They are all Jews. It comes to the same thing." London's own French-born wife, having been persuaded by the Communist authorities and hundreds of resolutions passed in factories and offices that her husband and his "fellow conspirators" had indeed been traitors, operating under directions of the former US Secretary of the Treasury, Henry Morgenthau, and the Israeli leaders Ben-Gurion and Moshe Sharett, wrote this letter to the President of the Czechoslovakian court trying the "Slansky plot": "As a Communist and as a mother I am happy that the treacherous gang was unmasked. I can only join all honest people in demanding a just punishment for the traitors."[37]

While in 1963 the Czechoslovak Communist Party repudiated the trial of Slansky and his colleagues as a "frame-up and violation of socialist legality," four years later, in the wake of the Arab-Israeli war in June 1967, some members of the Party, anxious to emulate the Polish example, forged letters allegedly written by Simon Wiesenthal, the Director of the Documentation Centre of the Federation of Jewish Victims of Nazism in Vienna, to prove that a number of Communists of Jewish origin, as well as a number of non-Jewish "Zionists," known for their advocacy of greater democracy and reforms, were planning a "counter-revolution" and "the restoration of capitalism."[38] The campaign of vilification, chiefly directed against Alexander Dubcek, First Secretary of the Party, was fed by the Kremlin, the Polish Communist Party and, most articulately, the Socialist Unity Party (Communist Party) of the German Democratic Republic (the Soviet zone of Germany). On August 25, 1968 *Neues Deutschland*, the Central Organ of the East German Communists, announced "Zionist forces have taken over the leadership of the Czechoslovakian Party." Among the Dubcek reformers was one man whose Jewish origin was established: Eduard Goldstuecker, a prewar Communist, prior to 1951 Czech ambassador to Israel and Sweden, sentenced to life imprisonment in 1953 as a follower of Slansky, rehabilitated in 1963 and appointed professor of German literature at Prague University. On September 4, 1968, *Izvestia* added another person alleged to belong to the "Zionist forces." It was Jiři Hajek, since April, 1968, Czechoslovakia's Foreign Minister. He was now identified not only as a Jew whose real name had been Karpeles, but as someone who "collaborated with the Gestapo during the German occupation so as to save his skin." *Izvestia* and *Neues Deutschland*, which reprinted the article in German, had, however, committed a first rate blunder. As the *Volksstimme*, the official organ of the Communist Party of Austria, on September 5, 1968 put it, "They caught the wrong Hajek. Foreign Minister Dr. Jiři Hajek was never a Jew nor ever called Karpeles. Quite apart from the intolerable style of

Izvestia with its strong anti-Semitic under (or over) tones, the other allegations are pointless, since the attacked Hajek is not the same as the Hajek they aimed at. There is really a Hajek, not Jiři but Bedřich, who was once called Karpeles. He changed his name in 1945 to Hajek. He did not collaborate with the Gestapo because, as a Jew, he had to flee and lived during the war in Britain as an emigrant, actively participating in the Czechoslovak liberation movement. Apart from having once been called Karpeles, *Izvestia's* allegations do not fit this Hajek either."

Meanwhile the curious accusation that "Zionist forces" had taken over the leadership of Czechoslovakia's Communist Party and thereby produced the "counter-revolution," prompted the USSR, Poland and the German Democratic Republic to anticipate the Brezhnev Doctrine and to invade Czechoslovakia with their armed forces. That *Neues Deutschland*, in the vein of the Hitlerian newspapers in the years 1933–45, indulged in a special "hate Israel" campaign, and endeavoured to discredit the Czechoslovak reform movement as a Jewish-inspired plot, cannot surprise anyone who knows that many leading functionaries of the Socialist Unity Party, and the administration of the German Democratic Republic, are former members of the Nazi Party and the Hitler Youth, who after World War II had received clearance from the Soviet Military Administration as "useful citizens", the Soviet term for informers and sycophants.

Between August 1968 and February 1969 about 35,000 persons fled from Czechoslovakia to Austria. Ten percent of them were estimated to be Jewish which means that Czechoslovakia, in which over one thousand years ago the Jews had organized solid communities, is nearly *"Judenrein."*

C. Southeast Europe

In Bulgaria, unique in Eastern Europe, nearly the entire Jewish community of 50,000 Sephardic Jews (i.e. Jews who in the fourteenth and fifteenth centuries had left Spain for the Turkish Empire of which Bulgaria was then a part) had survived the Nazi onslaught. A united front of Bulgarians, inspired by the monarchy and the Orthodox Church, succeeded in protecting the Jews. It was only after the Soviets occupied Bulgaria and transformed it into a Communist People's Democracy that some 45,000 Jews left the country for Israel.

In Hungary where Communists of Jewish origin, led by Bela Kun, presided over a short-lived Soviet regime in the spring of 1919, thereby giving rise to the view that Communism was a "Jewish conspiracy," the conservative regime of Admiral Horthy (while limiting Jewish applicants in university admissions to a strict quota system) protected the 725,000 Jews of Hungary until March 1944 when the Nazis occupied the country and shipped 400,000 Jews, over half of the Jewish population, to the gas chambers in Poland.[39]

Hungary was the first Soviet satellite in which the ruling Communists played on the anti-Jewish sentiments in the general population and accused a life-long Communist, Laszlo Rajk, a member of the Politburo of the Party and Hungary's Foreign Minister, of belonging to a "Trotskyite-Zionist group." In September 1949, in a show trial in Budapest Rajk and six co-defendants, three of them of Jewish origin, were sentenced to death as "agents of a world-wide Zionist conspiracy," operating under "orders of the American Intelligence Service." It must be added that Matyas Rakosi, the Communist leader who prepared the scenario of this trial, was himself biologically a Jew. The Jewish question played no noticeable role in the 1956 uprising of the Hungarians against the Soviet Union and its emissaries in the Party and secret police. The present Kadar regime tries to harmonize the spirit of the "Hungarian October" with the fact that on November 4, 1956 the Soviets moved their tanks into Budapest and other major Hungarian centres. The Jewish community of close to 80,000 members operates the only rabbinical seminary in the whole Communist world, a Jewish high school, a Jewish library and museum, kosher meat stores and various plants producing unlimited quantities of matzoh for Passover. In Budapest, where most of Hungary's Jews live, there are thirty synagogues.[40] It seems that their contacts with those Jews who had emigrated to Israel are not disturbed. In contrast to Communist Rumania which alone among the Soviet satellites refused to sever its diplomatic relations with Israel after the June 1967 war, Hungary, on June 12, 1967, broke off its relations with Israel. Nevertheless, while Janos Kadar denounced "Israeli expansion," he stressed that he was not "against the State of Israel or against the Jewish people." Since 1956, when Rakosi, one of Stalin's most trusted lieutenants, lost his power, governmental anti-Jewish discrimination seems to have been arrested.

In Rumania where anti-Semitic organizations such as the Iron Guard and the League of Christian National Defence anticipated the Nazis, some Jews seemed to have saved themselves from death in the Polish extermination camps by bribing their captors and finding refuge in Rumanian peasant communities which, however, were not immune to collaboration with the German Nazis and the Rumanian Iron Guardists. Some 300,000 Rumanian Jews (one-third of Rumania's Jewish population before World War II), however, succeeded in avoiding the trains of sealed box cars in which the Nazis shipped their human cargo to the gas chambers on Polish soil. After the return of the survivors of Auschwitz and other camps and the immigration of 50,000 Jews from Soviet-annexed territory, the Jews numbered almost 400.000. Of them about 200,000 left for the Jewish region of Palestine (since 1948: Israel) between 1946 and 1952.

In Rumania the Communist Party had been founded (in 1921) by two foreigners, a Bulgarian-born non-Jew, Christian Rakovsky, and a Russian-born Jew, Mikhael Katz. As Nicolai Ceausescu, who in 1965 became Secretary General of the Party and initiated a remarkable emancipation from Soviet domination, stressed at the Party Convention in 1966, the Comintern had "appointed leadership cadres from among people abroad who did not know the Rumanian people's life."[41] Among these Soviet-trained Communists, Ana Pauker, the daughter of a rabbi, ranked as the most powerful figure after 1945. When in 1952 she was removed from her offices as Secretary of the Central Committee of the Communist Party, Deputy Premier and Foreign Minister, her Jewish background may have been a contributory cause. But it must be noted that Pauker, herself, had been most active in the persecution of the Rumanian veterans of the Spanish Civil War, a number of persons of Jewish origin among them. The "Rumanization" of the Party and state administration led to the reduction of "alien elements" which affected also the Moscow-trained Communists of Jewish origin but did not single them out. At one time, in February 1959, the Rumanian government issued a statement accusing "Israeli and imperialist circles" of publishing that there was a mass migration of Rumanian Jews to Israel, ruining thereby the good relations between Rumania and the Arab countries. Subsequently, Jewish families were granted exit permit only if they gave as their destination Austria, a country which they and the Rumanian authorities knew quite well served as a stop-over for Jews on their way to Israel. The Jewish population of close to 110,000 remaining in Rumania appears to enjoy complete religious freedom. On June 9, 1967 when all other East European leaders, including Tito, signed in Moscow a statement condemning "Israeli aggression," Secretary General Ceausescu and Prime Minister Maurer of Rumania refused to give their signatures. During the following years, Rumania continued to maintain good relations with both Israel and the Arab states.

In Yugoslavia 61,500 Jews (out of a prewar Jewish population of 72,000) were shipped to concentration camps and murdered by the Nazis (with the collaboration of Croats and Bosnian Moslems, instigated by Haj Amin el Husseini, the former Grand Mufti of Jerusalem and founder of the Palestine Arab Higher Committee).[42] The rest took up arms and fought against the Nazis. Since the Yugoslav resistance movement was split into two camps, the Cetniks under the monarchist, Draza Mihailovic, and the partisans under the Communist leader Josip Broz Tito, some Jews fought in the ranks of the Cetniks, later wrongly accused by Tito as collaborationists, while the majority of them joined the Tito camp. Of the 10,000 Jewish survivors of the Hitlerian period, some 8,000 left for Israel. While after his expulsion from the Cominform in 1948, Tito may have hoped to build a Communist Yugoslavia free from the despotism of the Soviet Union, Yugoslavia remains a police state modelled in the image of the USSR. Although there has not been an officially trumped up "Jewish question" and "Zionist plot," the government of Yugoslavia has been as anti-Zionistic and anti-Israel as the Soviet Union, People's Poland and the German Democratic Republic (the Soviet zone of Germany).

III. Anti-Jewish Discrimination in the Arab States

As in the Soviet bloc countries in which anti-Jewish policies and laws must not be called "anti-Semitism," and frequently assume the dialectical disguise of "anti-Zionism," the Arab states, too, normally prefer their anti-Jewishness not to be known under this name. Some Arab spokesmen argue that Arabs cannot possibly be "anti-Semites" because they themselves are "Semites." Others insist that they cannot possibly be accused of "anti-Semitism" or "anti-Jewishness" since they do not consider the present-day Jews to be of Semitic-Jewish origin. Nevertheless, no sophistry of propaganda can cover up the fact that since the restoration of a Jewish statehood (Israel) in Palestine in May 1948, most Arab-Moslem states have considered the Jewish minorities in their midst enemy aliens and a potential fifth column of the State of Israel.

In Iraq (ancient Mesopotamia), where Jews had been living uninterruptedly for 2,500 years (since the Babylonian Exile), the monarchical government (following the defeat of the Iraqi troops in the 1948 war against Israel) ordered the sequestration of Jewish property and businesses, at the same time banning all Jewish emigration.[43] When, due to Western pressures, this ban was lifted in March 1950, 123,500 Jews, deprived of all their former material possessions (except for $140 per adult and $56 per child), emigrated to Israel in what became known as "Operation Ezra and Nehemiah."[44] Of the 2,500 Jews who remained in the country close to 2,000 fled after the June 1967 war. On January 27, 1969, nine Jews and on August 25, 1969 two Jews were publicly hanged as "imperialists and Zionist spies."[45]

In Syria the Jewish population of 30,000 in 1943 was reduced to 15,000 in 1947 by emigration to the USA, Lebanon and the Jewish area of Palestine because the civic equality which the Jews obtained under the French mandate (1920) had been severely infringed upon by first the Vichy Regime and later (1946) the government of the now independent Syrian Republic. A further exodus was caused by the 1947 pogroms in Aleppo and the terrorization of the Jews in Damascus. In the summer of 1948 the Syrian government issued a decree prohibiting the sale of Jewish property, followed by the freezing of Jewish bank accounts. After the union with Egypt and the founding of the UAR in 1958, the prohibition on the exit of Jews (decreed in 1953) was cancelled on condition that the emigrants accepted confiscation of their property by the government. In March 1964 this decree was revoked by a new law which prohibited Jews from travelling more than three miles beyond their place of residence.[46] During the June 1967 war, most of the 3,000 Jews who had remained in Syria were arrested. Only those Jews escaped persecution who lived among the Moslem Kurds who themselves were accused by both the Syrian and Iraqi governments of pro-Israel sentiments. As reported in the *New York Times* (April 14, 1974) following the slaying of four young Jewish women and two Jewish men attempting to cross into Lebanon in early March 1974, some 1,000 Jews of Damascus had taken to the streets asking the Syrian authorities to be allowed to leave Syria. In Damascus, the *New York Times* reporter writes, "On each street leading out of the ghetto Moslem shopkeepers keep a watch on comings and goings. They are believed to fulfill the role of police informers. At 10 P.M. curfew for Jews is strictly enforced. State employees and members of the military are reminded by notices displayed in many offices that they may not buy from Jews... Jews live largely by manufacturing in family workshops silver and copper articles that are sold in the souvenir shops of the souk, or ancient covered market, where they are not allowed to own shops themselves... When the head of a Jewish family dies, his property is forfeited to the state."

In the Hashemite Kingdom of Jordan (which in April 1950 replaced the kingdom and emirate of Transjordan, the area of Palestine east of the Jordan River) the few settlements of (orthodox) Jews were destroyed during the 1948–49 war by the Transjordanian legion (then under the command of a British officer, Brigadier General J. B. Glubb). Those Jews who did not succeed in escaping to the Jewish sector of Jerusalem were massacred. Synagogues and the ancient Jewish cemetery on the Mount of Olives were desecrated.[47]

In Egypt (since 1958 also known as the UAR) the Jewish population of 80,000 (before 1948) experienced active persecution ever since the establishment of the State of Israel. In the summer of 1948 bombs were planted in Jewish neighbourhoods, and Jewish businesses were looted. On "Black Saturday," January 26, 1952, most Jewish properties (estimated at 25 million dollars) were confiscated or destroyed. During the Sinai Campaign (November 1956), 3,000 Jews were imprisoned without charges. Of the 2,500 Jews who stayed in Egypt until the June 1967 war, all but 400 emigrated during the 1967–1972 period. Although there was no law passed to this effect, their exit permits were given on the condition that they would forfeit whatever possessions they still had to the Egyptian government.[48] Their passports were stamped with a red Arabic letter, stating they were Jews and enabling police and custom officials to pay special attention to them.[49]

In Libya, where Jews had settled 700 years before the area fell to the Arabs, the Jewish community of circa 36,000 members (before 1945) was victimized by street mobs in 1945 and again in 1948. Realizing that the monarchical government of King Idris I was either unwilling or too weak to protect them, 32,000 Jews left the country. The remainder (except for 220 Jews) fled after the June 1967 war since the new republican government refused to provide Jews with passports or any other documents for legal emigration.[50]

In Tunisia the Jewish community of 80,000 (as of 1956 when Tunisia gained its independence from France) faced no governmental discrimination. Habib Bourguiba as President (and leader of the Social Democratic Neo-Destour Party) was the only Arab leader who acknowledged the right of the State of Israel to exist and urged a peaceful settlement between Israel and the Arab states. The fact that most of the Jews left

Tunisia (chiefly for France) was motivated by their fear that the Bourguiba government would not be able to maintain itself for long and that Algeria in alliance with Egypt (and the support of the Soviets) would come to dominate Tunisia. While Bourguiba accused Nasser before and after the June 1967 war of having provided Israel with a *casus belli*, a wave of solidarity with the Arab states swept the Tunisian population. During the June 1967 war, Moslems set fire to the Great Synagogue in Tunis and publicly tore up the Torah Scrolls. After the war there was a new wave of emigration indicating that the Jewish community of Tunisia would shortly disappear altogether.[51]

In Algeria, conquered by the French during the 1830–1847 period, the Crémieux decree of 1870 had given citizenship to all Algerian Jews since there were no religious obstacles preventing the Jews from following French law. When the Fifth French Republic, in 1962, granted independence to the Algerians, the 150,000 Jews realized that the close to eight million Arab Moslems led by a Pan-Arab and pro-Soviet National Liberation Front could not be trusted to allow them civic equality within the new Algerian Republic. Panic-stricken, most Jews (like most other Algerians of non-Arab extraction) abandoned their possessions and left Algeria. It is estimated that 80% of the Algerian Jews settled in France which treated the Algerian Jews on an equal footing with the non-Jewish repatriates.[52]

In Morocco (which attained its independence from France in 1956) the Jewish community is in the process of liquidation. Of 199,156 Jews recorded in the 1951 census, but 31,000 were still in Morocco by 1972. Here, as in the case of Tunisia, the cause cannot be seen in any internal anti-Jewish policies on the part of the government (Hassan II since 1961). But the fact that it allowed Arab Moslem army units to be recruited in Morocco and to join the Eastern Arabs in the June 1967 war and again in the Yom Kippur war in October 1973, coupled with anti-Jewish demonstrations on the part of Arab nationalist groups in the leading Moroccan cities,[53] has, no doubt evoked in the Jews the desire to leave the country (in which their ancestors had settled before it was the Roman province of Mauritania).

In Yemen the Jews experienced for hundreds of years the full force of Islamic orthodoxy. Considered to be the Moslems' "deadliest enemies," "expelled by God from the community of human beings," "carriers of diseases and pests," possessed by a "criminal character which is the outcome of the Jewish faith" (these quotes were used by the Fourth Conference of the Academy of Islamic Research at Al Azhar University in Cairo),[54] the Jews of Yemen were forbidden to wear the same clothes as Moslems, to own property, to ride on animals and, of course, to make converts. In reverse, Arab Moslems were encouraged to convert the Jews to Islam, and Jewish orphans were converted by force. As soon as the Jews of Yemen heard about the Jewish colonization efforts in Palestine, they petitioned for the right to emigrate, and a number of them arrived there in 1910. By the time the State of Israel was proclaimed (May 1948), 18,000 Yemenite Jews had settled within its frontiers. In 1949 the entire Jewish community, numbering 46,000 persons, was transferred to Israel in what became known as "Operation Magic Carpet" (so named by the Yemenite Jews who had never before seen airplanes and were now flown to a land which their ancestors had been forced to leave by the Roman legions close to 1,900 years ago).

In Saudi Arabia there are no Jews. The Jewish communities which existed in the north and centre of the Arabian peninsula, particularly in Medina, before the rise of Islam were wiped out during the continuous tribal warfare in the area. By the time Ibn Saud, the rule of Najd, conquered Mecca (1924) and forced the Hashemite Sharif Hussayn to abdicate his leadership of the Hejaz, the entire Arabian Peninsula (except Yemen and the British Crown Colony of Aden) had become "*Judenrein.*" But the fact that the Jews of Israel defeated the Arab Moslem armies in 1948–1949, 1956 and again in 1967 caused special consternation to the Saudi dynasty which supports the Wahabi Ikhwan (Brotherhood) Movement, a puritanistic branch of Islam which believe that faithful Moslems will always prove to be militarily superior to Christians, Jews or other non-believers. A sort of *Jihad* (Holy War) spirit prompted the government of King Faisal (Ibn Saud's younger son who had followed his brother Saud on the throne) to send Saudi Arabian troops to the Syrian front against Israel in the Yom Kippur war in October 1973 and to heed the demand of the USSR which in supporting the Arab cause had asked the heads of the Arab oil producing states to stop all oil shipments to the USA, known to have supported Israel.

IV. Anti-Jewish Discrimination in Latin America, Spain, and Portugal

In Cuba, the 1959 revolution brought profound changes in the socio-economic structure of the country and practically wiped out the economic stability of the majority of Cuban Jews. As a consequence, most of the 12,000 Jews who lived in Cuba before 1959 have emigrated. Since the establishment in Havana of the Secretariat of the "Tri-Continental Organization" (the Latin American associate of the Palestine Liberation Organization), the Cuban press has become increasingly anti-Israel in tone. In the United Nations, the Castro government of Cuba has consistently supported the Arab position against Israel.

As of 1972, the Jewish populations of major Latin American states were:

Argentina	500,000	Colombia	13,000
Bolivia	2,000	Costa Rica	1,500
Brazil	150,000	Mexico	40,000
Chile	6,000	Uruguay	50,000
		Venezuela	15,000

These Jewish communities enjoy civic equality as well as religious and cultural freedom. Anti-Jewish propaganda (books, magazines, posters, etc.) stems chiefly from Leftist revolutionary groups, embassies of the Arab states and offices of the Arab League, as well as some professedly anti-Communist traditionalist Catholic organizations.

In Argentine, where the Jewish Colonization Association had in 1890 settled some 50,000 Jews from Czarist Russia in a number of agricultural villages (some 6,000 Jewish families are still farming),[55] the Argentine economist Beveraggi Allende, in widely distributed pamphlets (possibly financed by the Information Office of the Arab League), has accused the Jews of harbouring a secret plan to establish a Jewish state of Andina in Patagonia as an integral part of "the machinations of world Zionism to invade national territory."[56] In July, 1972, a pamphlet was widely circulated in church circles entitled. *The Jews, Israeli Citizens*, warning non-Jewish Argentines of the "Jewish danger." Since it was signed by the "Sisters of the Sacred Heart," the Delegación de Asociaciones Israelitas Argentinas, composed of representatives of the Jewish organizations of Buenos Aires and the provinces, contacted the Order to protest against the issuance of the pamphlet. The Sisters declared that their name had been used without authorization.[57]

In Brazil the "liberal" (but invariably pro-Soviet) weekly, *Pasquim*, used the attack on the Israeli athletes in Munich to liken Golda Meir to Hitler since both, according to the periodical, were responsible for the massacre of Jews – Hitler through genocide; Mrs. Meir because of her unwillingness to free the Arab terrorists in Israeli prisons.

A report on Brazil in the *American Jewish Yearbook* states, "There were no racial or religious problems in Brazil because the Jews, like all other ethnic or religious groups, had become an integral part of Brazilian society. However, some Brazilian publications, especially the so-called liberal ones, have been strongly anti-Israel and, consequently, somewhat anti-Jewish.... There has been a resurgence of Arab activities. The Arab League Office in Rio de Janeiro published, besides its monthly *Oriente Arabe*, anti-Israel pamphlets and apparently also sponsored books by Yasir Arafat, leader of the Palestine Liberation Organization, and other anti-Zionists, as well as a strongly anti-Jewish version of Karl Marx's *Zur Judenfrage*. An anti-Semitic book published in Portugal, *The Plot Against the Church*, has been distributed mainly in the cities of the interior."[58]

In Chile the so-called "Chilean Way to Socialism" under the leadership of Dr. Salvador Allende, elected President in 1970, caused some 6,000 Jews, most of them survivors of the Nazi Holocaust and immigrants from the Soviet bloc countries, to emigrate, partly to the USA, largely to Israel. Those Jews who stayed, facing an increasing polarization in Chilean politics, seemed to have been diversified in their preferences. Some favoured Allende who was credited with efforts to bring about an ecumenical spirit among the Catholics, Jews and Protestants; others supported the anti-Marxist coalition of the centre and right-wing groups for fear that ahead of Allende's "marxism" lay a full-fledged Communist totalitarianism.

In Mexico anti-Jewish sentiments had been nurtured for decades by the lower echelons of the Catholic clergy as well as various hate-mongering racist groups.[59] Former students of Patrice Lumumba Friendship University in Moscow, trained on North Korean soil in urban guerrilla warfare, joined by Arab students, have since 1967 been active in promoting anti-Jewish and anti-Israel propaganda on Mexican university campuses. After Mexico's establishment of diplomatic relations with the People's Republic of China, anti-Israel materials written in Spanish and produced in Mainland China have been arriving in ever-increasing quantities in the major cities of Mexico.[60] All this has prompted the government and the mass media to condemn any type of anti-Jewish discrimination. A centre of Ecumenical Studies organizes inter-faith youth meetings and seminars for both clergy and laity. At the suggestion of the metropolitan archdiocese (Cardinal Monsignor Miguel Dario Miranda), rabbis have been asked to explain Judaism and Jewish history in ecumenical meetings.

In Venezuela the left-wing groups, mainly the Communist Party and followers of Castroite Cuba, disseminate anti-Israel propaganda via forums, conferences, posters in streets and universities. As in the other Latin American countries, the word "Jewish" is deliberately inter- changed with "Zionist," "Israeli" and "Israelite," thereby reviving some deep-rooted anti-Semitic trends.

While the geographical proximity and political stature of the USA with the largest and most flourishing Jewish community of the world has, no doubt, a beneficent influence on the Latin American countries, their lack of a genuine cultural pluralism and political stability causes many Jews in them to view their own future without much confidence. As a consequence, Zionism is popular among the Jews of Latin America who since 1948 have constituted an important potential for emigration to Israel.

In Spain (9,000 Jews as of 1972) and *Portugal* (580 Jews as of 1972), the two European states which colonized Latin America, the Jews en- joy today legal equality and economic opportunity. Spain's religious liberty law of June 28, 1967 affirmed that the 1492 Order of King Fer- dinand and Queen Isabella banning the Jews from Spain was void. Under this law and various implementing ordinances, synagogues were constructed in Madrid and Barcelona which, however, are not permit- ted to have any visible identification since public religious display is the prerogative of the Roman Catholic Church. While the law of June 28, 1967 guarantees equal facilities for marriage and burial for non-Catho- lics as for Catholics, the right to choose the faith of one's children and a juridical recognition of non-Catholic associations, the pervasiveness of an orthodox Catholicism in Spain has preserved a number of pejorative stereotypes about Jews and Judaism. In Madrid, for example, the cult of Nino de la Guardia, venerated as the victim of an alleged Jewish ritual murder in the Middle Ages, is still in full force. While the Franco government must be credited with a number of efforts to assist Jews persecuted in Arab countries, it has so far refrained from curbing the overt anti-Semitism of the Junta Espanola Tradicionalista and the neo- Nazi "New Order" organization whose members are known to smear swastikas on Jewish homes and the synagogues.

In Portugal there are still numerous Marranos (persons whose an- cestors were converted to Catholicism by force in 1497 under the Edict of King Manoel II) with strong Jewish sympathies.[61] Whether the *coup d'état* of April 25, 1974 under the leadership of General Antonio de Spinola (promising "a democratic Portugal, truly free and with more social justice")[62] will remove the fear of religio-political reprisals and facilitate the reconversion of the Marranos to Judaism remains to be seen.

V. Anti-Jewish Discrimination in Western Europe

In the Scandinavian states, the Netherlands, Belgium, the Federal Republic of Germany (West Germany), Austria, Switzerland, France, Italy, Great Britain, New Zealand, Australia, India, the Republic of South Africa, Canada and the USA, political systems which profess constitutionalism (i.e. the Rule of Law, majority decision and minority protection of all persons recognized as citizens), there is no governmental anti-Jewish discrimination and on the whole the Jewish position in these countries is secure. This, however, does not prevent individual Jews and some Jewish organizations from expressing every so often their sentiments of insecurity concerning the various activities of anti-Semites able to use and abuse the freedoms of speech and organization guaranteed by democratic regimes even to their avowed enemies. Although the authentic Jew, aware of his Jewish background and consciously identified with it, will take it for granted that non-Jews see in him the spiritual, if not the genetic, descendant of Abraham, Isaac and Jacob, of Moses and the prophets and either trust him or distrust him because of his heritage, the assimilationist Jew, weakened by the adverse consequences of centuries of anti-Jewish persecution, identified by others as a Jew but personally inclined to escape this identification, is by far more sensitive *vis-à-vis* factual or imagined anti-Semitism. Anxious to avoid whatever traits the anti-Semite cares to stigmatize as "Jewish," the assimilationist frequently accepts the anti-Semite's accusations (that the Jewish faith is inferior or that Jews are not sufficiently patriotic or that Jews are too loud or too rich or too poor). Most Jews quite naturally demand complete freedom of religious life and civic equality (i.e. the same legal rights and duties which are granted to non-Jews). But they will, normally, have no objection to organizations of a private character which confine their membership to Catholics, Episcopalians or other denominations. The peripheral Jew, on the other hand, is prone to support endeavours which are to facilitate his access to groups and aspects of social life which otherwise would exclude him as a "Jew." He is, therefore, known to raise the banner of anti-Semitism whenever a private golf club or a dining club limits its membership to a Christian denomination. In his anxiety to belong, he forgets that there are all kinds of Jewish organizations (also clubs) of a similar nature.

In the Scandinavian states, Sweden (15,000 Jews as of 1972), Norway (800 Jews as of 1972), Denmark (6,000 Jews as of 1972), Jews consider themselves and usually are considered by the non-Jewish majorities as different from their neighbours in only one respect, their religious faith. Interestingly enough, most of the Polish Communists of Jewish origin, compelled by their fellow Communists to leave Poland after 1967, chose Stockholm and Copenhagen as places of refuge (100 of these expellees went to Oslo). A few of these Polish émigrés have revived the concept of the Polish Jewish socialist Bund (prior to the Nazi-Soviet occupation of Poland in 1939) of a national Jewish cultural autonomy in Europe based on the Yiddish language. Others who now admit that they were mistaken in assuming that "socialism will end all anti-Semitism," might have wished to go to Israel. They realized, however, that due to the political influence of the orthodox rabbinate on the government of Israel, their non-Jewish wives and children would not be accepted as Jews unless they would go through lengthy orthodox conversion rituals. Virtually all of these emigrés who account for nearly one-fourth of Scandinavia's Jewish population are atheists or at least religiously indifferent. In contrast to traditional Scandinavian Jewry, they, however, avow a cultural Jewishness expressed in Jewish secular organizations, emphasizing Yiddish literature, "Jewish" music and "Jewish" food. Unsullied by any anti-Semitic tradition, the Scandinavians have so far shown no animosity toward these newcomers.[63]

In the Netherlands (30,000 Jews as of 1972) and *Belgium* (40,000 Jews as of 1972) government and general population have consistently expressed their solidarity with Israel and world Jewry. Anti-Semitic utterances are immediately quenched by an overwhelming public opinion.

In the Federal Republic of Germany (circa 30,000 Jews as of 1972) and in *Austria* (circa 9,400 Jews as of 1972) the Jews who survived the Hitlerian Holocaust (including German and Austrian as well as East European Jews), are understandably most sensitive concerning any reappearance of anti-Semitic organizations and expressions.

In the Federal Republic of Germany anti-Jewish agitation, unequivocally discouraged by the government authorities, stems from two sources: the Communist and to a lesser extent socialist left which readily allies itself with Arab anti-Zionist and anti-Israel terrorism and derives its organizational strength from student organizations at West German

universities and right-wing extremists who oppose the trials of former
Nazi functionaries, deplore the reparations payments made by West
Germany to Israel and individual Jewish victims of Nazism, warn the
Germans that "Israel might drag them into another World War," etc.
While the National Democratic Party, as a power factor in West German politics, has declined (to 0.6% of the electorate in the Federal elections in 1969), its weekly organ, *Deutsche Nationalzeitung*, can still boast
of some 120,000 readers. On October 19 and November 2, 1973 its
headlines were: "Israel Rules America" and "Are We to Bleed for Israel?" The murder of eleven Israeli athletes in the Olympic Village by
Arab terrorists identifying themselves as members of "Black September" had been facilitated by the Communists whose youth organizations had familiarized the Arabs with the terrain and later argued
that the Munich act was not criminal since its causes were the existence
of Israel and the homelessness of the Palestine Arabs. In this connection
it must be noted that various groups of Turkish "guest workers," organized in the "National Socialist Turkish Workers' League" in West
Germany, have also sided with the Arab terrorists and the German
Communists. As a consequence, in March 1972 the executive of the
German trade union federation demanded federal and state action
against these Turkish "guest workers" for "nearly criminal racist agitation." Although the German authorities, for foreign policy reasons,
seem to be reluctant to react decisively in the case of anti-Semitism
practiced by foreigners, West German courts have on the whole been
willing to punish rather severely any Germans who were caught in the
desecration of Jewish tombstones or the publication of clearly anti-
Jewish statements. While the tiny Jewish community in the German
Democratic Republic is forced to affirm every so often its devotion to
the "Socialist Workers' and Peasants' State" and its support of the
Communist Middle East policy, the Jews of West Germany enjoy complete political and religious freedom.[64] In their support of Israel they
can count on the co-operation of influential segments of the general
population. In contrast to the Jewish community in the German Democratic Republic which is dying out;[65] the Jewish communities of West
Germany, worshipping in beautifully rebuilt synagogues, have actually
increased in number from 25,000 in 1968 to over 30,000 in 1972.

In Austria the neo-Nazis, like the Communists, prefer to attack Zionists and Israel rather than the "Jews." Nevertheless, as *Profil*, a Vienna
independent magazine, stated on November 9, 1973, both the neo-
Nazis and the Communists decry Israel as a "product of Jewish high

finance." When Chancellor Kreisky (himself of Jewish origin) ordered
the closing of the Soviet Jewish refugee transit camp at Schoenau Castle, former Nazis are said (by *Profil*) to have complimented him on the
"wise decision which at long last relieved Austria of its Jewish burden."
Other neo-Nazis allegedly favour the existence of Israel since they see
in it a chance of getting rid of the Jews without the "inhumanity of the
final solution." In the Socialist Party (which elevated Kreisky to the
chancellorship of Austria), a poll conducted by the Institute for Market
and Social Research between October 17 and 25, 1973, that is, shortly
after the outbreak of the Yom Kippur War, discovered that nineteen
percent of the members of this Party could be called "outspoken anti-
Semites" and fourty-four percent "moderate anti-Semites." Since the
poll, however, established no solid criteria for the meaning of "outspoken" and "moderate" anti-Semitism, it cannot be taken too seriously. Nevertheless, like the Poles (past and present), many Austrians demonstrate a widespread belief in ethnocentricity militating against Jews
as members of a minority faith and people whose loyalty to Austria is in
doubt.

In Great Britain (410,000 Jews as of 1972) anti-Israel agitation on the
part of the Communists and anti-Jewish agitation on the part of candidates of the National Front (in recent parliamentary and municipal
elections) have prompted the Association of Jewish Ex-Servicemen to
organize a research centre which is to identify every anti-Jewish group
operating in Great Britain and to make its findings available to the
police authorities and the press.

In France (550,000 Jews as of 1972), URSS, the official French language bulletin of the Soviet Embassy, appears to have been the chief
instigator of anti-Semitic campaigns. In September 1972 it published a
violently anti-Jewish statement stating i.a. that the Talmud propagates
racism and the despoilment of all non-Jews. In the same breath it
charged the Jews of Israel with teaching in their schools that all Arabs
must be exterminated. The Chief Rabbi of France, Grand Rabbin Jacob Kaplan, called the article "the vilest attack on the Jews of France
since the downfall of Hitler." Since it violated French law, prohibiting
incitement to racial or religious hatred, the International League
against Racism and Anti-Semitism (with headquarters in Paris)
brought civil suit against the Soviet Embassy. The French editor of
URSS argued that the article was merely a translation from the Russian and that the original was written in the USSR. Nevertheless, the
French court ruled that the article violated the law banning incitement

to racial and religious hatred and ordered the Soviet Embassy to print the text of the verdict in URSS.[66] Ever since the June 1967 war Paris and its suburbs have been the chief French localities for physical confrontation between Arab students, pro-Arab leftists (Trotskyites, Maoists, pro-Soviet Communists) and Jewish students and their non-Jewish (chiefly right-wing) sympathizers.

VI. Anti-Jewish Discrimination in the United States

In the USA (6,150,000 Jews as of 1972) the Anti-Defamation League (organized by the Jewish Fraternal Order B'nai B'rith) reported (New York Times, November 20, 1972) "a marked increase in the scope of American anti-Jewish activities, incidents, organizations and publications." The dangers to American Jews, the Anti-Defamation League stated, came from the "left" as well as from the "right," "from otherwise respectable sources and from extremist segments of minority groups." There can be no doubt that in the permissive climate of the USA the (circa 100) members of the American Nazi Party as well as Black Militants are openly advocating the cessation of all American aid to "the government of Israel," the "return of Israel to the Palestinians," etc. It is also an established fact that in Harlem, the former Negro ghetto of New York and still the quarter of the relatively largest concentration of Blacks in the USA, campaigns of Black militants (foremost among them the Black Muslims and the Black Panthers) against the "Jewish landlords" have overtones of anti-Semitism. In view of the efforts of the Federal government and a number of state governments to increase the student enrolment of Blacks and other minority groups, coupled with an "affirmative action" programme prepared by the Health, Education and Welfare Department of the US government for the hiring practices in colleges and universities, Jewish students and educators experience deep anxieties and fear the introduction of a quota system which would severely reduce the proportion of Jews in academic institutions. President Nixon's guidelines in this respect, stressing the need to avoid discrimination and placing greater emphasis on merit made it clear that equality of opportunity for members of minority groups (Jews are not considered a minority group in the USA) and women, did not imply any fixed quotas or proportional representation.[67] In their confrontation with Black militants, the Jews ideally conformed to the joking remark of a Philadelphia politician who raised the question, "Do you know what a conservative is?" and answered, "He is a liberal who got mugged the night before."

Since the end of World War II, most American conservatives and rightist groups have shown no inclination toward anti-Semitism. In sharp contrast to the Black militants and the radical left (particularly the American Communist Party and the Socialist Workers' Party), American conservatives and rightists have, on the whole, been staunch defenders of Israel.

Even in countries like the USA, Great Britain and Canada, generally devoted to socio-cultural pluralism, adherents of a pre-pluralist ideology of a "Christian America," a "Christian Britain," a "Christian Canada" become at times the sources of Jewish worries since these pre-pluralists feel called upon to indulge in campaigns to proselytize among the Jews. Spearheaded by evangelical bodies (e.g. The American Board of Missions to the Jews), a number of obviously well-financed "campus crusades for Christ" and appeals by the "Jesus Freaks" purport "to saturate the entire nation with the claims of Jesus Christ."

While some Jewish organizations, particularly orthodox Jewish youth groups, have preoccupied themselves with countermeasures to develop an increasing sense of identity among young Jews, there is little indication that the efforts of the Christian proselytizers have been successful.

In this respect it must be noted that anti-Jewish as well as anti-Israel manifestations do not originate only in the quarters of the relatively few non-Jewish extremists in the American population. Defamation of Jews and Judaism as well as the condemnation of Israel and Zionist efforts stem every so often also from persons of Jewish origin indulging in what has been labelled "Jewish self-hate." While the overwhelming majority of the Jews of the USA, affiliated with either the Reform Jewish congregations (about one million), Conservative Jewish congregations (about $1\frac{1}{2}$ million) or Orthodox Jewish congregations (about one million), a variety of Zionist and pro-Israel organizations and a vast number of charitable and education institutions, live a dignified Jewish religious and cultural life, a few Jewish-born outsiders to Judaism participate (and at times lead) in anti-Jewish and anti-Israel manifestations. A large proportion (in New York City about one-third) of the 4,000 to 5,000 "Students for a Democratic Society" (a deliberate misnomer to cover the terroristic activities of Communist and anarcho-Communist university students) are the sons and daughters of affluent Jewish parents who, themselves indifferent to Judaism, failed to transmit any Jewish values to their offspring. Characteristic of "Jewish self-hate" are also the anti-Jewish slurs, stereotypes and insulting innuendos in birthday cards, comic posters and "party books" produced and distributed chiefly by persons of Jewish origin for whom the Nazi Holocaust had admittedly no meaning whatever and who insist "it's fun to be Jewish."

VII. Inter-Denominational Discrimination in the State of Israel

In Israel (2,723,000 Jews as of 1972) anti-Zionist and anti-Israel activities are carried out by various Arab groups which enjoy the support of the Arab states and the Palestine Liberation Front (and its splinter organizations) with headquarters in the half-Moslem, half-Christian Lebanon, Syria and Libya. Discrimination against non-orthodox Jews can conceivably be attributed to the Orthodox rabbinate in the State of Israel which guarantees religious freedom to all faiths, but, ironically, not to all wings (or denominations) of Judaism. The Orthodox rabbinate which influences the government policies of Israel via Mafdal, the National Religious Party (frequently a member of the government coalition) and the Torah Religious Front has so far seen to it that neither the Reform rabbinate nor the Conservative rabbinate have been recognized in Israel. In view of the fact that there is no separation of church and state in Israel, Jews can, therefore, only be married (or divorced) by an Orthodox rabbi (or an Orthodox rabbinical court).

Of particular significance is the Orthodox stance with regard to the problem of "Who is a Jew?" since under the "Law of Return" Jewish immigrants are given access to immediate citizenship. Promulgated by the Knesset (Israel's Parliament) on July 5, 1950, to alleviate the plight of the homeless and stateless Jewish survivors of the Hitlerian Holocaust, this Law states, "Immigrant visas shall be issued to any Jew expressing a desire to settle in Israel, except if the Minister of Immigration is satisfied that the applicant acts against the Jewish people."[68] The exception was invoked by the Israeli Supreme Court when in 1962 a Catholic monk, Brother Daniel (Rufeisen), claimed to be a Jew entitled to immediate citizenship under the Law of Return. From the point of view of the Orthodox interpretation of Halakhah (a collection of oral teachings interpreting the written laws of the Torah), he was indeed a Jew since he was born by a "Jewish" mother and since Orthodox halakhaic interpretation does not absolve a Jewish convert to Christianity (or any other faith) from his obligations as a Jew. But the Israeli Supreme Court decided that "the Jewish people" does not consider a convert to another faith a Jew. By rejecting Brother Daniel's appeal to benefit from the Law of Return (allowing him the alternative way of becoming a citizen of Israel via normal naturalization procedures), it de-emphasized the biological component of a person's Jewishness (birth by a Jewish mother) and stressed the Biblical definition of the Jew (laid down in the Book of Ruth, Chapter I) as a conscious member of the Jewish faith.[69]

Since in the last decades a number of Jews in the USA and Western Europe have derived their Jewish identity not from birth but rather from conversion to liberal Judaism (Reform Jewish or Conservative Jewish congregations), their claim to be Jews might be jeopardized by the Orthodox rabbinate of Israel should they desire to emigrate to Israel. This, in turn, has engendered a movement among Jews inside and outside Israel demanding the recognition of the non-Orthodox wings of Judaism by the State of Israel and an Israeli version of the separation of church and state concept which without doing violence to prophetic Judaism would interpret the Jewish tradition in the light of the exigencies faced by modern Jewry in Israel and the Diaspora.

VIII. SUMMARY

Generally speaking, the phenomenon of anti-Semitism has always been as universal as the dispersion of those men and women who identified themselves as Jews. Whether it was their antagonism to the Hebrew-Jewish monotheism with its highest ethical expression in the prophets or to the Zionist concept of returning the Jewish people to an ordinary place among the nations, anti-Semites throughout the world have never lacked arguments – sometimes sophisticated half-truths, sometimes crude lies – to cast aspersion on the Jews. The fact that today "New Left" and "Old Left" radicals, including some men and women of Jewish origin, support Arab nationalists and the Communist-governed states in hammering at the bugbear of "Zionist imperialism", the further fact that Neo-Nazis, various "rightist" radicals steeped in religious and ethnic prejudices, and even some respectable commercial interests anxious to do business with Arab governments join them in denouncing Jews and Judaism, only attests to a deeply rooted irrationality for which Auschwitz was but a symbolic manifestation. In view of modern technology, it is, therefore, not inconceivable that the internationalism of anti-Semitism might result in an anti-Semitic international eager and able to trigger a new epidemic of anti-Jewish hate.

NOTES

1. In *Der Sieg des Judentums über das Germanentum* (published in 1871), Wilhelm Marr opposed the legal equality of Jews in Germany (enacted in 1872) and founded in 1879 the "Antisemitenliga." *Das Dritte Reich, Seine Geschichte in Texten, Bildern und Dokumenten,* Ed. Heinz Huber and Artur Mueller, Munich, Desch Verlag, 1964, p. 496. *Encyclopedia Judaica,* New York, Macmillan Co., 1971, Vol. 3, p. 87, calls Wilhelm Marr "the originator" of the word "anti-Semitism."
2. Prior to the outbreak of World War II, official Nazi policy was to deprive the Jews of their economic assets but to allow those to emigrate who could secure a visa to a foreign country. However, shortly after the Munich Agreement of September 29, 1938, the Nazi leaders began to contemplate the murder of those Jews who in the event of war would still be on German soil. On November 24, 1938 the SS organ, *Das Schwarze Korps,* declared that "the fate of such Jews as the outbreak of war should still find in Germany would be their final end, their annihilation *(Vernichtung).*" Quoted in Gerald Reitlinger, *The Final Solution,* New York, The Beechhurst Press, 1953, p. 8.
3. The Jewishness of a person was established on December 27, 1932 by the Soviet of People's Commissars which created the "single passport system" for the USSR (*Pravda,* December 28, 1932). It was their knowledge of Yiddish, their own religious faith or that of their parents which formed the criteria for the nationality of the Jews. William Korey, "The Legal Position of Soviet Jewry" in *The Jews in Soviet Russia since 1917,* Ed. Lionel Kochan, London, Oxford University Press, 1972, pp. 77–78.
4. Quoted in *Russian Jewry 1917–1967,* Ed. Frumkin, Jacob; Aronson, Gregor; Goldenweiser, Alexis; and Lewitan, Joseph. New York, Thomas Yoseloff Publishers, 1969, p. 172.
5. *Ibid.*
6. Joshua Rothenberg, "Jewish Religion in the Soviet Union, in Kochan, *op. cit.,* p. 165.
7. *Birobijaner Stern,* October 22, 1959. Quoted in *Jews in Eastern Europe: A Periodical Survey of Events Affecting Jews in the Soviet Bloc,* Ed. Emanuel Litvinoff, London, May 1960, p. 54.
8. Rothenberg, *op. cit.,* p. 164.
9. *Ibid.,* p. 179.
10. Quoted and criticized as an anti-Semitic libel in the official organ of the Communist Party of Italy, *L'Unita,* December 15, 1964.
11. *Jews in Eastern Europe, op. cit.,* November 1972, pp. 11–12.
12. An English translation of the text, quoted in *Church and State under Communism,* was prepared by the Law Library of Congress, Washington, 1964 (pp. 12–17 and pp. 18–24).
13. Richard Cohen, *Let My People Go,* New York, Popular Library, 1971, p. 146.
14. *Light on Soviet Jewry: Report of a Conference on Jews in the USSR,* Board of Deputies of British Jews, London, Woburn House, June 15, 1969, p. 41.

15. Boris Smolar, *Soviet Jewry Today and Tomorrow*, New York, Macmillan, 1971, p. 41.
16. *Figaro*, April 9, 1958.
17. *Réalités*, May 1957. Quoted in William Korey, *op, cit.*, p. 91.
18. *Pravda*, February 28, 1963.
19. *Pravda*, April 4, 1964.
20. Leonard Shapiro, "Introduction," in Kochan, *op, cit.*, p. 12.
21. Quoted in *Jews in Eastern Europe, op. cit.*, July 1969, pp. 12, 14, 16 and 17.
22. Quoted by Theodore Shabad, "Nazi-Zionist Link in War Is Seen in a Soviet Novel," *New York Times*, March 11, 1973.
23. Quoted in *Jews in Eastern Europe, op. cit.*, January 1970, p. 73.
24. Franz Mehring's biography of Karl Marx, University of Michigan Press, 1962, p. 3.
25. *Jews in Eastern Europe, op. cit.*, November 1972, pp. 15–17.
26. Quoted in *Light on Soviet Jewry, op. cit.*, pp. 30–31.
27. *Jews in Eastern Europe, op. cit.*, April 1971, pp. 78–85.
28. *Ibid.*, p. 29.
29. On September 3, 1972 a letter, signed by 200 Jews in Moscow, to the Supreme Soviet stated, "Expenditure for the higher education of a day student, including subsistence grants, amounted on average according to the national economy for 1961–1970 to 600 rubles or, over a period of $5\frac{1}{2}$ years' study to 3,300 rubles. The expenditure for a student of an external or evening department is between $\frac{1}{2}$ and $\frac{1}{4}$ of these sums. When entering a higher institution of learning, the student takes upon himself one obligation only: to work as assigned for three years as a young specialist. On completion of this term the person has the right to change his speciality or to go abroad ... Jews going to Israel are forcibly deprived of their USSR citizenship. In violation of the Resolution of the Soviet of Ministers of the USSR, No. 803, of September 22, 1970 they are required to pay a state tax of 500 rubles in spite of their not having made applications requesting to be deprived of their citizenship." Quoted in *Jews in Eastern Europe, op. cit.*, November 1972, pp. 50–51.
30. This document, published by *Samizdat*, the underground free press of the USSR, was reprinted in *Midstream: A Monthly Jewish Review*, New York, The Theodor Herzl Foundation, February 1974, pp. 40–41.
31. Rudolf Höss, the Commander of the Auschwitz concentration camp from May 1, 1940 to December 1, 1943, in a written statement admitted that during his Auschwitz commandership "three million Jews were killed in Auschwitz." He added that "400,000 Hungarian Jews were killed in Auschwitz alone in the summer of 1944." *Das Dritte Reich, op. cit.*, p. 534.
32. Abraham Duker, "Anti-Semitism," in *Handbook of World History*, Ed. Joseph Dunner, New York, Philosophical Library, 1967, p. 55.
33. *Trybuna Ludu*, March 20, 1968.
34. Paul Lendvai, *Anti-Semitism Without Jews: Communist Eastern Europe*, New York, Doubleday 1971, p. 108.
35. *Ibid.*, p. 91.
36. Reitlinger, *op. cit.*, pp. 492–493.
37. Artur London, *L'Aveu: Dans l'engressage du Procès de Prague*. Version française. Paris, Gallimard. 1968.
38. Lendvai, *op. cit.*, p. 274.
39. *The New Standard Jewish Encyclopedia*. Ed. Cecil Roth and Geoffrey Wigoder. New York. Doubleday. 1970, p. 935.
40. Lendvai, *op. cit.*, p. 321.
41. *Scinteia*, Bucharest, May 9, 1966.

42. *The New Standard Jewish Encyclopedia, op. cit.,* p. 2004.
43. On May 14, 1948, when the State of Israel was proclaimed, the criminal code prescribed imprisonment for seven years or death for those found guilty of "Zionism." Since the military courts, aware of the aim of the authorities to fill the government treasury with the money of the Jews, judged on the evidence of two witnesses, it was not difficult for anyone to label any Jew a "Zionist." *Encyclopedia Judaica, op. cit.,* Vol. 8, pp. 1452–1453.
44. *Ibid.,* p. 1453.
45. *Ibid.,* p. 1456.
46. *Ibid.,* Vol. 15, p. 646.
47. *The Synagogue Light,* Ed. Rabbi Meyer Hager, New York, Synagogue Light, Inc., Vol. 40, No. 7, April 1974, p. 7.
48. *Encyclopedia Judaica, op. cit.,* Vol. 6, p. 501.
49. *American Jewish Yearbook,* American Jewish Committee, New York, American Jewish Committee, Vol. 69, 1968, p. 136.
50. *Ibid.,* p. 138.
51. *Encyclopedia Judaica, op. cit.,* Vol. 15, p. 1448.
52. In 1965 Algeria's Chief Court declared that the Jews (fewer than 1,000 Jews had remained in Algeria) were no longer under the protection of the law. With one exception, all synagogues of Algeria were confiscated and converted into mosques. (*loc. cit.* Vol 2, p. 619.)
53. The Istiqlal Party published in 1965 the *Protocols of the Elders of Zion.* In 1967 it called on all Arabs of Morocco to boycott the Jews. (*loc. cit.,* Vol. 12, p. 346.)
54. The Fourth Conference of the Academy of Islamic Research, Al Azhar. Academy of Islamic Research, Cairo. General Organization for Government Printing Offices. Rajab 1388, September 1968, pp. 373, 527, 530 and 535.
55. *American Jewish Yearbook, op. cit.,* Vol. 74, 1973, p. 319.
56. *Ibid.,* p. 327.
57. *Ibid.,* p. 328.
58. *Ibid.,* p. 337. *The Plot Against the Church,* authored by Maurice Pinay (probably a pseudonym) made its first appearance in an Italian edition *(Complotto contro la Chiesa),* distributed among the delegates of the Vatican Council in Rome in 1962. The book was also published in German, Spanish and Portuguese translations. The English translation was issued by "Sons of Liberty," Hollywood, California, in 1967.
59. *Ibid.,* p. 353.
60. *Ibid.,* p. 354.
61. *The New Standard Jewish Encyclopedia, op. cit.,* p. 1554.
62. *New York Times,* June 1, 1974.
63. Bernard K. Johnpoll, "Polish Jews in Scandinavia," *Midstream: A Monthly Jewish Review, op. cit.,* Vol. XX, No. 2, February 1974, pp. 47–57.
64. The German Democratic Republic (GDR) rejected all restitution claims made by Israel in 1951. Its leaders argued that the crimes against the Jews were committed by the Nazis and not by the German people. Since all Nazis had been banned from public life (this assertion is contradicted by the Jewish Documentation Center in Vienna under the Directorship of Simon Wiesenthal), the GDR felt itself under no obligation to pay reparations. When after the conclusion of the Basic Treaty between the Federal Republic and the GDR in October 1972 the GDR became eligible for recognition by the Western nations, Israel renewed its claims for indemnification from the East German state as a partial successor to the Nazi regime. The GDR, again, rejected these demands and used its new membership in the U.N. to denounce Israel as a "colonialist,

racist and expansionist outpost of world imperialism." *American Jewish Yearbook*, *op. cit.*, pp. 470–471.

65. *Ibid.*, p. 472.
66. *American Jewish Yearbook, op. cit.*, p. 398 and Maurice Friedberg, "Soviet Jews After the Yom Kippur War," *The American Zionist*, New York, Zionist Organization of America, Vol. LXIV, No. 7, May 1974, p. 9.
67. These implementational rules were effected by University Centers for Rational Alternatives, a nation-wide organization of university professors under the chairmanship of Sidney Hook who not only led the struggle against the introduction of ethnic (racist) quotas but must also be credited with having pioneered the restoration of order and rationality in the American institutions of higher learning which in the 1960's had been politicized by American leftists and had witnessed a near collapse of their academic functions.
68. Quoted in Joseph Dunner, *Democratic Bulwark in the Middle East*, Grinnell, Grinnel College Press, 1953, p. 41.
69. Moché Catane, *Qui Est Juif?* Paris, Editions Robert Laffont, 1972, pp. 66–67.

Foreign Workers in West Germany

CLEMENS AMELUNXEN

DR. CLEMENS AMELUNXEN, a native of Munster in Westphalia, studied law and theology in Germany, Belgium, and the United States, where he held a Fulbright Fellowship at Duke University. After obtaining his Doctor of Laws degree in 1953, he entered the German judicial service and served as a judge in various district and circuit courts. Since 1962 he has been a judge of the High Court of Appeal (Criminal Division) of the Rhineland, and he is now a member of the Presidency of that Court. His other public-service activities include the chairmanship of the Youth Welfare Council in Leverkusen, serving as counsel to the Road Safety Committee at Dusseldorf, and membership on the Judicial Examinations Commission.

Dr. Amelunxen has published 13 books on various aspects of public law, among which the following may be mentioned (titles translated from German): *Criminality in Women* (1959); *Suicide* (1962); *The Child and Criminality* (1963); *Political Criminals* (1964); *The Small States of Europe* (1965); *The Victim of Crime* (1970); *Industrial Protection and Industrial Crime* (1973); *Society and Its Rights* (1974); and *The Individual in Modern Criminal Law* (1975). Dr. Amelunxen's decorations include the Commander Cross, Order of Merit, Principality of Liechtenstein, and the Pontifical Medal awarded by the Holy See.

Foreign Workers in West Germany

CLEMENS AMELUNXEN

This study deals with certain aspects of the life of non-self-employed foreigners working in the Federal Republic of Germany. Their situation will be examined in its legal, occupational, economic, social, family, and educational dimensions. Their relation to the national society in this country and the possibility of their integration will be explored. It will be seen that discrimination and prejudicial treatment exist in varying degrees. The causes, therefore, are partly predetermined by national idiosyncrasies, linguistic difficulties, and technological backwardness. But they are also attributable to deficiencies in legislation, mistaken political judgments, and personal prejudices on the part of the host country population.

I. INTERNAL EUROPEAN MIGRATION

Ten million Europeans have left their original areas of settlement. They come from the agrarian regions of the South and crowd into the metropolitan centres of the Northern industrial states. Many are recruited and many immigrate illegally; nearly all are employed as workers. They get their families to follow them when this is allowed. In Switzerland 25% of the industrial workers come from foreign countries, as do 12% of the industrial workers in France. The Benelux countries, Austria, and Sweden likewise absorb foreign labour into their economies. Even states that provide workers receive at the same time unskilled labourers from still less developed countries south of the Mediterranean Sea.

The migration of guest workers is caused by a lack of industrial labour demand in the countries of origin. It is also a symptom of overpopulation in agriculture, which is characterized either by great latifundia or by scattered plots so small as to be sub-marginal. Medium size farms that provide an adequate base of existence for a family, usually with many members, are seldom to be found.

The emigrants come from the backward marginal zones of their countries that have not been sufficiently influenced by regional development programmes: typically from Galicia and Andalusia (Spain), Sardinia and Sicily (Italy), Anatolia (Turkey), as well as Montenegro and Macedonia (Yugoslavia). The stream of migration also flows from the periphery into the centres of Europe. Where the centres of expansion within the home countries (Madrid, Barcelona, Milan, Belgrade, Salonika) do not provide sufficient employment, workers cross the border in a northerly direction.

The result of all this is that precisely the highest developed industrial societies of Central Europe are receiving a flow of people whose social, economic, and educational level is relatively low. Almost inevitably, this situation gives rise to "pluralistic" problems. As a step toward understanding them, the case of the Federal Republic of Germany (FRG) can be considered.

II. NUMERICAL DISTRIBUTION

The F.R.G. has about 60 million inhabitants, of whom about 27.5 millions are gainfully employed. The "Alien Quota" (percentage of foreign workers in the total numbers of dependent employees as of 1973) stands at 10.8%. Among the population resident in West Germany, about 5% are foreigners. A total of about 2.6 million foreigners are working in West German industry and business. Somewhat more than 30% of these are women. Nationalities (in the sense of citizenship) are represented as follows:

Turks	528,200	22.4%
Yugoslavs	466,100	19.9%
Italians	409,700	17.5%
Greeks	268,100	11.4%
Spaniards	179,500	7.9%
Portuguese	69,000	2.9%
Moroccans	15,300	0.7%
Tunisians	11,200	0.5%

In addition to these actively employed people there are about one million family members who do not work, including about 500,000 children and juveniles under sixteen years. Approximately 60% of the married male and 90% of the married female guest workers live with their spouses in the F.R.G..

According to the results of the 1970 census, 50% of all foreigners lived within only 4% of the spatial area of the Federal Republic. The "density of foreigners" is thus highly differentiated. In late 1972 it varied between 1.6% in the agricultural area and 24% in the districts of industrial concentration (Munich, Stuttgart, Ludwigsburg, Frankfurt, and West Berlin). The economically highly developed State of Baden-Württemberg has the highest quota of guest workers (16%), whereas the lowest quota (5%) is found in the state of lower Saxony. Fifty percent of the port workers in Hamburg are Portuguese.

The guest workers are distributed among the individual sectors of the economy as follows:

Iron and metallurgical industry	38.0%
Other manufacturing industries	24.0%
Building industry	18.0%
Service Enterprises	17.0%
Mining	2.5%
Agriculture	0.5%

Half of the male and approximately one-third of the female guest workers had already lived more than four years in the F.R.G. as of 1969. The numbers of returning immigrants are low, and many of those who have returned to their countries of origin come again to West Germany.

III. Law and Policy Concerning Aliens

The F.R.G. denies officially that it is an "immigration country," even though the facts of the situation would so indicate. The contention that West Germany is a non-immigration country enables the state to avoid accepting the legal consequences of the changed population structure, so far as the number and domicile of the guest workers are concerned. The laws are not directed toward the goal of making the guest workers citizens or even integrating them into West German society. They serve rather the purpose of preventing and obstructing such a development, in line with the old concept of a "compact and homogeneous nation."

The state doubtlessly has weighty reasons for this standpoint. As "permanent visitors" or even more as "new citizens," guest workers in the mass are very expensive for an industry that cannot count on continuous expansion. The desired homogeneity of the German population would also be endangered by the uncontrolled arrival and settlement of several million aliens.

The law of the Federal Republic concerning aliens, the main elements of which have been codified in the *Ausländergesetz (Law on Aliens)*, discriminates against the guest workers in comparison with the permanent population, as a part of which they do not count. As citizens of a country belonging to the European Community, the Italians are in the most favourable position. In West Germany they profit from the extensive freedom of movement within the territory of the European Community. Those affected most unfavourably, on the other hand, are the approximately 300,000 "illegals" (mainly Turks and Moroccans) among the guest workers, who without official recruitment and sojourn permit, often responding to the inducements of dubious private arrangers, arrive as "tourists" or are smuggled in.

A. Work and Sojourn Permits

Since November, 1972, entry by visa into the Federal Republic for purposes of employment has fundamentally been blocked. Job-seekers from the countries of recruitment (except Italy) can only enter with a work permit card issued by the foreign offices of the "Federal Labour Institute." This leads in practice, for example, to the situation in which the husband of a Turkish female guest worker, who enters West Germany as a tourist, is administratively expelled at the end of three months, even if housing and employment are available. As a result of the energy crisis, recruitment from non-European Community countries was stopped at the close of 1973.

The leeway for administrative discretion is broad, as the following provisions of the *Law on Aliens* demonstrate. The sojourn permit can be granted before or after entry into the country, but then may be declared invalid – even before entry. It can be granted without time limitation, for a limited time, or for a limited area; it may also be shortened or prolonged, as well as being made subject to conditions and requirements. An alien can be deported if he endangers the security of the Federal Republic or violates provisions of tax law – but he may also be denied permission to leave West Germany for the same reasons.

As a consequence, not even the appearance of a right to sojourn may arise. The foreigner is to a large extent the object of manipulation by the alien control authorities. He has, to be sure, the right to appeal to the Administrative Courts against arbitrary measures, but administrative "discretion" is in its nature difficult to probe through judicial control.

The linking of the sojourn permit to proof of housing suitable for a family and proof of specific employment – the conduct of an independent business or trade is not allowed – hangs as a permanent Sword of Damocles over the guest worker. It often subjects him to triple arbitrary treatment: at the hands of the employer, the landlord, and the official of the alien control authority. Complaints to the employer may lead to loss of the job; those to the landlord to loss of housing. Without work and a place to live, however, there is no longer any legal sojourn – and thus a vicious circle closes.

The sojourn permit is usually limited to one year. When application is made for an extension, the same factors as in the original granting of a permit must be examined all over again. There is also a one-year waiting period before a worker may have his family follow him, and even then permission is limited to the spouse and minor children; the admission of more distant relatives, to look after the children, for instance, is not permitted.

More than one million foreign workers, who have conducted themselves without reproach for between five and eight years, would meet the qualifications for a permanent sojourn permit (the so-called "little naturalization"). But only 7,000 of these have achieved this more or less assured status. Non-European aliens usually do not receive a permanent sojourn permit, even after marriage with a German woman. Not infrequently, this situation causes the break-up of families.

It should be pointed out that in other European countries the guest workers are in a better position so far as alien control law is concerned. In Sweden, where their percentage of the population approximates that in West Germany, they are officially classified as "immigrants" and receive the unlimited sojourn permit after the third year. In Switzerland too, the "permission to settle" and in France the "privileged residence card" is granted more generously.

B. Discrimination in Penal Law

The guest workers do not figure prominently in West German crime statistics. Generally, their share of crime is even lower than their proportional part of the population; the "paragons of virtue" with the lowest criminality are the Spaniards. This is all the more gratifying and praiseworthy since most guest workers are still in their younger years, at an age in which criminals are "active." They likewise often live in a perilous cultural and adjustment conflict, and they are more strongly exposed than the Germans to the temptations of an affluent society, which they are experiencing for the first time.

Exceptions to this favourable picture exist in connection with the crimes of disorderly conduct, assault and battery, and to some extent in the case of sexual offences as well as murder and manslaughter. The causes sometimes lie in the jealous passions of "hot blooded" Southerners, in unaccustomed imbibing of alcohol, in forced celibacy, in lack of leisure-time activities, as well as in traditional conflicts between different nationalities (Turks versus Greeks, Croats versus Serbs) or religious confessions (Catholics and Moslems).

German criminal justice does not always show the necessary understanding for these idiosyncrasies. Legal protection through court-appointed defence attorneys, interpreters, and experts is generally provided in a correct manner. But it is often to be read in the decisions of the lower courts that the punishment must be measured partially because the alien culprit has "abused the hospitality of the Federal Republic." The Federal Supreme Court has ruled against such considerations with refreshing clarity: such an "abuse" is not a permissible reason for increasing the sentence, since the equality principle of the Constitution also applies to foreigners.

Many lower courts also insist on "adjustment to German style and custom" (according to Franz), even in cases in which a mode of acting should at least not be reckoned as increasing the gravity of the offence, according to the national mentality of the foreigners concerned. The "deterrent function" of punishment is as a rule emphasized more strongly toward guest workers than toward Germans. In traffic offences involving alcohol, the driver's licence is often suspended for an unproportionately long time. The personal assessment of the alien accused in the explanation of judgment is likewise not seldom discriminatory.

Incarceration of guest workers during investigation often lasts longer than that of Germans suspected of the same offences. This can indeed be justified under the law on the grounds of increased danger of flight. It is a fact that the guest worker has no permanent personal or economic connections in the Federal Republic. He can return to his home country at any time without having to fear extradition. Indemnification for incarceration during investigation when innocent is only paid to aliens if the appropriate reciprocity is guaranteed by international treaty with the home country concerned.

In the penal institutions of the Federal Republic, the guest workers usually find themselves in total human isolation caused by the lack of any opportunity to communicate through speech. They consequently suffer considerably more severely from imprisonment than do their German co-prisoners. The foreign consulates make efforts for their compatriots within their capabilities, but even they cannot do much for the prisoners. Pastoral care is at best possible for Southern European Catholics but not for Turkish and North African Moslems.

After serving his sentence, the guest worker is usually expelled to his home country by the administrative authorities. But even minor offences punished with fines may lead to non-extension of the sojourn permit and thus practically to deportation.

IV. DISADVANTAGES IN VOCATIONAL AND ECONOMIC LIFE

On paper, the guest workers enjoy equal rights with their German colleagues in labour and welfare law. Theoretically, they receive "the same wage for the same performance." But things look rather differently in practice. The employment of guest workers has brought the West German economy two important advantages. Guest workers make it possible for German employees to rise to more qualified occupations and better paid positions. They also help to meet the need of the West German economy for regional mobility. Guest workers constitute the mobile shock absorber in the labour market, to be moved around as needed in times of boom and depression.

In practice, guest workers perform the most menial and dirtiest tasks. They drag tar spreaders, carry pig iron, clean toilets, and cart away the garbage of affluence. The public service departments of most West German municipalities would collapse were the guest workers to disappear overnight. These people are classified in wage groups – for the women there are special "light wage groups" – for which German workers can no longer be recruited. At least in the first year of their employment they are paid only what the wage standard calls for, while their German colleagues receive benefits over and above the standard. Aliens receive child allowances, but no separation allowances incident to split households.

Guest workers, who in the best case occupy places on the assembly lines of mass production, especially in automobile manufacture, suffer two and a half times as many accidents on the job as their German colleagues. This is not only attributable to lack of knowledge of accident-prevention rules, but also to the higher objective danger of their particular jobs. Preventive health measures called for under the Federal Social Aid Law are not available to guest workers.

If the average monthly gross earnings of the foreigners in 1972, which stood at DM 1280 per month, was only slightly less than the global German average of DM 1346, the reason was to be found in the fact that the generally lower paid guest workers performed more over-time hours and thus worked longer than the Germans. They also have less of their earnings at their own disposition, since they transfer approximately one-fourth of their net wages to their families in the home countries, thus contributing significantly to improving the balance of payments of lesser developed regions.

The savings offers of West German banks and credit institutions can be used by the guest workers only to an insufficient degree. The freeze periods on savings certificates, which alone provide a respectable rate of interest, are usually longer than the limited times the guest workers have to stay in the Federal Republic. Although the will to save and actual savings – about one-third of the wages – are considerable, the monies concerned are usually deposited in accounts with short with-drawal notice times and interest rates that are too low. Despite all this, approximately 800,000 guest workers maintain such accounts at German banks and savings institutes with total deposits of about 3.7 billion Deutsche Mark.

Other forms of saving assume minimal importance. The mere hoarding of money, senseless in times of inflation, looms closer in the consciousness of guest workers than making that money grow. Insurance savings for old-age annuities hardly occur to the guest workers, who are wholly caught up in the day-to-day battle for existence. Home-buyer's savings plans are likewise excluded for those guest workers who are interested in a house or an apartment in their own home lands; such objects are not financed by German building societies. Saving through purchase of securities finds no interest, for the lack of information if for no other reason.

The fact that every fourth male guest worker possesses his own car speaks at first glance for an elevated standard of living; the Italians have the highest ratio of motorization (27%), while the Turks (with 18%) are the least motorized. But the cars concerned are usually old jalopies purchased at second- or third-hand, which leave much to be desired so far as traffic safety is concerned. Guest workers have a disproportionately high involvement in traffic accidents. In the first year after a car has been purchased, the damage payments by insurance companies run 64% above the German average. The discriminatory but economically justifiable introduction of a "Malus" (insurance surcharge based on risk) for guest workers has been considered for quite a long time.

V. The Language Barrier

The number of guest workers who achieve upward social mobility is extremely low. Most of them begin as unskilled workers performing menial tasks, and even later they hardly improve their status. Only a few – and in this case again particularly Italians – rise from the lowest functions to achieve intermediate qualifications, not to speak of supervisory positions in the work force. The major obstacles to vocational careers are the lack of education and inadequate knowledge of the German language. The two handicaps are (according to Borris) inextricably linked with each other. Guest workers will remain a marginal group on the edge of West German society for as long as they fail to master or master only inadequately the German language. Without this communication skill they are underprivileged in their jobs and almost always isolated in their leisure time.

According to a questionnaire survey of the "Federal Labour Institute," the percentage of guest workers who are skilled workers only reaches the West German average of 40% in the case of those workers who classify their own knowledge of German as "very good." Four-fifths of the guest workers consider their knowledge less good, and are thus practically immobilized vocationally. The knowledge of German decreases with growing distance of the homeland from the Federal Republic. The "marginal inhabitants" from Turkey and Portugal show the smallest percentages classifying their knowledge of German as "very good": 7% and 5% respectively. Precisely these groups are under-represented in the higher and over-represented in the lower earnings brackets.

An older statistical analysis (cited by Salowsky and Schiller) provided a clear distribution as early as 1968. Knowledge of German on the part of male guest workers was correlated with their gross hourly earnings, with the following results:

Gross Hourly Wages

Linguistic Ability	Less than DM 4	DM 4 to DM 4.99	DM 5 to DM 5.99	DM 6 and Up	Unknown
Fluent German	10%	46%	29%	4%	11%
Broken German	14%	59%	19%	2%	6%
No German	24%	61%	10%	0%	5%

It is thus seen that only 10% of those possessing no knowledge of German achieved a gross hourly wage of DM 5 or more, whereas every third guest worker among those who spoke fluent German achieved that level of wages. Every fourth worker among those who spoke no German remained below the minimum wage of DM 4.00, but only every tenth of the fluent speakers of German were in that unfortunate category. These ratios have not changed until the present time, even though the general level of wages has risen.

Trade unions and evening schools offer a large number of language courses at low cost and even free of charge. Unfortunately, however, such classes have hardly any students. One reason for this is the physical and mental fatigue after strenuous manual labour and many hours of overtime, which hampers the pursuit of further education during the fragment of leisure remaining in the evening. Another reason, however, lies in the fear that permission to stay in Germany may be cut off before it is possible to learn good German or so soon that the effort will have been in vain.

The relations between deficient language skills and social discrimination will be examined later in this essay.

VI. The Housing Situation

The housing situation of guest workers in the FRG can only be described with the simple word "catastrophic." Discrimination against a marginal group becomes evident in this case with ominous clarity.

During the first few years after their entry, but also frequently longer, the majority of guest workers live in "plant accommodations" of their respective employers. These are mass quarters furnished with Spartan economy in the immediate proximity of places of work, and are administered by "house managers" with authoritarian disciplinary powers. In the second phase of the stay in Germany, often beginning with sending for the family, living quarters are rented on the free market. At this point, the guest workers become easy victims for sharp operators and rent gougers.

In a survey and analysis of empirical data covering the most populous West German State of North Rhine-Westphalia, Zieris studied the equipment, furnishings, and location both of plant accommodations and of rental housing. The following paragraphs summarize his findings.

One-third of the plant accommodations are old buildings erected before 1948. One quarter are temporary wooden buildings. In 46% of the cases, the accommodations do not provide the legally prescribed sleeping-room area of six square meters per person. The common living area (dayrooms, kitchens, washrooms, and toilets) comprises an average of 3.24 square meters per person. Twenty-four percent of the accommodations have no dayrooms, and 8% have no kitchens. Half the toilets are not provided with wash basins, and 17% can not be heated. In 75% of the accommodations the entry of persons of the opposite sex is forbidden or limited, and in 30% no visits at all are allowed. The rent charged is DM 8.58 per square meter of sleeping room. A total of 100,000 aliens – 16% of all guest workers employed in North Rhine-Westphalia – live in quarters of this kind, of which 23,000 figured in Zieris' survey.

In the case of rental living quarters, the picture is hardly any better. Their dimensions lie considerably below the otherwise usual standard for the native population. Sixteen percent of the foreign families inhabit only one room; 39% two, 28% three, and 15% four rooms or more. This means that 84% of the foreign families are confined to living quarters of three rooms or less. Almost half of these families live in quarters with less than thirty square meters living area; almost three quarters have living space less than forty-five square meters in total area. The average living space per dwelling unit is calculated at 38.5 square meters – compared with 60.1 square meters as the average for all rental living quarters in North Rhine-Westphalia. The living quarters of aliens thus lie 36% below the normal average living area.

A look at the average number of persons per room makes the disparity even more apparent. The aliens are housed at an occupancy rate of 1.53 persons per room or ten square meters per person, while according to the average for North Rhine-Westphalia 0.79 persons live in one room and each person enjoys 22.9 square meters of living space. The density of occupation of dwelling units inhabited by aliens is thus twice as high as the general average. The rent charged, on the other hand, lies one-third *above* that average. Dwelling units where aliens live have only 0.7 beds per person, and thus a deficit of 30% according to the usual standard that each person should have his/her own bed.

Forty percent of these dwelling units contain a kitchen, 21% no kitchen but a cooking alcove, and 35% have only a primitive place to cook. The last 4% have no cooking facilities at all. Only 32% of the foreign families have a toilet within their living quarters, while 10% do not even have a flushing toilet within the house. Only 16% have a bath tub and only 17% central heating in their living quarters. In 53% of all units surveyed, not all rooms could be heated, and 12% did not even have running water within the house. One percent of the dwelling units lacked electric lights.

In contrast to these figures 81% of all rental dwelling units in North Rhine-Westphalia have a toilet; 69% have a bath tub, 28% central heating, and 95% their own running water. While 41% of all rental dwelling units in North Rhine-Westphalia are located in new buildings (since 1948), 85% of the foreign families live in old buildings. Every fifth foreign family lives in "emergency" quarters (cellar, attic, shack, temporary wooden building, or garden house) that do not even meet the lowest standard for conventional housing within the State.

Since half of all dwelling units are rented furnished, enterprising landlords can raise their prices exorbitantly by valuing old pieces of junk as though they were costly antiques.

With regard to the size and equipment of dwelling units and their sanitary conveniences, the Spaniards and the Italians appear to enjoy the best relative situation. They are followed by the Yugoslavs and Greeks and – at a distance – the Turks and North Africans. In Cologne and West Berlin "compact" Turkish urban districts have developed in the meantime, in which guest workers' families – following the departure of the previous German inhabitants – are "among themselves." Houses ripe for the wrecking ball are crammed from cellar to roof with families of many children. Illegal influx is no longer to be controlled.

Here as well as in Munich a development is taking place that could be observed when Puerto Ricans began to stream in masses into New York City: the indigenous people flee when the foreigners come – a situation giving rise to the dangers of the ghetto and to foci of major social crises.

VII. School Problems of the Children

The approximately 500,000 children and adolescents who live in the FRG as family members of guest workers are subjected to considerable disadvantages in the school system. Compulsory school attendance was extended to them only then years ago. Until 1966 only about 70% of these children were in voluntary attendance at German elementary schools.

The Ministers of Education of the German constituent states had in 1964 considered the question whether special schools should be established for this growing minority. The decision was in favour of the German school. Considerations of cost were a major reason, but note was also taken of the unfavourable experiences of other countries that reported ghetto effects in connection with minority schools. There is, at the same time, no intention to engage in Germanization, such as was successfully practiced during the previous century in the case of alien children in the Ruhr mining area. The German school is not intended to and does not integrate the children of guest workers into German society.

Even the registration of children for school purposes is wholly inadequate. The children of guest workers are not indicated in the work contracts of their parents. Even the Inhabitants' Registration Offices lack sufficient information. Many guest workers are not interested in having their children enrolled in school. As a result, there are a large number of children who fail to attend school even today. They care for the smaller children at home, while their parents are away at work. From this group are recruited the 5% of totally illiterate people in West German society. School attendance also suffers from frequent housing and job changes by the parents, as well as from return to the home country on short notice.

There is no doubt that the children of guest workers are over-strained in the German school. They have to think in two languages simultaneously. While they do not forget how to speak, they often forget how to write their own languages, since writing is seldom practiced in the home. They acquire only insufficient skills in the German language, of which their German schoolmates already bring a fundamental knowledge. As a consequence of bilingual education in the school and in the family, they are subjected to considerably greater psychological and physical strains than are the indigenous children (Müller).

It is of course the policy, where possible, to establish special classes in which guest worker children are instructed during a transitional period, with particular assistance in improving their German. It is also planned to offer instruction in the geography, society, and culture of the country from which the children originate. In most cases, however, these well-intentioned policies exist only on paper. They are seldom carried out in the actual instructional programme. The special school activities are usually cancelled for lack of teachers. The guest worker children are frequently isolated in the German classes and find no contact with their indigenous schoolmates. The teacher fails to encourage them and take an interest in them. The process of instruction, taking place in a language that is new for them, remains an alien happening. Thus they sit their time on the school bench more or less without purpose. The parents are usually neither able nor willing to give any help with the children's homework.

The result is that a large proportion of the guest worker children do not complete their education. In North Rhine-Westphalia 40% of them are discharged from school without a certificate of completion when they leave at the end of their compulsory school attendance. Sixty percent of the children do not achieve an elementary-school completion certificate. Many even leave the schools before their mandatory school attendance has expired. More often than not they fail to complete the work prescribed for their grades, and they are frequently forced to repeat the same grade several times. Many, as a result, fail to advance beyond the fourth grade. Their resignation is expressed in frequent truancy, which makes their educational goals even harder to achieve. Instead of educating, the school actually functions as an instrument of social and human retardation. The inequality of their opportunities compared with German children is notorious.

A defensive and discriminatory attitude toward youthful guest workers is particularly apparent in the continuation schools (Berufsschulen: part-time schools attended by apprentices). These institutions, which all young people who have left full-time schools are required to attend during their first years of apprenticeship, are overloaded and overstrained to begin with. An absolute prerequisite for successful attendance is knowledge and mastery of the German language. Where these are lacking (because not learned in school earlier), or when young people of continuation-school age arrive directly from their home countries, the continuation schools have no facilities for helping them.

The situation is aggravated by the fact that the majority of foreign young people do not enter apprenticeship arrangements but are only employed as unskilled workers – like their parents. Only 2% of them receive apprenticeship training as skilled workers or in different branches of industry or business. Thus within the continuation school, the masses of young guest workers are assigned to the lower "general" classes, which are most severely hit by the shortage of German teachers. There is also a tendency not to admit young guest workers to the continuation schools or to "excuse them from attendance" because their educational difficulties are already anticipated. In a continuation school in Stuttgart, for instance, one hundred out of 170 foreign young people of the same age were "excused" from attending school because of insufficient knowledge of German.

It is therefore not astonishing that guest worker children hardly ever attend any advanced schools. Only 1% of them are to be found at modern secondary schools or university preparatory schools. With alien children constituting 7.5% of the school population in West Berlin in 1973, only forty-six aliens among 2,467 children transferred to a higher school. Among the causes – beyond those already mentioned – is the fact that the parents are mainly interested in having their children earn money as quickly as possible, and not in the advanced education of their children. An additional factor is that alien children have no legal right to state aid for training and further education.

In general, it can be concluded that the minimal opportunities for education and social advancement available to guest workers are not likely to show the least improvement in the next generation, now growing toward adulthood.

VIII. Social Discrimination

Vocational contacts between the West German population and the guest workers are limited to the necessary minimum. Non-vocational relations during leisure time and in human association are few and far between – and almost non-existent among neighbours.

The more fragmentary the German citizens' knowledge of their guest workers, the stronger their prejudices against them. These prejudices, in turn, hamper people getting closer together and thus getting to know each other better – so that a vicious circle results. The feeling of "being different" with which the nation delimits itself determines the discrimination. The inner attitude of a majority of the Germans, particularly those of the older generation, swings in a significant way between haughty arrogance on the one hand and feelings of envy and hate on the other. Where positive attitudes are to be found, they tend to emphasize the "utility" of the alien as a "useful servant." (Bingemann).

Ethnic chauvinism is expressed in the stereotype that guest workers are dirty, unkempt, underdeveloped, stupid and lazy. "Order and cleanliness," still a typical German idol, is generally denied as an attribute of guest workers. Guest workers are spoken to in the familiar form and by their first names. Their difficulties in adjusting to an industrial society are ignored. Since they do not understand the German language correctly, the Germans talk with them in a kind of primitive "Pidgin-German" thereby hampering them in gaining a better knowledge of the language. According to their nationality, they are labelled with discriminatory nicknames and derogatory appellations, such as "sheep thieves, camel drivers, Hottentots (North Africans), Mohammeds, Caraway Turks, Mussulmen (Turks), Partisans, bear trainers (Yugoslavs), spaghetti eaters, Macaronis, lemon shakers (Italians).

German inferiority complexes, on the other hand, find their expression in sexual envy and the fear of competition. The southern guest workers, Italians in particular, are said to have a strong erotic attractive force for German women as well as surplus capacity for sexual performance. Jealous aggressions are an obvious consequence. The criminality of the guest workers is greatly overestimated and is described in the German boulevard press in exaggerated and discriminatory language. The resulting distrust against supposed "knife heroes" and "moral rascals" is great. During economic crises and even in times of declining prosperity, foreign competition in the labour market is a target for attack. Guest workers are then denounced as "capitalist mercenaries, robbers of jobs, depressors of wages, and parasites," while unfounded opinions gain currency, such as: "If all we Germans should work one hour longer, we could send that whole foreign pack back where it belongs."

The EMNID Opinion Research Institute (news story in *Welt am*

Sonntag, January 6, 1974) asked citizens of the Federal Republic which guest-worker nationalities seemed particularly sympathetic to them. The survey indicated this hierarchy of popularity:

Yugoslavs	14%
Spaniards	9%
Italians	9%
Greeks	8%
Turks	5%
North Africans	3%
Portuguese	2%
No answer	54%

The position of the Yugoslavs at the top of the scale of popularity is probably attributable to the fact that they have, on the average, worked longer in the FRG than other guest workers and have achieved better language skills. It may also be the case that they are closest to the German national character. The relatively low popularity of the North Africans and the Portuguese is explainable, on the other hand, by their low occupational status and their low linguistic skills – factors that are closely interrelated. The fact that more than one-half the respondents gave no answer is significant in itself.

The guest workers themselves are greatly afflicted by and disappointed with the discrimination they encounter. Many of them come from countries – such as Spain and Turkey – that are linked with Germany through traditional political friendship. They now experience the fact that such "friendship" is in no way practiced in the realm of human contact within the Federal Republic. Nostalgic reactions and social frustrations are phenomena frequently observed among them. Rejected by the indigenous population, at best treated with indifference, they find their social communication limited to their own national groups on the job as well as in leisure time. On Sundays and holidays their home market place is replaced by the German railway stations: there is their rendez-vous, and their place for making deals and exchanging gossip.

Despite all this, 20% of the guest workers – and as high as 50% among those who live in the FRG with their families – would like to remain permanently in Germany. This is understandable for the single reason that the money earned in West Germany still exceeds by a considerable amount the low wages available in the home country. But it is likewise understandable that only 10% of the guest workers – the highest proportion being among the Italians and Yugoslavs, the lowest among the Spanish and Portuguese – are interested in full integration into the German population. But even in the latter groups there are occasional marriages between Germans and foreigners.

Where closer personal relationships are otherwise to be observed, the people concerned – as is the case, for instance, with Dutch, Austrians and Swiss – are members of higher occupational group, as well as independent business or professional people, with good linguistic skills. These groups, however, do not constitute "guest workers" in the technical sense, even though they are only temporarily employed in West Germany.

What the guest workers admire or at least recognize in the Germans is that bundle of national virtues of which the Germans themselves are still proud: order, diligence, punctuality, discipline, precision, and efficiency. What is criticized is precisely the overemphasis and exaggeration of these characteristics, as well as these constantly recurring faults: hardheartedness, pride, lack of humour, careerism, the chase after money, and lack of understanding for other people – which added together constitute social discrimination in all its forms.

IX. A Glance into The Future

The poet Max Frisch once wrote: "We called for manpower, but human beings came." The grave and depressing problem of the guest workers in the highly industrial societies of Central Europe must be approached with this point of view.

It must be emphasized that this problem has been brought into public consciousness as never before within the last few years. Churches, parties, and trade unions in the FRG, as well as the non-governmental welfare organizations and a large number of independent social scientists have laid their fingers on the wounds. This development opens the way to greater understanding for the foreigners on the part of the German population, and certain improvements can already be observed. But the problem cannot be solved with good will alone.

There is no doubt that the FRG is unable to get along without guest workers – not even in times of energy crisis or temporary economic recession. It should be possible to regulate the influx of foreigners through a system of "rotation." Under such a system guest workers would be recruited for a predetermined shorter tour of duty and then replaced with guest workers from the same country – perhaps even from the same family. Many social and human difficulties that occur precisely in the case of employment for periods of medium length could thereby be avoided. This would indeed be no purely economic solution. A related question of urgency is whether a small nucleus of foreign "core workers" should not be permitted to achieve integration into the German population – something they themselves desire after a lengthy sojourn in Germany – through legislative measures and vocational as well as liberal education.

The Federal Republic of Germany can no longer close its eyes to the fact that it has become an "immigration country" for all practical purposes.

REFERENCES

Bingemer, Meistermann, and Neubert, *Leben als Gastarbeiter. Geglückte und misslungene Integration* (The Life of Guest Workers. Successful and Unsuccessful Integration). Cologne, 1970.

Borris, M., *Ausländische Arbeiter in einer Grosstadt* (Foreign Workers in a Large City). Frankfurt, 1973.

Bundesanstalt für Arbeit (Federal Labour Institute), *Ausländische Arbeitnehmer. Erfahrungsberichte* (Foreign Employees. Reports of Experience). Nuremberg, annually since 1961.

Bundesministerium für Arbeit und Sozialordnung (Federal Ministry for Labour and Social Order), *Ausländische Arbeitnehmer in der Bundesrepublik Deutschland* (Foreign Workers in the Federal Republic of Germany). Bonn, 1971.

Cinnani, G., *Emigration und Imperialismus*. Munich, 1970.

Dörge, F. W., "Gastarbeiter – Europas Proletariat?" (Guest Workers – Europe's Proletariat?). *Gegenwartskunde* (Contemporary Studies), 1968, No 3.

Föhl, C., "Stabilisierung und Wachstum bei Einsatz von Gastarbeitern" (Stabilization and Growth in Connection with Employment of Guest Workers). *Kyklos*, International Journal for the Social Sciences, Vol. XX, 1967.

Franz, F., "Fremdenrecht und Fremdarbeiterpolitik" (Alien Law and Alien Labour Policy). *Deutsches Verwaltungsblatt* (German Administrative Gazette), 1973, p. 662.

—, "Zur Reform des Ausländer-Polizeirechts" (Reform of Police Law Concerning Aliens). *Deutsches Verwaltungsblatt* (German Administrative Gazette), 1963, p. 797.

Klee, E., *Die Nigger Europas*. Düsseldorf, 1971.

Koch, H., *Gastarbeiterkinder in deutschen Schulen* (Guest Worker Children in German Schools). Königswinter, 1970.

Lehmann, R., "Deutschland – Deine Neger!" (Germany – Your Niggers!), *Kontraste*, No. 4, 1971.

Marplan-Forschungsgesellschaft für Markt und Verbrauch (Marplan Market and Consumer Research Society), *Gastarbeiter in Deutschland* (Guest Workers in Germany). Frankfurt, 1971.

Müller, H., *Gutachten zur Schul- und Berufsbildung der Gastarbeiterkinder* (Advisory Opinion on Schooling and Vocational Training of Guest Worker Children). Published by: Verband Bildung und Erziehung (Federation for Culture and Education), Cologne. Bochum, 1972.

—, *Rassen und Völker im Denken der Jugend.* (Races and Peoples in the Thinking of Youth). Stuttgart, 1960.

Pagel, U., *Das Problem der sozialen Anpassung im Licht einer vergleichenden Socialpsychologie* (The Problem of Social Adjustment in the Light of a Comparative Social Psychology). Kiel, 1967.

Salowsky, H., and Schiller, G., *Ursachen und Auswirkungen der Ausländerbeschäftigung* (Causes and Effects of the Employment of Foreigners). Cologne, 1972.

Schloesser, W., *Arbeitsplatz Europa* (Europe as a Place of Employment). Cologne, 1966.

Senate of the city of Hamburg, *Bericht über die wirtschaftliche und soziale Lage der ausländischen Arbeitnehmer in Hamburg* (Report on the Economic and Social Situation of Foreign Workers in Hamburg). Hamburg, 1971.

Weber, R., "Isolation oder Assimilation," *Der Arbeitgeber* (The Employer), 1965, Nos. 11 & 12.

Wieczerkowski, W., *Frühe Zweisprachigkeit* (Early Bilingualism). Munich, 1965.

Zieris, E., *So wohnen unsere ausländischen Mitbürger* (Thus Live Our Foreign Compatriots). Düsseldorf, 1972.

Zwingmann, C., *Ein psychologisches Problem ausländischer Arbeitskräfte* (A Psychological Problem of Foreign Workers).

Spain: Regional, Linguistic and Ideological Conflict

MANUEL MEDINA

MANUEL MEDINA was born at Arrecife de Lanzarote on the Canary Islands, and studied at the University of La Laguna, Canary Islands, where he graduated as a *Licenciado en Derecho* (Bachelor of Laws) in 1957. He continued his studies at the University of Madrid, where he was awarded the Doctor of Laws degree in 1961, and at Columbia University, New York, where he earned the degree of Master of Comparative Law. He has been a Professor of International Law and Relations at the University of Madrid since 1958, and was promoted to full professor (Catedrá-tico) in 1975. He held a Fulbright Fellowship for study in the United States in 1961–62, and was Danforth Lecturer in the United States in 1967. He is Assistant Director of the *Revista de Instituciones Europeas* and an editor of the *Revista Española de Derecho Internacional*. He is the Secretary-General of the Spanish Branch of the International Law Association, and has served as Vice-Dean and Acting Dean of the School of Political Science at the University of Madrid.

Professor Medina's publications include (titles translated from Spanish): *The United Nations Organization: Its Structure and Functions* (Madrid, 2nd edition, 1975); *The Theory of International Relations* (Madrid, 1973); and *The European Community and Its Constitutional Principles* (Madrid, 1974).

REGIONS AND
LINGUISTIC MINORITIES
IN SPAIN

Catalan
Basque
Galician

Spain: Regional, Linguistic and Ideological Conflict

MANUEL MEDINA

I. The Inter-Group Spectrum of Modern Spain

During the second half of the 15th Century, Spain became a modern State with the unification of the Crowns of Castile and Aragon and the expulsion of the Arabs from the Iberian Peninsula. Queen Isabella of Castile and King Ferdinand of Aragon, after conquering the Moslems of Granada, decided to expel the Jews and set up the Inquisition to preserve the religious unity of Spain. During the 16th and 17th Centuries, the "Moriscos" (Arab minority) were also expelled, and the country became fully unified from a religious and racial point of view. Only the Gypsies have survived as a racial minority group in today's Spain. While most members of this racial group have been assimilated by the majority, groups of Gypsies do still survive in some areas of Spain, carrying out their traditional activities of cattle-trading, singing or peddling in the country-side or the cities. Distrust for this ethnic minority is still widespread in Spain. Due to the lack of appreciation by Spanish society towards their cultural values, their extinction as a group might take place in a very short time. Economic development, education, increased communications and other factors of modernization appear to make the survival of this group impractical, as it becomes more feasible for their individual members to integrate into the majority group under the present trend for urbanization.[1]

Due to the marginal character of the Gypsies in modern Spain, a study of inter-group relations will have to centre on the existence of two kinds of minorities: the regional-linguistic groups and the religious-ideological dissidents. Regional minorities have survived five centuries

The author wants to state his appreciation to Miss Adriane Allen for her help in drafting the English version of this paper. The final version has been written by the author alone, however, and any deficiencies in it are of his sole responsibility.

of political unification and centralization. Indeed, they appear to be stronger today than ever before. Religious and ideological dissidence developed in Spain since the 19th Century, under the influence of the French Revolution and as a consequence of political and economic modernisation. While class and sex discriminations are important in modern Spain, they offer a pattern familiar to most western countries. It is not possible to deal with the nuances of these problems in Spain in a paper of this nature. Relations between regional-linguistic groups will be considered first then the situation of the religious-ideological minorities. In any case, the two sets of issues are interrelated. Both the existence of a "regional problem" and a "non-conformism problem" are the natural consequence of an absolutistic conception of the unity of the State, incompatible with more flexible attitudes towards pluralism.

Spain is divided into fifty administrative districts called "provinces" *(provincias)*. The province, however, does not mean much in terms of cultural or socio-economic realities. Spaniards owe loyalty to their historical regions: *reinos* or *regiones*. Traditional regions enjoy partial recognition as administrative divisions in some areas: ecclesiastical provinces, university districts, military regions. All these are administrative subdivisions of the central Government, but no political recognition has been given to the reality of the region as centre of strong cultural and social attachments.

Regional identification does not always have the same intensity. Coastal and border areas appear to have a higher degree of regional sensitivity than the central areas. A strong basis for regional differentiation arises from the existence of regional languages. Thus, we can make a first distinction between Castilian-speaking regions and those where a different language is spoken. There are three main areas where Castilian competes with other languages: Galicia, in the Northwest, the Basque Country in the North, and the Catalan-speaking regions of the North-East. As to the Castilian-speaking areas, there is not complete uniformity among them. Some of these regions are separated by geography from the main core of Castilian-speaking Spain: the Canary Islands by 1,000 miles of ocean, and Asturias by the steep Cantabrian mountains. Arab traditions and a strong regional folklore separate Andalusia from Castile. Extremadura, Murcia and Aragon are differentiated by their history and some dialectal peculiarities from Castile. Castile itself is divided into three historical regions: Leon and Old Castile, in the North, and New Castile, in the South. Regional differences are also noticeable among the "non-Castilian" regions, as we will see next.[2]

In any case, the basic element for regional differentiation is the survival of a local language. Although the local languages find themselves today in a difficult position, and in many places are only spoken by a minority, they have served as a point of reference for regionalist demands and aspirations. The situation varies from one region to the next, and it is impossible to generalize. This is why it is necessary to consider with some details the specific problems of each of the three areas with minority languages: Catalonia, the Basque Country, and Galicia.

II. The Catalan "Empire"

Catalan (*català*, in vernacular) is a romance language. Although some parallels have been made with the *langue d'oc* or *provençal* of southern France, it has independent origins and a strong personality of its own.[3] It was the distinctive language of the Principality of Catalonia, one of the Christian kingdoms that were set up in the early centuries of the Spanish struggle against the invading Arabs (8th to 10th centuries). Catalonia eventually merged with the neighbouring Kingdom of Aragon, and the united Kingdom of Aragon-Catalonia flourished in the latter part of the Middle Ages. Castilian was spoken in Aragon and the inland territories of Valencia. Catalan was the language of Catalonia, most of Valencia, the Balearic Islands, the Roussillon (now a part of France), Andorra, and the small area of Alguer, in the island of Sardinia. When King Ferdinand of Aragon married Queen Isabella of Castile, the eastern territories of the Peninsula preserved their political and legal institutions, their customs, and their native language. The Spanish-speaking territories of Aragon were easily assimilated into the more populous and politically powerful Castile. The Catalan areas, on the other hand, have always been a thorn in the side of unitarian Spain. This is in part a consequence of their comparative economic strength. In 1640, the principality of Catalonia proclaimed itself an independent state, with French support. The revolt had to be suppressed by the Castilian armies. In 1700, the Catalan areas supported the Hapsburgs against the Bourbons in the Spanish War of Succession. After the triumph of the Bourbons in Spain, the Catalan regions were deprived of their self-government and public laws (*fueros*). The kingdom of Valencia was even deprived of its own private laws. In 1931 and 1934, the

Spanish republican Government put down renewed attempts by Catalonia to set up an independent state.[4]

Thus, the history of modern Spain has been marked by almost constant friction between the central government and Catalan centrifugal trends. Several factors have contributed to the survival of a Catalan national conscience after five centuries of national unity. The language is the most important basis for preservation of a Catalan identity. The Catalan language has a wealth of literary tradition and is still widely spoken in Catalonia, the Balearic Islands and most of Valencia. There are dialectal differences in spoken Catalan. The Catalan of Valencia is much closer to Castilian than the Catalan of the other two regions. Local dialects also exist in the three regions. Written Catalan has been unified through the efforts of learned societies. Castilian is spoken in most Catalan areas. As the official language of the country, it is compulsorily taught at all levels of education. This means that even in Catalan areas, most people write in Castilian, while they are unable to write in Catalan. Castilian is the only language that can be used by the administration and in courts of justice, and it enjoys the official protection of the state.

Catalan has fought an uphill battle to survive as a native language. To a large degree, this battle has been won, but not without casualties. The future remains uncertain, due to the continued pressure of the official language and the afflux of Castilian immigrants into the Catalan areas. In large areas of Valencia, especially in the larger cities, Catalan has become a second language, only spoken by the lower classes and barely written. In Catalonia and the Balearic Islands, Catalan is still generally spoken among people of Catalan ancestry, and written by the upper classes. However, it can always be displaced in transactions involving Castilians. Many of the Castilian-speaking immigrants are unable to speak or understand Catalan, and have not been assimilated into the Catalan culture.[5]

The Catalan-speaking areas of Spain are wealthy and densely populated. With only twelve per cent of the national territory, they contain twenty-five per cent of the Spanish population. The Principality of Catalonia has the highest per capita income among the Spanish regions. Valencia and the Balearic Islands also enjoy a higher-than-average per capita income. Barcelona and its hinterland have always been a prime manufacturing centre. The city of Barcelona has maintained through the centuries its reputation as the first port for business and trade in the western Mediterranean. The economic strength of Barcelona has been a source of strength as well as a source of weakness for regional identity. On the one hand, only a region like Catalonia that has been ahead of the rest of Spain for several centuries could have maintained an attitude of defiance and independence towards the central government. Regionalist movements have always been more moderate and less self-assertive in the Islands and Valencia. On the other hand, Catalonia has become a centre of attraction for migrants from other Spanish regions. The process has accelerated since the end of the Civil War. While only a few thousand Spaniards emigrated to Catalonia in the thirties, some 30,000 immigrants were arriving in Catalonia in the forties and fifties. In the late sixties, annual immigration was over 50,000.[6] Finally, in the seventies, Spanish migratory capacity appears to have been exhausted, and Catalonia is now importing Moroccan migrants in large numbers.[7]

The accelerated pace of immigration has prevented in many cases the assimilation of the immigrants, especially in the industrial belt of Barcelona. While children and grand-children of most immigrants who arrived in Catalonia before the Civil War have been assimilated into the Catalan society, and can speak Catalan, many of the post-1939 immigrants live in Castilian-speaking "ghettos" of manual workers or middle-class employees. Under these circumstances, the ability to speak Catalan is a symbol of prestige and a tool for social promotion. In some cases, Castilian-speaking immigrants can reach the higher social and economic positions without full assimilation, but the existence of social differences between established Catalans and newly-arrived Castilians has prompted accusations of discrimination against the immigrants. There is some basis for these accusations in the attempt by the Catalans to preserve their cultural heritage against the continuous interferences by the central government. The "in-circles" in Catalonia are basically formed by business and civic leaders born in the region and who have learned Catalan as their mother-tongue. Those who are unable to speak Catalan may be called "murcianos", i.e., natives from the southeastern region of Murcia, from which many of the immigrants come, regardless of their geographical origins. An attempt to learn Catalan at a later stage would be futile.[8]

The situation is somewhat different in the other two regions, due to their less marked degree of economic development. In Valencia, social prestige has been tied, for a long time, with the ability to speak Castilian, while Catalan was spoken mainly among peasants.[9] In the Islands, Catalan purity had been preserved by geographic isolation.

Since the Civil War, the situation has partially changed. With some degree of success, a new intelligentsia strives to make Catalan a distinguished cultural instrument among Valencians. There is also a revival of Valencian regional pride.[10] In the Balearic Islands, tourism has brought economic prosperity and closer association with Catalonia. For the first time, Castilian-speaking immigrants are needed in the Balearic Islands, and social cleavages similar to those existing in Catalonia are now appearing in the Islands. Inter-group differences in the Balearic Islands are softened, however, not only from the lack of strong political regionalism, but also from the presence of large numbers of foreign tourists, who either speak their language or communicate in Castilian.[11]

Political regionalism has played a very important role in Catalonia since the end of the 19th century. The republican government of 1931–1939 granted autonomous status to Catalonia. In turn, the Catalan-speaking areas supported as a rule the republic against the "National Movement" of General Franco. The new regime imposed heavy restrictions on the utilization of Catalan since 1939. It penalized with fines and or imprisonment the utilization of Catalan in government facilities and public places. Only in 1943 did the Franco government allow publication of Catalan classics. In 1946, living authors were also allowed to publish in their regional language. Today, some six hundred titles are published every year in Catalan. These include translations from other languages and scientific books. There are no Catalan daily-papers, but several periodicals are published in this language.[12] Without any specific television or radio stations, some programmes are broadcast in Catalan.[13] Catalan theatre has revived in Barcelona. Modern Catalan songs, with social and political contents, are well represented in the Spanish record market. There are very few movies with Catalan soundtrack. Even though the Government has allowed the teaching of the regional language in public schools, Castilian is still the dominant language in education. Concessions by the central government in these and other matters have somewhat channelled regional demands into the cultural sector.

It cannot be said, however, that the regional problem has been solved. Catalans demand greater support from the central government in the preservation of their "regional heritage." Some demands refer to the need to facilitate the assimilation of immigrants into the local society through compulsory education in the Catalan language and culture. These demands are somewhat similar to those of the Flemish in Belgium and the French-Canadians in Quebec. Recognition of linguistic monopoly of a minority language within a region may be the only way to preserve regional identity in the homogenizing circumstances of our times. Political regionalists are not satisfied, in any case, with piecemeal concessions by the central government in the cultural domain. Following the precedent of the Spanish republic, political regionalists are intent upon obtaining some kind of federal arrangement, under which some governmental functions could be transferred to the region. Under the present regime, such a demand cannot be fulfilled, because the Government is committed to a policy of rigid centralization. Only through a process of democratization at the national level could a federal formula be approved. Thus, the Catalan regional problem is closely intertwined with the political process in the nation as a whole. It appears that only a more democratic central government would be willing to recognize some degree of regional autonomy.

While there is discrimination against the Catalan language, the Catalans themselves enjoy recognition in Spain for their entrepreneurial spirit and good sense of economics. On the other hand, Catalans might be accused of being thrifty, devious and overbearing, with too much concern for money. The cultural level of Catalonia is well above the national average. Barcelona is not only the centre of the Catalan culture, but also the core of an important sector of the Castilian-speaking culture. Catalans have a certain sense of superiority towards the other regions, with the possible exception of the Basque country. For them, Castilians are lazy, bureaucratic, and intolerant. Relations between the three Catalan regions are somewhat complex. Valencians have ambivalent feelings towards their wealthy neighbours, and are equally divided in affections – and disaffections – towards the two larger metropolitan areas of Madrid and Barcelona. The relative isolation of the Balearic Islands has allowed the *Balear* people to preserve their regional culture without too much difficulty. Nevertheless, while Catalonia acts as a powerful magnet for the Islands, resentment against Castile is practically non-existent here.

III. GALICIA: SOCIAL AND LINGUISTIC PROBLEMS

With slightly less surface than Catalonia and only half its population, Galicia is just the opposite of Catalonia in many respects. It is primarily an agricultural region. Its population has decreased during the last twenty years, due to massive emigration. Nevertheless, Galicia has some affinities with Catalonia due to the survival of another Romance language in that region. Galician (*Galego*, in vernacular) was formed in the early Middle Ages in the Western parts of the Iberian Peninsula, including what is now Portugal. Later on, Portugal became an independent kingdom and developed its language. Galicia, on the other hand, became a part of the kingdom of Castile and was subject to its cultural influence.[14] Thus, Galician has been strongly influenced in its later development by the Castilian language. In any case, even today, Galician and Portuguese cannot be considered as two different languages, but rather as dialects of the same romance language.[15] The Galician literary language had an early start. Under Alfonso the Tenth of Castile (the "Wise King", 1252–1284 A.D.) it was used as a literary language of the Castilian court. With the political and economic decline of the region, Galician ceased being used as a literary vehicle after the 15th Century, and the upper classes dropped it in favour of Castilian. Galician literature was only revived in the 19th century, mostly through the works of the poetess Rosalía de Castro (1837–1885). A "renaissance" of literary Galician took place between the publication of Rosalía's first book, in 1863, and the beginning of the Civil War in 1936.[16] Here, as in Catalonia, the new government imposed several years of silence on regional literature. Between 1936 and 1945 only one book was published in Galician. Since 1946, however, Galician literature has slowly developed, first in poetry, later in fiction, and finally in philosophy, science, sociology, and economics. Publications in *Galego* are small in number by comparison with Catalan books. The melodic, soft tones of the Galician language seem to make it extremely appropriate for poetry, and this is the area where Galician literature has been most successful.[17] Nevertheless, a new generation of Galician regionalists is intent upon the strengthening of their native culture through the wider use of the language in more "serious" areas of intellectual endeavour: essays, novels, and the sciences. There are some cultural reviews published in Galego, but no daily papers. The regional language is not taught in school, and one of the fundamental claims of the Galician regionalists is the teaching of the local language: "Galego na escola". A new intelligentsia centered around the University of Santiago de Compostela has managed, however, to restore to Galician some of its lost lustre.[18]

The problems of the Galician language are similar in many respects to those of the Catalan language in Valencia. Optimistic calculations estimate that eighty per cent of the population of Galicia can speak Galego.[19] It is a fact, however, that we find a true diglosia here as in Valencia. Galician is the language of the peasants and fishermen, while the upper and middle classes only speak Castilian. Valencian regionalists can point with pride to the economic and cultural success of their northern neighbours in Catalonia. Galicians cannot claim the same degree of success of their southern neighbours of Portugal, even though Portuguese is spoken by some seventy million people. There is no apparent superiority of the Galician-Portuguese civilization over the more widely spread Castilian culture. The use of Galego is associated in the Iberian Peninsula with the impoverished *braceros* who are hired as seasonal workers in other parts of Spain. In Latin-America, *"Gallego"* is a spiteful name given to all Spanish immigrants. The sensitive, sentimental Galician appears to the haughty Castilian as ignorant and submissive, although rather guileful. The Galicians feel that their region has been marginalized, and that the economic prosperity that Spain has enjoyed in the last twenty years has not profited them at all. Hence, Galician regionalists direct their demands to the central government today in two main areas. First, Galego should be protected as a valuable contribution to national culture in the same vein as Catalan or Basque. It should be taught in public schools as a means to prevent further erosion. But protection of the language is not enough. Galician regionalists feel that more should be done to promote the economic development of the region. Although self-government has always been a demand of Galician political regionalism, Galicia never obtained autonomous status as Catalonia and the Basque country did during the Spanish republic. Therefore, without underestimating the prospects for a rebirth of strong political regionalism, present Galician demands are centered in the cultural and economic sectors. The moderation of these demands, however, has not made it easier for the central government to placate regional grievances. The persistence of emigration from Galicia is the best evidence of the relative underdeveloped stage of the regional economy and society.[20]

IV. The Basques: Ethnics, Language, and Politics

The Basque country is a small area at the western end of the Pyrenees. It comprises four Spanish provinces (Navarre, Alava, Guipúzcoa, and Biscay) and the French department of the Basses-Pyrénées. Technically,

the province of Navarre constitutes an independent region, as it formed
for centuries one of the Iberian Christian kingdoms, but its inhabitants
are from Basque origins and consider themselves Basques. Only in the
16th Century was the kingdom of Navarre annexed to the united kingdom of Castile and Aragon. The other three Spanish provinces from
the "Provincias Vascongadas" or "Región Vasca", and they have been
attached to Castile since the Middle Ages.[21] The population density is
relatively high in the Spanish Basque-Navarre area, with very high
figures in the provinces of Guipúzcoa and Biscay. With a surface smaller than the region of Valencia, the Spanish Basque area has a population of two and a half million.

The Basques found their claims to a differential treatment in two
important elements: the existence of a Basque race and the survival of a
language that cannot be related to any other living language. As a
matter of fact, the successive invaders of the Iberian Peninsula never
managed to establish absolute control over the Basque country, and
this has allowed the survival till modern times of a people with
specific anatomical and cultural peculiarities.[22] On the other hand, the
Basque country was not absolutely free from foreign presence. The
larger provinces of Navarre and Alava appear to have been rather assimilated into the Spanish national culture, and the Basque language
has barely been spoken in these provinces for several centuries. The
Basque language has survived till the 20th Century in Biscay and Guipúzcoa, which form the core of the Spanish Basque country. Here
again, the impressive industrial development of these two provinces has
eroded the language and the regional culture.

As in Catalonia, immigration from Castilian-speaking areas has been
intense, especially since the Civil War, and it is difficult to ascertain the
percentage of non-Basque blood in the area at present. Intermarriage
between Basque aborigines and immigrants has been frequent, and it is
difficult to find pure racial types in the cities. The Basque language is
still spoken in the rural areas, fishing villages and smaller industrial
cities, but very few people use it in the larger cities of Bilbao and San
Sebastian. As in Valencia and Galicia, the upper classes abandoned
their regional language, which was a second-rate language, used only
by the lower classes. Here again the phenomenon of diglosia is found,
with prevalence of Castilian over the native language. Political regionalists and the new Basque intelligentsia have to revive the language,
which never had an important literary tradition. This task appears to
be formidable, due to the intricacies of Basque, a non-Romance lan-

guage and, therefore, very different from the Castilian spoken by the cultivated sectors of Basque society. Unlike Catalonia, knowledge of the regional language is not required to progress on the social ladder, and this facilitates the assimilation of Castilian immigrants.[23]

Although Basque is today a minority language within the region, and notwithstanding the importance of immigration, the Basque country still offers peculiarities of its own. Basques are reputed to be hard-working, honest and reliable. Within Spain, people from Navarre are differentiated as being somewhat tough, hard-headed and intolerant, but they also enjoy the reputation of being serious and hard-working. Basques or Navarres are not noted for their sense of humour, but one of the most famous Spanish *"fiestas"* take place in Pamplona, the capital city of Navarre. Non-Basques are called *"maquetos"* in the Basque country, and this word carries connotations of unemployed, opportunistic migrants who come to the Basque country to take advantage of the prosperous economy and good nature of the Basques. The Basques have been especially suspicious of the central government. The history of the Basque country has been marked by civil wars and uprisings against central political power.[24]

During the 19th century, most of the Basque country supported the dynastic claims of the Carlist branch of the Spanish royal family against the Madrid government of Queen Isabel the Second. Carlism combined dynastic claims and political conservatism with regionalist demands. Their motto *"Dios, Patria y Fueros"* ("God, Fatherland and Regional Rights") expresses well the political aspirations of this movement, which found support in the rural, mountain areas of the Basque provinces and Navarre. It should be noted, however, that the main cities remained loyal to the liberal government of Queen Isabel. After one and a half centuries, Carlism is still a political force in Spain, with strong roots in Navarre. During the Civil War, the Carlists sided with the national movement against the republic, in the provinces of Alava and Navarre.[25] The more developed and industrialized provinces of Biscay and Guipúzcoa sided with the republic, which granted autonomy to the "Euzkadi" region during the Civil War. Since the end of the 19th century, the Basque Nationalist Party (*"Partido Nacionalista Vasco"*, or *P.N.V.)* had incorporated the demands for regional autonomy without the conservative overtones of the more traditional Carlist movement, and it fell upon this party to provide a political backbone to the republican resistance in the Basque area. After bloody feats of war, which included the bombardment of the city of Guernica by the

German air force, the Basque autonomous government was defeated by the Franco troops. The autonomous government was abolished, the privileges of Biscay and Guipúzcoa were withdrawn, and the new government undertook the task of erasing Basque nationalism. The Carlist movement was merged with the *"Falange"* party, although it survived *de facto* as *"Comunión Tradicionalista"*.

Thirty-five years after the end of the Civil War, a revival of Basque nationalism is patent. Besides the traditional formations of the Carlists and the Nationalist Party, an extreme-left group has been formed with the initials *E.T.A.* (*"Euzkadi ta Askatasuna"*, "Fatherland and Freedom"). *E.T.A.* attempts to set up an independent Basque state encompassing Spanish and French territories. While the *Partido Nacionalista Vasco* has concentrated in "political", non-violent means to secure some degree of autonomy for the region, *E.T.A.* parallels the Carlists in the use of force. The new organization appears to be to the left of any other political formation, combining nationalist tenets with radical proposals for social and economic change. For almost ten years, the Spanish newspapers have reported the activities of *E.T.A.* These have culminated in the assassination of Admiral Carrero Blanco, President of the Government, in December 1973, which has been claimed by an *E.T.A.* commando unit. Finally, it should be mentioned that notwithstanding its far-fetched political and economic goals, the *E.T.A.* appears to find support among some sectors of the Basque Catholic Church, and this gives the *E.T.A.* some resemblance to the "Irish Republican Army." Actually, it has been alleged that the Irish and Basque extremists have been co-operating in their political and military actions.

Thus, the Basque political problems appear to be more acute than those of Catalonia and Galicia. The right to use the language is only a marginal question. The Basque country does not face today the complex social problems which immigration and emigration have posed to Catalonia and Galicia. Nevertheless, immigration is threatening the survival of the Basques as an ethnic and cultural group. The Basque language has become a minority language within the region itself, and it has ceased being an effective barrier against immigration. This may partly explain why cultural issues have been superseded by politics. Carlism and Basque nationalism appear to be two faces of the same coin, one looking to the right and the other to the left. The exaggerated centralism of the present regime has nurtured an even more explosive reaction represented by the *E.T.A.* movement. Once more, authoritarianism and rigid centralism have not proved effective in dealing with a deep social and cultural conflict which probably cannot be solved with simplistic solutions.[26]

V. RELIGIOUS MINORITIES AND IDEOLOGICAL DISSIDENCE

Spain is an overwhelmingly Catholic country, where almost every citizen is baptized, married and buried by the Church. The number of members of other religions or sects is estimated at 30,000, that is, less than one per thousand of the total Spanish population of 34 million.[27] There are no figures available for atheists or other persons not affiliated with a given religious denomination. The high degree of religious homogeneity can be explained with the persecution of religious dissidents by the Inquisition between 1492 and 1812. Even after the abolition of the Inquisition, religious intolerance has been a dominant feature of Spanish life. Proselytism by non-Catholic sects and any freethinking was severely curtailed during most of the 19th century. Only since 1868 have Protestants been allowed to practice their cults and to proselytise in Spain. The 1931 republican constitution provided for the separation of Church and State and the widest margin of freedom of conscience. After the Civil War of 1936–1939, a new wave of religious intolerance settled, and the rights of religious minorities were again curtailed. Under the 1945 "Bill of Rights" ("*Fuero de los españoles*"), the Catholic religion was proclaimed the official religion of the state. Public manifestations by other religious denominations were not permitted. Non-Catholic denominations lacked juridical personality and could not own any properties. Thus, they had to be registered in the name of private persons. Outdoor signs of non-Catholic temples were absolutely forbidden. Discrimination against non-Catholic was established by law for education and government jobs.[28] In the late sixties, the legal position of non-Catholics was considerably improved. The "*Fuero de los españoles*" was amended, and new laws provided for some degree of religious freedom. In any case, non-Catholic sects and their members have to be registered in a special registry of the ministry of justice, proselytizing by non-Catholics is still subject to severe limitations.

Spontaneous discrimination against religious minorities has also decreased in the last decade, partly as a consequence of the more tolerant attitude of the Second Vatican Council. Discrimination is still in existence, however, in the rural communities of central Spain. The coastal areas and the cities appear to be more tolerant towards religious dissidence. Non-Catholics are also more numerous in the peripheral regions and in Madrid than in the hard core of Castile. In any case, legal obstacles and social prejudice have combined to prevent the formation of strong dissenting communities in any given city or region. With the possible exception of the Jewish community, recruitment into religious minorities usually takes place only among the working class and the lower middle class. There is no social advantage to be gained by joining a minority sect, and in many cases it may mean social isolation. The Spanish intelligentsia has not been attracted by non-Catholic sects. Instead, following a pattern familiar in Latin countries, intellectuals, professional men and industrial workers withdraw from the Church into positions of agnosticism, atheism and anticlericalism, rather than seek a new religious experience. Actually, the largest group of dissidents is formed by the agnostics. Figures are not available, because most members of this group maintain their official allegiance to the Catholic Church, and are baptized, married and buried in the Church. From indirect sources on religious beliefs and attendance to mass, it can be inferred that only fifty per cent of the Spanish population maintains real allegiance to the Catholic Church.[29] Intellectuals and liberals in general are sympathethic to Protestants and other religious minorities, but this positive feeling is not accompanied by any admiration. Spanish leftists consider religious affiliation of any kind as a form of escapism from the social and political problems of the country. Catholics, on the other hand, have evolved from an attitude of hostility to one of understanding, coupled with the realization that religious minorities do not constitute a threat to their established way of life. Spanish Protestants tend to look toward Catholics as irreligious, opportunistic and intolerant.[30] Religious minorities have, in a sense, some features in common with regional minorities, but they lack the aggressive, dynamic spirit of the regionalist movements. This may partly explain the lack of appeal to the Spaniards of minority sects. Spaniards tend to identify at present with more aggressive movements. Thus, the Jehovah Witnesses have done considerably well in their proselytising campaign in Spain in recent years, due to their obstinate refusal to be inducted into the military service. Conscientious objectors are increasing in numbers, notwithstanding the imposition of severe penalties by the government. Only recently has the government felt the need to soften these penalties, after the failure of an attempt by the government to introduce a more liberal legislation on conscientious objection.[31]

Active membership in the Catholic Church or some of its lay organizations is usually advantageous for social promotion. The expression of religious or ideological dissidence may disqualify for high positions in government and business. In turn, this compels ideological dissenters to be prudent in the statement of their personal views. A greater degree of tolerance is noticeable today in intermediate positions, including teaching and government jobs. In general, the intense process of modernization that Spain has experienced in the last twenty years has toned down ideological conflicts and has widened the margin of tolerance. Religious and ideological differences are generally accepted in business and family relations, and in many cases they are not an obstacle to marriage. Nevertheless, generalizations are always dangerous in this area, due to the importance of social and regional differences within Spain.

VI. Conclusions: Prospects for a Pluralistic Society in a Unitary State

The present Spanish constitution is based upon the principle of political and ideological unity.[32] The unitarian structure of the state does not allow much of an elbow room for political regionalism. While most Spanish regions seem to have accepted centralized government as a fact of life, some of the more dynamic and politically conscious regions, such as Catalonia and the Basque country, have not resigned themselves to this state of affairs. Actually, since the Civil War of 1936–1939, a recurrence of regionalist feelings can be appreciated in areas which had been traditionally quiescent: Valencia, Galicia, and the Balearic and Canary Islands. Linguistic, geographic, economic and racial arguments have been presented to justify the need for some degree of regional autonomy. Some regionalist movements (like E.T.A.) demand outright separation from Spain. Although it appears that there is no imminent threat of a break-up of the national state, political regionalism has to be taken seriously, as proved by history and recent events. On the other hand, the majority of the Spanish population is still located in areas where the need for local autonomy is not felt. For many of the supporters of the present regime, any demand for political or cultural autonomy is tantamount to high treason, and anti-regionalist feelings may run as high as regionalist demands. Some key sectors of the Spanish society are deeply attached to a unitarian concept of the state. This is especially the case of the civil servants and the military, but other sectors of the Spanish society are also opposed to any concessions to regionalism. This conflict between regionalist demands and centralist intransigence is a latent one, which might easily precipitate under crisis conditions.

The republican constitution of 1931 had found a compromise formula to accommodate regional demands with the principles of the unitarian state. Spain was to be "a unitarian state, compatible with the granting of autonomy to some regions, provinces or cities."[33] Under this formula, Catalonia was granted autonomy at an early date, and the Basque provinces of Biscay and Guipúzcoa became the autonomous Euzkadi region in October 1936, after the initiation of the Civil War. Autonomous status was also being considered for Galicia and Valencia before the outbreak of the Civil War, but these projects never materialized. In spite of the moderation of this semi-federalist formula, the republic encountered tremendous obstacles to overcome. On the one hand, the concessions did not entirely satisfy regionalist demands, and the republican government had to use force to put down two Catalan revolts, in 1931 and 1934. On the other hand, the concessions went too far for the taste of the centralists, and they prompted disaffection for the republic among the military and the Castilian middle classes. The whole experiment in political regionalism ended with the blood-bath of the Civil War. It would be an over-simplification to say, however, that the issue of regionalism *versus* centralism was a clear-cut one even at the time of the Civil War. Thus, the Catalan and Valencian right-wing parties were in favour of regional autonomy, but opposed social change. At a moment of crisis, they sided with the nationalists against their leftist countrymen. In the Basque country, the Carlists joined the Franco forces, while the *Partido Nacionalista Vasco* supported the republic. In consequence, Navarre and Alava joined the national movement in the first days of the war, while Biscay and Guipúzcoa supported the republic till militarily defeated.

Similar contradictions are found when moving from the political to the cultural front. A Catalan or a Basque culture can only survive through the assimilation of immigrants. Otherwise, the regional culture will be drowned in a Castilian cultural flood. This has partly happened in the Basque country, where the regional language is only preserved in parts of two provinces, while the great majority of the Basque population have lost their ancestral language and culture. This may explain the political activism of the *E.T.A.* movement, which cannot invoke an already non-existent cultural identity. In Catalonia, the regional language has fared better, but the large number of immigrants have reduced Catalan in some areas to a language of the upper and middle classes, while the working class speaks Castilian. This is why the new regionalism demands protection from the central government to preserve regional culture, even to the degree of requesting compulsory teaching of the regional languages in public school. Compliance with demands of this kind might eventually lead to the tracing of linguistic frontiers as in the case of Belgium and Canada. It is unthinkable that the present government would be willing to yield to these demands, which would be to the detriment of the Castilian-speaking minority in Catalonia. Here, like in the Basque country, preservation of the regional culture becomes problematic in the long-run without some degree of political autonomy.

Valencia and Galicia have many elements in common. The main factor of crisis for the regional culture in these two regions lies in a situation of diglosia, where the upper classes have given up their native language in favour of Castilian. In both areas, the regionalist movements seek a cultural revival of the local language among the upper classes and some form of compulsory education in Catalan and Galician in public schools. Political regionalism is not too aggressive at present in either region, but the development of regional feelings among the elites may forebode a greater role for regional politics in both areas. Moreover, Galicia has been shaken in recent years by widespread industrial action which appears to prove right the assertions of the new regionalists in the sense of a need for greater economic development in the region. Nevertheless, the lack of strong political regionalism in Valencia or Galicia makes it unlikely that a serious crisis may occur in either region. The absence of strong political regionalism also means that the central government does not feel too great pressure to satisfy the demands of these regions, by contrast with the almost constant concern for Catalonia and the Basque country.

The present regime has attempted to soften the impact of its centralist policies through some limited concessions. Many of the restrictions imposed upon the use of regional languages have been lifted, although none of them has received official recognition for legal or administrative matters. The government has authorized the teaching of Catalan in public schools, but Castilian maintains a privileged position as the official cultural language. In the area of economic planning, the government has encouraged the development of industrial plants in Galicia, the Balearic Islands and other peripheral regions. It has also

sponsored studies about regional economic development. More recently, the government has divided the national territory into a certain
number of zoning areas, which reproduce substantially the traditional
regions. The zoning division has been viewed as an attempt to meet
some of the demands for political regionalism. In any case, these actions by the central government do not go beyond a modest decentralization of some administrative tasks. They have not been accompanied
by political democratization, which would allow for a more effective
participation of regional minorities in the governmental process.

There are some promising signs of ideological liberalization. Non-
Catholic denominations enjoy at present official recognition and a certain degree of freedom. The process of economic development and
modernization has lifted the mental horizons of Spanish society. Limitations on the freedom of thought and expression still in force are now
directly related to the preservation of the existing regime, without
attempting any more to impose a comprehensive, totalitarian philosophy. In this respect, it appears that the government is yielding to social
pressure for more freedom and tolerance. This pressure is clear to the
initiated observer of Spanish mass media, which have developed a jar-
gon of their own to designate their demands for social and political
change. The newspapers are constantly asking for "*apertura*" ("open-
ing-up") and "*participación*" ("participation"), which are euphemisms
for "democratization". When a political commentator or a government
official talks about projects for "political associations", everybody
knows that he is talking about political parties, which are banned as
such by present Spanish laws. Public opinion polls taken in recent
years show a marked change of attitude of Spanish society about extra-
marital sexual relations, divorce, religious beliefs and morals in general.[34]

The progressive outlook of Spanish society does not warrant, however, a prompt or easy solution to intergroup relations. First of all, only
through effective political democratization will it be feasible to grant to
the regional minorities some of their demands concerning self-government. Secondly, even a democratic regime would face serious tensions
arising from regional and ideological conflicts in Spain. After centuries
of strongly centralized government, Spanish society has lived in a political culture of centralized discipline and intolerance towards minorities. It would be unrealistic to expect a rapid thaw of latent social conflicts just by virtue of a constitutional change. Under conditions of political democratization, transition towards a pluralistic society would

require very competent statemanship. It is to be hoped, however, that the relatively high degree of economic development achieved by this country in the last decade will facilitate a peaceful transition to a modern, pluralistic society.

NOTES

1. R. Lafuente, *Los gitanos, el flamenco y los flamencos*, Barna, Barcelona, 1955; Jan Yoors and André López, *The Gypsies of Spain*, New York: Macmillan 1973.
2. J. Caro Baroja, *Los pueblos de España. Ensayo de etnología* Barcelona: Barna, 1946; S. Tax Freeman, *Neighbors. The Social Contract in a Castilian Hamlet*, Chicago: University of Chicago Press, 1970.
3. Josep Melià, *Informe sobre la lengua catalana*, Novelas y Cuentos, Madrid, 1970.
4. Antoni Rovira i Virgili, *Resum d'historia del catalanisme*, Barcelona: Barcino, 1936; Pierre Vilar, *La Catalogne dans l'Espagne moderne*, 3 vols., Paris, 1962; J. H. Elliott, *The Revolt of the Catalans*, Cambridge University Press, 1963; Maximino García Venero, *Historia del nacionalismo catalán*, 2 vols., Madrid: Editora Nacional, 1966–1967; Ferrán Soldevila, *Síntesis de historia de Cataluña*, Barcelona: Destino, 1973.
5. Antoni M. Badia i Margarit, *Llengua i cultura als països catalans*, Barcelona: Edicions 62, 1964, and "La integració idiomàtica i cultural dels immigrants. Reflexions, fets, plans", *Qüestions de vida cristiana*, No. 31, 1966, pp. 91–103; Francesc Vallverdú, *Ensayos sobre bilingüismo*, Barcelona: Ariel, 1972.
6. Josep Vandellós, *La inmigración en Cataluña*, Fundación Patxot, Barcelona, 1935; Ernest Liluch and Eugeni Giral, "La poblacció catalana", Appendix to the Introduction of the Catalan translation of A. Sauvy, *La poblaciò*, Barcelona: Edicions 62, 1964; *La immigració*, *Qüestions de vida cristiana*, No. 31, 1966; Antoni Jutglar *et al.*, *La inmigración en Cataluña*, Barcelona: Edición de Materiales, 1968.
7. In 1972, the number of north-African workers in the province of Barcelona was estimated at between 25,000 and 40,000, and most of them were working illegally. *Cf.* Margarita Sáenz-Díez, "Falta mano de obra en Cataluña", *Informaciones*, November 10, 1973, p. 3.
8. Francisco Candel, *Los otros catalanes*, Madrid: Ed. Península, 1965; Joaquim Maluquer i Sostres, *Poblaciò i societat a l'àrea catalana*, Barcelona: Edit. A.C., 1965; Antonio Figueruelo, *Cataluña: Crónica de una frustración*, Madrid: Guadiana, 1970.
9. Joan Reglà, *Aproximació a la història del País valencià* Valencia: L'Estel, 1968.
10. Joan Fuster, *Nosaltres, els valencians*, Barcelona: Edicions 62, 1962, and Castilian translation, *Nosotros los valencianos*, Barcelona: Edicions 62, 1967; Vicent Miquel i Diego, *L'església valentina i l'us de la llengua vernacla* Valencia: L'Estel, 1965; Rafael Lluís Ninyoles, *Conflicte lingüístic valencià*, Barcelona: Edicions 62, 1969; and *Idioma i prejudici*, Palma de Mallorca: Editorial Moll, 1971; Emili G. Nadal, *El País valencià i els altres. Peripècies i avatars d'una ètnia* Valencia: L'Estel; 1972. Cf. review by M. M. Azeoedo of the two books by Lluís Ninyoles in *Language. Journal of the Linguistic Society of America*, 49, 1973, pp. 733–36.
11. Josep Melià, *Els mallorquins*, Palma de Mallorca: Daedalus, 1967.
12. J. Torrent and R. Tasis, *Història de la prensa catalana* Barcelona: Bruguera, 1966; "Un diario en catalán", *Cambio 16*, No. 96, September 17, 1973, pp. 21–25.
13. Jordi García Soler, "Radio i televisiò en català", *Serra d'or*, August, 1969, pp. 19–20.

14. Manuel Murguía, *Historia de Galicia* Imp. de Soto Freire, Lugo, 1865; Emilio González López, *La insumisión gallega: mártires y rebeldes*, Buenos Aires: Citania, 1963; Vicente Risco, *Manual de historia de Galicia*, 2nd ed., Vigo: Galaxia, 1971.
15. F. M. Sarmiento, *Estudio sobre el origen y formación de la lengua gallega*, Buenos Aires: Nova, 1943.
16. José Luis Varela, *Poesía y restauración cultural de Galicia en el siglo XIX*, Madrid: Gredos, 1958.
17. R. Carballo Calero, *Historia da literatura galega*, Vigo: Galaxia, 1962.
18. Ramón Piñeiro López, *O linguaxe e as linguas*, Vigo: Galaxia, 1967; Jesús Alonso Montero, *Realismo y conciencia crítica en la literatura gallega*, Madrid: Editorial Ciencia Nueva, 1968.
19. Valentín Paz Andrade, *La marginación de Galicia*, Madrid: Siglo XXI, 1970, pp. 93ff.
20. Alberto Míguez, *Galicia, éxodo y desarrollo* Madrid: Edicusa, 1967; Carmelo Lisón Tolosana, *Antropología cultural de Galicia* Madrid: Siglo XXI, 1971.
21. Julio Caro Baroja, *Los vascos*, 3rd. ed., Madrid: Istmo, 1971; Philippe Veyrin, *Les basques*, 2nd. ed., Paris: B. Arthaud, 1955.
22. Arturo Campión, *Orígenes del pueblo euskaldun. Celtas, iberos y eúskaros*, Pamplona: Imp. y Lib. de J. Garcia, 1927.
23. *Cf.* William A. Douglass, *Death in Murelaga. Funerary Ritual in a Spanish Basque Village*, Seattle and London: University of Washington Press, 1970; Spanish translation by Eduardo Estrada, *Muerte en Murélaga: El contexto de la muerte en el País vasco* Barcelona: Barral, 1973.
24. Caro Baroja, *op. cit.*
25. Ramón Oyarzun, *Historia del Carlismo*, Madrid: Alianza Editorial, 1969.
26. Among recent books on Basque political problems: José-Domingo de Arana, *Presente y futuro del pueblo vasco*, Bilbao: Ercilla-Libros, 1968. José Miguel de Azaola, *Vasconia y su destino*, 1st vol., Madrid: Revista de Occidente, 1973.
27. Juan Estruch, *Los protestantes españoles*, Barcelona: Nova Terra, 1968, pp. 37ff.
28. About religious intolerance in Spain: Adolfo de Castro, *Historia de los protestantes españoles (y de su persecución por Felipe II)*, Càdiz: Imp., Librería y Litografía de la Revista Médica, 1851; Claudio Gutiérrez Marín, *Historia de la Reforma en España*, Mexico: Casa Unida de Publicaciones, 1942; Claudia B. de Wirtz and Winifred M. Pearce, *Spanish Harvest*, Spanish translation by Adán Sosa, *Cosecha española*, Mexico: Casa Unida de Publicaciones, 1949; Jacques Delpech, *The Oppression of Protestants in Spain*, London: Butterworth's Press, 1956; J. A. Monroy, *Defensa de los protestantes españoles*, 2nd ed., Tangiers: Luz y Verdad, 1959. A delightful personal account of the problems of Protestant proselytism in 19th century's Spain, in George Borrow, *The Bible in Spain*, Everyman's Library, No. 151, London: J. M. Dent & Sons, 1961. and William I. Knapp, *Life, Writings, and Correspondence of George Borrow*, 2 vols., New York: G. P. Putnam's Sons, 1899.
29. *Cf.* data at *Comentario sociológico*, vol. I, No. 2, June, 1973, pp. 193 ff.
30. Estruch, *op. cit.*, pp. 197 ff.
31. Jesús Jiménez, *La objeción de conciencia en España*, Madrid: Edicusa, 1972; Luciano Pereña Vicente, *La objeción de conciencia en España*, Madrid: Propaganda Popular Católica, 1972.
32. *Cf.* Jorge de Esteban Ed., *Desarrollo político y constitución española*, Barcelona: Ariel, 1973.
33. Constitution of 1931, Article 1.
34. Data on attitudes towards premarital sex, contraception, work by married women, divorce, etc., in *Comentario sociológico*, vol. I, No. 3, September, 1973, pp. 251 ff.

Peru: Portrait of a Fragmented Society

GEORG MAIER

GEORG MAIER was born in Karlovo, Bulgaria, but has spent most of his life in the United States and in Latin America. He attended the University of Florida, obtaining his Bachelor of Arts degree in 1959. His graduate study, in Political Science, was at Southern Illinois University at Carbondale, where he obtained his M.A. in 1962 and his Ph. D. in 1965. Until recently, Dr. Maier was Associate Professor of Government at Southern Illinois University at Edwardsville, but he is now managing a business in Quito, Ecuador, and serving as advisor on planning and administrative matters to the Ecuadorian Ministry of Government.

Dr. Maier's academic fields of interest are comparative politics, Latin-American politics, political sociology and political parties and interest groups. He has published two books: *The June 2, 1968 Presidential Election in Ecuador;* and *Historical Dictionary of Ecuador* (with Dr. A. W. Bork). Some of his articles deal with the boundary dispute between Ecuador and Peru; the structure and political role of Ecuador's functional chambers; the political role of the Catholic Church in Ecuador; and social structure and social mobility in Ecuador. His field of academic enquiry and teaching is not limited to that country, however, but has extended over all parts of Latin-America.

Peru: Portrait of a Fragmented Society

GEORG MAIER

Peru, which became an independent republic in 1821, has the third largest land area of any nation in South America, with roughly 14 million inhabitants. Physically, the country is divided into three well defined areas which are nearly parallel to one another: the coastal region or *costa* on the West, the highland region or *sierra*, and the forested *montãna* area extending from the eastern slopes of the Andes down to the jungle. Ninety-five percent of Peru's population is rather evenly divided between the *costa*, which was occupied by the Spanish conquerors, and the *sierra*.

The Country's social organization reflects a complex, diversified society with closely integrated local cultures. One of the major characteristics of the country, which is also present in its neighbours Bolivia and Ecuador, is the extent to which it is permeated with Indian culture. Peru is also one of the most conservative countries in Latin America, still retaining many Spanish colonial patterns. A small upper class, composed mainly of the white descendants of the Spanish conquerors, has controlled the wealth of the nation and in conjunction with the armed forces and the Roman Catholic Church, it has formed an alliance by which it has controlled the political process as well. This arrangement has resulted in the creation of an elite dominated society with class stratification resembling a pyramid, which in turn rests upon a broad base made up mostly of thousands of disparate Indian communities. In fact, one may speak of two societies: a stratified Peruvian society the members of which share certain basic traits that give them cultural homogeneity;[1] and a non-stratified Indian society which is characterized by little concept of ethnic unity and only a dim awareness of the Peruvian national superstructure.

Additional ethnic elements living in Peru other than the white-*mestizo* and the Indian include Asiatics, Negroes, Levantine Arabs, and

European immigrants, many of whom are Jews. Sporadic immigration to Peru began in 1830 but the first sizeable number – about 87,000 – consisted of Chinese who were brought in as coolies between 1849–74. The Japanese began to immigrate about 1899 and by 1933, when most immigration tapered off, there were about 20,000. Unlike the Chinese, the Japanese have tended to resist complete assimilation into national society. Negroes were brought to Peru in relatively small numbers prior to Independence. They have mixed with the white-*mestizo* and Indians creating the *zambo*, a term designating anyone of mixed blood, but with predominant Negroid features. Mulatto (white-Negro) types are rare in Peruvian society. Of the European immigrants, the Italian colony is the largest, followed by Americans, Spaniards, British, Frenchmen, and Germans. Only the Italians, Spaniards and Germans have intermarried with Peruvians, and some of their names grace Peru's social register. Arabs and Jews have remained ethnically isolated and there appears to be more resentment of these two groups than towards the European or North American immigrant.

The various ethnic minority groups constitute but a small fraction (one percent) of Peru's total population. In 1970, there were about 12,000 Chinese, 18,000 Japanese, 35,000 Negroes and roughly 90,000 resident aliens.[2] They are concentrated in urban areas with most of them living in Lima. With the exceptions of the Negroes and Chinese most are engaged in various business enterprises and are relatively well off.

Peruvian attitudes towards these minority groups are numerous, confused, and subject to change. This situation is reflected in popular vocabulary, where the terms used to designate stereotypes have changed over time in line with the social attitudes of the moment. Nevertheless, racial prejudice has, according to Frederick B. Pike, proved an obstacle to Peru's becoming a nation.[3] He states:[4]

In preventing the emergence of a Peruvian nation, racial considerations have been more important determinants than geographic and regional problems. While it is unscientific to speak of racial considerations, especially in any sense implying superior and inferior races, or to attribute the characteristics of a people to race rather than to traditions, experiences, environment and cultural background, it is necessary to consider race in all of its most unscientific connotations in trying to understand Peru. Here there persists a strong tendency to attribute to different elements of the population characteristics, good or bad, and potentials, high or low, that are allegedly the result of racial determinants.

If Peruvians tend to accept that racial factors influence the habits and determine the potential of the diverse ethnic groups that comprise their population, they disagree heatedly as to whether specific racial influences have been positive or negative. There is absolutely no consensus among intellectuals as to whether the Indian "race," the Spanish "race," the Negro "race," the Oriental "race," and the mixed-blood *mestizo* or *cholo* "race" are good or bad races. Unhappily, each of the "races" has had skilful enough detractors to spread the belief in the inferiority of that particular race among a wide cross-section of the populace. On the other hand, champions of a Peruvian "race" have tended to praise exclusively only one of the five "races" said to make up the national population (Indian, Spanish, Negro, Oriental, mixed-blood) and disparage the other four. Many Peruvians have as a result come to accept the monotonously-repeated message that at least four of the country's "races" are bad. Not infrequently they have reached beyond this to the conclusion that all five are without redeeming features.

The Indo-American caste of Peru is reflected in the languages spoken in the country. Over half of the population speaks Spanish which is the official language and the one used in the instruction in public schools. Spanish is the predominant language in most of the nation with the exception of the Southern Andes region where Indian tongues prevail. Peruvian Spanish is often referred to as the most "correct" Spanish spoken in South America. This is partly attributed to the relative isolation of Peru during much of its history. More important, however, is the fact that during the colonial era the Spanish-speaking aristocratic minority made every effort to retain and perpetuate its language in the prevailing prose form, "feeling that the supremacy of the language could be threatened by inroads of Quechua or other Indian dialects."[5]

In addition to Spanish there are a number of native languages, most prominent of these being Quechua and Aymara. Originally, Quechua did not have as great a distribution as it has today. It was spread as far north as Colombia and as far south as Chile by the Spanish clergy who deliberately taught Quechua as a sort of Indian *lingua franca*. Of lesser importance is Aymara, which is spoken between Arequipa to the north to the southern basin of Lake Titicaca in Bolivia and into the north Chilean highlands. Both Quechua and Aymara are the majority languages where they are spoken. Moreover, the use of Quechua appears to persist by its widespread use. In 1969 the government undertook steps leading to the introduction of Quechua and Aymara as languages of instruction in some *sierra* schools. The survival of both languages, however, is at present assured because of the low mobilization of the Indian rather than to this new government policy. At present, the status of Quechua and Aymara reflects the prevailing attributes of Peruvians towards Indian culture, with one group advocating their preservation and the other their suppression.

Roman Catholicism is the dominant faith of Peru and its primacy has rarely been questioned. Introduced at the time of the Spanish conquest, the Church has exercised great influence in the nation's affairs and continues to receive constitutional protection from the State. Religious tolerance is likewise constitutionally protected but other religions, Protestant sects in particular, have fared poorly. By 1971, the total Protestant community numbered some 100,000, or less than 1 percent of the population.[6]

Religion forms an integral part of the Peruvian's life and has an equally great impact on Peruvian society. While the Church has supported the present government and approved of its revolutionary reforms, traditionally she had contributed to the ideological reinforcement of upper class superiority by supporting conservative governments and defending *Hispanidad*, i.e., all that is Spanish. Thus, from the beginning there existed a tie between the Spanish elite and the Church hierarchy. To the Indian masses on the other hand, the Church has served a different purpose. As Mariategui states:[7]

The external trappings of Catholicism captivated the Indian, who accepted conversion and the catechism with the same ease and lack of comprehension. For a people who had never differentiated between the spiritual and temporal, political control incorporated ecclesiastical control. The missionaries did not instill a faith; they instilled a system of worship and a liturgy, wisely adopting them to Indian customs. Native paganism subsisted under Catholic worship.

While regional, social, ethnic, racial, and religious issues have acted as divisive forces in Peru's drive towards national unity, the single largest factor undermining the search for national consenses lies in the country's social classes. A study of Peru's social and political realities can be best understood and analyzed within the context of class categories and the conflict of social classes.[8] In this chapter, an attempt is made to outline Peru's class structure and show how inter-class conflict and exploitation of one class by another has fractionalized Peruvian society and undermined attempts to unite the nation.

I. Social Stratification in Peru

The present-day population of Peru is divided rather evenly between two social groupings: one deriving from the Hispanic-Peruvian culture which constitutes Peruvian society and the other from the indigenous Peruvian cultures. The co-existence of these two groups, the white or Creole and the Indian, has been marked by great social distance resulting from the significant disparities in levels of achievement and participation in the social, political, economic, and cultural life of the nation.

The group which identifies with the Hispanic-Peruvian culture is composed of the urban and most of the rural population of the *costa*, practically the entire urban population of the *sierra* and the *montaña* and a small percentage of the rural population of the same two regions. In the rural region where members of Peruvian and Indian society meet, a caste division has traditionally prevailed. The individual criteria for membership in this group, as contrasted with Indian society, are: adequate knowledge of the Spanish language though not necessarily literacy, and Western European style of dress, usually including factory-made clothing and shoes. Indian features will bar easy rise in this group but are not sufficient to preclude entrance.

The Hispanic-Peruvian element is stratified along lines of social class, while Indian society forms a caste separated and isolated from the class system of Hispanic-Peruvian society. At first glance there seems to exist, furthermore, a tripartite division along socio-cultural and economic lines, Whites, *cholos*,[9] Indians. Both the class-caste separation and the socio-cultural and economic division conform, however, to the ideology of the ruling groups which favours social distance from Indian society and racial pride in 'white' status. The actual interpenetration of all Peruvian social groups is, however, stronger – especially under conditions of incipient social change – than admitted by the ideology of the ruling groups. This ideology, therefore, takes the classical form of false consciousness.

In an objective, impartial view Peru's social stratification seems to be based on the distinction between five groups. Four of these – the Creole upper class, the industrial upper class, the middle class, and the urban lower class or *cholo*[10] identify with the Hispanic-Peruvian society. The rural lower class or Indians, identify with indigenous cultures.[11]

A. The Creole Upper Class

While the word Creole *(criollo)* has many connotations, it is here used to identify the descendants of the Spanish conquistadores and a few European immigrants who were the heirs of Spanish power in the towns of the *costa*. The land in the *sierra* was left to their *mestizo* counterparts, the offspring of the original white landowners and Indian women, who became the ruling class of that region. This situation led to an early division of the upper class. The "Creole upper-class,"or "coastal plutocracy," constitutes one-tenth of one percent of Peru's population.[12] It is composed of a conglomeration of landowners, financiers, bankers, real estate developers and private entrepreneurs whose wealth came from harvesting guano and producing fishmeal. The basis of wealth of the *sierra* aristocracy, on the other hand, was predominantly agrarian. Members of this group, in particular, place great emphasis on one's family name or *apellido* which reflects on the social status a person enjoys in society. To have the right *apellido* is a virtual assurance of finding all doors open, a privilege not even money or political aggressiveness can buy.

Most of the coastal plutocrats make up "national elite" whereas most members of the highland aristocracy form part of the "provincial elites." At the highest level of wealth and status the latter have merged with the former through intermarriage or membership in the most prominent social clubs. In contrasting these two groups it is possible to say that the coastal plutocracy is more flexible because its members have varied social origins and derive their wealth from different sources. The highland aristocracy, on the other hand, is more closed and drawn from a smaller number of families whose ancestors settled in Peru during the viceregal era.

The national elite has generally been considered synonymous with "the oligarchy" that has dominated the national scene since colonial days. They have served as a liaison between the country and the outside world and have been able, in collaboration with the military establishment, not only to rule the country but effectively oppose social and economic reform. The military junta, which took over in 1968, has assumed a decidedly leftist stance, however. While President Juan Velasco Alvarado is not a Marxist, he has, for the first time in Peruvian history, characterized his government as Socialist. The government's objective is to adopt a purely domestic philosophy "in order to emerge from the capitalist world without falling into Communism."[13] The properties of the national elite have been nationalized and many of them have left the country. The highland aristocracy, on the other hand, has remained relatively intact.

B. *The Industrial Upper Class*

During the past two generations Hispanic-Peruvian society has experienced changes similar to those witnessed in other Latin American nations. The population has increased substantially with a steady influx of people from rural communities to urban centres. Industrialization has likewise increased leading to the incorporation of powerful industrial elements into the upper class. Members of the industrial upper class form four-tenths of one percent of Peru's population.

One of the major characteristics of the Peruvian industrial upper class is that their wealth is not based on ownership of land although they may acquire some for prestige purposes. Secondly, this group is composed of relatively recent arrivals, primarily Europeans, Levantine Arabs and East European Jews. As William Whyte concluded,[14] "in any gathering of industrialists in Peru today, you will find a large proportion of immigrants and sons of immigrants." The reasons for a scarcity of traditional Peruvian families who have founded and developed industrial enterprises lies in the fact that manufacturing has not been part of the cultural pattern.

The number of members of this class that have gained access to the social circles of the traditional upper class is relatively small and has not resulted in a fundamental transformation of the basic social structure. The industrial sector accounts for roughly twenty percent of Peru's gross national product and a large portion of the industry has been in the hands of foreign interests. Consequently, absorption into the traditional upper class circles has proved to be of little inconvenience for the latter.

Whether the industrial upper class will increase in size and attain greater political power is, at present, difficult to answer. The 1968 military government has nationalized many foreign economic interests and has also induced members of the traditional upper class to engage in industrial enterprises. The latter has been accomplished by nationalizing their estates and giving them the option of either being paid in long-term government bonds with a low interest rate or immediately buying into an industry. While some have become active in industry, many others have preferred to retire and live on the interest. To a large degree this is attributable to situations which exert negative effects upon economic progress: "(1) those who exert effort and apply their abilities do not obtain the rewards; and (2) those who obtain the rewards do not put out effort nor apply their abilities."[15] Unfortunately, both of these situations exist in Peru today. They inhibit economic growth and thereby the expansion of the industrial upper class which would allow it to act independently from the Creole upper class.

C. The Middle Class

For purposes of identification the term middle class is used here to describe that social grouping which is excluded from the elite but maintains a considerable social distance from the masses. As Francois Bourricaud states, "the denomination 'middle' indicates that those who apply it to themselves; or to whom it is applied, are principally concerned with their relationship towards those above or below them."[16] Socially, this class, which accounts for about 20 percent of the total population, represents the lowest level of those Peruvians who like to think of themselves as 'whites'. There is little awareness of class cohesiveness among the various segments that make up this group, and there are great ranges of variation, not only between the urban and rural middle class according to regional differences, but also within the urban middle class.

The middle class has traditionally included professionals, intellectuals, members of the bureaucracy, and military officers. Certain educational criteria and professional qualifications tended to serve as guidelines for admittance to this group. Members of this old middle class lacked political, social, and economic independence, and their multiplicity of interests exhibited little conformity and potential for united action. Unable to exert any political muscle, they assured themselves a measure of acceptance by emulating upper class standards and value judgements.

A marked tendency towards urbanization and subsequently industrialization[17] in this century has led to the inclusion of a new element into Peruvian middle class society. This new element, urban and business oriented, developed a loud political voice under the dictatorship of Augusto B. Leguía (1919–30) whose great desire was "to rule Peru through the support of the only socio-economic sector he understood and admired."[18] Leguía's main positive contribution during his rule is generally considered to be his fostering of a "new, self conscious and sensitive middle sector, as a counterveiling force to the aristocracy."[19]

In spite of Leguía's support and the continued industrial growth of the country, the middle class has remained small and divided. The members come from diverse backgrounds: ethnically from white to Indian, economically from self-employed to salaried employees,[20] and socially from traditional elements to newcomers. This results in a large socio-economic gap between the upper and lower classes. The well-off members, who have traditionally emulated upper class habits, have, in fact, been co-opted by the latter. On this subject, Carlos A. Astiz notes that "the Peruvian upper class... has provided for the absorption of the old middle-class segment into the civil and military bureaucracy and even into private organizations such as corporations and banks, thus subsidizing the middle class and giving it a clear stake in the status quo."[21] The arrangement, he continues: "consists in a *quid pro quo* of at least passive acceptance of the system in exchange for stable employment and a dignified role in society."[22] As long as the existing socio-political system has been able to meet these requirements, the middle class had contributed to the maintenance of the status quo.

D. The Lower Class

The lower class, which comprises about 80 percent of Peru's population is divided into urban and rural segments, *cholo* and Indian, respectively. The *cholos*, who account for 30 percent of the lower class, are primarily found in urban areas (Lima, Arequipa, Callao, Chimbote) and coastal plantations. Racially, the *cholo* is overwhelmingly Indian or *mestizo* in background, although in the *costa* it also includes persons of Negro and Oriental ancestry.

The Cholo

Cholos are those Peruvians whose life-style represents a fusion of urban and Indian ways of life. Members of this class are labelled *cholos* in the ideology of the upper class stratum. Like the Indian, the *cholo* works with his hands, but unlike the Indian, he has enjoyed some sort of formal education which makes him aware of the social conditions which surround him. Not accepted by the "whites" and too proud to side with the Indians, the *cholos* represent the expanding, socially most restless and politically most alienated force in society. As a consequence, they have followed any political movement or leader that has promised them a better future. They have also provided the cannon fodder for the many revolutionary uprisings in the country's history and the bulk of the present Peruvian army.

The urban lower class is made up of the industrial labour force, artisans, taxi cab drivers, barbers, tailors, petty tradesmen, – anyone employed in a low-status occupation usually involving manual labour.[23] A large segment of this class is made up of personal service workers and more specifically, household servants. These people have become accustomed to industrial conditions and have usually two generations of experience in urban living. They have well-defined aspirations toward a better standard of living and improved education for their children.

The *cholo* class in the rural areas, also called the "old-style" *cholo*, is composed of small farmers, plantation workers, small-town craftsmen, and traders. The status of the rural *cholo*, who mediates between the regional upper class and the Indian, is determined by the patronage they receive from the rural elite in return for performing services that the elite find degrading to perform.

The *cholo* differs from the Indian by his identification with national Hispanic-Peruvian culture and social consciousness. He comprises a transitional element which has abandoned, at least temporarily, Indian lifeways and attendant cultural problems, in favour of economic improvement. In contrast to the Indian, the *cholo* is more aggressive and politically and economically more astute. Accordingly, the *cholo* has been referred to as the "new Peruvian race," and the "new middle sector" of Peru. At present however, the *cholo*, much like the middle class and the Indian, lack political and worker organizations as the means to force their "entry into politics."[24]

The Rural Lower Class

The rural lower class in Peru consists of most of the *sierra* agricultural population part of the coastal plantation workers and the primitive Indian tribes in the eastern jungles. Together they make up the remainder of Peru's lower class. They also find themselves at the bottom of the country's social pyramid.

The rural population of highland Peru consists of Quechua and Aymara-speaking Indians who constitute the largest sector within indigenous Peruvian society. It is a massive submerged sector which bears the classic stigmata marking its members as among the dispossessed peoples of the world: poverty, illiteracy, a high infant mortality rate, low participation in local and national institutions, and low self-esteem.[25] The last is accompanied by an instinctive even fatal passive resistance towards anything that suggests progress. If possible, the Indian produces enough to survive and maintains the role of an indifferent spectator *vis-à-vis* the world around him. The Indian's indifference and resistance is supported by the members of Hispanic-Peruvian society who look upon him as a child if not a brute. In fact, the Indian is addressed as *hijo* or child by them. Social boundaries are sharply drawn and social distance is relatively strict between the two societies. The elite believe that the Indian is inferior and that his status is fixed.

Within Indian society all organization is limited to the community. Inter-community relations take the form of trade and cycles of fairs or *fiestas*. The basic elements of the *fiesta* system, whose origin can be traced back to early colonial days, has furnished a large segment of the rural population, both Indian and *mestizo*, with a language with which to express social relations.[26] This social relationship is not formalized, however, and the differences among communities sharing the same cultural heritage and placed in the same cultural context can be fully as significant as similarities among them.

On the community level, there appears to be little reciprocity in the pattern of mutual obligation between Hispanic-Peruvian society and Indian society. The Indian supplies the manpower, pays tributes (in crops and animals) to support the appointed officials, carries out public works, and sustains the *fiesta* cycle. In return, the Indian presumably receives "protection" from outside encroachment and spiritual satisfaction from church services. Little comes his way in the form of economic and educational benefits. Indians who make demands upon their landlord, demands which they are entitled to by law, have been subjected to retaliation in the form of harassment, beatings, and threats against their lives.[27] Whatever the Indians have achieved has been the result of a protracted struggle and if they appear fatalistic and conservative to the outsider it is because changes that have taken place in Peruvian society have brought them more harm than benefits.[28] Consequently, in their attitude toward most of the outside world, including Peru as a nation, the Quechuas could be described as indifferent and the Aymara as apprehensive.

A gradual improvement of the Indians' lot has, nevertheless, occurred, particularly since World War II. Indian communities, with encouragement from politicians and labour organizations, have organized and have sought a greater role in national society. The Indian's cause has been supported by (1) political parties such as APRA[29] and Popular Action,[30] (2) Marxists and church-oriented peasant movements, (3) revolutionary leaders such as Hugo Blanco and Héctor Béjar[31] and (4) Peruvian intellectuals who have championed the Indian's cause. The latter, called *indigenistas* are led by Luis E. Valcarcel, Manuel Gonzalez Prada, Hildebrando Castro Pozo, and the Marxist intellectual José Carlos Mariátegui. The *indigenista* group has often provoked reaction which is sometimes heard among their adversaries. Carlos A. Astiz states: "A strictly racial approach is often found, with the explicit or implicit suggestion that the Indians should be transformed and acculturated forcibly or eliminated."[32]

Probably the greatest effort in incorporating the Indian into national society has been made by the revolutionary government which assumed power in 1968. This has been accomplished not so much by new programmes but by the general outlook and attitude of the government. Entire Indian communities in the *sierra* have abandoned their traditional way of life. Some have received expropriated land or work it on a cooperative basis and others have moved into the expanding urban labour markets. More importantly, those Indians who have left their communities have sustained their ties of kinship and community loyalty, thereby establishing a viable line of communication between the city and the rural area. Whether these efforts of the government will effectively end self-imposed isolation developed over centuries of exploitation, however, remains to be seen.

II. Social Values and Social Mobility

A. Social Values

Given the relatively rigid social stratification, social values tend to differ not only between Hispanic-Peruvian society and the indigenous

Peruvian societies, but also among the various strata of the former. Traditional Hispanic and Catholic values, rooted in the forms of medieval thought, have not only guided and bolstered national society but have continued to the present to act as a catalyst on the nation's culture. Some values have remained unchanged since colonial days, whereas others have acquired a distinct Peruvian flavour.

From earlier infancy the Peruvian individual is exposed to a socialization process which acquaints him with the behaviour patterns appropriate to the different class statuses. Consequently, the adult Peruvian is well conditioned to operate within the class to which his birth has assigned him and to possess the appropriate attitude of superiority or inferiority concommittant with his class status. Particularly in highland Peru where the traditional value system has remained intact human relationships reflect an underlying inequality of men.[33]

This value system includes four essential elements: hierarchy, paternalism, authoritarianism, and status. In any social relationship the hierarchical situation is of primary concern. This is followed by paternalism, which implies the superiority of one actor over the other. This, in turn, is followed by authoritarianism which those in a superior position develop on a permanent basis as a result of their status. Finally, there is the all important element of status. Because every relationship is a status relationship, strictly contractual relations are not possible in Peru. To enter into a binding contract implies that the parties involved were previously independent of one another and equal.[34]

To these aristocratic values can be added the value attached to the ownership of land which not only assures one of economic well being but, more importantly, guarantees a certain measure of social prestige and political power. Owning land guarantees a person rights traditionally exercised by a landlord or *patrón* and denied to the landless or *peón*. Being a landlord a person is provided with "a measure of security, status in the community, freedom to act and speak freely, some degree of influence in government and a share in its benefits."[35] Land also ensures access to educational opportunities for one's children and farm credit or irrigation improvements. Because land ownership directly relates to such benefits, every strata of Peruvian society, including the Indian, has been preoccupied if not obsessed with a desire of owning a piece of land. It is with such an acquisition that they begin to plan for the future.

Because of the significance attached to the ownership of land, it has never been easy to acquire. The traditional landowners have always considered ownership of land an exclusive right. This has allowed the large landholder to preserve the socio-economic and political status quo. It has also had some positive as well as negative effects on the whole society. Peru remained relatively isolated from the outside world until the turn of the last century allowing the development of the collective goal of *"Peruanidad"* which is often formally expressed in the patriotic attitude. At the same time, there was no significant immigration to Peru until the 1930's. In describing 19th century Peruvian history, Watt Stewart states:[36]

... the great landowners were, aside from the church, perhaps, the strongest power in the country. It was very difficult for any government to refuse to listen to their demands. This fact, more than any other, explains the failure to attract the European immigrant. The great sugar or cotton planter did not want the land settled by the modest husbandmen, the genuine colonizer. He merely wanted hands to work his broad acres, and at a good profit – which meant cheap labour.

Accordingly, the only sizeable group of foreigners that was brought to Peru were Chinese coolies during 1849–75. According to Stewart, they were not only exploited but the "Asiatic problem" ranks as one of the most serious social and economic problems Peru has faced since Independence.[37]

Individual values[38] stressed in Peruvian society and instilled particularly among elite youths include male exemption from work. Only the *cholo* and the Indians are expected to work, whereas the rest of society's male population devoted their efforts to social pastimes only to be occasionally interrupted by work. Secondly, the pursuit of manliness *(machismo)* is thought as desirable and ideal. While *machismo* may be best defined as intransigence it also involves sexual conquests and the ability to maintain a succession of mistresses. A third value inculcated into Peruvian youth is the desire for dominance. However great this urge to dominate may be it is considered impolite to manifest aggression. The latter, nevertheless, motivates individual behaviour. To get ahead in life, it is important to have these values or, more important, to have members of one's peer groups believe that one has them.

Values held by members of the national society are in sharp contrast to those shared among the Indians. The ingrained patterns of dominance and submission to which they are born and later conditioned is elevated by them to the level of a supernatural principle. In terms of material values, the Indian, much like his members of Peruvian society, seeks to acquire wealth and prestige. Outward display of wealth and holding a position of commercial authority provided an Indian with a measure of social status *vis-à-vis* the rest of the community.

B. Social Mobility

In spite of the possibility of upward mobility, social stratification remains rigid. The rigidity of the social structure is especially pronounced in the rural areas of the *sierra* where the traditional economic system has experienced the least amount of change and has provided the least opportunity for upward social mobility. Ownership of land determines status in a manner whereby the more land he owns the higher is the social prestige of the owner. Small land holders, on the other hand, incapable of securing additional land, are relegated to a low social status. Upward mobility is particularly difficult in the countryside because of a lack of opportunities to become educated or change jobs. Thus, the peasant is forced to leave his rural environment in order to move up socially.

The Indian population is particularly handicapped in moving upward socially because Indians, in most cases, own no land. Ethnic differences and resistance to becoming acculturated provide further roadblocks. Within his own community the Indian may enjoy certain status if the local priest appoints him to serve as a sponsor *(prioste)* for a fiesta or to various other local offices. In general, however, he must leave the city, change his native ways and hope to improve his social and economic status. The system of obligatory military service is often the most effective method of incorporating the Indian into Peruvian society.

Social mobility assumes greater significance in the large cities, and, in particular, in the Lima-Callao metropolitan area, where the greatest changes of a socio-economic and political nature have taken place. There, a light skin, economic fortunes, appointment to high public or private offices, and exercise of a profession have contributed to upward social mobility, solidified the middle class, and brought about some additions to the upper class. Members of the urban lower class have experienced greater difficulty in moving up the social scale, not to mention those Indians who have migrated to the cities. Cultural and psychological limitations have proved to be difficult barriers to becoming part of the urban economic process.

Downward mobility is also possible when a family loses its fortune. The downward passage from middle class to *cholo*, however, does not occur frequently since it involves a drastic change of habits and behaviour and is not just a change in social relationships and economic status. From *cholo* to Indian (rural lower class) the transfer is relatively simple, where horizontal mobility is possible, and fundamentally involves little more than a change of residence.

In summary, it may be stated that social mobility in Peru is difficult given the fairly rigid class structure. When it does take place it is usually the result of the modernization process the country has undergone in the last decade. This implies that social mobility is greater in the urban areas and that the middle and upper classes have benefited the most from it. Insofar as the members of the indigenous-Peruvian societies are concerned, however, social mobility has only been of relative significance.

III. CONCLUSION

In colonial days, Peruvian society was largely a reflection of its feudal economy. Class status was ascriptive and social mobility, therefore, accidental. The wars of independence and the subsequent proclamation of the Republic brought about the demise of the Spanish-born upper class or *chapetones* and the rise in social status of the *criollos* and military officials who rose socially through the armed forces. This minor realignment in the class structure did not change ostensibly during the 19th century.

The first phase of social mobilization in Peru took place between 1890–1940 and, more specifically, after 1930. During this period Peru's export economy was developed and modernized, major highways were built, the mining industry diversified and expanded, and Lima was transferred from a slumbering colonial capital into a modern city.[39] Rather than to rely on the importation of coolies, labour during this period was provided by the native labour force which led to an ever-increasing migration from the rural to the urban centres and from the *sierra* to the *costa*. This, in turn, led to the formation of a rural working class and industrial working class and an expansion of the middle classes. The emergence of these new classes coincided with the entry of APRA into Peruvian political life in the 1920's. APRA, generally considered a populist-type coalition, represented the first attempt to express an ideology of social transformation in Peruvian politics.

During the second period, 1950–60, Peru entered a phase of extended social mobilization characterized by a movement of people of all the regions, the *sierra* in particular, toward the metropolitan area of Lima-Callao. This internal migration coincided with increased industrial expansion and a population explosion. Between 1940 and 1961, the population of the Lima-Callao area has almost doubled with one half of the new immigrants coming from the *sierra*. The total urban population has increased from 35.4 percent in 1940 to 47.4 percent in 1961 and 51.0 percent in 1967.[40] This rather massive influx into the urban centres which continues to the present day is motivated primarily by economic considerations. But economic expansion has not kept pace with the migration and few of the new urban dwellers have been economically absorbed.

In spite of these migrations and continued efforts of the present military government to integrate the Indian into the national sector the social mobilization of the Indian has remained relatively low. As David S. Palmer says: "... these attempts may be met with apathy or, at most, an attempt to further their own particular interests rather than the objectives desired by the innovators from the national government."[41] Thus, in the context of a pluralist society, the Peruvian nation continues to reflect the model of "devisive pluralism" on the Hispanic-Peruvian level and "segmented pluralism" on the indigenous level.[42]

Because a large segment of the population falls into the "segmented sector" and because this implies low social mobilization, the military government has considerable "political space" within which to govern. Adding to this the unchallenged position the military enjoys within the "divisive sector," the military government may well succeed in establishing itself as a reformer rather than as a mere restorer of stability.[43] Whether the intended reforms will be brought to fruition and lead to a unified Peruvian nation, however, remains to be seen. Good intentions against such staggering odds may not be enough.

NOTES

1. These common traits are: use of the Spanish language, the Hispanic political and Catholic religious systems, a European economic system in the process of transition form pre-industrial to industrial, and a shared tradition of participation in the vicissitudes of Peru as an independent nation.
2. Peru, *Area Handbook for Peru* (Washington, D.C.: U.S. Government Printing Office, 1972), pp. 70–1.

3. Frederick B. Pike, *The Modern History of Peru* (New York: Frederick A. Praeger, 1967), p. 5.
4. *Ibid.*
5. Peru, *Area Handbook*, p. 73.
6. *Ibid.*, p. 97.
7. José Carlos Mariátegui, *Seven Interpretive Essays on Peruvian Reality*, trans. by Marjory Urquidi (Austin: Texas University Press, 1971), p. 135.
8. Carlos A. Astiz, *Pressure Groups and Power Elites in Peruvian Politics* (Ithaca: Cornell University Press, 1969), p. ix.
9. There is no agreement among sociologists and anthropologists regarding the moment when an Indian becomes a *cholo* since the difference is socio-cultural and economic rather than racial. *Ibid.*, p. 72.
10. The *cholo* is found predominantly but not exclusively in an urban setting, whereas the Indian lives primarily in a rural environment.
11. This classification is employed by Astiz. *Ibid.*, p. 67.
12. The percentages used here are, likewise, from *Ibid.*
13. Quoted in Peru, *Area Handbook*, p. 77.
14. William F. Whyte, "Cultural, Industrial Relations, and Economic Development: The Case of Peru," *Industrial and Labor Relations Review*, XVI (July, 1963), 585.
15. *Ibid.*, 592.
16. François Bourricaud, *Power and Society in Contemporary Peru*, transl. by Paul Stevenson (London: Faber and Faber 1970), p. 58.
17. Carlos A. Astiz notes that as in other Latin American countries, "Peruvian urban growth preceeded industrial development . . . and has led to the foundation of a middle class of limited usefulness, with a limited role in the type of society which still exists in the country." Astiz, *Pressure Groups*, p. 67.
18. Pike, *The Modern History*, p. 217.
19. *Ibid.*
20. There exists a well-marked distinction between the traditional usually self-employed and rural middle class and the *empleados* or salaried employees who compose the lower middle class and live in the cities.
21. Astiz, *Pressure Groups*, p. 67.
22. *Ibid.*, p. 68.
23. Magali Sarfatti Larson and Arlene Eisen Bergman, *Social Stratification in Peru* (Berkeley: University of California Press, 1969), p. 54.
24. Richard Stevens, *Wealth and Power in Peru* (Metuchen, N. J.: The Scarecrow Press Inc., 1971), p. 150.
25. While there are no precise universal criteria by which Indians are defined, identification in Peru varies with geographic location and social status. The following may be considered as social traits denoting "Indian-ness:" (1) being poor, (2) going barefoot or wearing sandals, (3) speaking primarily Quechua or Aymara, (4) being illiterate, (5) wearing distinctive dress and practicing certain dietary habits, (6) living in or identifying with an Indian community, (7) living on subsistence agriculture, (8) chewing coca. While not all these traits are necessary to be classified as Indian, traits 3, 4, 6, and 7 are the most essential. See Sarfatti Larson and Eisen Bergman, *Social Stratification*, p. 39.
26. Hans C. Buechler, "The Ritual Dimension of Rural-Urban Networks: The Fiesta System in the Northern Highlands of Bolivia," in William Mangin, ed., *Peasants in Cities: Readings in the Anthropology of Urbanization* (Boston: Houghton Mifflin Company, 1970), p. 62.
27. For an excellent treatment of this subject, see F. La Mond Tullis, *Lord and Peasant in Peru*, Cambridge: Harvard University Press, 1970.

28. Astiz, *Pressure Groups*, p. 86.
29. APRA stands for Popular American Revolutionary Alliance and was founded in 1931 by Victor Raúl Haya de la Torre. This political movement has attempted to effect a more equitable land distribution and to ameliorate the living conditions of the lower classes and Indian society. For an excellent study of this political movement, see: Harry Kantor, *The Ideology and Program of the Peruvian Aprista Movement*, Berkeley: University of California Press, 1953.
30. AP stands for Popular Action and was organized in 1956 by diverse elements that had supported Fernando Belaunde Terry in the elections of that year. AP's platform was similar to that of APRA with agrarian reform and integration of the Indian into national society being of high priority.
31. The guerillas attempted to bring about a fusion of the most combative peasants and the revolutionary petty bourgeoisie. See: Héctor Béjar, *Peru 1965: Notes on a Guerilla Experience*, trans. by William Rose (New York: Monthly Review Press, 1969), p. 11.
32. Astiz, *Pressure Groups*, p. 45.
33. Stevens, *Wealth and Power*, p. 41.
34. *Ibid.*, p. 42.
35. *Ibid.*
36. Watt Stewart, *Chinese Bondage in Peru* (Durham: Duke University Press, 1951), p. 8.
37. *Ibid.*, p. 1.
38. Stevens, *Wealth and Power*, pp. 85–6.
39. *Ibid.*, pp. 33–4.
40. Quoted in *Ibid.*, p. 127.
41. David Scott Palmer, *"Revolution from Above:" Military Government and Popular Participation in Peru, 1968–72* (Ithaca: Cornell University Press, 1973), p. 13.
42. In "divisive pluralism" the sub-units of society are "in contact with government and with each other," and in "segmented pluralism" the sub-units of society are "in contact neither with government nor with each other." *Ibid.*, p. 15.
43. "In order to achieve these ambitious goals once in the seats of political power, the military has proceeded to change the rules of the political game. At the same time, they have worked to alter the structure of interaction in such a way as to permanently displace the oligarchy from its past position as the 'peak' group in the basically divisive pluralistic framework society. The new rules of the political game provide for the following changes: (1) An increase in the role of the State in the economy, society, and culture, (2) Ubiquitous military control at the key points of the top and intermediate levels of the political apparatus of the state; i.e., the government bureaucracy, (3) Conscious efforts to make the political parties and pressure groups less relevant and less legitimate by systematically cutting out their access as institutions to the political system, (4) Conscious efforts to rechannel demands and supports of citizens onto the output side of the political system, (5) An expansion in the definition of political participation to include involvement in economic and social arenas and at the local level, while at the same time reducing (at least for the nonce) citizen participation in the authoritative allocation of values at the national level, (6) Conscious efforts to subordinate all groups to the military elite, by forcing them – through withdrawal of power resources and by effective control of key access points – into a clientelistic relationship with the military. The groups can choose to withdraw from the larger clientelist network, if they wish, (7) Encouragement of qualified and visible civilians to accept roles within the government apparatus *as individuals* in advisory or policy implementation positions." *Ibid.*, pp. 258–9, 61.

Endangered Cultures: The Indian in Latin America

MIGUEL LEÓN-PORTILLA

MIGUEL LEÓN-PORTILLA, a native of Mexico City, received his M.A. in Philosophy (1952) from Loyola University of Los Angeles, California, and his Ph. D. (1956) from the National University of Mexico. He has been a Professor on the Faculty of Philosophy and Letters, National University of Mexico, since 1957 and he was Director of the Inter-American Indian Institute from 1960 to 1966. He is presently Director of the Institute of History of the National University of Mexico.

Among the many professional organizations to which professor León-Portilla belongs, several pertinent to the present study are: Société des Americanistes, Paris; Sociedad Mexicana de Antropología; Permanent Committee of the International Congress of Americanists; Patronato del Arte Indígena Oaxaqueño; Sociedad Defensora del Tesoro Artístico de México. He was awarded the Elías Sourasky Prize in 1966 and a Guggenheim Fellowship in 1969. Among his published books are *Aztec Thought and Culture* (Spanish 1956; English, University of Oklahoma Press, 1963); *The Broken Spears, Aztec Account of the Conquest of Mexico* (Boston, Beacon Press, 1962); *Precolumbian Literatures of Mexico* (Spanish 1963, English, University of Oklahoma Press, 1969); and *Aztecs and Navajos, A Reflection on the Right of Not Being Engulfed*, 1975.

Endangered Cultures: The Indian in Latin America

MIGUEL LEÓN-PORTILLA

Three attitudes in the Sixteenth Century affected the destiny – survival or disappearance – of the indigenous peoples and cultures of the New World. The first attitude is that of Fray Bartolomé de las Casas. Many passages from his works could be cited, but his *Brevísima relación de la destrucción de las Indias* (Very Brief Account of the Destruction of the Indies) reflects the essence of his thought:

So many, so frightful and so great have been the deaths and cruelties inflicted on the Indians – and they are still being inflicted – that all these lands will be devastated and lost. If Your Majesty does not find a remedy, everything here will soon come to an end, and there will be no Indians to live here and all these countries will remain uninhabited and barren.[1]

The central thought of Fray Bartolomé sprang from the experiences he had gone through, since his stay on the Caribbean Islands, on seeing the gradual extinction of the indigenous populations. His attitude, nevertheless, was not fatalistic. Las Casas believed such cruelties, deaths and destruction of the Indians could still be prevented, and herein lay the reason for his struggle and why he became the passionate defender of the Indians.

Another point of view on the fate and possible presence or disappearance of the Indians was offered by another Dominican who, like Fray Bartolomé, also lived in conquered Mexico, Fray Domingo de Betanzos. He also observed how the native population was diminishing, and was alarmed to such an extent that he became persuaded that they were to be totally extinguished. He journeyed to Spain to make his view public in the Imperial Court, incurring the enmity of some and the praise of others. In a letter he expressed concisely what he thought:

The Indians will soon be gone, a statement which I repeat, sign and confirm. And this will have happened before this generation passes. I say that by the time that those who are yet children in our own (Spanish) nation attain the age of seventy, at most, they will see the end of the extinction of all the Indians, and it is possible that this will occur before forty years have passed. And this is true, perfectly credible, even though some may think I blaspheme.[2]

A highly different approach is revealed by a remarkable scholar and investigator of the Aztec pre-Hispanic past, in all probability the most distinguished of all, Fray Bernardino de Sahagún. He likewise had knowledge of the decreasing number of Indians, due to forced labour and epidemics. Though well aware of the situation, he saw reality with optimism. In one stark sentence in his *Historia General de las Cosas de la Nueva España* he expressed his view:

It is my opinion that there will always be large numbers of Indians in this land.[3]

These were the three attitudes of awareness, all thought-provoking, from three individuals who could be called Spanish representative minds of the Sixteenth Century in the New World. Betanzos typified that of the pessimist. Sahagún was the optimist who believed that the Indians would not disappear. Las Casas, deeply conscious of the threatening dangers, devoted his existence to fight for respect for the dignity of the Indians and their cultural values. Today, after three centuries of Colonial life and one and a half of independent life in Latin America, it can be recognised that Las Casas was right in his struggle and that Sahagún did not err in his prophetic affirmation.

I. Forms of Indian Presence

In the Latin American world of today various forms of the presence of the Indian can be found. There are the remains of ancient cultures revealed by archeology, the wonders of art, the codices which have survived and the native texts. These constitute an Indian presence which aids the evaluation of his partially vanished world, revealing much of his "face and heart", as the natives themselves express it. This first form of the indigenous presence undeniably confers elements of individuality upon the cultures of the Latin American peoples, particularly in countries such as Mexico, Guatemala, Peru, Ecuador and Bolivia.

A second form of the Indian presence is to be found in the culture elements of the native world, which, incorporated in the ways of life of the rest of the people of Latin America, continue to exist. Many native words have become a part of the Spanish national language in Mexico, Guatemala and other countries: Nahuatlisms, Mayisms or, in Peru, Bolivia and Ecuador Quechuisms; likewise, the obvious influences on foods and popular arts. All this, and a great deal more, turns those who have come to settle in Latin America into mestizos by culture. Several studies have been made on indigenous contributions to universal culture. Alfonso Caso[4] and Juan Comas[5] have published valuable works on this subject. (Caso 1946; Comas 1957) If these writers have shown that numerous elements of the pre-Columbian world have been integrated by universal culture, it is obvious that all individuals living in the American Continent have received even more. Thus, this second form of the Indian presence is a reality.

The main interest, however, is the living presence of millions of Indians. In the Latin American continent today there is a population of some two hundred and eighty million inhabitants. Within this growing number the percentage of Indians is large enough to be significant, and far too large to be ignored. How many there are, where they live, how to determine who should be considered as Indian, since these questions have been studied from an anthropological point of view, many criteria have been used. Probably the most acceptable answer is that an Indian group is one which, in its culture, in its ways of life and social organisation, preserves a high proportion of the characteristics and elements of pre-Columbian origin. The speakers of native languages in the case of Mexico, for instance, are approximately ten per cent of the total population, that is rather more than five million. Of these some three and a half million are bilingual in varying degrees, the rest are monolinguals, speaking only their native language.

If language is an obvious trait, there are others which are less apparent, such as the structure of a community in which there survive, as in some areas of Mexico and Guatemala, elements of the ancient *calpulli* or ward system, or in the Andean zone the organization of the *ayllu*. In those communities there are functions which are distributed in the traditional manner for the celebration of fiestas throughout the yearly calendar. In these cases often more important than the structures imposed by the political organization of the modern countries (for instance, the municipal system) there exist patterns of traditional government, far stronger and deeper-rooted. Other cultural elements of ancient origin are also visible in the lineage relationships, in the internal administration of justice, in religious and magical practices, in the artistic creations, in the labour systems, in the manner of exploiting the earth and in the traditions which are at the basis of the community's sense of identity.

To be sure, there coexist in these communities a number of traits of pre-Columbian origin with others of Colonial and modern proveniences. These groups, preserving to different degrees a variety of ancient cultural elements, constitute what can be called the indigenous presence. Within the Latin American world, therefore excluding Canada and the United States, four major areas of the Indian presence stand out.

II. Major Areas of Indian Presence

One area is Middle America or "Mesoamerica", i.e. a large part of Mexico and adjacent portions of Central America. In the case of Mexico, the northern zone of the country has to be excluded since it does not fall within the geographical-cultural concept of Middle America. This portion will be dealt with later. In Middle America, where the ancient high cultures flourished, the Indian presence is felt today, in some regions with great intensity. A good example of this is given by the Mexican state of Oaxaca where there are more than fourteen different native groups, and more than half of the total population is indigenous. In Chiapas, Yucatan, Campeche and Guatemala the Maya Indian presence is also strong and obvious.

In Middle America there have existed and exist today intense processes of cultural and ethnic rapprochement between the Indians and those of predominantly European culture. On the other hand some indigenous groups have maintained a kind of "insular" character, that is to say, communities such as those of the Mixe, Tzeltal and Tzotzil, Cora, Huichol, and others, are more or less isolated entities that have managed to preserve a strong sense of identity.

A second and very different zone is that of the Andean area, whose heart-land is Peru, Bolivia and Ecuador. To these should be added some regions in southern Colombia and in the north of Chile and Argentine. It comprises a vast area, with heavy concentrations of Indians since pre-Columbian times and throughout the Colonial and modern periods. Within it there has been a limited mixture with the Indian.

On the coast of Peru, for instance, lie the Spanish cities and in the mountains the great conglomerates of natives.

A fact is that there exist today in some of those countries persons who feel proud of the absence of indigenous ethnic traits in their families. On the other hand, in the mountains and the highlands live millions of natives. It probably would not be an exaggerated estimate to state that the Quechua speaking people of Peru, Bolivia and Ecuador in their totality are more than fifteen million at this date. The majority have lived a wretched existence though certain projects, in some ways positive, have been initiated among them.

To consider a third area, again different: take that of the Indians of the tropical forests. Small groups, preserving what has been described on occasions as a "primitive culture" live there. On the margins of a river, at certain spots, may be found, for instance, five hundred Indians; at other sites there may exist smaller clusters, or slightly larger ones, though never numerous. Regarding the aboriginal population of Brazil, for example, opinions vary widely.

How many dwellers of the tropical forest live in the Amazon basin, along its tributaries, or in the Orinoco basin? Doubtlessly there are many groups with their distinct languages and differences. Usually they live near the great means of communication of the jungle, which are the rivers. Perhaps, in their totality, the South American dwellers of the forest do not number more than half a million. Several of these communities are very seriously endangered, that is, on the verge of disappearing.

There remains the fourth area, inhabited by a different kind of Indian, to a certain extent also marginal, such as the Chibcha in Colombia, the Araucanian of Chile and several small Argentinian groups in Patagonia and in the Chaco. It is also important to mention, in the north of Mexico, the Papago, Seri, Yaqui, Mayo and Tarahumara, all of whom have preserved a deep rooted sense of identity. Actually, the native population of the northern portion of Mexico, outside of the Middle American area, numbers approximately one hundred and fifty thousand persons. The largest group is the Tarahumara, the others being much smaller.

In a word, the Indian presence in the Latin American world is particularly manifest in both Middle America and the Andean area. These two nuclear zones since ancient times were the scene of high cultures and civilizations with dense human conglomerates and elaborate social and economic patterns.

It seems pertinent now to raise some questions about the meaning of the indigenous presence today. Do the native communities constitute a specific social class? Clearly the majority of the Indians are poor, victims of different forms of exploitation. Nevertheless they do not necessarily form a specific social class. Actually it can be noted that sometimes within an Indian group different social and economic levels are manifest. Are they to be identified with the peasantry? Only to a certain extent, since some Indians are only partially devoted to farming. Basically, today's natives are groups with a different culture and, to a large measure, ethnically distinct from one another. Compare, say, the ways of life of a Mixe Indian from Oaxaca with those of a Maya from Yucatan. Physical anthropology has shown that there exist marked differences even within a single linguistic family, such as that of the Maya. (Comas 1966) On the other hand, it is worth remembering that the influence of Spanish culture has served often as a homogenizing factor since the Sixteenth Century. Undeniably a good number of elements of European origin have infiltrated in the indigenous cultures, dress was modified, a number of institutions became implanted to some degree, various forms of syncretism were born as is shown, for instance, by contemporary religious beliefs and practices. Nevertheless, as Fray Bernardino de Sahagún had foreseen it, the fact remains that "the face and heart" of the Indian – his cultural presence and sense of identity – has survived in large areas of the New World.

III. Contemporary Attitudes vis-à-vis the Indian in Latin America

The first attitude, with a long history behind it, is that of "Indianism". It was born as something more or less romantic. The idea was that "something should be done for the Indians". In the case of Mexico this point of view became manifest since the days of Independence and in some instances it was associated with the understandable purpose of turning the new country's attention inward, toward that which is autochthonous, which gives individuality, which is opposed to the former dominators with whom the old Colonial ties had been broken. The new laws and constitutions of most Latin American countries established equality among the Indians, mestizos and descendants of Europeans, but this did not solve the problem. In reality the Indians did not enjoy the same benefits. Many years were to pass before Indianism could become more operative, better qualified.

Indianism, with an anthropological approach, actually originated with the consummation of the Mexican revolution of 1910. Dr. Manuel Gamio was one of the first to conceive this approach. He demonstrated how anthropological science should become a guide to adequate forms of action and how it should lead toward an understanding of the cultural differences. Only thus could be conceived projects to reach the Indian communities, as it was useless to base action upon the pretence that one was living in a country with a culturally homogenous population. These ideas, promoted by Dr. Gamio, resulted in an Indianism ever tending toward the basic principles of the social sciences. This Indianism became institutionalized on the founding of a number of organisms. Thus was born, in 1941, the Inter-American Indianist Institute whose first director was Dr. Gamio.

Affiliated organizations were created with the same purpose of serving as advisory boards to the various Latin American governments in a type of action which would take into account the Indian presence in a definitive and integral manner. Some of the National Indigenist Institutes were endowed, in different measures, with the means of achieving their function as advisory boards and their direct work in the zones of native population. It must be admitted, however, that in several cases these indigenist organisms, created to comply with an international commitment derived from the Convention of Patzcuaro (1940), subscribed to and ratified by the majority of American nations, were simply new bureaucratic offices producing meagre practical results. And again we must insist on the magnitude of the problems confronted by the native communities citing as examples the cases of Bolivia and Peru. In the first of these countries, out of a total population of about four million, almost three million are Indians in many respects destitute and deprived of the possibilities of participating in the socio-economic life of their country. In Peru, out of some fourteen million inhabitants probably more than nine are Quechua and Aymara.

In Mexico, the National Indianist Institute was created as an organism for action. Based on anthropological research of an integral nature, its end was that of planning and making possible the changes which it was thought opportune to introduce in the different Indian zones. With this in mind, this institute organized a series of "Coordinating Centres." Some of these Centres, such as that which functions in Chiapas among the Tzotzil and Tzeltal, have experienced considerable development. Others exist in diverse parts of Oaxaca, Yucatan, Guerrero, Puebla, Veracruz, Michoacan, Nayarit and Chihuahua. It must be recognized however, that the work of the National Indianist Institute due, among other reasons, to the limitations of its economic resources, has affected a relatively small percentage of the groups that form the Indian population of Mexico, perhaps only about a 25 per cent. Concerning the philosophy which has guided the action of this Institute it may be said briefly that its aim has been to foster the socio-economic betterment of the indigenous groups thus to achieve their effective integration into the life of the nation. (See: Caso 1955 and 1962).[6]

There exits, on the other hand, a different attitude on the part of several Latin American governments. This may be described as derived from criteria prevalent among some experts in the economic sciences. The central idea of this way of thought is that the native groups have to be assimilated by the culture of the majorities as a consequence of the application of development programmes carried on a national level. These programmes must affect also the Indians and in the end must transform their traditional ways of life. In consequence, in the minds of those who hold these views, it is superfluous to undertake other specific forms of action, such as those held by Indigenism which takes into account the cultural differences and values of the native communities.

Finally there are the attitudes which several anthropologists have expressed in the last years on re-examining these problems. In their judgment, respect for different culture values, which obviously is one of the tenets of the anthropological sciences, cannot be reconciled with action directed toward "bettering" the Indians through processes described as "Mexicanisation", "Bolivianisation", "Peruvianisation" and so forth. What is important, according to this criterion, is to strengthen the native groups, fostering their economic development, but favouring at the same time the conservation of their ethnic and cultural identities, in order to achieve the constitution of autonomous nationalities, such as those existing in the Soviet Union. Those holding such a point of view sometimes add that this will only become a reality through radical changes in the social and political structures of the respective Latin American countries. (See, for instance, Bonfil 1969[7] and Stavenhagen 1966–67[8]).

Closely related to the above position are the statements often expressed by some anthropologists when talking about the social scientists' responsibility and the possible development of Indian groups. For example, the Mexican Daniel Casés writes:

Problems of responsibility have faced the Mexican anthropologist from the very beginning of his education. The principles, which guide the work among Indians of the Mexican government Indigenous National Institute, are based on the same goals listed for efficient administration and the introduction of middle class values. ...

The term "internal colonialism", used by some of our social scientists to refer to this situation, is adequate up to that point, although it is unfortunate for the understanding of the total system of which our countries are a part; among other things it implies a dualism and even a pluralism which do not exist.[9]

Regarding the latter attitude various critical comments could also be quoted, such as those in the paper written by Professor Alfonso Villa Rojas entitled, "On the new ideological tendences of some anthropologists and Indianists",[10] and published in *América Indígena*, the quarterly magazine edited by the Inter-American Indian Institute.

An interesting approach to these same questions has been offered also by Dr. Rudolf A. M. van Zantwijk,[11] a Dutch ethnologist who has delved deeply into the situation confronting some of the Indian groups of Mexico and other American countries. Concisely he states:

A solution is only possible if autochthonous and therefore often "undeveloped" aspects of the (native) culture are no longer confused with identity. Active government support of the Indian cultural identity need not be a check on economic progress and integration. Government support of some aspects of the Indian social identity will even allow them to remain functional. Such support is not likely always to promote a so-called "free" capitalistic development. ...

Just as all aspects of the social Indian identity should be respected, in the same manner the different historical processes in the diverse regions or countries of Latin America must also be taken into account as a necessary antecedent. In the case of Mexico it is clear that the historical processes of three centuries of existence as "New Spain" and a century and a half of independent life have led irreversibly to the formation of a mestizo nation. The consequence in this case has been, to a considerable degree, a cultural and ethnic fusion through the assimilation of values, mainly Indian and Spanish. Mexico has thus amalgamated its autochthonous legacy with many other cultural elements acquired in later periods. But accepting that this is the situation of the majorities, it is undeniable that there remains a question to be answered: what will be the fate of the surviving aboriginal groups, those who, in various forms, have preserved a culture which differs from that of the mestizo majority?

On the other hand it is worth remembering that, for instance, in the Andean area the historical processes have been different. There the fusion of the Quechua and Aymara with the people of European origin has been relatively smaller. Obviously, to evaluate the prevalent attitudes regarding the Indian presence in the diverse Latin American countries one must keep in mind that their historical processes have not been the same. Thus it has been said, for example, that in a country like Mexico, whose evolution has consisted of and is a cultural and ethnic fusion, any attempt to invent "autonomous nationalities" sounds unrealistic.

IV. The Idea of Participation

What must be striven for above all is the *participation* of the Indian groups in the integral reality of the life of their respective countries. It used to be said: "the Indians must be incorporated in our culture". Then was maintained "they must be totally integrated" as if the intention were that of doing away with their identity and values. The idea of *participation* means something else. In order to attain the participation of the Indian groups in the integral reality of the life of their respective countries, in no way is it necessary to attempt to destroy their culture values nor their sense of identity. There will be groups who, on counting on the means of participating in the national economy, will also assimilate the culture of the rest of the mestizo population. Others, however, will decide (and they should be encouraged) to *participate*, preserving their cultural identity.

It is important that it should be recognised in every type of action, state or privately sponsored, that there are the Indian groups with different cultural backgrounds which will participate in the economic life, and in the development of the nation if obstacles are removed. These often consist of forms of social injustice which have taken deep roots through the centuries. Anthropological Indigenism deserves credit for growing awareness of the fact that, when groups of different cultures are involved, governmental action must be based on the recognition and acceptance of cultural and ethnic plurality.

Furthermore, the distinction established here between what used to be called "the integration of the indigenous groups" and the concept of *participation* is not merely a play on words. It may be necessary to repeat that the idea of fostering this participation by means of specific development projects would be invalidated if it would lead to infringement of the different cultural values or of the right to maintain a distinct sense of identity.

Obviously no country exists on earth which is not interested in having all the groups in its population, including those which constitute minorities, participate in one way or another, in its political, social and economic life. This participation, in reference to the often neglected Indian presence, is a necessity which cannot be postponed in the context of the Latin American countries. The fact that many groups, even in our times, remain marginated or exploited, barely eking out their subsistence, is so inhumanly grave, that no one can afford to ignore it.

NOTES

1. Fray Bartolomé de las Casas, "Brevísima relación de la destrucción de las Indias" in Lewis H. Hanke and Manuel Jiménez Fernández, eds., *Tratados*, Mexico, Fonda de Cultura Económica, 1965, Vol 1, pp 191, 192.
2. Fray Domingo de Betanzos, *Memorialda . . . el Consejo de las Indias*, original manuscript Archivo Histórico Nacional, Madrid, 1535, Diversos 18.
3. Bernardino de Sahagún, *Historia general de las cosas de Nueva España*, 4 Vols, Angel M Garibay, ed., Mexico, Editorial Porrúa, 1956.
4. Alfonso Caso, "Contribución de las culturas indigenas de Mexico a la cultura mundial", in *México y la Cultura*, 1946, México, Secretaría de Educación Pública. "Que es el I.N.I.?" (What is the National Indigenous Institute?) Mexico, 1955. "Los ideales de la accion indigenista", in *Los centros coordinadores indigenistas*, Mexico, 1962.
5. Juan Comas, *Principales contribuciones indigenas precolombinas a la cultura universal*, Mexico, Instituto Indigenista Interamericano, 1957.
6. See note 4.
7. Guillermo Bonfil, "Reflexiones sobre la politica indigenista y centralismo gubernamental de México", American Society of Applied Anthropology, XXVIII Meeting, Mexico, 1969.
8. Rodolfo Stavenhagen, "Seven Erroneous Theses about Latin America", *New University Thought*, vol. 4 no. 4, Detroit, 1966/1967.
9. Daniel Casés, "A comment on New Proposals for Anthropologists", *Current Anthropology*, vol. 9, no. 5, Chicago, 1968.
10. Alfonso Villa Rojas, "La responsabilidad de los científicos sociales. En torno a la nueva tendencia ideológica de antropólogos e indigenistas", *América Indigena*, vol. 19 no. 3, Mexico, 1969.
11. Rudolf A. M. van Zantwijk, *Servants of the Saints, The Social and Cultural Identity of a Tarascan Community in Mexico*, Assen, 1967.

The New Indianism and the Menominee of Wisconsin

VICTOR J. HANBY

VICTOR J. HANBY received his Bachelor of Arts degree with Honours from the Department of Government of Essex University (England) in 1968. The following year he completed an M.A. in Political Science at McMaster University in Hamilton, Ontario. He is currently a candidate for the Ph.D. in government at Essex University. He has held a Social Science Research Council Research Studentship (1969–71) and a Fulbright Fellowship (1973–74) and has served as a visiting Professor in the Departments of Political Science and Racial and Ethnic Studies at Michigan State University. Since September 1972 he has held the post of Lecturer in Political Sociology and Race Relations in the Department of Sociology at the University of Stirling.

Mr. Hanby is the author of a number of research papers and articles, including "Embourgeoisement and Labour Elites," a paper presented to the British Sociological Association Annual Conference at Stirling in July, 1973; "Thresholds of Representation and Thresholds of Exclusion: an Analytic Note on Electoral Systems," (with D. Rae and J. Loosemore), *Comparative Political Studies*, January 1971; and "Clientele Markets, Organizational Dynamics and Leadership Change," *Electoral Studies Yearbook*, March 1975. He is the editor of *The American Indian: A Social and Political Re-Evaluation*, to which he contributed two articles: "The New Reservation" and "Wounded Knee and After." Mr. Hanby is currently engaged in a study of the urban Indian community in Lansing, Michigan (to be conducted from June 1975–September 1975).

The New Indianism and the Menominee of Wisconsin

VICTOR J. HANBY

The seizure on New Year's day, 1975, of the former Roman Catholic novitiate in Gresham, Wisconsin by an armed group of Menominee Indians is again testimony to the new mood of militancy which has gained increasing hold among some sections of the U.S. Indian community over the past years. By 4 February, when the occupying group finally relinquished control of the building, all the various forces, factions, and trends of contemporary movements within the United States Indian population had emerged in sequence to the fore of this conflict. In so doing they presented a detailed picture of differences evident within Indian society. The Gresham "affair" in its genesis and resolution, is in a fundamental sense a microcosm of the contemporary social and political firmament of the United States Indian. Yet it is also partly reflective of a problem which, while not exclusively Menominee, being shared by a handful of other tribes, produced in the period 1953–1973 a particular set of difficulties for the Menominee tribe. The actions of the militants at Gresham are in part, therefore, a legacy also of the specific experiences of that period.

I. THE POVERTY SYNDROME

The national, social and economic scenario against which Gresham took place is one where poverty seems the rule rather than the exception. Disparate sources[1] report the following statistics for the Indian. Life expectancy is now 44 years, about 20 years lower than for the general population. The natural death rate of Indian children of between 1 and 14 years is almost three times greater than for other United States children. The incidence of suicide among Indian males between 15 and 45 is roughly four times that of white males of the same age category. The tubercular death rate is six times that of the rest of the population. Alcohol abuse affects one in three of adult Indian males. Middle ear infections, rarely seen in other groups in the population, are a major medical problem on Indian reservations. In addition, the 1970 United States Census[2] found that 39% of the Indian population was living below the United States Government designated "poverty-level" whereas the proportion for the remainder of the population was 14%. The median income of Indian families was $5832, for the United States as a whole it was $9590. Nearly 70% of all Indian males are employed within the following four occupational groups – operatives, labourers, service workers, and craftsmen. Only 9% could be located within professional or technical categories. More than one out of five Indian men had less than five years' schooling. About one-third of all Indians over 25 had completed secondary education as compared to 53% of the total population. Patently, the Menominee had contributed their share to these statistics. By the end of 1973, social conditions amongst the Menominee had sunk to a new low. Around 50% of the tribe was regularly collecting welfare and 41% of the families had incomes below the federal poverty level. Roughly 50% of their homes had electricity, and considerably fewer than this had piped water. Anaemia and tuberculosis were rife, and with alcoholism as a third health problem, the inadequate health services could merely make ineffectual efforts to treat a problem which could only benefit from wholescale injections of finance and trained medical personnel.

Though such figures seem disturbing enough, it is what they are reflective of which requires most consideration and to understand this and hence in essence the Menominee problem, one is required to refer back to an ill-conceived social experiment of the United States Government in 1953 called "termination".[3]

II. "Termination" and its Effects on the Menominee

In that year, Congress passed two measures designed in effect to bring the Indians into the mainstream of United States society, by firstly (Public Law 280) extending under law the right of certain states to apply their civil and criminal jurisdiction to the reservations. Secondly, (Concurrent Resolution 108, 83rd Congress) by rescinding the hitherto special legal and social status of reservations and tribal governments. Such measures while designed primarily to terminate the special relationship between the Federal Government and reservation tribes, were not performed out of vindictiveness or bad faith by the Federal Government. Public Law 280 was a response to a request by particular tribes for external help in policing the reservations, when tribal funds could not support a responsibility which was by law exclusively tribal. Concurrent resolution 108 was designed to achieve a position of "full freedom" for the Indian in United States society. Nevertheless, as both were implemented without consultation with the Indians and even if they were solely additional reflections of the traditionally naive and uninformed attitudes of the agencies responsible for Indian affairs, yet their effect was to cause consternation among all Indian groups and for the Menominee and a small collection of other tribes the net result was disastrous.

By the early 1950's the Menominee had become one of the most prosperous tribes in the United States; with a well-stocked forest and a sawmill to process the wood, at the beginning of Eisenhower's first administration the tribe was making enough money to pay the Federal Government for the cost of its administration. Profits from the mill supported a health service contracted for at a local hospital as well as allowing the tribe to pay off the Bureau of Indian Affairs for roads, schools and a power plant constructed on its behalf. In addition, $10 million was deposited with the United States Treasury in the tribal account, and about $8 million of this came from a law suit in 1951 against the Bureau of Indian Affairs for mismanagement of the forest and tribal land. It was in an attempt to distribute part of this that the Menominee troubles began in earnest. Because of federal trusteeship, this money could only be allocated through Congressional appropriation. In early 1953, the tribe petitioned Congress for a proportion of their monies, firstly to allow them to give each tribal member $1500 in a self-help housing programme as well as secondly make financial provision for improvements to the sawmill, expand community services, and modernise and reconstruct part of the hospital. The Congressional answer was relatively pointed; the Menominee could have all their money if they agreed to termination. A condition which to the bulk of Congressmen didn't seem unduly unfair as the Menominee had enough money, seemed sufficiently assimilated into American society and were clearly satisfactorily profit-oriented. A hasty tribal vote, with only a 5% participation rate, effectively predetermined by the need for ready money among the bulk of those participating, produced a vote for termination. In 1954 the Menominee Termination Act was passed, the tribal role was closed and the tribe was required to produce a termination plan. The precipitate manner in which the termination decision was taken produced impractical deadlines, polarised opinion within the tribe, and succeeded in forestalling final acceptance of termination plans until 1961. Then a corporation, Menominee Enterprises Incorporated, was created to manage the tribal assets, individual Menominee were given 100 shares of stock each in the corporation, the Bureau of Indian Affairs discontinued all their services, the former reservation became a separate county of Wisconsin, and the tribal council was dissolved.

It soon became clear that the assets of the new corporation were insufficient to meet the costs involved in providing county-wide services. As a county, the Menominee were legally obligated to meet minimum State standards in all the services they provided. The payments in 1954 to tribal members and an additional payment in 1955 when the B.I.A. discovered another error in their accounting procedures, meant that by 1961 only about $2 million was left in the tribal budget. The bulk of these monies soon disappeared through the cost of up-grading the saw-mill to Wisconsin industrial standards. The Indian school and local hospital were closed as insufficient funds were available to meet necessary expenses, again to provide State standards of service. Most local children then had to travel to neighbouring Shawano county for their education and health and dental care became minimal. The profits of the mill which before termination provided finance for nearly all these services, were rapaciously devoured by the new property taxes on the former tribe's forest lands. Federally protected Indians do not pay tax on their land.

In a bid to alleviate some of these problems the Federal and Wisconsin governments in the period 1961–1973 provided subsidies to the new county of about $19 million, merely to keep the county functioning as an administrative unit. Even so, basic services continued to be seriously lacking. Some primary education was supported and a child-care centre was established in the former Bureau of Indian Affairs building. The only high school in the area remained in the neighbouring county, however, and Menominee secondary school graduation figures in 1973 were still only 22%. A community school was founded for high-school dropouts, though with mainly volunteer teaching. Health care remained a serious problem. Two community-health nurses and a part-time National Health Service doctor provided the sole sources of medical care. Hospital cases had to be moved to neighbouring counties occasionally by bus or car, as the inadequate county ambulance service oftentimes couldn't meet demand. Private ambulance services from neighbouring counties would often refuse to make the trip into Menominee county.

By the early 1970s the financial situation of the county was so fragile, that the Menominee corporation had been forced to sell areas of land bordering the local lake, Legend Lake, for holiday home development in order to infuse a degree of solvency into the county treasury. The economic and social conditions of the Menominee continued to deteriorate. The outdated mill, the main source of employment, installed limited automatic machinery which reduced the operating payroll but contributed to the number of unemployed and increased the burden on the welfare services. Migration to the cities of Milwaukee, Chicago and Minneapolis continued to grow, so that at the end of 1973 the bulk of the two and a half thousand inhabitants were either below sixteen or over sixty.

While the Menominee were gradually succumbing to the economic and social pressures that developed after termination, new social movements were beginning to make their presence felt within the Indian population at large. At first such movements were very much in the Indian tradition, tribal not pan-Indian in emphasis but as more and more instances of such behaviour manifested themselves, the emphasis gradually changed until specific grievances of specific Indians were supported as catalysts against the old order by Indians from disparate backgrounds. Particular grievances thus became symbols of general problems facing the Indian in United States society and were projected accordingly.

III. The Rise of Indian Protest

Two incidents in the late 1950s seem to provide bench mark dates for the new Indianism. In 1958 some 350 Lumbee Indians, a non-reservation rural community in North Carolina, attacked and fired on a Ku Klux Klan meeting, routing the meeting to such an extent that Klan members had to be rescued by state police. The same year, the Tuscarora of New York, after tribal members had been arrested in demonstrating against a hydro-electric development to be built on their state reservation, "invaded" the Department of the Interior in Washington, intent on making a citizens' arrest of the Indian Commissioner or the

Secretary of the Interior, whoever was available. Though neither the "arrest" nor the opposition to the development were successful, the leader of this opposition group, Wallace "Mad Bear" Anderson, capitalized on the publicity his movement aroused and subsequently emerged as the leading force behind United Nations of American Indians, a movement designed to unite the separate tribes on New York's eight State reservations along the lines of the old Iroquois Confederation.

The early years of the 1960s was a period of sporadic violence between Indian groups and various agencies of the Federal and State Governments, but two developments in the mid-1960s established the basis for the widening of the conflict from an exclusively tribal to a fundamentally pan-Indian nature. Over the period 1964–1966, the State of Washington conservation agency had been continually arresting tribesmen who insisted on their traditional treaty "right" to fish wherever they wished. Out of this situation had arisen the Survival of American Indians Association, designed to link the several small tribes of north-west Washington in their fight against the state. In 1961, a group of university-educated Indians in an effort to nationally articulate the problems of the Indian, weaken the hold of tribal governments which were perceived as repositories of patronage for vested interests and dismantle the existing bureaucratic methods of dealing with Indians, had formed the National Indian Youth Council. The N.I.Y.C., while its adherents were of a basically radical nature, was opposed to the more open militancy of some of the "fishing rights" proponents, yet it saw in the Washington problems the essence of the Indian situation in society at large. In 1966, therefore the N.I.Y.C. organised a "fish-in" where the tactics of passive resistance were copy-book civil rights. Several arrests were made, and there was limited violence! The "fish-in" however was important in two respects for the new Indianism. As an example of Indian determination and spirit its impact reverberated across the United States in a number of similar disputes over both fishing and hunting rights, and in the establishment of numerous, local, self-help organisations. More importantly, and this seems clearly one of the N.I.Y.C. ancillary aims, it demonstrated the fundamental differences in attitude and behaviour which existed *between* and *within* Indian tribes.

Many tribal governments actively opposed demonstrations of the "fish-in" type as numbers of them had actually concluded agreements with the Washington authorities. In effect, the bigger tribes agreed to police their members in return for certain fishing rights, thus making conditions almost impossible for the smaller, financially-restricted tribes. In actuality, the continued fishing of the smaller tribes and their rejection of state policing of fishing threatened to re-open the problem for all tribes. The tribal governments of the larger tribes, therefore, put considerable pressure on the smaller groups to observe State laws and not activate disputes which were, and should remain local problems. The N.I.Y.C. demonstration produced a salutary lesson for many councils of the larger tribes. Large numbers of younger tribal members ignored their councils' advice and actively supported those smaller tribes who were actually involved in the dispute. This was to be a lesson many later tribal councils also had to learn to their cost. However, even though the Washington confrontations had laid important groundstones for subsequent developments, the majority of elected Indian leaders and their moderate supporters saw such developments as nuisances rather than forerunners of future events. They were in addition esoteric complaints, the significance of which might not be obvious to non-reservation or non-rural Indians.[4]

A. Mobilization of Urban Indians

The escalation of the Vietnam War in 1966–1968 and the gradual erosion of community action and the War on Poverty programmes changed the locus of Indian dissent. Like many Menominee, increasing numbers of Indians had been attracted to urban areas, some as a result of another ill-advised Federal programme "relocation",[5] some in voluntary search of education and employment. The attrition of the War on Poverty programme had a very significant effect on this urban element, and a particularly strong impact on younger Indians who had been prime beneficiaries of job training programmes.

The immediate response was the emergence in west coast cities of the United States of dissident Indian groups making loud claims to "red power",[6] restoration of traditional lands and drastic action to redress the present situation in which Indians felt themselves. The emotional tone in which much of their "red power" rhetoric was couched, touched a sympathetic cord in many reservations too, particularly amongst the young, though rarely among tribal councils. The spirit of the appeal though in many instances more loud than coherent, was clearly

expressed in terms of Indianness or Indianism, and not with particularly tribal referents.[7]

Such appeals found many adherents in the mid-west cities of St. Paul
and Minneapolis. In particular in 1968 in a predominantly Indian part
of south Minneapolis, a police-activity mon:toring unit formed itself
into the American Indian Movement,[8] with radical aims and clearly
expressed ambitions. As the A.I.M. founders see themselves and their
supporters as a movement rather than a formal organisation, there is no
manifesto, simply a statement of goals. "The situation regarding the
native American people will not be changed by words or polite petition.
This has been tried for all as 'the plight of the Indians'. It is the responsibility of the Indian people themselves to change realities around
them, when these realities are repressive and oppressive. We are simply
doing these things that all Indians agree should be done, in a way that
will be effective."[9] Its initial impact was primarily in the cities in which
it was founded but swiftly the movement expanded to other mid-west
cities as more and more Indians became appraised of its aims and, more
specifically, of its ability to produce positive results out of essentially
negative conflict situations. Basically an urban-based movement, the
initial concern of A.I.M. was to help the situation of those whose predicament was to a large part forgotten by the bulk of the United States
population, that large group of Indians who through living in urban
areas were denied Federal support. It is not generally recognised that
only Federally recognised tribes, i.e. tribes recognised in formal treaties,
or those living on Federal trust lands, are eligible for Government
grants or to be serviced by the Bureau of Indian Affairs. The Federal
authorities do not recognise Indians on state reservations, in rural communities or in urban conurbations.[10] Yet the 1970 census reveals that
only 28% live on reservations, 24% in rural areas and the remaining
48% in urban areas. As the A.I.M. increased their membership and
gained support through the mid-west with local groups forming A.I.M.
chapters for the support the association affords, the movement began
to escalate both its education programme and its impact in the Indian
community generally. Increasingly the A.I.M. investigated complaints
arising from reservation groups, and began to play a larger intervening
role in disputes which arose in that particular domain. In particular
it began to side with dissident reservation groups against elected tribal
leaderships, a scenario which ultimately was to come to fruition at
Wounded Knee in 1973.

B. *Urban-Reservation Collaboration and Wounded Knee*

The urban oriented movement in the western coastal states had, as a means of highlighting the urban Indians' neglect, attempted in 1965 to force the Federal government to return the island of Alcatraz to the Indians under the terms of an 1868 treaty. Though this effort was unsuccessful, Alcatraz remained as a symbolic focus for grievances for most Indians in the San Francisco and the Los Angeles areas. In 1969, when the conditions there of the urban Indians had seriously declined, the problem was exacerbated in San Francisco with the burning of the Indian centre, thus removing the sole source of Indian aid and guidance in the area. Discussions on how to alleviate the local Indian situation led to the formation of a pan-Indian group, Indians of All Tribes and the subsequent seizure of the old symbol, Alcatraz. "Culturally... mixed, they were seeking to develop a new source of Indian identity, drawing its strength from pan-Indianism and the commonalities of urban experience and its strategies from college campuses and the urban guerilla movements."[11] Most important of all, the Alcatraz group provided a model to be emulated in other urban areas, most notably in the Seattle area with the creation of another multi-tribal, urban-rural group, United Indians of All Tribes.

1972 saw the emergence of the first collaborative effort between reservation- and urban-based movements. A Trans-America caravan called the "Trail of Broken Treaties" was planned, which would link Indian opinion across the nation as the caravan made its way from the west coast to Washington, D.C. Though the bulk of the caravan were reservation residents and represented most points of non-urban Indian opinion, the impetus and guiding hand were clearly urban, if not specifically A.I.M. based. After its stop in St. Paul, Minnesota, the caravan became the vehicle for the transmission of the Twenty Points,[12] a programme of reform supported by the Indians, designed to reformulate their total relationship with the Federal government in particular, and United States society in general. After its arrival in Washington, the "Trail of Broken Treaties" achieved a notoriety it had not intended. Confusions led to the occupation of the Bureau of Indian Affairs, greater confusions fired by fear of police reprisals led to serious damaging of the building in creating fortifications. Elected Indian leaderships for the most part, criticized the "Trail" and the subsequent sit-in, saving most of their vilification for the A.I.M. participants. Moderate

Indian opinion was seriously alienated by the violence of the occasion. The A.I.M. leaders vowed policies of no-surrender without concessions from the Federal Government in future actions, when the Nixon administration finally rejected the Twenty Points following the end of the occupation. Demarcation lines were clearly drawn within Indian opinion over the activism within its midst, producing on one side particularly, deep senses of frustration and desperation, and on the other, feelings of anger and anxiety about, and opposition to such volatile tendencies.

The various factions, both urban and non-urban, which had been growing up or coming to prominence from the late 1950s, and which constitute the new Indianism, saw the trends that they started, collide at Wounded Knee in February 1973. Though this incident has gone into the folklore of Indian culture along with Custer's Last Stand, in essence it was little more than a local problem which in its escalation reflected many of the dominant trends in Indian society, not all of which are complementary. Dissident groups on the Sioux Pine Ridge Reservation in South Dakota, unhappy about the nature and spirit of the tribal government's treatment of many group elements on the reservations had coalesced support around Russell Means, a Sioux member who'd spent most of his life in Minneapolis and who was instrumental in the early successes of the A.I.M.. Means had returned to the reservation ostensibly to study traditional religion, but with the clear additional purpose of some day running for tribal chairman, and an even stronger purpose of solving the outstanding problems which were besetting the tribe, not least of all the high-handed and authoritarian practices of its chairman. It also became clear that Means and other A.I.M. leaders who were also in the area following disputes they'd helped settle in Nebraska and neighbouring South Dakota areas, did not simply become leaders of this movement to remove the incumbents but also provided a focal point for all discontentment present on the reservation. It was relatively heavy-handed and brutalistic efforts by the tribal leader, his followers and local police to prevent A.I.M.'s participation on the reservation that provoked the occupation of Wounded Knee[13] village. Subsequent developments catapulted A.I.M. into the forefront as spokesmen for Indians everywhere, whose problems were certainly basically reflected in the actual occupation and preceding events, but who also looked at Wounded Knee as a symbol of both past and present injustices. The battle-lines were clearly drawn within the Indian community nationally, on the one side stood the Federal Government, the Bureau of Indian Affairs, the Tribal Council, village elders and conservative Indian opinion everywhere, from Sioux wives and mothers to the ultra-conservative National Tribal Chairmen's Association. On the other side stood A.I.M., young Sioux members and assorted radical Indian support from disparate groups in the country, both urban and reservation. Wounded Knee was the first major example of conflict between the old and new orders of Indians, between the new Indianism and the traditional. When the incident was over, even groups which had not supported A.I.M. and its Indian nationalist viewpoint were aware of the existence of a new movement within the Indian community. A movement which while not encapsulated in one group or one individual as even A.I.M. was simply one strand within it, did seem to express one coherent idea, a belief in Indianness not solely in being Sioux, Chippewa, Menominee, etc.

C. Menominee Protest – The Gresham Episode

The incipient strands of self-determination and self-help that arose in the Indian community in the 1960s and which received their most extreme expressions at Wounded Knee, had not passed by Menominee County. In 1969, D.R.U.M.S. (Determination of Rights and Unity of Menominee Stockholders) was formed to act as a pressure group on the Menominee Board which was perceived by many Indians as a white puppet structure. Five members of the Board of nine were by statute outsiders, and membership of this Board was in fact determined by a "voting trust" not by direct tribal vote. In addition, the control and administration of shares owned by minors and mental deficients was held by a private bank. Block voting by the bank tended to have a significant impact on what was said and done in the county. One of DRUMS' first moves was to attempt to stop the sale of Menominee land for holiday development. This was successful, but only after three years of struggle and the eventual loss of 5000 acres. Gradually DRUMS moved into a more overtly political role, actively supporting a drive to get the tribe "de-terminated" and calling for support for that position. As this movement gained strength in 1971, a joint effort was consolidated between DRUMS and the voting trust which in effect was the primary decision-maker in the Menominee structure, to take their case to Congress. This move was spearheaded by Ada Deer, Chairwoman of the voting trust, who in time became the main lobbyist for de-termination. In December 1973, the Federal Government restored the Menominee reservation, disbanded existing structures and authorised the Menominee Restoration Committee to map out the details for the future, with Ada Deer as head of the new tribal government.

In the period 1971–1974, other changes were being effected which also had direct consequences for the Gresham affair of this year. Small numbers of Indians began to filter back to Menominee county from the urban areas they'd migrated to earlier. Some were merely being drawn back to the home roots they had originally left in search of a better life, with a desire and sense of responsibility to help. This group became particularly active in various community programmes but found the paucity of financial resources to overcome the fundamental social and economic problems, an increasingly provoking frustration. Other groups in a trickle at first, returning from the cities where they had remained unemployed, merely returned on the premise that if unemployment was to be their lot, it might as well occur among friends and relatives. This group in the main in the 18–25 age category had oftentimes come into contact with A.I.M. and similar groups, while they'd languished in the cities. They therefore comprised a politically, potentially potent group whose disaffection with white society was intense and whose patience with the controlling tribal authorities was limited. Their impact on the younger element which had remained on the former reservation was, as might be expected, considerable.

The re-establishment of the Menominee as a Federally recognised tribe had increased expectations across the board considerably, both among the radical and conservative groups. Certainly the sawmill returned a profit for the first six months of 1974, and more money began to flow back into the community. However, the wheels of federal authority turn very slowly, and the conditions on the reservation on re-establishment were worse than they had ever been. Large infusions of personnel and finance were immediately required but even had this been the case, it would have been several months before conditions improved measurably. As it was, such help did not come, and conditions improved only marginally. Education remained in a very basic state and health services stagnated, both waiting for new B.I.A. funds to upgrade old buildings or construct new ones. The new tribal government now had to counsel moderation and patience, and to many young Indians, its former activist leaders now began to sound very much like apologists for the status quo. In 1974 therefore, the material condition of the Menominee did not seem to represent much of a change from that under termination. To a generation whose major external, political stimuli had been "fish-ins", Alcatraz, the Trail of Broken Treaties, Wounded Knee, A.I.M. and sundry other groups, this situation was more than unsatisfactory.

As a symbol of the surplus which existed in white society and the comparative deprivation which prevailed among the Indians, the Menominee militants pointed to an empty 64 room, well-equipped, deserted, novitiate in a neighbouring county. Indian claims that the building was on land "stolen" from them by treaty violations, and hence should be handed over for use as a new medical facility, were unsurprisingly ineffectual. The occupation was the immediate response, to dramatise the financial and social conditions of Indians everywhere, and to draw attention to the plight of the Menominee, and their particularly depressing experiences.

The group which lead the occupation, the Menominee Warrior Society, had only existed for a short period prior to the occupation, and drawing its support mainly from the disaffected young groups on the reservation, though not formally, a chapter of A.I.M. clearly drew its inspiration from their brand of Indianism. Michael Sturdevant, the spokesman of the Warriors Society, defended the occupation in forthright terms. "If I knocked on somebody's door and said 'Hey, my people need a hospital,' how many of those people's doors would have been opened to me? I would not have received an audience." Not only did the occupation reveal the narrowness of alternatives which many Indians feel face them in their dealings with the remainder of United States society, but it also again demonstrated the conflicts which exist between the new Indianism and established tribal governments and moderate Indian opinion. The American Indian Movement swiftly established observers close to the occupation site, providing moral and material support, and also chief negotiators for the Indians. The participation of the A.I.M. functioned clearly as an important militant "imprimatur" for the stand of the Menominee and the Gresham activists were quickly supported by demonstrations and marches in Indian communities across the United States.

Within the Menominee reservation, however, opinion was far from one-sided. The immediate response of the governing body was to denounce the action as "unethical" and as "anarchy", a decision which swiftly provoked a march in a neighbouring town amongst Indians not involved in the occupation, in support of the activists. Some three weeks later, the tribal government had reassessed the situation sufficiently for Ada Deer to redefine their position as one of sympathetic understanding for the principles underlying the action, by viewing the sit-in as run "by frustrated people with the best of motives" who through their lives "had been engulfed by racism".[14]

Yet, while Gresham is a relatively good example of the pervasive impact that nationalist pan-Indianism has achieved in contemporary United States society and the incisive strains to traditional Indian structures that such movements portend, it is far from clear whether the coming year will see the emergence of Indian militancy as a united national force. Certainly the past few years are testimony to the increasing frequency of activist outbreaks among Indian communities; from the legalistic emphases of the Passamaquoddy and Penobscot of Maine, the Havasupai of Arizona, Aleut and Inuit of Alaska and many others who are claiming through the courts former tribal land; from the urban self-help emphases of the Chicago based Native American Committee, and American Indians United, plus some of the other urban-based movements mentioned earlier, with their boycotts of discriminatory government programmes, their advisory capacities to authorities, State and Federal, on Indian education, their establishment of Indian centres in urban areas and their projection of Indians and Indian sponsored candidates for government office, to the more publicly aggressive outbursts, of the Mohawk band who in May 1974 seized a former Girl Guide Camp in New York State and laid claim to an additional nine million acres, of the Onandaga of New York State also, who in August 1974 physically ejected every non-Indian from their reservation, this a long seated grievance had received new impetus just prior to the evictions through the presence on the reservation of Russell Means and other A.I.M. followers, as well as of course to Alcatraz, the Bureau of Indian Affairs sit-in, Wounded Knee and Gresham.

IV. Cross-Currents in the Indian Community

While these are all testimony to the spirit of the new Indianism, and a belief that Indians should come forward to help their own regardless of basic tribal affiliations, some activities within the Indian community have actually been in the opposite direction. In 1972, urban interested groups began to lay the foundations for the formation of a national urban movement to be called National American Indian Council, A.I.M. seemingly the pre-eminent movement among all urban groups, rejected the NAIC programme and refused to join. At present the Hopi and Navajo are engaged in a lengthy and protracted series of litigations over control of 3000 acres of Arizona with rich mineral deposits at stake. This dispute now publicly being fought in the courts had smouldered for three years in a series of sporadic burning and destruction of respective tribal properties. While such incidents do not themselves threaten the continued existence of the new Indianism, they do rather point to the uneasy balance of conflicting loyalties and jealousies which underly it.

Neither is it clear that these tribes which like the Menominee suffered the bitter experiences of "termination" will provide the new "Greshams". For most of the small tribes, the experiences were uniform, their land was sold and the people scattered. At least two of the major tribes, the Siletz of Western Oregon and the Klamath of Southern Oregon, unlike the Menominee didn't retain their lands as a single unit but sold large tracts of it for development and National Forest. Only about three hundred Siletz remain in their traditional region, and around six hundred Klamath, though there are scattered groups throughout Oregon. Embryo movements have begun for reinstatement as federally recognised tribes, though the problems are overwhelming. Many different groups now hold title to the lands, the Federal government having purchased the last Klamath lands as late as early 1974 as additional National Forest land, and the majority of both tribes are scattered outside the state.

Yet what is clear is the fact that even if such efforts at de-termination are unsuccessful, any movements they inspire will be well served by the ground swell of opinion created by the new Indianism of the past decade or so. Indeed *any* local Indian problem will be so served but it is doubtful whether such actions will constitute the emergence of a national Indian movement. The most likely possibility would point to one or more existing group movements asserting their positions within the loose framework that exists at present and striving to increase their impact within the Indian community at large, at once paradoxically unifying and dividing.[15] Whether the parthenogenesis of a unified, national Indian movement will be manifested in 1975 or not, United States society may once again expect images on their television screens of Indians brandishing guns, as at Gresham, and once again demanding "the Indians shall not be required to remove from the lands that are hereby set aside for them",[16] even if for many of them for the moment, such lands are more figurative than real and may as much concern an Indian centre in Chicago or Detroit as a reservation in Wisconsin or Oregon.

While, following Gresham, there can be no doubt of the overall impact that the new Indianism has wrought amongst certain segments of the Indian population in U.S. society, there can also be little doubt over the basic tactical weakness of the Indian's present position. As a group the Indian is politically and socially weak.[17] His dispersion across the U.S. in relatively small aggregations effectively means that oftentimes only drastic action of the type discussed in this paper may bring immediate results. Yet such action itself often subsequently details the extrinsic weakness that the Indian experiences in his general dealings with the remainder of U.S. society. At Gresham the National Guard arrived, in the first instance, not with the express intention of forcibly removing the militants but of interposing themselves between the Indian militants and local police authorities and armed, white vigilante groups who were threatening to take the law into their own hands. Indeed, this interpretation seems well supported by an analysis of the terms under which the Indians relinquished control of the Novitiate. Apart from the general prescriptions relating to the future Indian use of the Novitiate, a major plank of related Indian demands was that their safety be guaranteed by the National Guard against actions by the police and local groups. It seems therefore that the capacity of the Indian to help his position in society might be as much conditioned by his efforts to invoke sympathetic responses from the dominant society as by his own attempts to change the situation through the impact of the new Indianism. Clearly this seems to call for finely attuned, diplomatic insight in both the creation and resolution of future confrontations. Even disregarding for the moment such refined, tactical, connotations, while the new Indianism and its present mixture of pragmatism, bravado and cultural symbols, can patently at present create such confrontations, the real question in essence, seems to be but for how long successfully? The new Indianism needs soon to begin to translate itself into a vehicle for significant social change and make positive gains while the opportunities present themselves. The tolerance by the dominant society of actions of the order of the Wounded Knee and Gresham must surely be a calculus of finite magnitude.[18] Time and history, are not with the Indians and most of all, not with the new Indianism.

NOTES

1. *Current Population Reports*, United States Bureau of Census, No. 46, Washington D.C., 1972. *Ten State Nutrition Survey* (1968–1970) Health Services & Mental

Health Administration, U.S. Department of Health, Education & Welfare, Atlanta, 1972. *We Americans: Our Education*, United States Department of Commerce, United States Bureau of Census, 1973. T. W. Taylor, *The States & Their Indian Citizens* (United States Department of the Interior Bureau of Indian Affairs, Washington D.C., 1972). See also Richard M. Nixon, Presidential Message to Congress on Indian Affairs, July 1970.

2. *American Indians*, United States Bureau of Census Report, June 1973.
3. On termination see especially, W. E. Washburn, *Red Man's Land, White Man's Law* (Scribners: 1971); G. Orfield, *A Study of Termination Policy* (National Congress of American Indians: no date). Though termination was repudiated as a policy by President Nixon in 1970, in effect it has never been repealed by legislation. The Menominee restoration legislation referred solely to the Menominee.
4. The fishing rights dispute dragged on in sporadic fashion through the U.S. courts until in February 1974, a federal court decision issued in Tacoma, Washington, reaffirmed the right of Washington State Indians to fish in off-reservation areas and hence effectively recognised the supremacy of federal treaty law over state law concerning the treatment of Indian populations. The decision however was circumscribed with many conditions and qualifications that tribes were to meet, if their rights to fish and regulate their own members fishing, were to become operative. Further while recognising Indian prior rights, the courts also accepted the state's right to regulate the Indian in matters of conservation. Thus the present peace seems to have been achieved on the basis of a certain judicial ambiguity and its continuance seems to rest on a series of negotiations being conducted between sceptical Indian leaders and intransigent state officials. The future outcome is therefore unclear.
5. See W. Washburn, *op. cit.*, also F. Svensson, *The Ethnics in American Politics: American Indians*, (Burgess, 1973).
6. See A. Josephy, *Red Power* (McGraw Hill: 1971); G. Steiner, *The New Indians* (Harper & Row: 1968); V. Deloria, *God is Red* (Grosset & Dunlop: 1973).
7. Van Nostitz rather misses this point in his interpretation of red power as a "vigorous tribal pride". See S. Van Nostitz, "The Tragedy of the Red Indian – Genocide or Nemesis", *Plural Societies*, Spring 1974, Vol. 5, No. 1.
8. See especially R. Burnette & J. Koster, *The Road to Wounded Knee* (Bantam Books: 1974).
9. From the A.I.M.'s statement of goals. See *Akwesasne Notes*, Winter 1973, p. 15. A.I.M. was in fact founded by Denis Banks & George Mitchell. Mitchell the main theoretician of the movement was soon overshadowed by the more flamboyant activities of Banks & Clyde Bellecourt an early member, and resigned the movement immediately following the Bureau of Indian Affairs sit-in. His position as theoretician and main strategist had however been increasingly devolving, in the period 1968–1972, to Russell Means.
10. In 1974 the U.S. Supreme Court ruled however that the B.I.A. had to service the needs of an off-reservation Indian who had brought a case against the B.I.A.. The B.I.A.'s refusal however to see this as a precedent establishing case, to be translated into general policy means that large numbers of Indians will continue to be denied services and support. A resolution to change this situation was proposed by Governor Milliken of Michigan at the 1974 Republican Governors Annual Conference and the resolution was to be communicated to the U.S. President for action by the Chairman of the Conference, the then Governor of Arizona, J. Williams.
11. F. Svensson, *The Ethnics in American Politics*.
12. See especially V. Deloria Jr., *Behind the Trail of Broken Treaties* (Dell: 1974);

V. Deloria Jr., *God is Red*, (Appendix IV); R. Burnette & J. Koster, *The Road to Wounded Knee*, Ch. 9.

13. See the Spring 1973, October 1973 & Spring 1974 issues of *Akwesasne Notes* for the most comprehensive reports of Wounded Knee and its aftermath.

14. See International Herald Tribune, January 22, 1975, for a reported interview with Ada Deer.

15. The somewhat schizophrenic strains produced within the Indian community by the new Indianism and the tension between the old tribal order and the new pan-Indian militancy are nicely encapsulated in statements made at his swearing in by the present Commissioner of Indian Affairs, an Athabascan Indian, Morris Thompson. Thompson's initial statement that he would prefer to, if possible, rely on the advice of such conservative groups as the National Tribal Chairman's Association, was only marginally balanced by his equivocations over whether he thought A.I.M. was a help or hindrance in creating a favourable metamorphosis of the Indian condition.

16. From an 1854 Wisconsin Treaty.

17. Van Nostitz is only partly correct when he says "the young Red generation is in general aware that it cannot reach its goals by force, being too small a segment of the whole population", but it surely also over-simplifies matters to continue, "so its leaders are trying to come nearer to their goals by working on public opinion or by means of legislation and jurisdiction". The current generation of leaders merely see force as incidental but oftentimes necessary, to the movements of self-determination and self-help that characterise the new-Indianism and legislation itself is only a means to this end, unlike their fathers for whom a belief in legislation was an end in itself in many instances. Public opinion too is a tool to be utilised if the opportunity presents itself and not to be courted for its own sake, a fact which clearly distinguishes the leaders of the new pan-Indian movements from the present senior generation of Indian leaders as in for instance the National Tribal Chairman's Association and their belief that "good" publicity needs continually to be solicited from the behaviour of the Indian population. See Van Nostitz, *op. cit.*, p. 23.

18. This in essence alludes to the possibilities envisaged by B. Smith & J. Barrington-Moore particularly, concerning the dysfunctional effects of violence upon its perpetrators in terms of long term reactions by society, and that violence even when a tool incidental to the overall strategy needs to be realistically related to the attainment of particular ends. See B. Smith, "The Politics of Protest: How Effective is Violence" and J. Barrington-Moore Jr. "Thoughts on Violence and Democracy" in R. Connery (Ed.), *Urban Riots: Violence & Social Change* (Academy of Political Science; 1968); also R. M. Fogelson, *Violence as Protest* (Anchor: 1971), Ch. 1.

Discrimination in the Arab Middle East

COSTA LUCA

COSTA LUCA is the pseudonym of a prominent Arab business man who travels extensively in the Middle East and knows the problems discussed in this article from personal observation as well as scholarly study.

Discrimination in the Arab Middle East

COSTA LUCA

The Arab countries of the Middle East were parts of the Ottoman Caliphate until the end of the first world war.

The first duty of the Caliph was to safeguard the interests of the Moslem people and to apply the Moslem Law (Shari'a). The main sources of Moslem law are the Koran which is the literal word of God by divine revelation, expressed in Arabic, and the Hadith which is the pronouncements and daily behaviour of the Prophet Mohammad during his lifetime. Moslem scholars have throughout the ages interpreted and propounded the Koranic verses and the Hadith for the believers to follow.

Needless to say, there were various and sometimes differing interpretations.

In 1869, the Ottoman Caliphate codified the Shari'a in a single compilation called "al Majalla" for easier handling and reference by the jurists and courts. The Majalla remained the law of the land in most of the countries of the Middle East long after the dissolution of the Caliphate. This did not preclude other interpretations of the Shari'a being applied in other parts of the Arab lands such as in the Arabian Peninsula. However, regardless of the various interpretations, the Arabs have lived with the Shari'a as their law for fourteen centuries.

When these countries gained independence after the second world war, most of them elaborated constitutions and civil codes which were imbued with the spirit of Islam.

Regarding the subject of discrimination in non-Moslem countries, it is perhaps enough to refer to the laws enacted and the social attitudes of the people. This is also true of the Arab countries except for an additional factor. Alongside the secular laws and preceding them, and sometimes conflicting with them, is the Shari'a anchored in divine revelation and offering its followers a complete system of thought and conduct. It regulates in detail the daily behaviour of the individual, his relations with other individuals of his faith and other faiths, and his spiritual life. And although many Moslems, especially of the younger generations, do not perform the religious practices such as prayer and fasting as ordained by the Shari' a, still they are no less imbued with Islamic values and fundamental beliefs which condition their whole life and attitudes.

The degree of secularization of laws, institutions and attitudes in the Arab countries may be demonstrated by the following comparison. If we borrow from Christian analogy and terms, we find that the Arabs are now hesitantly feeling their way towards a "scholastic" era of accommodation between religious teachings and modern science. Their thought processes are as much confined by religion as the West was in the age of scholasticism. The Arabs are probably still ages away from raising the searching questions about religion and its relevance to modern society which the Renaissance thinkers raised four centuries ago.

What is still more indicative, is that contemporary Arab thinkers have not dared discuss the philosophical concepts related, among other things, to personal freedom of choice, which their predecessors, the Mutazila, publicly discussed in Baghdad and other Moslem cities more than one thousand years ago.

Alongside the Arab "scholastics" is a probably stronger traditional trend which in the past, defeated the Mutazila, and which continues to deny the need to amend or modify or by-pass the religious teachings as they are now interpreted, in order to deal with modern problems. The traditionalists claim that Moslems in the past have achieved a high degree of civilization and that Islam now has the ability to generate a superior civilization to that of the Western world if Arabs only follow their religion.

In developing a modern, civilized and free society, Islam may not need to evolve in the same manner or pass through the same stages of "scholasticism, the Renaissance and the secular state" as Christianity did. It may follow other roads and arrive at a completely different type of civilization. What is important for our discussion is that we should not lose sight of the influence of Islam on the laws of the countries concerned and on the behaviour of individuals and the state as regards discrimination.

In fact, we are left in no doubt of that influence since it is embodied in the constitutions and laws of all the Arab countries except Lebanon.

The following is a brief review of some constitutional provisions emphasizing the role of Islam in present Arab societies:

Saudi Arabia has no constitution because the Koran *is* the constitution. The Shari'a is applied in spirit and letter.

Article one of the Provisional Constitution of Yemen says:[1] Yemen is an Arab Moslem state... Article 3 says: Islam is the official religion of the state and *Islamic Shari'a is the source of all laws.*

Article 2 of the Kuwait Constitution says:[2] The religion of the state is Islam and Islamic Shari'a is a primary source of legislation.

Article 2 of the Jordanian Constitution says:[3] Islam is the religion of the state...

Article 1 of the Provisional Constitution of Qatar says:[4] (The religion of Qatar) is Islam, and Islamic Shari'a shall be a fundamental source of its legislation. Article 7(b) says: The state shall endeavour to instil proper Islamic religious principles in society...

Article 3 of the Provisional Constitution of Syria says:[5] Islamic jurisprudence shall be a principal source of the legislation.

Article 4 of the Provisional Iraqi Constitution says:[6] Islam is the religion of the state.

Article 2 of the Permanent Constitution of Egypt says:[7] Islam is the religion of the state and the principles of the Islamic Shari'a are a primary source of legislation. Article 19 says: Religious (Moslem) teaching is a basic course in the educational curricula.

All this emphasis on Islam being the religion of the state and of Islamic Shari'a or jurisprudence being the source, or a primary source, of legislation, is meant to prevent the enactment of laws contrary to the spirit of the Shari'a and to thwart any movement for an independent social and political development emancipated from the hold of religion.

So, when we encounter a constitutional provision or a law, such as the principle of equality before the law for all citizens, in direct opposition to the religious principles, which exclude Christians and Jews and others from that equality, as we shall see later, it would be wise to keep in mind that the Koranic text has a stronger hold on the mind of the Arab than the constitutional article. Therefore, the Arab is more likely to conduct himself according to what the Glorious Koran says, and what he has been taught at home and in school since early childhood rather than according to what the law of man tells him to do.

It should also be mentioned that to a devout Arab, this whole discussion is irrelevant. He would maintain that there is no discrimination whatsoever in Islam because God has organized Arab Moslem society in the most perfect manner. Discrimination embodies a sense of injustice and God cannot be unjust. Therefore, when we discuss in the following pages the status of women, of non-Moslems, of other Moslem sects and conclude that they are being discriminated against, we would be committing a wrong by criticizing an order of things which God in his wisdom has ordained for the well-being of all mankind including those who suffer discrimination according to our light.

I. The Concept of the Moslem State

It is useful at the start of our discussion to acquaint ourselves with some principles explaining the nature of Moslem Arab states which are different from Western concepts.

The first observation is that the central concern of the Moslem community has been and still is, not the structure of government, but the preservation of the system of religious and moral duties elaborated in the Law. Whatever rights were recognized for the Moslem were conditioned upon the performance of these duties.[8]

The Koran, therefore, did not deal at length with state organization or with the relations between the state and the individual or with economic matters. The Koran enjoins on the Moslems to obey God, the Prophet and "those of you who are in authority."[9] This is interpreted, with verse 83 of the same Surah, to mean that obedience is due to Moslem rulers but not to rulers of other faiths who may conquer the Moslems. Another brief reference in the Koran says that (public) "affairs are a matter of counsel."[10]

From these and other pronouncements, the theory evolved that political power is inseparable from divine revelation and cannot be shared. The Moslem's spiritual universe is full of the single strong and simple idea that there is no legality except in the service of God. Therefore, political power is sacred for whoever exercises it. It is taken or lost as a whole in its entirety and does not tolerate heresy, deviation or dissension.

Professor H. A. R. Gibb says: "...both in theory and in practice, all civil and judicial authority was derived by delegation from the ruler, and was revocable by him at will. The civil duty of the subject was obedience, his religious duty to use his influence for good within his sphere of action; only when the Faith itself was in danger had he the right, under proper leadership, to resort to force."[11]

In this system, the ruler enjoys absolute power of life and death, of repression and confiscation against anything that is foreign or in revolt against the prevailing order at the moment. The ruler also exercises these powers in ordinary peaceful life so long as he manages to convey the impression that he is not contravening the Shari'a. And since it is hazardous for anybody to claim otherwise, many Caliphs and rulers exercised and still exercise absolute and arbitrary authority over their subjects.

It should be noted in this respect that the Moslem concept of the state and of authority emanating from the ruler, is in direct opposition to the Western democratic concept. In the West, the theory is that the people are the source of all political power and that all authority is derived from them through the electoral process. Furthermore, the rulers are responsible to the people for their political actions.

The Moslem concept of the state is also in contradiction with the modern Arab constitutions which have copied the Western tradition and included the principle of the people's political supremacy.

The failure of the democratic parliamentary experiment in most of the Arab countries may be partly explained by this conflict from which the Moslem concept emerged victorious.

Another important feature of the Moslem concept of the state is the persistent Islamic ideal of social unity, with its emphasis on the evils of division.

From these general principles, the following practices can be observed:

In the long history of the Arabs after Islam, there have been uncounted rebellions, political assassinations and military coups. All have been motivated by the efforts of various Moslem sects and factions to seize political power. But there has never been a single revolt by the nobles or the common people protesting against a despot or demanding to share in the government of their country. There is nothing in Arab history comparable to the revolt of the nobles in Britain which culminated in the Magna Carta of 1215, or to the French revolution and the subsequent European revolutions of the 19th century seeking freedom for the individual from arbitrary and despotic rule.

The situation did not change in more recent times. In the last three decades, there have been many military coups and political assassinations and plots in the various Arab countries. But not in a single case do we find the people revolting against the despotic rule of a dictator or of a party. All the upheavals were the work of small groups of army of-

ficers motivated by selfish or idealistic aims as the case may be. Popular
participation came always after the fact but never preceded or accompanied it. One may conclude that freedom and political liberty, as
understood in the West, do not preoccupy the ordinary Arab.

This is emphasized by the fact that the ordinary Arab is not a passive
creature and is capable of revolt if he is sufficiently aroused by a cause.
He has repeatedly revolted against the foreigners of a different faith
who ruled the Arab countries after the first world war until all foreign
powers were ousted from Arab lands. But the Arab has never revolted
against a despotic ruler of his own faith.

The ideal of social unity and abhorrence of division within the Moslem community works against the practice of democracy based on the
political party system. It is ingrained in the Arab consciousness that
partisanship is wrong and that all Moslems should be brothers. Therefore, no viable political party has ever evolved in any Arab country,
with two reservations: When a political party led the fight against a
foreign occupying power; and when the party seizes power. It then becomes the ruler to whom obedience is due, and ceases to be regarded as
a political party dividing the people in competition with other parties.

The concentration of political power in the hands of the head of
state be he king, president of the republic or a military officer is all too
apparent regardless of any constitutional provisions to the contrary. If
we review the history of the Arab countries which practiced parliamentary democracy before military coups overthrew the system (Iraq,
Syria, Egypt, Sudan), and the countries which still adopt the parliamentary system (Lebanon and Jordan), we cannot find a single case in
which a government fell because parliament voted against it. All government changes, and they were and still are frequent, occur because
the head of state desires a change. Parliament feels that a no confidence
vote reflects on the standing and authority of the head of state although
the Constitution absolves him of political responsibility.

The ideal of social unity has been transferred by some Arabs to the
political field to form the basis of pan-Arab unity and Arab nationalism. This is not the place to discuss the inter-relations between Arab
nationalism and Islam. It is however our considered opinion that to the
vast majority of those who profess to be Arab nationalists, the two
terms are inter changeable.

Under the label of Arab nationalism, the revolutionary Arab countries of Iraq, Syria, Egypt, Sudan and Algeria, have adopted the socialist system. They have been able, especially in Egypt under Jamal

Abdel Naser, to interpret the Moslem Shari'a as favouring socialism. They even invented the term "Islamic Socialism." The reason for these mental acrobatics was the feeling of the revolutionary rulers that they have to remain within the bounds of Islamic legality. They also felt that in this way, socialism would be more acceptable to the populace.

On the other hand, the same Shari'a is interpreted in Saudi Arabia and other countries as precluding any interference by the government in private ownership and enterprise except in exceptional cases.

The conclusion is that in the areas where the Shari'a was not particularly interested, namely in the political and economic spheres, the various regimes whether capitalist or socialist, find ample room for interpretation and manoeuvre each claiming that its system is aligned with the Shari'a.

All this emphasises again that Islam is an all-embracing creed, which totally fills its adherents' lives. The present Arab rulers are aware of the fact and act accordingly.

II. Discrimination against Women

Women all over the world are striving to achieve equality with men in various fields such as the accessability of jobs, equal pay, promotion opportunities, political appointments, equal rights in family relationships, etc. In the more advanced Arab countries, women associations are working for these ends in the same way as in other underdeveloped countries. As these problems are world wide, we shall concern ourselves with two areas of discrimination against women, particular to Arab Moslem societies, namely, polygamy and divorce. They deserve discussion because their practice is widespread, they determine the character of family life which is described in all Arab constitutions as the basic unit of society, and because as a result, they influence the type of Arab society as a whole.

These two fundamental aspects of social relationships are not the subject of ordinary legislation which can be easily modified, but are directly regulated by Islamic teachings handed fourteen centuries ago by divine revelation in the Koran. Hence, no change can take place except by "interpretation," and "interpretation" in this case has been very narrow indeed.

We shall also review briefly the political rights of women to vote and be elected wherever there are elective institutions in the Arab countries of the Middle East.

Before dealing with these definite subjects it should be stated that Moslem men enjoy a preliminary sweeping privilege over women for God has ordained: "Men are in charge of women, because Allah hath made the one of them to excel the other, and because they spend of their property (for the support of women), so good women are the obedient, guarding in secret that which Allah hath guarded. As for those from whom ye fear rebellion, admonish them and banish them to beds apart, and scourge them. Then if they obey you, seek not a way against them. Lo! Allah is ever High, Exalted, Great."[12]

And since the word of Allah should be believed in its literal meaning by every devout Moslem, and since these religious teachings are still the law in some Arab countries, and are "the primary source" or "a primary source" of legislation in all the others, the discrimination against women is instilled in the spirit of the individual even when the law tries to soften the inequality in conformity with modern trends.

A. Polygamy and Divorce

The Moslem's right to have four wives at one time is based on two Koranic verses:

"And if ye fear that ye will not deal fairly by the orphans, marry of the women, who seem good to you, two or three or four; and if ye fear that ye cannot do justice (to so many) then one (only) or (the captives) that your hand possess. Thus it is more likely that ye will not do injustice."[13] Verse 129 of the same Surah says: "Ye will not be able to deal equally between (your) wives, however much ye wish (to do so)..."

Thus the only restriction against polygamy is the inability to do justice to many wives. But the measure of the ability is left to the sole judgment of the husband.

Similarly, the right of divorce is granted to the husband at his sole discretion. Only the way in which the divorce should be pronounced is regulated. This however, does not affect the absolute right of the husband to divorce his wife or wives by his own will for any trivial matter or for no matter at all, and at any time he chooses.[14] Needless to say, the wife does not enjoy any such rights.

The right to marry and to divorce, was exercised without limitation for hundreds of years and is still unlimited in certain Arab countries. This means that theoretically a Moslem can have a new bride every day on condition of divorcing one of his four wives. The two acts are only limited by whether he is financially able to pay the dowry of the new bride and the financial dues prescribed in the marriage contract with his divorced wife, and by his own appreciation that he can treat his several wives equally in all respects.

In the past three decades, a few Arab countries in the Middle East tried to regulate the abuses of polygamy and uncontrolled divorce by a stricter interpretation of the Koranic verses.

Syria issued the Personal Status Law of 1953. This law did not contest the Moslem's right to polygamy and divorce, but sought to regulate the financial aspect of the problem. Article 17 decreed that the husband should demonstrate his ability to support a new wife and that this ability should be ascertained in court.

Article 27 decreed that if the husband abuses his right to divorce, the court would accord the wife financial compensation. Otherwise, the right of the husband remains unimpaired and the wife exercises no similar right.

The Personal Status Law of 1971 of the *Democratic Republic of Yemen*, carried the same provisions for polygamy and divorce as the Syrian law.

The Iraqi Personal Status Law of 1959 is more strict.[15] No new marriage can be concluded without court authorization. This is subject to two conditions: a – The husband's financial ability to support more than one wife, b – A legitimate need for the new marriage. This is left to the discretion of the court.

Divorce in Iraq also becomes effective only after a request is presented to the court by the husband and a court decision is issued.[16] The wife does not enjoy a similar right.

In Egypt, the absolute right of the husband to polygamy and divorce remains unimpaired. No new legislation regulating these rights has been issued during the three decades of the Naserite "progressive revolution."

In the countries of the Arabian Peninsula: Saudi Arabia, Yemen, Kuwait, Bahrain, the Union of Arab Emirates, Qatar and the Sultanate of Oman, the Shari'a remains supreme unfettered by new legislation or interpretation in the field of polygamy and divorce. The same applies to Jordan.

In Lebanon, the Shari'a is also supreme among Moslems. The Lebanese parliament, with a majority of Christians, does not see fit to legislate for the Moslems in such matters closely related to religion. The various Lebanese religious communities apply their own laws regarding the personal status of their adherents.

Among all the Arab countries, only Tunisia outlaws polygamy and penalizes the husband who commits it. It also makes divorce subject to a demand presented to the court by either the husband or the wife.[17]

Still the Tunisian law is careful to specify that these measures have been taken with full respect of the Shari'a and with a liberal interpretation of its terms.

The above-mentioned areas of discrimination against women are apparently contradicted by constitutional provisions in certain Middle East Arab countries which specifically decree the equality of the sexes, or in more general terms, the equality of all citizens before the law (presumably including women).

In Iraq, the provisional constitution of July 16, 1970 affirmed the principle of equality. Article 19(a) says: The citizens are equal before the law, without distinction as to race, origin, language, social environment or religion.[18] Article 30(a) says: Equality in the occupation of public posts is guaranteed by law.[19]

Law no. 139, 1972, establishing the General Federation of Iraqi women proclaimed the right of the women to exercise equal political, social, economic and cultural rights as men.

Article 3 of law no. 63, 1972, organizing the General Federation of Iraqi Youth declares that the Federation "strives to establish complete equality in all responsibilities and rights between all Iraqis without regard to sex."

The Egyptian Permanent Constitution of Sept. 11, 1971 provides in article 11: The State undertakes to coordinate between the duties of the woman towards her family, her work in society and her equality with man in the political, social, cultural and economic fields, *without violating the principles of Islamic jurisprudence.*[20]

Article 40 says: Citizens are equal before the law without distinction as to race, origin, language, or creed.[21]

The Provisional Constitution of Qatar of April 2, 1970 provides in article 9: All persons shall enjoy public rights and shall be subject to equal duties without distinction on grounds of race, *sex* or religion.[22]

The Permanent Constitution of Syria of Jan. 31, 1973 (approved by referendum on March 12) says in article 25(c): All citizens are equal before the law regarding their rights and duties. Article 45 says: "The state guarantees for women the opportunity to participate fully and actively in the political, social, cultural and economic life (of the country). The state strives to remove all the restrictions which prevent the development of women and their participation in building the Arab socialist society."

The Jordanian Constitution of Jan. 1, 1952 says in article 6: Jordanians are equal before the law without discrimination as to race, religion or language.

The same provision is embodied in article 29 of the Kuwaiti Constitution of Nov. 11, 1962.

How do we accommodate between these constitutional provisions and the text in the same constitutions about the Shari'a being "the primary" or "a primary" source of legislation in the countries concerned; or between equality before the law and the Moslem's right to polygamy and arbitrary divorce also guaranteed by law? A Moslem learned man may provide a scholastic interpretation, but for the layman the best that can be said is that the constitutional provisions express a hope of things to come, a trend which may be realized in the distant future, while the reality remains anchored in the Islamic Shari'a and jurisprudence.

A report by the General Federation of Iraqi Women presented to the Seminar on the Status of Women in Arab Laws, held in Beirut in May 1974, aptly describes the situation. After referring to the efforts of the Iraqi government to remove discrimination against women, the report said:

However, there is a wide gap between these laws, and what the Iraqi revolution and Ba'ath Party believe in, and the actual conditions in Iraqi society; between the legal framework which abolished discrimination against women, and the social facts and values prevailing in Iraqi society which consider women in a different light, especially in the countryside and in the poor urban districts.

B. The Effects of Polygamy and Arbitrary Divorce on Arab Society

We said that we have chosen to discuss polygamy and divorce as examples of discriminatory practices against women because of the type of family they help to produce. In fact, Moslem society is greatly concerned about family life and this concern has found expression in the various constitutions which agree on considering the family the basic nucleus of society deserving every encouragement and protection by the state.[23]

However, this concern is vitiated by the atmosphere of insecurity and instability felt by Moslem women because of the husband's right to re-marry up to three other wives and to divorce his wife at his sole dis-cretion.

Care however, should be taken not to portray polygamy as the ge-neral order of things in Arab Moslem society. With the spread of mod-ern education, and greater contacts with advanced societies, polygamy becomes scarce among the better educated classes. Furthermore, the urban middle and lower middle classes are financially unable to afford the costs of a second wife. There remains the majority of the population composed of the urban poorer classes, the peasants and the nomads who still find that more wives, and consequently more children, en-hance the prestige of the father, do his work for him in the city streets or in the fields and provide him with an extra income as the children are put to work at an early age alongside the women.

But if the urban woman in the more affluent classes and in the middle and lower middle classes, feels more or less secure from the danger of having another wife sharing her husband with her, she feels no such safety from the husband's arbitrary right to divorce her. Hen-ce, the Moslem woman is eternally concerned with two things: How to prevent her husband from remarrying, and how to avoid divorce.

Chief among the ways used is to entice the husband to over-spend so that he will not be able to afford a second marriage or a divorce and the attending alimony. Another way is to give birth to as many children as possible so that the husband will be constantly occupied financially and emotionally. A third device is to hoard as much money and pos-sessions from the husband's property as the wife can possibly lay hands on, as security against a possible alienation or divorce.

The great number of polygamous marriages testifies, however, that the success of these devices is limited. The woman remains with a total feeling of insecurity and dependence. When she lives in the same house with another wife or wives, or even in separate houses, wifely intrigues against each other are common place. The children born to the same husband by different wives grow in this atmosphere of insecurity, in-trigue and bitterness. They learn to dislike and distrust their half brothers as competitors for their father's affection and generosity. They may grow to dislike their own father if he shows more affection to the children of a more favourite wife.

Thus, the formation of the family (where there is polygamy and arbitrary divorce), considered as the basic unit of Arab society, is characterized from its inception by disunity, mistrust among its members and dislike of each other. These characteristics of the basic social unit are reflected on society as a whole and this in turn partly explains the lack of social cohesion in Arab society.

C. Women's Political Rights

In the Arab East eight countries have parliaments or national assemblies. They are Lebanon, Jordan, Syria, Bahrain, Kuwait, Qatar, Egypt and Sudan. The Palestinian organizations have a national council which acts as a sort of parliament. Women have the right to vote and be elected in Lebanon, Syria, Egypt, Sudan, and the Palestine National Council.

In Bahrain, Kuwait, Jordan and Qatar, women do not vote and are not eligible for election. In the remaining Arab countries of the Middle East, there are no parliaments or representative assemblies. The women and the men are equally deprived of political rights.

The following table gives the number of women in the various parliaments and assemblies:

Parliaments	Total Membership	No. of Women Members
Lebanon	99	none
Syria	122	5
Egypt	300	9
Sudan	250	12
Sudan (Regional Southern Council)	57	3
Palestinian National Council	150	2
Tunisia	90	4

III. DISCRIMINATION AGAINST NON-MOSLEMS

The inhabitants of the Arab countries of the Middle East, outside the Arabian Peninsula, are Moslems, Christians and Jews. Christians have been banished from the Arabian Peninsula after Islam, and Jews, who managed to remain in Yemen until recently, migrated to Israel in the last two decades.

Professed atheists, if they exist, are not tolerated in the Arab countries, have no legal status and are not recognized by the state. They are therefore outside the scope of this article.

The case of the Jews in the Arab countries has been embittered by the Palestinian problem and it will be discussed more fully in another article. So we restrict our discussion to the status of Christians.

The subject is important only in the countries which have a sizable Christian population, namely, Syria, Jordan and Egypt, where Christians constitute 12–18 percent of the population. It is less important in Iraq where Christians are only 5–7 percent. In the Sudan, there are no native Christians in the North and those in the south are distinct racially from the Arab north and present a different problem.

The status of Christians and Jews is determined by a few Koranic verses. God, addressing the Moslems, tells them: "Ye are the best community that has been raised up for mankind. You enjoin right conduct and forbid indecency; and ye believe in Allah. And if the People of the Scripture (Christians and Jews) had believed it had been better for them. Some of them are believers; but most of them are evil-livers."[24]

Another verse says: "Fight against such of those who have been given the Scripture... and follow not the religion of truth (Islam), *until they pay the tribute readily, being brought down.*"[25]

These two verses firmly establish two very important concepts: First, that the Moslems are far superior to any other religious group. "Ye are the best community that has been raised up for mankind...," and secondly, that the Christians and Jews who had not accepted Islam should be conquered and "brought down" i.e. humiliated, and subjected to the payment of the "tribute" or poll tax which is a special tax not payable by Moslems.

At the same time, the Christians and Jews were barred from sharing in the government of the country and from the service in the army but they were granted "protection" of life and property by the Moslem state for paying the "tribute." They also retained the right to manage the affairs pertaining to their personal status (marriage, divorce, etc.) according to their own laws but they enjoyed no political rights of any kind.

So, from the birth of the Moslem state to the fall of the Ottoman Empire fifty years ago (a period of about fourteen centuries), the Christians and Jews of the Middle East lived officially and in law, as second class citizens, tolerated sometimes, often abused depending on the fanaticism of the Ruler, but never full citizens equal to the Moslems in rights and duties. Certain outstanding persons among them reached high position in government and administration under the Arab Caliphs and Ottoman Sultans, but they were employed for their ability and because they had no political ambitions. They were never employed because their communities had a right to share in the government of the country.

Other religious teachings accentuate the divisions within Arab society. The Moslem can marry Christian and Jewish women but a Christian or a Jew cannot marry a Moslem woman until he is converted to Islam.[26] The prohibition was confirmed in verse 10 of Surah IX (She That Is To Be Examined).

The Koran praises the Christians in some verses and attacks them in others but it has no good word for the Jews. It says: "Oh Ye who believe! Take not the Jews and the Christians for friends. They are friends one to another. He among you who take them for friends is (one) of them."[27]

Belittling their military abilities, the Koran says: "They will not harm you save a trifling hurt and if they fight against you they will turn and flee...".[28]

Equating the Christians with disbelievers (although in other places Christ and Mary are highly praised), the Koran says: "They surely disbelieve who say: Lo! Allah is the third of three; when there is not God save the One God. If they desist not from so saying a painful doom will fold on those of them who disbelieve."[29]

The Jews come for a much harsher verdict: "...And humiliation and wretchedness were stamped upon them and they were visited with wrath from Allah. That was because they disbelieved in Allah's revelations and slew the prophets wrongly."[30] And again: "Ignominy shall be their portion wheresoever they are found... They have incurred anger for their Lord, and wretchedness is laid upon them...".[31]

The reader is entitled to ask why bother with these old teachings when all the recent Arab constitutions decree the equality of all citizens before the law without distinction as to religion, race, etc.?

The reason is found in the character of Islam which is a religion and a state at the same time. Furthermore, Islam has such a strong hold on its adherents who believe implicitly in the word of God in the Koran, that whenever there is contradiction between the text of the constitution or law and the religious precepts regarding Christians and Jews, the contradiction is always resolved by administrative and practical measures in favour of the Shari'a and against the "equality" embodied in the law especially regarding political rights and appointments. In cases which cannot be "adjusted" by administrative measures, the modern law simply adopts the Koranic text such as that forbidding the marriage of a Christian or Jew with a Moslem woman.

This has led, over the years, to the establishment in every Arab country with Christian, Jewish and other Moslem minorities, of closed and separate communities who coexist but do not inter-mingle. The communities lead their own separate lives according to their own traditions. Their members do not mix except when business or need demands it. There is almost no social intercourse among the bulk of the population except in the higher hierarchies of business and administration, and an ordinary Christian or Moslem may spend his whole life without once receiving a guest in his house of the other faith.

In general, the Christians in every country of the Middle East other than the Sudan, are above the average in the level of education, in the mastery of crafts and technical skills, in the professions and in academic accomplishments. Historically, this is due to the fact that Christian communities developed their own primary and secondary free school systems way ahead of the Moslems. The foreign Christian missions later helped this development. When 65 percent of the Syrians and Jordanians remain illiterate to this day, and more than this percentage in Iraq and Egypt, we find that illiteracy among the Christians in these countries is quite low.

But perhaps more important than the ability to read and write is the receptivity to modern knowledge. The Christian has no inhibitions against modern civilization which is basically Christian. The Moslem Arab, on the other hand, starts with two deep convictions. God has told him that he is a superior being "Ye are the best community that has been raised up for mankind..." and he cannot but believe the word of God. Yet, he cannot help but realize that Christian societies are far more advanced and that there is not a single Moslem country in the world within the ranks of the "advanced and industrialized" countries. This apparent contradiction between the word of God and the actual facts bewilders the Arab Moslem. Many try to rationalize the situation by bringing out the worst features of Western culture and civilization. They therefore start with a negative attitude towards the values of modern civilization and remain forever foreign to it and unable to assimilate it. This is especially true of many Arab students who study in the West and who return with a deep feeling of bitterness and frustration.

The other conviction is related to pride in the glory of past Arab conquests and culture, and humiliation in the present decline. The Arab has been singularly unable to analyse the internal causes of this decline and to examine critically the cultural and social values and traditions inherited from a past nomadic society built on conquest and exploitation of the conquered, which have ceased to be relevant to the present age for at least the last two centuries. The refusal to review and criticize these traditions and values may be due to the fact that they are related to religious beliefs.

Therefore, the Arab blames external forces for his decline, chief among them the Crusades and "colonialism," both Western factors hostile to Islam.

These feelings accentuate the divisions within Arab society to which we referred. The various sects continue to evolve in separate closed communities with different outlook, habits and customs.

Naturally, social cohesion remains lacking and a national feeling is difficult to cultivate. The Christian, although above the average level in education and ability finds the roads blocked in the way of his advancement in public functions and the army. In many countries, advancement beyond a certain officer's grade is automatically blocked. In all, the number of Christians in the top grades of the administration is way below their numerical proportion, their academic standing or their proven ability. The obstacles placed in their way are contrary to the law, but there seem to be always administrative stratagems which achieve the desired results.

To take Syria as an example, before the military coups started in March 1949, and under the constitutional republic, there was not a single Christian among the eleven governors of the districts; there was one sub-governor among 21; there were three ambassadors out of almost forty, and there were two general secretaries of the ministries among seventeen. When Syria joined Egypt in the United Arab Republic in 1958, President Jamal Abdel Naser was scandalized because there were still a few Christian ambassadors in the foreign service. He felt that the Christians could not be trusted to hold state secrets. He soon rectified the situation. Egypt, for twenty years under Naser, never had more than one Christian cabinet member out of 22 and sometimes 28. In Iraq it is rare for a Christian to rise above the rank of captain or major in the army. The same applied to Egypt under Naser except for one or two exceptions.

To offset this picture which is unacceptable to foreign observers, all Arab countries seize on a Christian personality who has achieved prominence in public life and present him as an example of non-discrimination. But such rare cases do not hide the discrimination practiced against thousands in the lower ranks of the administration, army and the judiciary. And in all cases, the underlying sense is that the Christian is not an equal citizen but a conquered alien who pays "the tribute." He can be used to fill a post if needed, but he does not share in the government of his country as of right.

In the Sudan, where the Christians are concentrated in the south, a long war of resistance forced the Moslem north to agree to a sharing of political power with the Christian south. An agreement was concluded and promulgated on March 3, 1972 which gave the Christian south a wide range of local autonomy and fair share in the central government.[31]

The Christians of the Arab countries have not been inactive in their efforts to achieve full and equal citizenship in a truly modern national state. Before the first world war, they were prominent in the secret societies which operated in the Ottoman Empire and abroad for the establishment of an independent Arab national state separate from the Moslem Caliphate. And when Hosain, Sharif of Mecca, proclaimed his revolution in 1916 against the Ottoman Caliphate for establishing a national Arab state, the Christians of the Middle East rallied to his support. When his son Faisal became King of Syria on March 8, 1920, many of his top advisers were Christians from Syria, Lebanon and Palestine.

Later on, in 1936, Antun Sa'adeh of Lebanon, a Greek Orthodox Christian, organized the first avowedly secular party in the Arab countries calling for the union of Syria, Lebanon, Jordan, Palestine and Iraq. Sa'adeh was executed in 1949 but his party is still active in Lebanon.

Later on, another Christian, Michel Aflaq, in Damascus, started the Ba'th party also on secular bases. The party is now in control of Iraq and Syria. In Syria, it has enabled the Alawites, an oppressed minority Moslem sect, to gain power over the hostile Sunni majority.

The Christians are also prominent in the various local communist parties where religion is not a factor against them.

Among the Palestinian guerrilla organizations, the two extreme

Marxist groups (and therefore secular and non-religious) are led by two Christians. Dr. George Habash leads the Popular Front for the Liberation of Palestine, and Mr. Nayef Hawatmeh leads the Popular Democratic Front for the Liberation of Palestine.

The traditional Moslems describe these Christian activities as a sabotage of Moslem society and an attempt to break it up and destroy it. The Christians, on the other hand, feel that they are only striving to gain self respect, equal citizenship and rights in their homeland. Both sides seems to be telling the truth from their point of view.

IV. Discrimination against Moslems of Other Sects

In the Arab countries of the Middle East only Yemen, Bahrain, Iraq and Syria have sizable communities of different Moslem sects. Saudi Arabia, South Yemen, Kuwait, the United Arab Emirates and Qatar have on the whole, homogenous Sunni populations. So does Jordan, Egypt and the Sudan. The Sultanate of Oman has a large majority of the Abadite Moslem sect.

Yemen is divided into the mountain tribes of the Shi'a Zaydi sect (followers of the fifth Imam) and the Sunnis who inhabit the coastal plains. After the withdrawal of the Ottomans, power was seized by the Zaydi Imams who exercised it over the whole country. The Egyptian military intervention of 1962 was mainly opposed and made ineffectual by the opposition of the Zaydi tribes. At present there is a sort of compromise and agreement acceptable to both communities with the balance of power favouring the Zaydis.

Bahrain is also divided into Sunnis and Shi'a. The independence of the country was proclaimed on Aug. 14, 1971 and elections for the constitutional assembly were held on Nov. 30, 1972.[33] The assembly is composed of 44 members of whom half are elected and half appointed by the Ruler. Among the elected members fourteen are Shi'a and eight Sunnis. Although the elections were not conducted on religious or sectarian basis, the results may reflect the proportion between the two communities. The ruling family is of the Sunni sect and there are as yet no complaints of sectarian discrimination.

The Iraqi situation is quite different. The population is divided into about sixty percent Shi'a (followers of the twelfth Imam), thirty five percent Sunnis and about 5–7 percent Christian. The Shi'a are all Arabs while the Sunnis are divided into Arabs and Kurds. The Kurds number about 25 percent of the whole population, hence only 10–15 percent of the Iraqis are Arab Sunnis.

Under the monarchy the Sunnis, both Arabs and Kurds, claimed to represent more than half the population, and with the backing of the royal family, which was also Sunni, ruled the country. The Shi'a did not object because they greatly venerate and obey the descendants of the Prophet of whom the royal Hashemite family is a scion on the Sunni side. In exchange, the monarchy gave the Shi'a more positions and government posts than they were accustomed to under the Ottomans who maltreated them. So, on the whole, there was a sort of balance which satisfied the various religious and social communities.

The whole set-up was upset by the 1958 revolution. Abdel Karim Qasem (a Sunni), removed the monarchy and thereby alienated the Shi'a. His second in command, Abdel Salam Aref, also a Sunni, proposed an immediate union with Syria and Egypt already united in the United Arab Republic. This alienated the Kurds who consider any union with an Arab country an attempt to reduce their power by submerging them in a larger Arab majority. War ensued with no side able to achieve a military solution.

With the Kurds who constitute 25 percent of the population, in mortal conflict with the present regime, and with the Shi'a constituting 55–60 percent of a population in silent opposition, the Ba'th is now ruling Iraq with the backing of about 15–20 percent, if all the Sunnis supported the regime, which they do not. Consequently, a small Sunni minority of probably 10 percent is ruling by brute force the majority of Iraqis. The age-old religious hostility and bitterness of the Shi'a against the Sunnis which started on the death of the Prophet and was exacerbated by the killing of Ali's grandson, Hosain, in 680 at Karbala near Kufa, (Mesopotamia) at the hands of the Sunni Omayyads, is still as acute and vivid as if the injustice was done a few months back and not thirteen hundred years ago.

This injustice felt by the Shi'a is not only religious but had, from the start, a political colour. The Shi'a feel that they were cheated in the succession to the Caliphate after the death of the Prophet and they continue to be cheated in the modern age by the Ba'this who, even as a small minority, have seized power and deny it to the Shi'a majority.

Syria has an abundance of religious sects both Moslem and Christian. Besides the Sunnis, there are five Shi'a and twelve Christian sects. The population in 1949 was 3,177,175 of whom 69% were Sunnis, 15.7% Shi'a of the various sects and 15% Christian.[34]

The Shi'a sects were mentioned as follows:

Shi'a (followers of the 12th Imam)	13,708
Ismaelites (followers of the seventh Imam)	32,804
Druzes (an offshoot of the Ismaelites)	100,554
Alawites or Nosayris	355,468
Yazidis	2,889
Total	505,423

After 1949, the government statistics ceased to mention the various Moslem sects separately and lumped them with the Sunnis under the single heading of "Mohammadans." The Moslem minorities saw in this stratagem an attempt to forestall their demands for sharing in political power and they resented this and similar subsequent measures.

Homogeneity among the Syrian Moslems could not be achieved by changing the name of the group, especially that, among the various Shi'a sects only the members of the first group in the above table are recognized by the Sunnis as Moslems. The Ismaelites, Druzes, Alawites or Nosayris and Yazidis are referred to as "gholat al Shi'a" that is "the Shi'a extremists" and are not accepted as Moslems. Some of these sects have a secret faith of complex origin, do not hold the Koran and do not practice any of the prayers or do the duties required of the believer. Religiously speaking, the Sunnis hold the People of the Scripture, that is the Christians and Jews, as nearer to them than those extreme Shi'a sects.

It is very significant that in the Arab world, Christians have lived in mixed villages with Sunnis, Ismaelis, Druzes, Nosayris and Shiites. But Sunnis and Druzes do not co-exist in the same villages, nor Sunnis and the other extreme Shi'a sects, nor the extreme Shi'a sects one with the other.

Furthermore, inter-marriage between the members of these Moslem sects is almost unknown and social intercourse is very rare, so that each sect forms a separate and closed community within the social body.

The monopoly of political power exercised by the Sunnis lasted as long as the Sunnis officers, who held power directly or behind the scenes, were in effective control of the armed forces. However, the demographic composition of the country had special features which helped to end the Sunni monopoly and transferred it to the Alawite minority.

For whereas the Christians are dispersed in the Syrian towns and villages all over the country, the Shi'a sects occupy strategic geographical areas.

The Ismaelites occupy a region east of Hama in central Syria. But their offshoot, the hardy Druzes, occupy a mountain range in southern Syria near the Jordanian border.

The Alawites occupy the whole mountain range along the Syrian coast extending from Alexandretta to the Lebanese border.

These last two groups, secure in their regions where almost no Moslem lives, pursued their own separate development. They were estranged from the central government which looked down upon them and neglected their villages and towns in respect of roads, education, health and public services.

Feeling that they have no hope of escaping discrimination under Sunni rule, they flocked to join the political parties which advocated a secular state and attacked religious fanaticism. These were the Syrian Popular Party of Antun Sa'adeh and the Ba'th party founded by Michel Aflaq and Salah Bitar, mentioned previously. In fact, the rank and file of these two parties has always been largely constituted by members of minority groups such as Christians, Ismaelis, Alawites and Druzes. The Syrian government liquidated the Syrian Popular Party in 1950, so the Ba'th remained the only political organization which gave the minorities a hope of social and political justice and equality.

In another development, the young men of the Shi'a sects, poorly educated and unable to compete in the liberal professions or technical skills with the city dwellers, found in the army a source of income and a respectable profession shunned by the more affluent Sunnis. Gradually, the lower ranks of the army officers corps were filled with Alawites and Druzes of the Ba'th party. Some of them reached higher positions. So when the Ba'th party seized power in 1963, the Moslem minorities felt that their turn had come. But soon they fell among themselves and the Alawites liquidated the Druzes within the army in 1966.

Now, behind the façade of the Ba'th party and the Ba'th slogans and ideology, the Alawites hold effective political power, and the Sunnis complain of a type of discrimination they have practiced for centuries.

We are sure that the Syrian Ba'th party would object to this analysis and would like instead to put forward its programme of "progressive revolutionary socialism" applied without distinction to all Syrians. This is not the place to contest this programme. Suffice it to say that it has worked for the benefit of the dispossessed minorities because all the nationalizations and economic "reforms" have in fact improverished the wealthy and middle classes which are mostly Sunnis and partly Christian. By destroying the economic base of the Sunnis, the Ba'th and its Shi'a members have in effect deprived them of the means to regain political power. Moreover, no one can contest that the best armed units in the Syrian army are under the firm control of Alawite officers, or that the Alawites are found in every administrative department all over the country in numbers and positions unwarranted by the size of the community or the academic standing of its members.

V. DISCRIMINATION AGAINST NON-ARABS

Non-Arab minorities in the Arab countries are found in Iraq, Syria and Sudan.

In Iraq and Syria, the leading racial minority are the Kurds. There are also Armenians, and Assyrians who form part of the Christian population. There is also a small group of Circasians in the Golan Heights of Syria. The Iranians in Iraq are integrated with the Shi'a community and almost undistinguishable from it.

The Kurds in Iraq constitute 25 percent of the population as we said, and occupy the Kurdistan mountains in the north east of the country. No Iraqi government has ever succeeded in subduing them by force of arms. So when after numerous plots and coups, the Ba'th party, whose first principle is comprehensive Arab unity, seized power in Iraq on July 30, 1968, friction with the Kurds was bound to increase. In the end, the Kurds revolted in their mountains and the Iraqi army was unable to put the revolt down by military means. On March 11, 1970, the two sides arrived at a compromise whereby the Kurds gained a large measure of local autonomy within the Iraqi state.[35] But the participation of the Kurds in the central power in Baghdad was delayed. The Ba'this who hold all real authority in the Revolutionary Command Council refused to admit Kurds into this council or to vest real power in a democratically elected parliament in which the Kurds could be legally represented. War broke out again in 1973 and is dragging on with inconclusive results.

The Ba'th ideology insisting on the supremacy of everything Arab may be blamed for the present mortal struggle. But it should be remembered that after the monarchy, similar clashes between the government in Baghdad and the Kurds occurred. Some blame may rest with the Arab inability to accept political compromise in any situation.

The extreme territorial demands of the Kurds and the fear that they may seek total separation in the future may be another cause of discrimination. In any case, the inability on both sides to coexist while respecting the legitimate interests of each other, is inflicting deep wounds which would be difficult to heal and is making a compromise solution ever more difficult.

In Syria, the Kurds in Damascus and in the region of Aleppo who had migrated in Ottoman times, have been assimilated and do not present a problem. There was another migration of Kurds from Turkey to the Jazira region of north Syria along the Turkish border after the first world war. These Kurds number about 200,000. They occupied the lands, largely uninhabited at that time, extending roughly from Tal Kojak on the Iraqi border, along the Turkish frontier to Ras al Ain, a distance of about 200 kilometers long and thirty kilometers deep. They built villages, tilled the land and lived according to their tribal affiliations and traditions. There was very little friction or conflict between them and the nomadic Arab tribes to the south.

When the Ba'th party, with its pan-Arab ideology, seized power in 1963 it did not like the Kurdish concentration along the Turkish border. It claimed that the presence of a non-Arab population there is unsafe, although the Kurds are more hostile to the Turks than the Arabs. Syria was also apprehensive about possible connections between Syrian and Iraqi Kurds.

So the Syrian government began up-rooting the Kurds from their lands and pushing them south to the more arid regions. The policy has been completed and the Kurds dispossessed without compensation. Their lands have been turned into state farms run by government employees. This process was begun and terminated by administrative, arbitrary action without basis in law.

The Kurds have been effectively dispersed and reduced to poverty. Some migrated to the towns and to Lebanon to become unskilled labourers. Others remained in the more arid lands. In Syrian eyes, the Kurdish problem has been solved!

When Sudan gained its independence, it had to deal with the southern problem. The southern Sudanese are racially different from the northerners and are either Christians or animists. Khartoum sought to Arabize them and exercise direct rule over them. They revolted and took to the bush. The war was waged intermittently for sixteen years. The southerners suffered many hardships and casualties, their villages and homes were devastated and about half a million of them sought refuge in neighbouring African countries. The war was also a great drain on the manpower and resources of the north. Finally with the good offices of Ethiopia and others, accommodation was reached between the two sides in March 1972. The south achieved a large measure of self-rule and an adequate share of political power in the central government.[36] The agreement seems to be working smoothly and no complaints about discrimination is heard from either side.

VI. Discrimination when Moslems are a Minority

Lebanon is the only Arab country in which the Moslems are a minority. It is also the only Arab country in which the Shari'a has no sway and parliament does not have to be inspired in its legislation by Islamic jurisprudence.

Like Syria, Lebanon consists of many religious sects. But unlike Syria, it does not submerge these sects under a general label so as to deprive them of their identity. On the contrary, the Lebanese Constitution recognizes these diverse communities and guarantees them just representation in the cabinet and in public office.[37]

Furthermore, a tradition evolved alongside the constitutional provisions for the division of key political positions and administrative posts among the different communities. Thus, by general concensus, the president of the republic is always a Maronite Christian, the speaker of the house a Shiite Moslem, the deputy speaker a Greek Orthodox and the prime minister a Sunni Moslem. The distribution reflects the numerical strength of the main religious communities.

Likewise, the administrative posts from the highest ranks to the lowliest are distributed in a similar manner. The system may not produce an efficient administration but it serves the purpose for which it was designed. It satisfies the various communities in so far as the public office is a source of revenue for its holder, and represents actual participation of the community in running the country.

By another consensus, parliamentary elections are always run on the basis of six Christians of various denominations, to five Moslems of the various sects. Therefore the number of deputies in parliament is always a multiple of eleven. The present parliament consists of 99 deputies. A previous assembly had 66 an still and older house had 121 members.

In the Lebanese system, it is officially recognized that the unit of the body politic is not the separate individual but the individual organized within a religious community. The Lebanese system discloses openly what other Arab systems in the Middle East camouflage at the expense of their religious minorities.

A particularity of Lebanon in the Middle East is that many Sunnis have not reconciled themselves to the amputation operated by France in 1920 whereby Sunni regions in the north have been separated from Syria and incorporated into Lebanon. These Sunnis feel that they have been transformed by an arbitrary decision of a foreign power, from a ruling class in Syria into a political minority in Lebanon.

In recent years, Sunni Moslems assuming the leadership of all the Moslem sects, Shi'a, Druzes, etc., and claiming to speak in their name, began sounding the following grievances:

1. The extremists among them claim that the Moslems have become the majority through birth increase and should therefore hold the presidency of the republic. In support of their claim, they demand a population census.

The Christians counter that they agree to a census but in that case the Lebanese expatriates who still hold Lebanese citizenship should be counted. There are about three million Lebanese expatriates of whom more than 80 percent are Christians; hence a stalemate.

2. The Sunni Moslems claim that real authority is in effect exercised by the president of the republic and that the Sunni prime minister is almost a figurehead.

The Christians answer that these are constitutional powers agreed upon by all the Lebanese upon the formation of the state and designed to instil a sense of security among the Lebanese Christians who live surrounded by a Moslem sea.

3. The Moslems complain that there are key posts in the army and civil administration always held by Christians. They demand a share of these posts.

The Christians intimate that in every conflict between Lebanese interests and those of any other Arab country, or with the Palestinian organizations established in Lebanon, the Sunnis have consistently stood with the foreign Arab country and with the Palestinians against Lebanon. If the Sunnis had held the key posts in the army and administration, Lebanon would have long lost its identity and independence under the encroachments of the late Jamal Abdel Naser or Syria or the Palestinians. What is needed first, the Christians say, is a demonstration of loyalty to Lebanon on the part of the Sunnis.

In a more general context, the Christians wonder whether the Sunni clamour for equality and power-sharing embodies a readiness on the part of the Sunnis to deny the Koranic verse which makes the Christians second class citizens in relation to Moslems. As no denial of this superiority creed has ever been contemplated by a Moslem leader, Lebanese Christians cling to the present arrangements for fear of gradually falling to the status of the Syrian or Egyptian Christians.

In the last three years, the Shi'a Moslems under Imam Musa al Sadr, began dissociating themselves from the Sunni leadership to pursue the interests of their particular community. They realize that they are more numerous than the Sunnis and yet, because they were less developed socially and educationally, their share of the higher administrative posts went to the Sunnis. They are now agitating for more development projects in their parts of the country and for more say in the government. These demands, if satifsied, would be in good part at the expense of the Sunni population.

Unlike many Sunnis who still look to Syria as their true homeland and who have not yet accepted the reality of Lebanese independence, the Shi'a, because of the discrimination against their various sects in Syria, look to Lebanon as their only homeland in the region. Their loyalty has been repeatedly demonstrated in their conflicts with the Palestinian organizations along the borders with Israel.

Under these circumstances, one might suspect that the Lebanese Christians look favourably to a shift of political power from the Sunnis to the Shi'a.

REFERENCES

Mohammad Marmaduke Pickthall, *The Meaning of the Glorious Koran. An Explanatory Translation*, New York: New American Library.

Statistical Abstract of Syria, 1950, Damascus: Syrian Ministry of National Economy.

Record of the Arab World, Beirut, Lebanon: Research and Publishing House.

The Arab Compilation of World Constitutions, in Arabic, Cairo: General Directorate of Legislation and Fatwa, 1966.

J. Harris Proctor, ed., *Islam and International Relations*, London: Pall Mall Press.

NOTES

1. Provisional Constitution of Yemen elaborated by the Khamar Congress and issued by Presidential decree on May 8, 1965.
2. Kuwait Constitution of November 11, 1962.
3. Jordanian Constitution of January 1, 1952.
4. Provisional Constitution of Qatar of April 2, 1970 – Record of the Arab World, April–June 1971, p. 1965.
5. Permanent Constitution of January 31, 1973.
6. Provisional Constitution of July 16, 1970 – Record of the Arab World, July 1970, p. 4449.
7. Permanent Constitution of September 11, 1971 – Record of the Arab World, June–Sept., p. 2847.
8. H. A. R. Gibb, Islam and International Relations, p. 12.
9. Surah IV (Women), verse 59.
10. Surah XLII (Counsel), verse 38.
11. H. A. R. Gibb, *Op. cit.* p. 10.
12. Surah IV – Women – verse 34.
13. Surah IV – Women – verse 2.
14. Surah II, The Cow, verses 226–232.
15. Article 3, par. 4.
16. Personal Status Law 1959, art. 39.
17. Personal Status Law of 1956, art. 18.
18. Record of the Arab World, July 1970, p. 4449–54.
19. *Ibid.*
20. Record of the Arab World, July 1971, p. 2848.
21. *Ibid.*, p. 2850.
22. Record of the Arab World, June, 1971, p. 1965.
23. The Egyptian Constitution, art. 9; the Iraqi Constitution, art. 11; the Constitution of Qatar, art. 7(a); the Syrian Constitution, art. 44(a).
24. Surah III, The family of Omran, verse 110.
25. Surah IV, Reprentance, verse 29.
26. "This day are (all) good things made lawful for you... the virtuous women of those who received the Scripture before you (are lawful for you).." (Surah V, The Table Spread, verse 5) "...and give not your daughters in marriage to idolaters till they believe..." (Surah II, The Cow, verse 221).
27. Surah V, the Table Spread, verse 51.
28. Surah III, The Family of Imran, verse 111.
29. Surah V, The Table Spread, verse 73.
30. Surah II, The Cow, verse 61.
31. Surah III, The Family of Imran, verse 112.
32. Record of the Arab World, 1972, Vol. I, pp. 562–70 & Vol. II, p. 1342.
33. Record of the Arab World, 1972, Vol. I, pp. 51–2.
34. The Statistical Abstract of Syria, 1950, Table 3, p. 17.
35. Record of the Arab World, March 1970, pp. 1804–21.
36. Record of the Arab World, 1972, Vol. I, pp. 562–70, Vol. II, p. 1342.
37. Constitutional Law of Nov. 9, 1943, art. 95 zs amended.

Suffering and Struggle of the Kurds

L. M. VON TAUBINGER

LASZLO M. VON TAUBINGER was born in Töre, Upper Hungary, where his father owned a farm estate. After a secondary education at Premonstrant and Benedictine schools, he attended the University of Budapest, where he obtained his Doctor of Laws degree in 1941. He made a particular study of the economic and political problems of the Danubian region, and gave numerous lectures on this subject before audiences of the Royal Hungarian Foreign Policy Association, to which he belonged. Shortly after the occupation of Hungary by the Germans in March, 1944, Dr. Taubinger was arrested by the Gestapo. After the end of the war, Dr. Taubinger was sent to Prague by the Hungarian Red Cross to administer refugee matters, but soon learned that the Communists planned to arrest him. He fled to Munich. In the American Occupation Zone of Germany, he became the leader of the Hungarian refugee organization, as well as Bavarian representative of the British Red Cross Hungarian Welfare Committee. His organization provided relief and legal protection for more than 650,000 Hungarian refugees, and assumed general responsibility for Hungarian affairs in West Germany. While undergoing medical treatment in Switzerland between 1951 and 1956, Dr. Taubinger acquired proficiency as a political journalist, and was soon writing on East European affairs for various newspapers and magazines. He returned to Austria in 1956 to take charge of relief for Hungarians escaping after the unsuccessful revolution in their country. When this task had been completed he settled in Vienna as a foreign policy correspondent, dealing with Eastern Europe, world Communism generally, and problems of national minorities. His political analyses are broadcast by the Swiss Radio and appear in Swiss and English newspapers.

Suffering and Struggle of the Kurds

L. M. VON TAUBINGER

Following an interruption of four years, Iraqi Kurdistan has again been the scene of bloody fighting since March 12, 1974. The violence was precipitated by the socalled "Law Concerning Self-Administration of the Kurdistan Region," proclaimed unilaterally by the Iraqi Government on March 11. The courts refused to accept this statute of autonomy, giving as their reason that it not only failed to fulfil their demands for self-administration but also deviated in part from the undertakings formally agreed to in the armistice concluded in March 1970.

I. DEMOGRAPHIC AND HISTORICAL BACKGROUND

Kurdistan is only an abstract idea for most people. It includes, by and large, the wild mountainous area that surrounds the junction of the borders of Turkey, Iraq, and Iran as well as the northeastern corner of Syria. Exact statistics on the number of Kurds in each of these states are not available. The governments of the countries concerned tend to minimize the number of Kurds, insofar as they admit that Kurds exist in their countries at all. The Turks, for instance, deny that any Kurds live within the territory of their state, having simply declared that this minority consisted of "Mountain Turks." The Kurds speak of approximately 5.5 million compatriots in Turkey, 4 million in Iran, 2.2 million in Iraq, 500,000 in Syria, and about 60,000 in the USSR. They also claim that an additional 300,000 Kurds live outside Kurdistan. According to these figures, there are about 12,560,000 Kurds in the world.

According to United Nations statistics published in 1968, approximately 2 million Kurds live in Turkey, 5.4 in Iran, 1.7 million in Iraq, and 59.000 in the USSR. This calculation does not include a figure for Syria. The United Nations statistics would add up to a total of 9.16

million Kurds, not including the part of the Kurd people living in Syria.

The Kurds have a long history. From the very earliest times they have lived in the areas where they still live today. Their language is related to Persian and belongs to the Indo-Germanic language group. Reputable scholars believe that the Kurds are descendants of the Medes. It is known that their ancestors captured Nineveh in the year 612 B.C., thereby destroying the Assyrian Empire. The kingdom of the Medes thus established had a short history, however, it was overthrown by the Persians under Cyrus (550 B.C.). Since that time the Kurds have lived in various tribes, geographically separated from one another, which – according to the Greek author Xenophon – refuse to submit to rule either by the Persians or by the Armenians.

In the seventh century A.D. the Kurds were converted to Islam. The Sultan Saladin, one of their greatest leaders, united under his rule the territories of present Syria and Egypt as well as parts of Mesopotamia in the late twelfth century and the early thirteenth century. The majority of the Kurds later fell under Turkish rule, and parts of the areas they inhabited were incorporated in the Ottoman Empire; other parts were annexed by the Persians. A revolt in 1842 under the leadership of the Bedir Khans brought about the establishment of the short-lived independent Kurdish state which embraced the territory between the lakes of Van and Urmia.

After the Ottoman Empire had disintegrated in 1918, a Kurdish delegation lead by Sharif Pasha was among those invited to the peace negotiations at Sèvres near Paris. The third section of the Peace Treaty concluded with the Turks contains various provisions effecting the Kurds. Article 62, for instance, calls for a three-person commission appointed by the governments of France, Great Britain, and Italy, which was to work out a plan for the autonomous administration of those parts of the former Ottoman Empire that were inhabited by the Kurds. The commission was to work in Istanbul and have the plan ready within the following six months. The area involved was that east of the Euphrates, south of the Armenian border and north of the Syrian, Iraqi, and Turkish border. Article 63 of the Peace Treaty obligated the Turkish government to execute the proposal developed by the commission within 3 months.

Should the Kurds living in the afore mentioned territory apply to the League of Nations for the establishment of their own state after the elapse of one year, however, and should the League of Nations consider them capable of governing themselves, the Turkish government was then obliged by Article 64 of the Peace Treaty to execute the decisions of the League of Nations in this regard.

After the Turkish monarchy had been overthrown and the Republic proclaimed on November 1, 1922, however, the new regime under Kemal Ataturk refused to recognize the provisions of the Treaty of Sèvres. Ataturk concluded a new agreement with the Allies in June, 1923, in Lausanne. This new agreement made no more mention of the Kurdish question.

Since then, the areas inhabited by the Kurds have been divided among five states – Turkey, Iraq, Iran, Syria, and the USSR – with no regard to ethnic, economic, or cultural principles. The Kurds, however, have refused to be satisfied with their situation. They continue to fight for their freedom in constantly recurring revolts, sometimes limited to particular tribes.

II. KURDS IN TURKEY

As already indicated, the largest compact group of Kurds lives in Turkey. Kemal Ataturk, however, refused to recognize the existence of a Kurdish people within the new Turkey that he had constructed. He ordered the Kurdish schools to be closed, and publications in the Kurd language were prohibited. Many leaders of the Kurds were arrested, and for a while consideration was even given to a plan to resettle the Kurdish tribes into Anatolia, in order to prevent any contact between them and their fellow Kurds living in neighbouring countries.

As a measure of defence against such plans, an armed rebellion broke out in 1925 under the leadership of Sheik Seiid. Within a short time, the Kurds had succeeded in gaining control of major portions of the districts in which they lived. The Turkish government, however, suppressed this rebellion in a series of bloody battles with massive engagements of troops, and had the rebel leaders publicly executed on June 28, 1925.

The oppression that followed led to the outbreak of a new revolt in 1930: perhaps the bloodiest and longest Kurdish rebellion. This time the leader of the rebels was Isan Nuri. It took the Turkish army two years to regain control of the situation. From that time on, there have officially been no more Kurds in Turkey. They were renamed as "Mountain Turks" in order to hide the very fact of their existence.

The Kurds living in Turkey revolted for a last time in 1936, in order to protect themselves against the policy of the Ankara government. But this rebellion too, was drowned in blood within a year. An entire series of villages inhabited by Kurds was levelled to the ground. Those villagers who failed to flee into the mountains while there was still time were slaughtered. The leader of the revolt, Sayed Reda, was sentenced to death and hanged as a "Bandit".

Although the years since World War II have been marked by profound political, social, and economic changes in all parts of the world, this is not the case in the part of Turkey inhabited by the Kurds. Their districts are the most backward areas within the country, in which the people are denied the most basic national and human rights, including the official use of their own mother tongue.

III. THE KURDS OF IRAN

The situation of the Kurds living in Iran is today somewhat better than that of their compatriots in neighbouring countries. This is true despite the fact that the Iranians deny the existence of the Kurdish people as such and designate the Kurds simply as Persians on the basis of the similarity of languages.

The first major Kurdish revolt broke out in 1920 under the leadership of Simko. The rebellious tribes succeeded in bringing the major part of the areas where they lived under their control, and in defending these areas against the Persian troops for five years. Sporadic fighting broke out from time to time after the rebellion had been repressed. Simko was murdered in 1930; by the Persians according to the Kurdish version, and by his rivals as Iranians tell the story. The death of Simko deprived the Kurds of their leader, and they submitted to the government. A brutal terror followed. Teheran attempted to intimidate and denationalize the Kurds. The result of this policy, however, was the strengthening of national awareness.

The Democratic Party of Kurdistan (DPK) was founded in August, 1945, under the leadership of Ghazi Mohammed. This party was the successor of the Committee for the Rebirth of Kurdistan (*Komala i Zhian i Kurdistan*), which had been formed in the late 1930's through the merger of various Kurd political groups. Mohammed had previously visited the premier of the Azerbaijan Soviet Republic, Baghirov, to discuss with him the possibility of establishing a Kurdish state in those territories of Iran then occupied by the Soviet Union. Baghirov recommended to his guest that this committee be reorganized after the model of the Azerbaijan Democratic Party, which had been founded in the Russian-occupied areas of Iran and which was allied with the communist-oriented Tudeh Party. Mohammed followed this advice and organized the DPK. The party was founded on a purely national basis; it demanded equal rights and autonomy for the Kurds as well as introduction of the Kurdish language as an official language in Kurd-inhabited districts.

In January, 1946, the Kurd leaders in the Russian-occupied parts of Iran decided to establish the Kurd Republic of Mahabad. But the new State had only a very brief existence. Upon the withdrawal of Soviet troops, Mahabad was occupied by the Persian armed forces and its leaders were arrested and executed. It was of no help to Ghazi Mohammed and his followers that the leader of the Kurdish rebels in Iraq, Mullah Mustafa al Barzani, came to their aid with his troops. The Iranian armed forces consolidated their occupation in 1947 and perpetrated a considerable blood-bath among the population. Barzani's forces continued to fight the Persians for a while, but were finally forced to admit defeat and flee into the Soviet Union. This event marked the death-knell of the short-lived Republic of Mahabad.

Since that time, the Iranian government has carried out various reforms in the areas where the Kurds live, ameliorating their conditions of life. It can therefore be asserted that those Kurds who live in Iran enjoy an appreciably more favourable situation than their brothers in neighbouring countries. Not the least important reason for this is the tension existing between Iraq and Iran. In order to weaken the government in Baghdad, Teheran accords benevolent support to the freedom struggle of the Kurds in the neighbouring country under the leadership of Mullah Mustafa al Barzani.

IV. KURDS IN SYRIA

What is today the Republic of Syria was a French Mandate under the League of Nations from the collapse of the Ottoman Empire until 1946. The new boundary between Syria and Turkey was formed by the track of the "Baghdad Railway" built by the Germans during World War I. This border created three Kurdish enclaves on Syrian territory. The Kurds in these districts were the best off until 1946, as they were free to publish their own newspapers and use their own language. But after Syria had achieved its independence, the new regime in Damascus withdrew from the Kurd minority even these minimal rights.

The situation of the Kurds deteriorated even more after 1958, when Syria merged with Egypt for the first time. The consolidation was followed by mass arrests and deportations among the Kurds. The traditional place names of Kurd villages were transformed into Arabic and the regime had a programme drawn up for the "final solution" of the Kurd question. This programme was summarized by an important Baath official, Mohammed Talab Hilal, former secret police chief in the Kurdish province of Djesireh and later Minister of Food, in 12 points under the title "Studies Concerning the Djesireh District from National, Social, and Political Points of View."

By 1963 the Syrian government began to carry out the plans recommended by Hilal. The first step was the annulment of the Syrian citizenship of more than 120,000 Kurds. This measure was, among other things, intended to destroy Kurd families. Thus in typical cases parents were classed as Syrians but their children as aliens, with neither the right to attend school as children nor to accept employment as adults. The next step initiated was the forced resettlement of Kurd peasants from their farms within a strip of between 10 to 15 kilometers in width along the borders, and the settlement of Arabs within this strip. The purpose was to separate the Syrian Kurds from their fellow-Kurds living in Turkey and in Iraq. There followed the land reform, in which the farms of Kurdish peasants were confiscated and distributed among Arabs. The next step was a 1965 decree that prohibited, under the threat of criminal penalty, the building of new houses and the rehabilitation of existing buildings within the Kurd district. When a Kurd nevertheless made repairs on his house to save it from collapse, the authorities immediately ordered it to be razed to the ground.

The Kurds in Syria live today under the most brutal oppression, without enjoying any national rights whatsoever. They therefore cast hopeful glances at the struggle of their blood brothers in Iraq, the outcome of which may well have an effect on their own future.

V. The Kurds of Iraq*

Immediately after World War I, the Kurds in Iraq seemed destined to enjoy a comparatively favourable situation. This was a consequence of the fact that Iraq came under British control. The British allied themselves with Sheikh Mahmud Ben Sulaimaniya during the war, at the end of which they supported the formation of a Kurd-populated mountain region with at least partial autonomy. But this policy did not last long, since the Kurds were dissatisfied with the arrangement and demanded full autonomy for the entire Kurd territory between the Great Sab and the Sirwar. Several Kurdish revolts ensued. The Kurds proclaimed Sheikh Mahmud King of Kurdistan on September 14, 1922, but this revolt was crushed in 1924. Mahmud and his most loyal retainers fled into the mountains, from which in 1932 he launched a renewed but likewise vain effort to liberate the Kurd-populated regions from Iraqi rule.

A. The Kurd Policy of Independent Iraq

After obtaining its independence in 1932, the Iraqi government promised the League of Nations it would guarantee the Kurds free use of their language, introduce Kurdish as a second official language in addition to Arabic in districts where Kurds lived, and would also establish Kurdish schools. But Baghdad failed to keep these promises, and a series of new revolts followed.

After these battles proved unsuccessful, the Iraqi Kurds passed through a period of bitter oppression. Their situation was aggravated by inter-tribal rivalry. The leadership of the struggle for autonomy passed into the hands of the Barzani clan, where it remains to this day. The fate of the Iraqi Kurds is thus linked with that of the Barzanis.

The oppression and persecution led finally to a new Kurd revolt, which broke out in 1943, and was led by Mullah Mustafa al Barzani. After two years of fighting Barzani moved his operations to Iran in 1945, so as to assist the newly established Kurdish Republic of Mahabad against the Persians. He was appointed a general but was forced into exile with his loyal followers after the Iranian armed forces had

* _Editor's Note:_ This section was completed before the events of 1975 that culminated in Iraqi-Iranian accommodation at the expense of the Kurds.

occupied Mahabad and dissolved the Republic. He withdrew with his troops into the Soviet Union.

What Barzani did during his Soviet exile until 1958 is not known to us. In any case, he learned for a while under Stalin, as did many of his followers, who learned about prisons from the inside. He was later treated as someone who might be useful in the future as an instrument of Soviet power politics. But he never became a Communist and the Communists have never regarded him as one of their comrades. They still mistrust him today.

Comparative quiet prevailed in Iraqi Kurdistan during the decade and a half following Barzani's 1945 departure. After young officers had overthrown the monarchy and proclaimed the Republic of Iraq under Abdul Karim Kassem in 1958, Kassem summoned the Iraqi politicians living in exile, including Mullah Mustafa al Barzani, to return home.

The Kassem regime followed a pro-Kurd policy at the outset, for which reason Barzani and his followers supported it. Paragraph 3 of the new Iraqi Constitution declared explicitly: "Iraq is a Republic of Arabs and Kurds." The Democratic Party of Kurdistan was legalized at this time. The Kurds were granted schools and Kurdish newspapers and magazines were permitted.

But this period of Arab-Kurd co-operation proved short. As soon as the new rulers in Baghdad felt strong enough they lost interest in living up to the promise they had made to the Kurds. They began to arm the Kurd tribes hostile to Barzani and to incite them against the general. Mullah Mustafa therefore failed to return to Bagdad after a trip to Moscow, remaining in the mountains instead. With the support of friendly tribes and the DPK, he sent a 1961 ultimatum to Kassem from his mountain retreat. This document contained the demands for which the Kurds are still fighting today. Kassem answered Barzani's ultimatum by marching troops into Kurdistan and attempting to take the Kurd leader prisoner. Barzani reacted to this action of the Iraqi government by issuing a proclamation to all Kurds on September 11, 1961, urging them to take up arms. This was the hour of birth of the freedom struggle which the Iraqi Kurds, with only brief interruptions, have waged from 1961 until today.

The Iraqi troops were hardly in the same class with Barzani's Peshmerga (irregulars). The Kurd general limited himself to guerrilla warfare, withdrawing into the mountains in the face of frontal attacks. This tactic forced the attackers to fan out, whereupon his forces would wipe them out group by group. He thus succeeded in eliminating entire Iraqi brigades. His prestige grew in time to the point where entire police and army units defected to him with their weapons. Barzani never attempted, however, to hold villages and cities. These therefore became the preferred terrain for operations of the Iraqi units, who vented their brutality against the civil population, whom they were permitted to plunder almost without limit. Hundreds of thousands thus became civilian victims of the Kurdish War.

B. Kurd Affairs Under the Baath Regime

Kassem's dictatorship was overthrown on February 8, 1963, by a coup d'état conducted by Baathist officers, and Kassem himself was executed. The new regime declared its willingness to accord the Kurds legal equality and grant them full autonomy. Barzani's demands centered around the creation of a new central government in Baghdad with Kurd participation, as well as a regional government in Kurdistan. The ministerial chairs in the central government were to be distributed in a ratio corresponding to the Arab and Kurd sectors of the population. The Kurdish regional government was to have the following powers within Kurdistan: legislation, jurisprudence, Home Office affairs, education, public health, agriculture, and decision on regional matters. Barzani also demanded for the Kurds a percentual participation in the national income, particularly the oil profits. The Iraqi troop units should, furthermore, be withdrawn from the Kurdish districts. The Iraqi government rejected these proposals and made counter-proposals to the Kurds.

A new situation came about when the Baath Party also assumed power in Syria on March 8, 1963. Iraq and Syria soon began negotiations for the formation of a federation with Egypt. Barzani demanded the participation of a Kurd delegation in these talks and dispatched Jalal Talabani to Cairo at the head of a delegation. But Talabani was refused admission and forced to return home. Shortly thereafter, the Iraqi government arrested without warning the Kurd delegation that was in Baghdad negotiating about the conditions for autonomy. Simultaneously, the Iraqi armed forces opened a major offensive against Barzani's troops. They employed tanks, heavy artillery, and napalm bombs to attack Kurdish villages, thereby causing renewed heavy losses among the civilian population. The Syrian Al-Yormouk Division also took part in the offensive. Mass executions among the civilian population belonged to the regular order of business in the Iraqi-occupied cities. At the same time, however, the Iraqi government declared to the outside world that the Kurd problem had long since been solved and that the north of the country was entirely at peace.

C. Civil War and Attempted Reconciliation (1964–1970)

The Baath regime in Baghdad was overthrown on November 18, 1963. The new State President, Abdul Salam Aref, made the Kurds new promises that their demands would be fulfilled and concluded an armistice with them in February, 1964.

In the meantime, however, disunity erupted within the Kurdish camp. In the summer of 1964 the left wing of the DPK, under the leadership of Ibrahim Ahmed and Jalal Talabani, set about deposing Barzani as party chairman. Barzani responded to these efforts in July by calling a special meeting of the Central Committee of his party at Gûala Dixa, at which all the military leaders pledged their loyalty to him. Ahmed and Talabani as well as 12 additional Soviet-supported members of the DPK's Politbureau were expelled from the party. Talabani thereupon organized an armed revolt against Barzani; but the latter succeeded in crushing it. Barzani's enemies fled to Iran, where they were granted asylum in Hamadan. Certain ideological supporters of the rebel leaders switched over to the Iraqi camp. Ahmed and Talabani set up a counter-politbureau of the DPK in Hamadan; a substanstial schism in the party continued until 1970.

The peace between Barzani and the Iraqi government again proved of short duration. Fighting was resumed by the government troops in April, 1965. The offensive by the Iraqi army led to tension with Iran, since the battles extended to the Iranian border. Like his predecessor, President Aref denied that the war had been reopened and contended that his army was only being deployed to combat highway bandits in the north. This time, however, the renewed outbreak of fighting had its good side for the Kurds. Most of the rebels reconciled themselves with Barzani and returned from exile to rejoin his troops. Only Ibrahim Ahmed remained irreconcilable. Talabani stayed abroad, but canvassed for international support for the Kurdish cause. He and his associates did not return home until 1970, after the armistice of that year.

When President Aref was killed in an airplane crash on April 14, 1966, his brother Abjul Rahman Aref assumed power in Baghdad. The

Kurdish War was suddenly interrupted. But again the peace did not last very long. Only a month passed before the fighting broke out anew. This time the Iraqi troops suffered their heaviest losses since 1961, so that the government felt obliged to open negotiations with Barzani. On July 29, 1966, Prime Minister Al-Bazzaz published a 12-point programme for the solution of the Kurdish problem. His offer consisted of the following points:

1. Recognition of the Kurdish nation in the Iraqi Constitution;

2. The decentralization of administration must be carried out with due consideration for the interests of the Kurdish nation;

3. Introduction of the Kurdish language as an official language in Kurdistan;

4. The calling of parliamentary elections before the end of 1966;

5. The Kurds should be admitted to all public offices in proportion to their fraction of the total population;

6. Inclusion of Kurds in the General Staff, award of fellowships, as well as cultivation of the Kurdish language and culture at the University of Baghdad;

7. Public officials in the Kurdish provinces must be Kurds;

8. The Kurds have all rights that are proper for citizens in a parliamentary democracy – including the right to publish their own newspapers;

9. The granting of a general amnesty for all Kurds, and the re-employment of Kurd civil servants and workers in their former positions;

10. Those Kurds who defected from the Iraqi army or police to Barzani's troops are to return to their former units with their weapons within two months;

11. A Ministry for the Reconstruction of the North is to be created;

12. The government will make the return of the refugees to their homes its concern.

This programme was to be carried out within six months. But nothing came of it since the new Prime Minister, Nadji Taleb, refused to fulfil the obligations assumed by his predecessor. Barzani thereupon addressed an ultimatum to the government in March, 1967, warning against once more overtrying Kurd patience and threatening to reopen hostilities in June if the obligations undertaken by Al-Bazzaz should not be fulfilled.

But in the meantime the Six-Day-War between the Arabs and the Israelis broke out in June, 1967. The Kurds failed to take advantage of this development, and Barzani refrained from carrying out his threat. The government took advantage of his restraint by attempting to stir up Talabani, who had fallen out with Barzani again, to conspire against the latter. Scattered armed skirmishes took place between rival groups of Kurds. But Talabani's supporters were too weak to raise any serious danger for Barzani.

The pause in Iraqi-Kurd hostilities lasted until the summer of 1968. When the Baath Party came to power again on July 17 of that year, the new regime launched another offensive against the Kurds. After the government troops had suffered various defeats, Baghdad finally – in late January, 1970 – declared its willingness to negotiate a peaceful solution of the conflict. The government proclaimed at the same time a general amnesty for all Kurds and promised them autonomy. One of the major reasons for this conciliatory attitude was the fact that the new regime of Hassan Ahmed Al Bakr found itself facing both internal and foreign political pressure. It was forced to admit that it was in no position to win the war against the Kurds under then existing conditions.

An agreement between the government and Barzani's representatives was finally concluded on March 11, 1970. It consisted of the following points:

1. The Kurds are to obtain full political rights as well as autonomy within the Iraqi republic. The autonomous Kurdish region will be established on the basis of a census;

2. Recognition of the existence of the Arab and Kurd nations shall be anchored in the constitution;

3. A Kurd shall be designated as Vice-President of the Republic;

4. The Kurds shall participate proportionally in the legislative branch;

5. Kurds shall be included in the government and in the ministries without discrimination;

6. Appointments as public officials in the Kurdish regions shall be limited to Kurds or persons having complete mastery of the Kurdish language;

7. Kurdish and Arabic are both official languages in the Kurdish regions; Kurdish is the language of instruction in these regions;

8. All former military personnel and civil officials, students, and workers of Kurdish nationality are to be reinstated in their former offices, jobs, or fields of activity;

9. Cultural and educational advantages shall be extended to the Kurds;

10. The Kurds may establish their own trade unions and student, youth, women's, and teacher organisations, which shall be affiliated to the corresponding Iraqi national organisations;

11. Resettled Arabs and Kurds shall be returned to their former domiciles;

12. Measures of assistance shall be carried out for those Kurdish citizens who suffered most severely from the wars;

13. A Kurdish Development Commission shall be established and charged with formulating a development plan for the Kurdish region;

14. The agrarian reform shall be expedited in the Kurdish districts;

15. The central authorities shall have jurisdiction over the exploitation of natural resources within the autonomous region;

16. The secret Kurdish radio transmitter and all heavy weapons of the Kurds shall be turned over to the government during the final stage of implementation of these provisions.

The agreement contained six secret clauses, of which one provided for a census to be conducted before the end of 1970. Those districts of the country in which the Kurds constitute 51% of the population were to be unified, to constitute the autonomous region of Kurdistan. The non-fulfillment of the secret clauses, about which little has actually become known, may well have played an important role in the onset of the current tensions and the outbreak of the new war.

D. The Kurdish Struggle for Freedom Since 1970

The new phase of the Kurdish struggle for freedom in Iraq began on March 12, 1974, a day after President Bakr had officially promulgated a statute of autonomy for Kurdistan. This statute had been determined unilateraly by the Iraqi government, and was not accepted by the Kurds. Since then, fighting has continued without interruption in Iraqi Kurdistan, although Baghdad publishes no information whatever about this war. Such information as is available comes from the radio transmitter of the Kurd partisans, and is often reprinted in the Persian newspaper "Kayhan International," which is published in English. Some information comes from journalists who visit Barzani's headquarters.

The reason for the renewed revolt lies in the fact that during the four years that elapsed between the signing of the agreement on March 11,

1970, and the issuance of the autonomy statute on March 11, 1974, the Iraqi government did little or nothing to strengthen Kurdish confidence in Baghdad. A government-sponsored assassination attempt against Barzani took place on September 30, 1971, costing the lives of several in his entourage. The measures for achieving autonomy specified in the March, 1970, agreement, such as determination of the Kurdish districts by census and decision as to jurisdiction over Kirkuk, the capital of the northern oil-field district, were never carried out.

Because of the worsening internal political situation, however, Al Bakr promised the Kurds during the summer of 1973 that he would fulfil the agreements negotiated with them within a year. He achieved formal compliance with this promise by proclaiming the unilateraly determined autonomy statute on March 11, 1974, demanding at the same time that Barzani accept the statute within a two week deadline and lead his KDP into the so-called "National Front," consisting of the Baath and the Communist Party. Barzani rejected this demand and answered the government's ultimatum with a counter-ultimatum: should Baghdad fail to fulfil the Kurdish demands within two weeks and fail to guarantee them a share in the Iraqi oil income, he would renew hostilities against the government.

The Kurdish radio station summoned the people to mobilize. Barzani's followers withdrew into the mountains, and the Iraqi army was deployed on the Mosul-Erbil-Kirkuk line. Soon thereafter, the Iraqi armed forces began the merciless bombardment of Kurdish villages.

The Kurdish rejection of autonomy as proclaimed by Bakr is understandable if account is taken of the fact that the new statute represented a greatly shrunken version of the agreement negotiated in 1970. While the new statute also gives the Kurds local legislative and executive organs, their powers are limited in favour of the central authority of the state to the point where little more than an advisory function remains. The delimitation of the Kurdish region is likewise highly restrictive in character. Through deliberate drawing of boundaries in contentious districts in many localities, Arab majorities can easily be manipulated into existence. This is particularly true for the district of Kirkuk, where more than 80,000 Kurds were forcibly resettled for the benefit of the Arabic sector of the population. The statute no longer has anything to say about proportional participation of the Kurds in the central government, and the important command positions in the army are likewise to remain in the hands of Iraqi officers.

The Kurds, however, demand an autonomy that would permit them to manage their own affairs by themselves and to secure their region against any military encroachment by the Arabs. For this reason, they also insist on having a fixed share in the oil income of Iraq, justifying this demand by pointing out that Kirkuk is a Kurdish city and oil produced there is Kurdish oil.

The city has long been a bone of contention. During the last fifteen years Baghdad has, as already mentioned, done everything possible to replace the Kurdish tribes who live around Kirkuk with Arabs by means of resettlement measures. The clause of the agreement concluded in 1970 promising the return of the forcibly resettled Kurds to their domiciles was never fulfilled.

The political circles surrounding Barzani were also shocked by the procedure employed by President Bakr for enacting the autonomy statute. He did indeed turn to Barzani and the KDP under Barzani's leadership for advice at the beginning. But when these Kurdish leaders made difficulties for him, he summoned his own Kurdish assembly, composed principally of Barzani's former opponents belonging to the left-wing of the KDP, such as Jalal Talabani, Hashem Aziz Akrawi or Abdullah Ismael. This pliant body voted as expected for the autonomy statute elaborated by the government.

The abyss between the government on the one hand and Barzani and the KDP on the other has become too wide to be bridged. The tension has been made even more acute by the attempts of Baghdad to promote a Kurdish quisling party under Aziz Akrawi, whom the Kurds consider the most loathsome kind of renegade. Another defector, Muhi Eddin Maruf, was made Vice-President of the Republic. But the government appears to have been unable to persuade more than eighty Kurds to serve in the Baghdad-sponsored Kurdish Regional Assembly and thus provide a counterfoil to the KDP.

Kurds who remain loyal to Barzani have fled in masses out of the cities and into the mountains. According to information broadcast by the Kurdish radio more than 100 physicians, 100 jurists, 300 engineers, 500 teachers, 5,000 civil servants, 4,000 soldiers and police, the entire faculty of the Kurdish University of Sulaimaniye and 10,000 students fled into the districts controlled by the "Peshmerga" during the first few weeks. The government set a bounty of 100,000 dinars on Barzani's head and ordered extensive bombardments of the Kurdish villages, including the use of napalm. Between the onset of hostilities and the middle of May, 1974, 74 villages were razed to the ground by bombardment and numerous other villages were partly destroyed. The number of victims of these bombardments among the civilian population was 2500 dead and 4000 wounded up to the end of July, 1974, according to information released by Barzani's headquarters. The number of refugees rose to 250,000. They are presently living in caves and tents along the Iraqi-Persian border, partly on Iranian territory. The United Nations, otherwise so ready to support African guerrilla movements, left unanswered the Kurds' appeal for help that was hand-carried to Secretary-General Waldheim by two emissaries from Barzani, the former ministers Abdul Rahman and Muhsen Dizayee. Despite its official concern for peace and freedom, the international organization appears to take little interest in what happens to the victims of this war.

The Kurds are nevertheless hopeful. Barzani has under his command about 45,000 "Peshmerga" troops as well as a militia of about 60,000 with inferior training and weapons. The Iraqis have mobilized a force of 100,000 for the war and have deployed the most modern Soviet weapons, such as T-22 supersonic bombers, Zukhoi-20 bombers, a large number of helicopters, and T-62 tanks. According to Kurdish information, Soviet officers are engaged in operational tasks at the airbase near Kirkuk, from which the Iraqi Air Force launches its bombing sorties against Kurdish villages.

In an interview given to the *Washington Post*, Barzani called upon the United States to give him military as well as political support. "If the Americans protect us against the wolves in Baghdad, we can entrust the exploitation of the Kirkuk oil fields to an American company," the Kurdish leader indicated. These initiatives of Barzani however, have remained unsuccessful so far. From no quarter has he received heavy weapons comparable to those of Iraq, except for a few old anti-aircraft guns from World War II, and can only count upon a limited amount of help from the Iranian government.

The renewed outbreak of the Kurdish war in Iraq happens to be most inconvenient for the USSR. It is no secret that Russian weapons once more are being employed against a people fighting for its freedom. This compromises the USSR in the eyes of the peoples of the Third World, at a time when Moscow is actively bidding for their favour. The Russians also fear that new defeats of the Iraqi army will endanger the internal stability of the country, presently considered Moscow's closest ally in the Middle East, with the possible result of a new coup d'état in Baghdad. The Soviet government has therefore attempted to mediate, but the previous experiences of the Kurds made them distrustful of Moscow. They know that the Russians are mainly interested in Iraqi petroleum and in their strategic position on the Persian Gulf, which they obtained through their treaty of friendship with Baghdad. Two emissaries of Moscow, Boris Ponomaryov, candidate of the Politburo and Secretary of the Central Committee of the CPSU, and Rostislav Ulyanovsky, deputy chief of the International Division of the CPSU Central Committee, called on Barzani, in November, 1973, after a visit in Baghdad. Butt hey could not persuade the Kurd leader to bring the KDP into the National Front and to change his demands against the Iraqi government in the sense desired by Baghdad.

This diplomatic fiasco explains the fact that Soviet propaganda as well as communist propaganda generally has attempted to discredit Barzani ever since. The Kurd leader is now branded as "the paid lackey and mercenary of Arab reaction, the oil trusts, zionism, and imperialism." From one day to the next, the formally lionized Barzani became in Moscows an "agent of the feudal lords, an enemy of land reform, and a traitor to the Kurdish cause" who is today "supported by all reactionary forces in the Near East." In point of fact, Barzani continues to enjoy the support of the majority of Kurds – peasants, mullahs, workers, students, and intellectuals – who flocked as volunteers to his banners to fight for the freedom of their people. His objective is not the total separation of the Kurdish regions from Iraq – as contended by Bagdad and by Soviet propaganda – but to achieve Kurdish autonomy within the Iraqi state. Such autonomy must, however, not be limited unilaterally by the Iraqi government and forced upon the Kurds, but must be such as to guarantee the national rights and self-government of the Kurdish people.

As things stand today, a long and bloody war is to be expected. Experience indicates that the Iraqi army will hardly be in a position to conquer the Kurds. Its failure to do so, however, can easily lead to the overthrow of the Baghdad government. In spite of all the bloodshed and all the political and economic losses that the Kurdish war has cost Iraq in the past, the rulers in Baghdad have not yet understood that the Kurdish problem cannot be solved by force. The love of freedom of this people is stronger than the power of the weapons with which Al Bakr would choke it in blood. The fate of his predecessors is therefore likely to overtake him before long.

The Kabyls: an Oppressed Minority in North Africa

EMMANUEL SIVAN

EMMANUEL SIVAN was born in 1937 in Kfar Hahoresh in the Palestine Mandate. He studied history in Jerusalem, where he obtained his B.A. degree in 1962 and his M.A. in 1963. He then attended the Sorbonne in Paris, where he received the degree of Docteur en histoire in 1965. Since 1966, Doctor Sivan has been a member of the history department at the Hebrew University of Jerusalem, where he was promoted to Associate Professor in History in 1972 and became head of the department in 1973. Professor Sivan's publications to date include two books: *L'Islam et la Croisade:* Paris, 1968; *Communisme et Nationalisme en Algérie coloniale:* New York: Columbia University Press, in press. Among his numerous articles, the following may be mentioned as pertinent to the present study: "Le caractère sacré de Jerusalem dans l'Islam aux 12e–13e siécles," *Studia Islamica XVII*, 1967; "Réfugiés syro-palestiniens à l'époque des Croisades," *Revue des Études Islamiques XXV*, 1967; "Jihad ideology during the Crusades era," *Proceedings of the Israeli Historical Convention* (Hebrew), 1968; "Islam and the Mediterranean – a Contribution to the debate on the Pirenne Thesis," *Proceedings of the Israeli Historical Convention*, 1970; "Modern Arab Historiography of the Crusades" *Asian and African Studies VIII*, 1972; and "Slave-owner mentality and Bolshevism" *Asian and African Studies IX*, 1973.

The Kabyls: an Oppressed Minority in North Africa

EMMANUEL SIVAN

The Kabyls are a part of the Berber-speaking peoples, who are to be found throughout North Africa (the Maghreb) and especially in Morocco and Algeria. The Berbers are the ancient inhabitants of the Maghreb, established there long before the Phoenician and Roman conquests. Their racial origin is Hamaitic, or to be more precise, of a mixed Hamaitic-Semitic branch. They continued to constitute the majority of the North African population even after the seventh century Islamic conquest, for at least five centuries. Their resistance to the conquest itself was the most severe the Arab troops had ever encountered; the post-conquest process of Islamization was particularly slow and fraught with difficulties, because of the Berber attachment to their pagan particularism. And even after Islamization achieved more substantial progress, Arabization was very marginal.

In the 12th and 13th centuries an invasion of Arab nomads, the Banu Hilal and Banu Sulaym tribes, swept the whole Maghreb. The nomads, driven from upper Egypt by the Fatimid empire, destroyed much of Maghribi agriculture, decimated a sizable part of the population and ultimately settled down. This massive Arab colonization – unparalleled in scope and durability in the previous five centuries – introduced a dramatic change in the racial balance of the Maghreb, although even today it is estimated that the majority of the Algerian and Moroccan population is of Berber origin. Following the change in the racial balance, the linguistic balance began to change as well, although much more rapidly. The Berber-speakers became a minority, even before the 19th century French conquest. It should be stressed that Berber particularism is defined today in *linguistic terms*. The Berber language, unlike the Kurdish, for example, is merely a spoken language, sub-divided into various dialects, each reigning in a different area, and sometimes highly differentiated from the next. Berber dialects persisted

– as could be expected – in regions where a high concentration of members of this race remained, and particularly in almost inaccessible mountain areas into which the Arab invaders barely penetrated.

In Algeria the major Berber concentrations are:

1. The Aures mountains in the south-east.

2. Mzab on the borders of the Sahara (where Berber particularism is also given a religious dimension; the inhabitants belong to the Ibadite heterodox sect.)

3. Greater Kabylia (or the Jurjura region), bordered by the narrow coastal plain in the North, the Mitidja plain in the West, (40 kilometers east of Algiers), Wadi Summam in the East and Wadi Isser in the South.

4. Smaller Kabylia (or the Babur region) with the same North-South border as in (3), and extending from Wadi Summam in the West to the Bejaya (Bougia) area in the East.

The most acute Berber problem in present-day Algeria is that of the Kabyls. For the sake of clarity and brevity, this study concentrates on Greater Kabylia (roughly corresponding to the Tizi Ouzou district). Small Kabylia (roughly the Setif district) is geographically open to the Constantine region (North-East Algeria), which is the most arabized region in this country; hence arabization in the former area is relatively high and, moreover, the economic elements which render the Kabyl question so acute are much less dominant there. Nevertheless, a part of this analysis will apply to it as well. The same will hold for the Aures region. Although lacking territorial continuity with Greater Kabylia and in spite of the fact that it is populated by nomads (while the Kabyls are sedentary) speaking a different Berber dialect (the *Chouia*) and even known for their age-long hostility to the Kabyls, the Aures have many economic, cultural and political problems analogous to those of the Kabyls.

The Berbers represent three to four million Algerians; one-fourth to one-third of the population. Estimates vary because of lack of clear linguistic census in mixed areas. Yet it is greater Kabylia which surpasses each of the other concentrations in size and in level of cultural-political consciousness. It is, moreover, very near to the capital, Algiers, and thus perceived as potentially dangerous to the government, and hence is more systematically muzzled and oppressed.

I. The Cultural-Racial Problem

The Greater Kabylia region remained more staunchly hostile to the Arab conquerors than any other part of the Central Maghreb (called Algeria only since the 16th century). Because of the mountainous nature of its topography, the Arabs usually held only two or three garrison towns while their presence was almost nonexistent in the Kabyl hinterland. The population, living mostly in tiny villages, maintained its traditional structure, based upon tight kinship relations. The Arabs – due to no sheer coincidence – called them *Kaba'il* (tribes). Hence the more vulgar name Kabyls. The Kabyls, no less significantly, call themselves, in Berber, *Amazigh*, i.e., freemen. The political authority in each village belongs to the *jama'a* (council of clan heads) and all contacts of the Muslim government with the local population are conducted through its mediation. Even after the Kabyls had finally embraced Islam, they managed to retain much of their particularism. Worship was conducted in Arabic and in Kabyl (a rare phenomenon in Islam where Arabic is given a monopoly), the Kabyl customary law continued to exist alongside the Muslim written law *(Shari'a)*; Islamic jurisdiction was carried out by the jama'a or by the jama'a together with Muslim *kadis* (judges), in the neighbourhood of the Arab garrison towns. In cases of conflict between the Kabyl and the Muslim legal systems, Kabyl customary law had the upper hand; thus Kabyl women enjoy a higher status than Muslim females (no veil, rights of husbands for unilateral divorce narrower, etc.).

If Kabyl autonomy was preserved under Arab and later Turkish rule, it became somewhat undermined after the French conquest (1830). The French, whose physical presence in Kabylia was much more significant than that of previous rulers, did not differentiate in the 19th century between Arabs and Berbers, both because of ignorance and out of an aspiration for administrative simplification. French and Arabic alone served for contacts between the natives and their colonial masters (which brought about an extension in the use of Arabic); Muslim law was officially incorporated into the French ruling system and was imposed according to its written version (hence curtailment of the influence of Kabyl customary law); and the Kadis were made French officials (hence a decline in the authority of the jama'a.)

In the late 19th century when the French acquired a better knowledge of local problems and disposed of their assimilation policy, they began to try to move the evolution backwards. Special institutes were founded to train officials dealings with Berbers, and a good many of the prerogatives of the jama'a and of the customary law were reintroduced. Assimilation – to the extent that it continued – now concentrated on the Berbers, trying to build upon the different (and more superficial) nature of Kabyl Islam. The same *divide et impera* policy explains the preference towards Kabyls (and Berbers in general) in education, in the army and in administration.

This *volte-face* came too late. French 19th century policy – imposed in an efficient and centralistic manner quite unmatched in the past – reduced the cultural gap considerably and it was difficult to widen it again. Whatever the tenacity of the new policy, it was to a large extent offset by the rise of 19th century Algerian nationalism based upon the common denominators of territory, Islam and the *Arab language*, at least as a written and sacred language (which Kabyl surely was not.) The reformist Ulama' (jurists) movement which was the main protagonist of modern day nationalism exerted itself tirelessly to create a new consciousness, common to Arabs and Kabyls alike, and narrowly attached to the Algerian framework; their koranic (elementary) schools, madrassas (religious colleges) and sermons in mosques served at the very least to counter the French effort and maintained the cultural gap at its relatively narrower stage attained late in the 19th century (paradoxically enough thanks to colonialism!).

While preserving much of their old particularism – but without the separatist bent of yore – the Kabyls thus came to be progressively converted to Algerianhood. Being a warlike race, jealous of its freedom (cf. their name) and known for its xenophobia, it is small wonder that the Kabyls even spearheaded the nationalist struggle which passed into more and more violent stages. Kabylia was in a situation of latent terrorism from 1947 (led by Krim Belkassem) and together with the Aures it constituted one of the two spots where the Algerian War of Independence broke out in 1954; it was in those two regions that warfare was waged ceaselessly and in the most ferocious manner until 1962.

As the war further consolidated Algerian national consciousness, one could no longer speak of a Kabyl consciousness as a *separatist problem*. The Kabyls problem today is primarily that of a cultural-racial particularism *within* the Algerian framework. For Kabyl continues to serve as the major spoken language of Greater Kabylia, Kabyl folklore proves itself to be of astonishing vitality, the rural social structures remain intact (albeit their political role is secondary and unofficial) and even in the Islamic domain Kabyl customs (some of them of pagan origin) are still predominant *de facto*. At the level of political consciousness two solutions are suggested amongst the Kabyls:

1. *Particularistic concept*: aspires to endow all Algerian region (and Greater Kabylia is particular) with a certain administrative autonomy, designed to let each region foster its cultural uniqueness while preserving the central role of the Arabic language. As a matter of fact, such a tendency already existed in the pre-independence national movement in the form of the *Parti Populaire Kabyle* (PPK) which split in 1949, in the name of these principles, from the major political party led by Messali Hadj. It is quite significant that while at the time many Kabyls (including guerilla leader Krim Belkassem) denounced the PPK, as objectively serving the French, after independence was achieved, Krim Belkassem and many other Kabyl nationalists came to espouse views similar to those of the party. In the framework of the post-1962 uncontested national unity, even good patriots could now advocate such regionalistic views.

2. *Berberistic concept*: Attacks the preferential status of the Arabic language, emphasizes the fact that the Berbers were the original inhabitants of Algeria and claims that they still constitute a majority of the population. The protagonists of this concept thus call for equal status for Berber with Arabic (if not downright precedence) systematic fostering of Berber consciousness common to all Berber-speakers with their unique history, and above all – institutionalized political expression for the *fait Kabyle* and the *fait Berbère* in general (i.e., a formula modelled on the community regime in Lebanon). Significantly enough, holders of this view tend to think in overall Berber terms both because of theoretical assumptions (concerning cultural and racial community of all those who speak Berber dialects) and out of the wish to put forth a united sizable bloc *vis à vis* the dominant Arab bloc.

II. THE DEMOGRAPHIC AND ECONOMIC PROBLEM

The ethnic and racial problems are rendered more acute, in the Kabyl context, due to the territory and population data. It is estimated that the inhabitants of Greater Kabylia, including temporary immigrants, is roughly 1,300,000 (i.e., 11% of the Algerian population). About 900,000 of them reside in that region itself (area: 6000 square kilometers.) Density is thus 150 per square kilometer, the largest density in a non-urban area in all of Algeria. In some rural sub-districts in Kabylia (such as Fort National) density is even as high as 548 per square kilometer. The natural growth rate (which is slightly greater than the Algerian average of 3.2 per thousand yearly) intensifies the problem in a spiraling manner.

The resources of Greater Kabylia are however, very poor. The mountainous nature of the terrain (on average 800 to 1,200 metres above sea level, with many ridges rising to 2,000 m. and more) and the fact that it is extraordinarily compartmentalized by ridges lying crosswise to one another, account for only 30% of the soil being cultivatable (including tiny pieces of land on the slopes). But for lack of sufficient anti-erosion operations, merely a small part of the cultivatable area is congenial for the profitable crops of Algerian agriculture: citrus, wheat, vegetables, and wine grapes. Kabylia's share in the gross Algerian product in these domains is a mere 15%. Most of Kabyl's crops consist of olives (55% of the agricultural product), figs, cherries and tobacco. The topographical data and the nature of the crop result in the predominance of small private holdings (89% of the cultivatable area comprise plots of less than 100 dunams.) Hence, persistance of traditional cultivation methods, low mechanization ratio and a tendency to family production units (both with respect to manpower employed on the farm and to goals of production), 40% of the product goes for self-consumption.

Kabylia is devoid of mineral resources (except for a lead mine in Palestro). But for the existence of relatively cheap energy resources – i.e., water flowing in the ravines – much of the production capacity of the local hydro-electric stations is used for other Algerian districts, due to the paucity of industrial plants in Kabylia. Full-fledged industry is virtually non-existent. Even the olive oil plants which satisfy 43% of Algerian consumption are termed large workshops, utilizing handicraft methods. No other industries for advanced processing of agricultural products (e.g., fruit conserves) can be singled out. Under the heading of non-agricultural industry one finds next to nothing unless the few metal and construction materials workshops can be termed industrial plants. This picture of under-development is further accentuated by the poor transport infrastructure: road density less than 0.1 per square kilometer, a single railway linking Tizi Ouzou with the Algier-Annaba (Bone) mainline. This poor infrastructure is not explained by topographical difficulties alone, as many possible roads are unpaved and utilized by animal transport alone.

This combination of high population density and economic under-development results in immigration (to other parts of Algeria and abroad) on the one hand and low level of income and employment for those who remain in Kabylia. It is estimated that if all the Kabyl work force remained in the region, it could be provided with a mere 45 work-days per person per year. About 200,000 Kabyls have migrated (on a temporary or a permanent basis) to Algerian coastal towns since the end of World War II. They are prominent in towns as manual labour-ers and small tradesmen. After 1962, Kabyls took hold of many aban-doned European commercial business. Even those Kabyls who settled permanently in the coastal towns maintain close relations with their native area where many of their clansmen still live. Inter-marriage in their urban communities or marriage with present-day inhabitants of Kabylia is the predominant pattern.

Probably 270,000 Kabyls work abroad (mostly in France, but also in Germany and Belgium). This is usually a young working force (males aged 20 to 25, bachelors or married men who go without their fami-lies.) They immigrate temporarily, though when they return they may not necessarily come back to Kabylia and may well settle in the coastal towns where they may find work in the industrial professions they have worked in in Europe. The money the immigrants send to their families is not only important for Algerian balance of payments, but it also en-ables those who remain in Kabylia to balance their family budgets. This need for balancing stems from the fact that the local population, 88% of whom are employed in agriculture, maintain a very low level of income and employment – due to the nature of the terrain as well as that of the crops – and that, even after the immigration which com-prises in many villages 60% to 80% of the young work force. But even with the immigrant revenues, the average income in Kabylia is much lower than the level of agricultural areas on the coast (e.g., the Mitidja valley east of Algiers, or the Cheloff valley to the west of the capital.) Income level in Kabyl towns (including Tizi Ouzou with its 55,000 in-habitants) falls much lower than that of the coastal towns. This close proximity of a poor area to wealthy ones results in an envy and an an-tagonism which do not exist in such intensity in other poor (and Ber-ber) areas such as the Aures, situated very far from the coastal centres.

III. The Kabyl Problem in Independent Algeria

The Kabyl problem today is both the product of the historical and economic data analysed above (with the modifications introduced by the Algerian war) and of a series of new political data created in independent Algeria (some of it connected, of course, with the war years).

The Algerian War of Independence contributed, as has already been stated, to blurring the Arab-Kabyl dichotomy. The present-day Algerian identity system is based on the fatherland, Islam (as a cultural tradition but not necessarily as a religion), and on the Arab language, as the historically sacred, Koranic language, and as a counter-weight to the assimilatory power of the French language. Kabylia lost a tenth of its population during the war, developed a deep attachment to national independence, and hence, separatistic tendencies have practically completely disappeared. Even Berberistic tendencies, which reject Arab cultural superiority, have weakened somewhat. With the spread of primary education (50% of school-age children, as opposed to 10% in 1954), Kabyl children are much more intensely exposed to Arabic. Kabyl youth are in the process of becoming bi (or tri) lingual, using both Kabyl and dialectical Arabic (Maghribi style). (As a written language, French is also used to a large extent.) The cultural renaissance characterizing present-day Algeria which expresses an aspiration to strengthen national identity and cut off the links with the colonial past, engulfs Kabylia as well: the new mosques, religious colleges and convents tend to facilitate the diffusion of the Arabic language.

A. Persistence of Kabyl Particularism

Nevertheless, the aspiration to preserve Kabyl particularism (within the national framework) is strong and vital. The war years themselves largely contributed to fostering it. Kabylia was throughout the war a special command (wilaya no. 3) where all officers and men were Kabyl (the first wilaya commander had been Krim Belkassem). The Kabyl command of the A.L.N. (Armée de Libération Nationale) operated in total autonomy (because of difficulties in communication, the nature of the topography and the *quadrillage* method employed by the French). Thus, a practically closed Kabyl circuit was created, speaking its own language and operating according to its own traditions; and that, not only militarily (the foundations of which had been laid by the Kabyls in 1947, long before the FLN was established), but also in the parallel political-administrative organization founded by the F.L.N. (Front de Libération Nationale), in order to mobilize the civil population for the war effort and to cut them off from the French administration. This virtually independent organ fought, of course, for overall national independence, but it served at the same time to fortify and to accentuate the specific Kabyl characteristics. This dichotomy is expressed today in the fact that the Kabyl language is extraordinarily vital, even among the youth, while Kabyl concentrations abroad serve as a powerful catalyst for it. Though Islam is penetrating much deeper into Kabylia it is assimilating in the process many local customs of animistic origin. The rate of Arab-Kabyl intermarriage – a most significant indicator – records only a slight rise, even in coastal towns where Kabyls and Arabs at best work together, (though Kabyls tend to live in separate quarters.)

That tenacious Kabyl adherence to their sub-culture is all the more astonishing when compared to the very rigidity with which the new Algerian regime (particularly under Boumedienne) tries to enforce homogeneity and Arabization. The myth of national unity (which for the Boumedienne junta implies predominance of Arabhood in all fields and at all levels) makes it regard any kind of ethnic or cultural particularism as a mortal danger to the Algerian entity; hence, no proclivity for concessions, be they designed to preserve (let alone foster) some degree of pluralism. Kabyl particularism is the most suspect of its kind, notwithstanding the clear attachment of its proponents to Algeria and their lack of ties with France or other "neo-colonialist" countries. The inherent tendency of Arab nationalism for *Gleichschaltung* and intolerance takes no heed of these considerations. The government's ruthless drive has been extended since 1969: radio programmes in Kabyl are phased out, projects to open teaching of Berber dialects in the universities have been "frozen", the use of Arabic in elementary schools has been made exclusive and the number of such schools is pushed up.

B. Economic Deprivation of the Kabyls

Kabyl economic deprivation has at the same time been greatly intensified. Under Ben Bella (1962–1965) this was mainly the upshot of the

war which had been waged with particular ferocity in Kabylia. Tobacco growing was largely destroyed, both because of the hostilities and
more deliberately due to FLN rejection of smoking as part of the colonial culture; some of the plantations were uprooted while the situation
of others deteriorated through years of neglect; the restoration process
was painfully slow. Wine grapes suffered from a structural marketing
crisis (Algeria consumes almost no wine now that the European population is gone, and France, its old time major market, is saturated by its
own production and feels no responsibility any more for its ex-colony).
It is true that by 1965/66, restoration of plantations was completed and
agricultural production reached the 1954 level. But by then Kabylia
began to feel the results of the ambitious industrialization programme
launched by Ben Bella and put into high gear by Boumedienne and his
Minister of Industry, Belaid Abd al-Salam. This is an economic
growth strategy focusing on the build-up of heavy industry (iron and
steel), especially near Annaba-Bone, and the processing of natural gas
and oil (in the Oran districts.) This policy implies:

1. Scant governmental effort for agriculture, where Kabylia's
strength lies; only 16% of the development budget is allocated for this domain as against 48% for industry and 36% for gas and oil production.

2. The regime's developing both heavy and light industry (such as
the processing of agricultural products) which is much more suitable
for Kabylia. As heavy industry is based upon a high automation rate,
it is unlikely to absorb all of the Kabyl workforce or to decrease the
latent unemployment (i.e., low level employment of the workforce);
hence, continued need for immigration abroad.

3. Perhaps the most crucial implication: the unbalanced situation
between the hinterland and the coastal plan which even under the
French had always been in favour of the latter, has become more emphatically so with Annaba and Oran as thriving industrial centres and
Algiers as political-administrative and commercial. Kabylia, being an
underdeveloped hinterland area near the coast, feels the resulting
strain and resentment most acutely.

C. Discrimination in the Sharing of Political Power

To these two series of agelong problems a new one was added, related
to the character of the post-independence Algerian regime. To this
very day the regime must be explained to a large extent in terms of the

conditions associated with its birth during the war years: above all, by the guerilla army operating within Algerian borders and the political and diplomatic leadership and ALN (Armee de l'exterieure), located in Morocco and Tunisia, which were cut off from each other by almost impenetrable French frontier defences. The internal army always contested the political predominance of the noncombatant "external forces"; the Kabylia were first and foremost in this demand because of their heavier weight – qualitatively and quantitatively – among those fighting in Algeria itself. Power in independent Algeria fell in 1962, however, into the hands of an alliance of political exiles (the Ben Bella group) and the "external" ALN (commanded by Boumedienne), having defeated the internal wilayas commander led by the Kabyl (third) wilaya, in operations waged mostly in Kabylia itself. Boumedienne usurped power in 1965 on the basis of a group of Army staff officers who spent the war on the Moroccan border (the Oudja group). It is true that he also relied on a group of 'internal' guerilleros (the Zbiri group) but promptly got rid of this group (which, among other things, was inclined to back the particularism of the Berbers of the Aures.)

No wonder the Kabyls have the feeling that they did not obtain their due share in power proportionate to their contribution to the War of Independence. They feel, justly enough, that they played a decisive role in launching the war, did much to lead it throughout and suffered a particularly high casualty rate. Indeed, when one studies the evolution of the Algerian political elite, one perceives a continuous process of a decline of the Kabyl role. In the Provisional Revolutionary government established in 1958, they had three (out of eleven) ministers, including Krim Belkassem as Minister of Defence and Deputy Premier. But already under Ben Bella, they usually had one (or at most two) ministers, in a much enlarged cabinet, none of them holding key portfolios. The present Boumedienne cabinet holds only one Kabyl minister, the Minister of Labour Mazouzi, a very mediocre figure owing his position not to local Kabyl roots, but to the fact that he had been the faithful agent of central power in Kabylia (as FLN party boss there in 1962–64 and later as district governor). The only representative Kabyl personality ever allied with Boumedienne, Bachir Boumaza (who had been Information Minister in his first cabinet), rebelled against him in 1966 and joined the opposition in exile. The 24-member Revolutionary Command (established in 1965 as substitute for the National Assembly which had been dissolved), included three Kabyls, all ex-wilaya commanders (Mohamad Ouel-Hadj, Youssef Hatib and Salah Boubnider).

All three resigned in 1967 and joined the opposition. Perhaps more significant is the fact that under Boumedienne they had also been the only Kabyls in the secretariat of Algeria's single party, the FLN, a post they resigned at the same date. No Kabyls were appointed in their place in either institution which are now exclusively in the hands of Arabs. Among medium-level party cadres one finds no Kabyls apart from the secretary of the Kabylia district, who like Mazouzi, is a yes-man (béni – oui-oui) "planted" by the central government.

A parallel erosion of Kabyl positions was seen in other centres of power. The army after 1962 was based from the very beginning upon the units which had been kept in Morocco and Tunisia and where very few Kabyls served, and not on the "wilaya's army", where the officers corps – even outside the third wilaya, was to a large extent, Kabyl. The process was further catalysed by purges following the Kabyl insurrection in 1963, the Boumedienne putsch in 1965, and the Zbiri uprising in 1967 – all resulting in a decline in the number of Kabyl officers. The last Kabyl senior officer, Colonel Mohamed Ou El-Hadj, left the army after the Moroccan war (1964). Today one can find at most a few Kabyl majors. If one takes into account that for a warlike race like the Kabyls, the army has always been an important channel for social and economic mobility, their frustration is all the more evident.

The main centre of power in present-day Algeria is unquestionably the administration and more precisely the technocrats filling its upper echelons. Here Kabyl presence is almost unfelt. Of all the district governors (préfets) only one is of Kabyl origin; Gozali, the Tizi Ouzou governor; he is a completely Arabized Kabyl educated in the Middle East. It goes without saying that such a préfet – true to the French model copied by the ex-colonised – is deemed an inspector of the central government in Kabylia and not a regional representative *vis-à-vis* the Algiers bureaucrats. It should be stressed that the present Algerian administration is even more centralist than its French predecessors. The Boumedienne regime is not ready to acquiesce like the French (who lacked manpower); with a tenuous hold over rural areas, it is much more sensitive to the issues of unity and homogeneity and is much more suspicious of regionalist tendencies (because of the wilayas tradition). All the more so as regards Kabylia, which is so close to the capital. The centralization process comes ineluctably into a head-on collision with the particularist Kabyl tendency, a tendency which the growing under-development (and widening gap with the Arab coastal towns) only serves to render more acute.

IV. KABYL OPPOSITION

Kabyl resentment and frustration are directly reflected in the prominent role Kabyls play in the Algerian opposition movement and in the particular interest this movement gives to the Kabyl and Berber problems. This does not necessarily mean opposition organizations are Kabyl-centered. Mindful of the danger that they might be labelled as playing into the hands of colonialism by fostering separatism, such organizations usually cloak their particularistic demands with the broad mantle of "regionalism" or "pluralism", and profess (sincerely enough) their adherence to national unity. They usually try to rally non-Kabyls to their cause and to present other (and wider) criticisms of the powers-that-be: corruption, dictatorship, pseudo-socialism, etc. All of this is expecially true of the two main opposition organizations, both of which may be termed particularistic rather than Berberistic.

1. Front des Forces Socialistes (FFS): led by the Kabyl Hussein Ait-Ahmed, co-founder (with Ben Bella) of the all-Algerian military underground organization in 1949, co-founder of the embryonic CRUA (future FLN) committee in 1954, taken prisoner by the French (with other FLN leaders) after his plane was intercepted in 1956. Ait Ahmed was a delegate to the Algerian Constituent Assembly and instigated an open armed revolt in Kabylia against Ben Bella (Summer, 1963). His deputy (and military commander) was Mohamed Ou El Hadj, the last commander of the Kabyl wilaya in the war. The rebels cashed in on the bitterness reigning in Kabylia because of administrative centralization and the hostility of Arab technocrats to Berber particularism, arbitrary nationalization of petty commerce, and lack of help to the agriculture ravaged by the war. The prime professed goals of Ait Ahmed were a more democratic (and socialist-minded) regime, and an end to the Ben Bella policy of subversion against neighbouring Maghribi regimes. Yet as secondary goals he called for wide autonomy in the various regions. For all his attempts to achieve a broader appeal by playing down his Berberism, it was evident that his followers were mostly Kabyl and particularly sensitive to the latter goal. There was even an outspoken minority group inside his movement, led by Mohammed Bessaoud, which advocated the fostering of the Berber unique characteristics in general (and the Kabyl ones in particular), by resuscitating the Berber alphabet and Berber-Kabyl culture. Bessaoud never

tired of declaring his opposition to Arabhood as a predominant component of Algerian identity.

The uprising was especially strong in the heartland of Kabylia. Yet even in the heartland it was greatly handicapped by the fact that Ait Ahmed refrained from enlisting the support of the prestigious Krim Belkassem, fearing that he might overshadow him. Thus the hardcore followers of Belkassem during the underground period gave him only partial support. Ben Bella, a consummate tactician, brought the revolt to an end by utilizing the outbreak of the border warfare with Morocco, in order to attract to his side Mohamed Ou El-Hadj and many of his soldiers, in the name of the Sacred Union slogan. After they were dispatched to this war, the rebellion – hitherto thriving and tenacious – collapsed. Ait Ahmed himself was captured, condemned to death, amnestied and later fled from jail. In 1965 he established the FFS in Spain and Germany as an exile and immigrant organization with the help of Kabyl intellectuals such as Mabrouk Si Hassan and Ben Abdallah and student circles in Paris under Ali Myal. In the beginning the FFS had a firm basis in Algeria stemming from the die-hard nucleus of uncaptured rebels who were to launch guerilla warfare under Colonel Si Sadou. But Boumedienne's ruthless policing operations extirpated these remnants. Being primarily a movement of intellectuals in exile, the FFS was unable to regain its hold in the fatherland through temporary Kabyl workers abroad, indoctrinated and trained and then returned to Algeria (a method used from 1926 to 1954 by Messali Hadj, the founder of Algerian nationalism and used today by the Krim Belkassem movement). The FFS was gravely hit when King Hassan of Morocco, who had in the outset financed some of its activities, later transferred his support to Krim Belkassem. The personal rivalry between the latter and Ait Ahmed (harking back to the underground years), was a further drawback, depriving the FFS of mass proletarian support and perpetuating its high-brow character. Even after the assassination of Belkassem (v. infra), hostility between the two movements persisted and further incapacitated the Kabyl opposition in general (notwithstanding attempts of mediators such as Si Herawi.)

2. Mouvement Démocratique du Renouveau Algérian (MDRA): the youngest and the most important opposition organization, founded in October 1967 by Krim Belkassem, who up to that date, because of his bitterness at being deposed from power in 1962, was active in private business only (first in Algeria and later in Paris). His comeback to the

political scene is to be accounted for by the severe crisis the Algerian regime had been undergoing in summer-autumn 1967 (following the Zbiri uprising), which appeared to be developing the dimensions of a revolutionary crisis. His organization enjoyed King Hassan's support and a benevolent neutrality of the French authorities (quite unlike their usual hostility to Algerian opposition groups), thus expressing the French consideration that Krim Belkassem might actually take power in the nebulous situation then reigning in Algeria. The hard core of the organization was made up of people who had constituted his entourage during the war and in the first months after independence, when he was still in power: Amirat Slimane, Mourad Terbouise, Lichani, Belhassan, Akli, etc., all Kabyls.

The platform of the MDRA repeated most of the criticism of Boumedienne voiced by other opposition organizations (dictatorship, an economic policy aimed at producing spectacular projects but not at solving basic problems, lagging fight against illiteracy, etc.). Yet it was significantly silent about the more resolutely Socialist measures usually demanded by other organizations (nationalizations, self-management, progressive taxations, etc.). Belkassem also never had recourse to the Marxist terminology so dear to the FFS and to the rest of the opposition. It is to be assumed that the economic and social options of the MDRA were much more conservative, which is well in line with what is known of the personalities of its leaders.

The MDRA is very outspoken regarding the need to recognize the ethnic and cultural diversity characterizing Algeria, and to foster this diversity and grant it a *political expression*. The organization denounces the discrimination of the hinterland in general and of Kabylia and the Aures in particular, both in cultural and economic fields. If the MDRA calls for better relations with the Maghribi states, an end to subversion and even for North African unity, it justifies all that in terms of the Berberistic common denominator in North Africa, which also differentiates it from the Arab east *(Mashriq)*.

Belkassem did not succeed in taking advantage of the Zbiri 1967 uprising centered in the Berber-populated Aures, because the latter was quelled when the MDRA was still in the process of organizing. It was then that he decided to launch several spectacular operations *in Algeria*, something that other opposition movements had never tried. Their aim was both to make their presence felt and to recreate a climate of instability, and probably crisis, in Algeria, through the vicious circle of resistance – repression – intensified resistance, etc. The first operation was an assassination attempt against a key (but very unpopular) figure of the regime, Kaid Ahmed, the FLN boss and ex-finance minister. The attempt, coordinated by Lichani failed, and all the conspirators (most of whom were immigrant workers who had returned to Algeria,) were subsequently captured.

In June 1968, Amirat Slimane entered Algeria secretly, and tried to reconstruct the network, but a denunciation by one of his men led to the dismantling of the whole structure with its main protagonists sentenced to life imprisonment (1969).

Henceforward the MDRA concentrated upon long-term educative efforts among immigrant workers in France and Germany, putting its trust in the future and waiting for its hour. As the Kabyl (and Berber) problem remained acute in Algeria, and as relations with the fatherland were kept through workers coming back, temporarily and permanently, the potential danger of the MDRA for the regime was not negligible. This was also, the estimate of the regime itself. Boumedienne proposed to Belkassem a return to Algeria and a job as minister (1969), but Belkassem rejected it out of fear that a trap was being set for him. In November, 1970, Algerian secret agents assassinated him in Frankfurt. This was a heavy blow to the MDRA and to the opposition in general. Yet Mourad Terbouche soon succeeded in taking command and in reorganising the movement.

Apart from the FFS and MDRA "particularism", there exists one clear cut "Berberistic" opposition group, the Paris-based *Académie Berbère*. Its founder and leader is Abd El-Kader Rahamani, a Kabyl who had served in the French army and who became a well-known figure when he headed a group of Muslim officers, who resigned from the French army during the Algerian war to avoid having to fight their co-religionists. They were subsequently imprisoned. In their publication, *Amazigh*, Rahamani and his deputies, Imazigan and Ould Slimane call for:

1. Broad autonomy for Kabylia as well as for other Berber speaking areas (Mzab Aures),

2. Adequate representation of Berbers in the various centres of power,

3. Systematic fostering by the government of the language, customs, folklore, historiography and the particular Islamic character of the Berbers (and especially of the Kabyls),

4. Granting the right to Berbers living in predominantly Arab concentrations to preserve their specifity (e.g., in the educational system),

5. Rejection of the definition of Algerian identity as exclusively Arab.

REFERENCES

Anon., Nationalisme et provincialisme Kabyles, *Magherb*, Paris, No. 33, May–June.

M. A. Bessaoud, *Le FFS, Espoir et trahison*, Ed. Sinbad, Paris, 1962.

G. H. Bovisquet, *Les Berbères: histoire et institutions*, Paris, Presses Universitaires, 1957.

Y. Courrière, *La Guerre d'Algérie*, 4 vols., Paris, Fayard 1968–1971.

E. Gellner, & C. Micaud, eds., *Arabs & Berbers: From Tribe to Nation in North Africa*, London, Duckworth, 1973.

R. Montagne, *The Berbers, their Social & Political Organization*, London Cass, 1973.

—, *Villages et kasbas berbères*, Paris, Alcan, 1930.

—, *Les Berbères et le Makhzen*, Paris, Alcan, 1930.

J. Morizot, *L'Algérie Kabylisée*, Paris, Peyronnet, 1962.

H. Reichard, *Westlich von Mohammed, Geschick und Geschichte der Berber*, Köln, Kiepenheuer & Witson, 1967.

W. B. Quandt, *Revolution & Political leadership: Algeria 1954–1968*, M.I.T. Press, Cambridge, Mass. 1968.

Discrimination in Pakistan: National, Provincial, Tribal

ALBERT E. LEVAK

ALBERT E. LEVAK was born in Pittsburgh, Pennsylvania in 1922. After service in the military he received B.S. and M. Litt. degrees from the University of Pittsburgh. He earned a Ph.D. in Sociology and Anthropology from Michigan State University in 1954. In 1963–64 he received a Post-Doctoral Award as a Senior Fellow, Urban Studies Center, Rutgers – The State University. He was a member of the faculty of Michigan Technological University, The Ohio State University, and Mississippi State University before joining the faculty of the Department of Social Science at Michigan State University where he is now a Professor. He also serves as an associate faculty member of the Asian Studies Center. In 1958 and 1959 he devised and carried out a training programme for the faculty of the two Academies for Rural Development at Comilla and Peshawar in Pakistan. In 1960 and 1961 he served as Advisor to the Academy at Peshawar. His research interests centre on race and ethnic relations.

Professor Levak has written numerous articles and bulletins on minority peoples, poverty, and community development. The most recent is "Provincial Conflict and Nation Building in Pakistan," in W. Bell and W. Freeman, eds', *Ethnicity and Nation-Building*, Beverly Hills, California: Sage Publications, 1974, pp. 203–221. His other publications related to Pakistan include monographs and articles on Social Research for Basic Democracies; Community Development in Pakistan; Underdeveloped Areas and Community Development; and Social Research in the Formulation and Implementation of Development Plans.

Discrimination in Pakistan: National, Provincial, Tribal

ALBERT E. LEVAK

> No matter to what community he belongs, whatever his
> caste or colour or creed, he is first, second and last a citi-
> zen of this state with equal rights, privileges and obliga-
> tions.
>
> Quaid-e-Azam

I. The Founding of Pakistan[1]

Before the British had departed as colonial rulers of the Indian sub-
continent, the Muslim populace perceived a degree of intolerance and
a false sense of superiority on the part of the Hindu majority. In his
presidential inaugural address to the Muslim League in 1930, Muhammad Iqbal proposed consolidation of the Muslim northwest-Ba-
luchistan, Northwest Frontier, Punjab, and the Sind – as a single political unit within an all-India federation. But, there was little support
for this proposal among the Muslim populace. They felt the Hindu
majority had shown little respect for the rights of minorities and they
wanted to preserve the culture of the Muslim communities against the
threat of ballot-box destruction.

Muhammad Iqbal and his ally Mohammed Ali Jinnah, a lawyer-
politician, came to the conclusion that the Muslims of the sub-continent could live in greater freedom and achieve their own destiny only
in their own sovereign state. Because of the persistence of these two
leaders, the Muslim League session at Lahore formally adopted on
March 23, 1940, the goal of independence for the Muslim areas of
India:

I would like to thank Clinton J. Lockert, Bibliographer, Michigan State University
Library, Dr. Etta Abrahams and Dr. William Bridgeland for their assistance in
preparing this manuscript.

Undocumented materials presented throughout the text have appeared in many
different acceptable sources. It would be a cumbersome task and reduce readability
if all were to be cited individually. There has been a frequency and a consistency of
the items in the following sources: *Asian Almanac, Asian Recorder, The Japan Times,
The* (London) *Times, The Guardian,* The *New York Times, Newsweek, Time Magazine*
and the *Washington Post.*

Resolved ... that no constitutional plan would be workable in this country or acceptable to the Muslims unless it is designed on the following principles, *viz.*, that geographically contiguous units are demarcated into regions which should be so constituted with such territorial readjustments as may be necessary that the areas in which the Muslims are numerically in a majority, as in the north-western and eastern zones of India, should be grouped to constitute "Independent States" in which the constituent units shall be autonomous and sovereign (All India Muslim League, n.d.: 1–2).

Seven years later, on August 14, 1947, Pakistan emerged on the international scene as a new state, different from most nations of the world. First, it was unique in its geographical dimension. East Pakistan consisted of the eastern half of Bengal Province, a portion of Assam and the tribal areas of the Chittagong Hill Tracts. It was separated from West Pakistan by one thousand miles of India. West Pakistan included the Sind, the Northwest Frontier, the Baluchistan Agency, the Punjab, and the small princely states of Amb, Bahawalpur, Chitral, Dir, Kahan, Las Bela, Makran, and Swat. Certain Pathan tribes also swore allegiance to the new government. Finally, the central government was located in Karachi in West Pakistan. Eventually, with the exception of Chitral and Swat, all these areas were merged into the four Provinces of Baluchistan, the North West Frontier, the Punjab, and the Sind. These Provinces were based upon ethnic identities. Pakistan thus became a multi-ethnic nation. All groups profess the same religion and uphold the same concept of Muslim nationhood, but maintain their distinct identities. They speak different languages, and their cultures and forms of social organization likewise differ.

The new state was established on the religious faith of Islam, and without Islam there would be no Pakistan. The Islamic faith formed the ideological basis for this country. Pakistan was founded so the Muslims of the sub-continent would be enabled to order their lives in accordance with the teachings and requirements of Islam as set forth in the Holy Quran. Pakistan would use law to reform the lives of the Muslim population and would reshape the social structure so as to translate the basic postulates of Islam into behavioural norms (Hoebel, 1965:43).

Islam is more than a religion in the Western sense; it is a total scheme for living and fails to accept the distinction between the secular and the religious spheres.

There is no hesitation regarding the validity of the universal truth of the Quran; and all followers accept its injunctions as permanent and unalterable. For virtually all Pakistanis Islam is the political doctrine of the nation. Nevertheless, from the first expression of concern in the 1930 Muslim League meeting and the final establishment of the State in 1947, there seldom has been any agreement on the practical application of Islam in the arena of politics and government. There has been a failure to achieve consensus on exactly what the words mean and imply. Meanings once accepted can change with changing cultural context through time. As Hoebel (1965:46) concludes:

Given the fact that there are significant basic differences between the Sunni and Shia Muslim sects and their associated laws in Pakistan, plus the fact that Sunni law has been interpreted and expanded by four major schools of Islamic jurisprudence, dating from the great period of Muslim expansion during the centuries immediately following Muhammed, it is clear that the finding of any authoritative body of Islamic postulates from the doctrinal literature is hopeless.

Along this line he reports a conversation with a Pakistani who said, "...Everyone has his say as to what is and what is not Islamic. And indeed there are fundamentalists, maybe two thousand ulema, who have no modern education but considerable influence among the people. Which view will prevail? Who can say?" (Hoebel, 1965:54)

This strong emotional tie to Islam and the freedom of interpretation poses the problems for present day Pakistan. Mohammed Ali Jinnah, Quaid-e-Azam (the Great Leader), had learned the meaning of liberal democracy and sovereignty of the people from the British. His image of Pakistan envisioned a modern democratic state, not a theocracy. From his perspective religion was a private affair. In the past twenty-eight years of Pakistan's existence, however, his image of Pakistan has been distorted by religious leaders and ignored by some politicians.

II. PAKISTANI POLITICS

The first twenty-eight years of Pakistan were not without serious problems. Jinnah, father of the nation, died eleven months after its establishment. Liaquat Ali Khan, the first Prime Minister, was assassinated in 1951. There were the continuing dispute with India over the Kashmir problem and the war with India in 1965. On eight occasions between 1947 and 1958, emergency rule or martial law was proclaimed in all or parts of Pakistan. In addition there have been at least five constitutions,[2] including the present one. The number of constitutions may not be too significant because some of those in effect earlier were as legal and as valid as the present one (Zafar, 1974:xxii). What has been significant has been the distribution of power both on the geographic plane as well as on the representative plane.

A. West Pakistan: The One Unit Scheme

Pakistan began with five Provinces, all of which were originally established on the basis of ethnic identity. In an attempt to bring unity to the nation, the Establishment of West Pakistan Act was passed on October 14, 1955. This Act consolidated the four Provinces of West Pakistan into one unit by governmental fiat. Those responsible defended the One Unit on doctrinal grounds, claiming that Islam stood for integration. There was immediate and continuing opposition to the One Unit plan – particularly from the leaders of the Sind, Baluchistan, and the North West Frontier. When the concept of the One Unit was being reconsidered in 1970, strong voices were heard on both sides of the issue. An editorial in a national newspaper published in Lahore declared: "...regionalism is repulsive to the idea of Pakistan nationhood. That is why regionalism must be held as severe and serious a threat to the Pakistan nationhood as atheistic socialism" (*The Pakistan Times*, 1970). Another paper published in the North West Frontier made the point, "We are now faced not only with the normal type of political differences between our different political leaders, but also with a fast and increasing religious fanaticism and intolerance generating hatred and antagonism between the different sections of our people" (*Peshawar Times*, 1970).

After a long seige of martial law, President Yahya Khan felt the nation should return to democracy, which had been an important force in its establishment. To this end he issued the Legal Framework Order, 1970, which was more appropriately described as the Constitutional Framework Order. He dissolved the One Unit scheme and re-established the five Provinces.

B. East Pakistan

Pakistan government officials conceded that East Pakistan, with approximately 58 percent of the total population, had been consistently discriminated against in the political, economic, and social spheres for the first twenty-three years of Pakistan's existence. A review of the events of those years makes it obvious that the West wing dominated the government bureaucracy, the industrial and commercial community, and the military to the disadvantage of the East wing. The East Pakistanis were aware of their position, but they did not revolt. They were tied to the Government by the ideology of Islam. But, they sought greater autonomy through the political process under the leadership of Sheikh Mujibur Rahman, the leader of the Awami League. (Levak, 1971:210-211)

The actions of the President through his issuance of the Legal Framework Order, 1970, and his desire to return to democracy and his willingness to allocate seats in the National Assembly on a one-man-one-vote basis was encouraging to the East Pakistanis. The first universal adult franchise election since 1954 was held on December 7, 1970. The Awami League, which was to the right of centre, won 167 of the 169 seats allocated to East Pakistan in the National Assembly. On the other hand, the Pakistan People's Party led by Zulfikar Ali Bhutto, which was to the left of centre, won only 82 seats of the 144 allocated to the West.

Immediately following the election there was optimism in East Pakistan. Its leaders had gained national power through the democratic process. Since the Awami League victory centered on the Six Points[3] advocated by Mujib there was much less optimism in the West wing. Since the new National Assembly would be responsible for the development of a new constitution, President Yahya Khan and Bhutto, advocates of a strong central government, preferred to negotiate.[4] These negotiations continued through March 25, 1971, when the Army moved into East Pakistan in a surprise attack to destroy the Awami League and eliminate the cultural and political leaders of East Bengal. Neither the President nor the military understood the depth of the ideological commitment to the concept of Bengali nationalism on the part of the East Pakistan populace. The fact that in Indian West Bengal people with similar language and cultural characteristics, even though Hindu in religious beliefs, supported the resistance of East Pakistan, demonstrated the strength of ethnicity over religion alone as a binding force. (Levak, 1971:214)

On April 17, 1971, three weeks after the initial attack, East Pakistan was named Bangladesh (Bengal Nation). The revolution raged on for nine months. On December 3, 1971, the Indian army became involved. And on December 17 the final surrender documents were signed when Pakistan accepted a cease fire. This resulted in the dismemberment of a Moslem nation. In its stead emerged Bangladesh, and a new Pakistan which would require reformation.

C. *Reforming Pakistan*

What is to be learned from the lesson of East Pakistan? An obvious conclusion is that religion alone is not a sufficiently cohesive force to hold a nation together. However, because the Muslims were able to gain their freedom from domination by the Hindus they felt they could organize themselves into a model Islamic state. Even though this failed in East Pakistan, the idea of Islam still evokes a strong emotional response. Today the problem facing the new Pakistan is the lasting integration of the constituent peoples – Baluchis, Pathans, Punjabis, and Sindhis. Each of these reflect differing economic, political, linguistic, cultural, and societal interests. To integrate these constituent peoples into a nation cannot be achieved by emotional appeals to Islamic doctrine alone.

III. Ethnicity and Nation-Building

Rupert Emerson (1960:27) discusses nationalism's role in creating pressures on the lives of peoples who have been traditionally bound by the norms of their ethnic identity. He referred to Pakistan's split from India where "...even the assertion that there is a single nation embracing the peoples concerned may be successfully denied." And the Center for International Studies (1960:203), in reviewing social, economic, and political change in developing countries, came to the conclusion that in achieving independence old ethnic, linguistic, religious or tribal loyalties come to the fore once again. "This had been particularly true wherever the new national leadership has been recruited largely from a single region or ethnic group..." The dissolution of the One Unit by the Establishment of West Pakistan Act coupled with the secession of East Pakistan has been followed by a sharp rise in parochial and provincial bitterness over linguistic differences, competition for resources, and distribution of power.

Weiner (1972:64) points out that public policy is affected by patterns of social organization in plural societies. While he enumerates four patterns, two may be appropriate for Pakistan. First, a country in which a single group is dominant in authority and in numbers. Second, any combination in which one or more minorities cross international boundaries. While presumably all four provinces are of equal stature, the Punjab appears to contain the dominant ethnic group, while the others are in a subordinate position.

A. The Predominance of the Punjab

The Punjab has always been a major source of food supply and has the world's largest irrigation system. It also has the highest degree of urbanization in Pakistan. It has played a significant role in Pakistani development. The leadership of the Muslim League came from this area. The Muslim League formally adopted the goal of independence here. Lahore, the principal city of the Punjab, is known for its educational institutions – Punjab University, the Civil Service Academy, the Administrative Staff College and other schools are located here. It is a centre of journalism and literature, a cultural centre, a commercial centre, and a headquarters for government.

This concentration of modernized elements has affected the perception of the Punjabi's role in the development of Pakistan. At the time of the dissolution of the One Unit rule it was reported (*Peshawar Times*, 1970) that seventy prominent members of the High Court Bar Association of Lahore, including some former Presidents of the Association and some retired High Court judges resolved:

Whereas the disintegration of the One Unit has been brought about at the behest and desire of the people of units (other than the Punjab), it is just and fair that the servants who belong to other regions should be reverted to their respective units irrespective of the fact whether the posts or assignments equivalent to their ranks and statuses are available or not.

The Punjab which has in the past made tremendous contributions and sacrifices for the collective overall interest of the country should not be made to suffer and its exchequer burdened with expenditures of officials who do not belong to it.

The constituents of Baluchistan, the North West Frontier, and the Sind have long resented the dominance of the Punjab. There is a threat to the authority of the Punjab on the basis of a political coalition of these Provinces. In the election of Bhutto to the Presidency, only two of the 82 votes he received were from outside the Sind and Punjab. These two were from the North West Frontier in an area peripheral to the Punjab. In addition, at the time of the negotiations with East Pakistan the representatives of the North West Frontier and Baluchistan supported Sheikh Mujib in his struggle for provincial autonomy.

In regard to the second point, i.e., minorities cutting across international boundaries, the most troublesome politically have been the Pathans of the North West Frontier and the Baluchis. While each of these Provinces has a principal city, large proportions of their population live under forms of tribal organization. In both instances these tribal organizations cross international boundaries into Iran and Afghanistan.

B. Public Policy and National Integration

There are essentially two public policy approaches for achieving national integration: (1) the destruction of distinctive cultural traits of the minority community and the assimilation of this community into some version of a national culture; or, (2) the establishment of national loyalty without destroying the minority sub-culture; that is, the policy of unity in diversity. In practice, the national leadership usually adopts elements of both policies (Weiner, 1972:64). The leadership may, however, lean more toward one approach than the other. In Pakistan, Bhutto's major problem has been how to maintain centre-provincial relations despite his preference for a strong central government. This problem focusses on two of the provinces in particular – Baluchistan and the North West Frontier. How to keep them in the union without having to share his authority and power? Some of his actions appear to be arbitrary and thus discriminatory. In addition, other forms of discrimination are evident as one views the actions of the Provinces and the tribal organizations towards members of their populations.

IV. BALUCHISTAN AND THE NORTH WEST FRONTIER

Since the establishment of Pakistan, these two Provinces have been advocating provincial autonomy. Only a few weeks after the secession of

East Pakistan the issue of separatism began to come to the fore. This issue may, however, be stirred up more by forces external to the boundaries of Pakistan than by the leadership of either Province. The new regime in Afghanistan is not sympathetic to Pakistan. According to one source (*Economist*, 1973:47) Afghanistan is renewing its demands that the Baluchis and the Pathans have a right to a new political arrangement "in accordance with their hopes and aspirations." It is believed that Afghanistan is looking for an outlet to the sea. As Syed reports (*Orbis*, 1973:971) Pakistani officials have referred to evidence that demonstrates the role of the Soviets in attempting to build secessionist movements in Baluchistan and the North West Frontier.

A. *The Pathans*

Most of the people living in the North West Frontier are Pathans (Pashtuns, Pakthuns, Afghans). They are a large self-aware group inhabiting both Pakistan and Afghanistan. At the time of Partition the Pathans were voicing their demand for a "Pakhtoonistan" as a reflection of their ethnic identity. The key argument for the creation of a "Pakhtoonistan" is on an ethnic plane. The Pathans feel they are different but equal in regard to all other Pakistanis and should have a separate homeland.

While there are those who advocate secession it is unlikely that it will become a major movement. As Barth (1969:117) points out, the Pathans are "generally organized in a segmentary, replicating social system without centralized institutions." The fact that they are a large population extended over ecologically diverse areas and in different regions in contact with different cultures makes organizing them for secession a problematical affair. Nevertheless, the leadership can be expected constantly to seek greater autonomy but within the framework of the Constitution. While there has been some violence it appears they will continue to pursue greater autonomy primarily through non-violent means.

B. *Baluchistan*

The Province of Baluchistan as a geographical area consists of vast ranges of mountains, huge expanses of desert, and rich plains in need of irrigation systems. The mountainous region, because it is in a virtually inaccessible place and at a great distance from Pakistan's major cultural centres, has remained almost completely isolated and out of reach of modern civilization. The total length of surfaced roads in this largest Province is less than 2,000 miles. In addition, the systems of communication, particularly in the hinterlands, are quite limited. As a consequence, life in the interior is very difficult and the people live at a subsistence level raising sheep and tilling small parcels of land.

A large portion of the Baluchistan Province is tribal in character. There are more than a dozen main tribes in Baluchistan, and they have a number of clan subdivisions. Some of these clans extend into the southeastern part of Iran. The three major tribes of Baluchistan are the Bugti, Marri, and Mengal. It is estimated that there are 300,000 Baluchis in this area.

As in any tribal society loyalties are to the family, lineage, caste and ethnic group rather than to the nation as a whole. These Baluchistan tribal groups are not inclined to regard their parochial orientations as being inconsistent with allegiance to the nation-state. They believe that so long as they remain Islamic, they are patriotic. They fail to distinguish between religion and politics.

While the Pathans are also organized into a tribal society there is a difference between them and the Baluchis. The Baluchi tribes are based on a contract of political submission of commoners under sub-chiefs and chiefs. (Pehrson, 1966:27) The Baluchi tribes are centrally led and as a consequence are more capable of pursuing long range strategies. The Pathans, on the other hand, are organized through fusion and *ad hoc* councils. In recent years the Baluchi tribes have been moving northward and extending their ethnic boundaries. As a result there is a flow of individuals from Pathan groups to Baluchi groups (Barth, 1969: 124).

In the past the Government has been reasonably conciliatory toward tribal groups because it was aware of the depth and resiliency of tribal roots. Early in the British Indian government rule it was recognized that to deny the tradition of tribal independence could well create problems of frontier unrest. This cultural sensitivity appears to be lacking in modern day Pakistan.

Political Discrimination. Jinnah said (Isa, 1973:i) on February 14, 1948 while addressing the Darbar at Sibi:[5]

The history of political reforms for Baluchistan is connected with the history of the struggle of the Mussulmans for freedom ... and now that I have the honour to be the Governor-General of our great country – Pakistan, it is natural that the question of reforms and securing for the people of Baluchistan an adequate say in the administration and governance of this province should be constantly in my mind. ...

Twenty-five years later, President Bhutto, by Presidential Rule on February 15, 1973, dissolved the elected coalition provincial government of Baluchistan and ordered the dismissal of the governors of Baluchistan and the North West Frontier. Both governors were members of the National Awami Party (NAP). Prior to this ouster there had been incidents of violence and lawlessness by the tribesmen. Government troops had been sent into the Province by central Government. The official reason given for the ouster of the Provincial government was the failure of the Awami's to put down the large scale violence in Baluchistan.

Mr. Bugti Akbar, leader of one of the most powerful tribes in Baluchistan, was appointed Governor. However, his appointment appeared to be tokenism; an attempt to appease the tribal dissidents since Mr. Bugti was a tribal leader. Because he was not transferred to the post of Chief Minister, which is more powerful than the governorship under the new Constitution, he resigned, effective January 1, 1974.

In addition to the removal from office of the elected NAP officials, hundreds of Party workers were also jailed as were some of the elected personnel. Another element of dissonance involved the communications media. A number of daily and weekly newspapers were shut down, and a number of editors were removed or jailed for failing to obey the order for precensorship of news and editorials.[6]

In a democracy the voice of the people is heard through their elected representatives and the perspective of the electorate is affected by differing viewpoints expressed in a free press. In the area of public policy it appeared as though President Bhutto opted to move in the direction of destroying distinctive cultural traits of the minority by imposing considerable restraints on the National Awami Party. This is particularly evident when one considers that the NAP has been the dominant political party in Baluchistan as well as the North West Frontier.

Economic Discrimination. There is no denying that Baluchistan has been denied an equitable share of the programmes for economic development. The British placed an emphasis on creating infrastructural facilities for their own strategic purposes. The single sheet budgets of the Agency of Baluchistan indicated, on an average, an insignificant annual outlay of Rs one crore[7] and a half for non-development purposes, social overheads, and so-called development projects.

The first meaningful attempt to establish development programmes was made after creation of the One Unit. These development funds were insufficient because of the lack of a field organization. On the average the expenditure for development in Baluchistan for the period 1960–61 to 1969–70 was only Rs. 3.50 crores. Even though this was a small amount, only 50 percent was used and the remainder was returned. The above-average annual development allocation was achieved primarily because, during the latter part of that decade, the Governor of West Pakistan was from Baluchistan. As a consequence, higher allocations were made to Baluchistan during his term of office. (Khan, 1973:iii)

During 1970–71, the first year of the creation of the Province of Baluchistan, a budget of Rs. 7.00 crores was set aside for development purposes. For 1971–72 it was Rs. 12 crores; and, for 1972–73 it was Rs. 40 crores. In addition to the constant increase another positive act was absolving Baluchistan of its debt of Rs. 39.63 crores to the Federal Government on April 7, 1973. At the same time a debt of Rs. 67.92 crores owed by the North West Frontier was removed (*The Pakistan Times*, 1973: 10).

For the first time, development programmes were aimed at improving the lot of the masses. But as the Governor of Baluchistan commented: "We cannot be too optimistic. The past neglect has left Baluchistan far behind in the race for development." (Khan, 1973:IV)

V. The Problem of Stratification

A. Inequality in the Allocation of Resources

Past economic planning in Pakistan has been characterized by discrimination. Maddison (1972:136–137) discusses this social policy of "functional inequality" at some length. Those favouring this policy argue that a high degree of inequality is necessary in the early stages of capitalist development. This inequality is required to promote savings and create entrepreneurial dynamism. This can be determined by the following statement from the Second Five-Year Plan:

Direct taxes cannot be made more progressive without affecting the incentives to work and save. The tax system should take full account of the needs of capital formation. It will be necessary to tolerate some initial growth in income inequalities to reach high levels of saving and investment. What is undesirable is a wide disparity in consumption levels. Tax policy should, therefore, be so oriented as to direct a large part of high incomes into saving and investment rather than consumption.

The Third Plan claims, "What is basic to Islamic Socialism is the creation of equal opportunities for all rather than equal distribution of wealth." Maddison also reports a statement from the Fourth Five-Year Plan which reiterates the old creed with some doubt. It states, "We cannot distribute poverty. Growth is vital before income distribution can improve." He refers to the conflict between the necessary economic dynamism and social justice, and possibly some second thoughts about the policy position. Maddison shows scepticism about the proposition that social justice could be better served at a later stage of development when the economy is richer. The possibilities of usefully redistributing assets at a later stage, he asserts, actually diminishes.

Griffin (1965:603–613) also made the observation that the economic planning was consciously designed to skew the distribution of increased income in favour of those already wealthy. Papanek, an advisor to the Pakistan Planning Commission, writes (Papanek, 1967:241–243) that these inequalities were directly related to the pattern of Pakistan's development. He goes on to say, "The problem of inequality exists, but its importance must be put in perspective... the inequalities contribute to the growth of the economy, which makes possible a real improvement for the lowest income groups... Great inequalities were necessary in order to create industry and industrialists..." And, finally, another author (MacEwan, 1971) writing on Pakistan's development states, "...in my opinion, one cannot speak of planning as independent of social organization."

There were twenty-two families in control of the industry and commerce in Pakistan. If, according to the Plans, the money was diverted to the development of industry, these families should have been involved. Yet when Bhutto became President of Pakistan one of his first moves was to threaten these twenty-two families with prosecution unless they returned the foreign exchange held abroad.

Twenty years of economic planning have taken place and yet Baluchistan has the lowest per capita income of the four Provinces. Is Baluchistan now receiving increased development funds because of

economic planning or is it because of Baluchi leaders prolonged opposition to the existing system? As Braibanti (1974:32) suggests, "...it
may not be demonstrable for many years to come, that the question of
cultural sensitivity is probably just as important as the question of economics in holding a state together."

B. Social Stratification

While no research on ethnic stratification has been carried out in Pakistan, conventional wisdom clearly points to the Punjab as being the
dominant Province. Baluchistan, because of its form of tribal organization and distance from the major centres of Pakistan, would probably
be lowest on the scale.

This difference can result in discrimination as is made apparent in an
editorial (*Peshawar Times*, 1970) which speaks of the non-locals settled
in either Baluchistan or the North West Frontier on a permanent or
temporary basis. They choose to live in urban areas and make no attempt to integrate, learn the native language, or cultural mannerisms.
They prefer to live as superiors, a privileged elite; they look down on
the "natives" and according to the concept of "colonialists" appropriate privileges which rightfully belong to the natives.

The same editorial describes a condition in Pakistan where there has
been a wholesale granting of scholarships and seats in professional colleges to outsiders – scholarships and seats which were meant for the
locals. The editorial also speaks of the same practice occurring in re-
cruitments for the services, especially the superior services.

As in most societies of the world, Pakistan's social structure contains
a clearly defined system of social stratification. Such a system ranges
from caste to class, depending upon the awareness and acceptance of
the system by the populace, the degree of rigidity, and the degree of
distinctiveness. Saghir Ahmad (1965: 82–83) examining class and
power in a Punjabi village, reviewed the literature on class and caste in
the subcontinent and found the exponents of caste to have three basic
positions. First, those scholars who believe that the genesis of caste is to
be found in Islam (Levy, 1960). Second, those scholars having a position similar to the first position, by holding that caste among Indian
Muslims is both a product of history and cultural contacts with Hindus; this position has been reinforced by Islam (Weber, 1946; Ansari,
1960; O'Malley, 1932; Karim, 1956; Slocum, 1959). The third cate-

gory of scholars define caste structurally and support the position that stratification, at least in certain regions of Pakistan, is amenable to such an analysis (Barth, 1960; Bailey, 1964; Berrman, 1960; Ibbetson, 1916).

Culturally, caste has been viewed as a phenomenon associated with Hinduism; structurally it has been perceived of as an extreme form of class rigidity. As Honigman (1962:54) has written, "Anyone familiar with Islam will understand... that Muslims do not employ the term 'caste' with the meaning it possesses for the Hindus." Rather it refers to a social bond which links people who have the same surname and essentially the same rank. As Pakistan moves through the process of modernization, some of the rigidity will disappear. Nevertheless, modernization has not yet reached the tribal regions of the nation.

In the stratification system of the Marri tribe the major division lines correspond to significant distinctions in the political system. A convenient index to this overall system of rank has been expressed in monetary terms. For those who have been wrongfully killed there is a traditional compensation paid to the rightful heirs. The Marri tribal council has defined a scale of blood compensation for men (Pehrson, 1966:28) as follows:

	Rupees
Sardar, or other member of dynastic lineage (Bahawalanzai)	8,000
Waderas, mukadams, motabar-e-mard (other prominent men)	7,000–4,000
Kaum-e-mard (Marri commoners)	2,000
Seyyal (Pathan, Brahui, Baluch, tribesman)	2,000
Kamin kum-asil (Pathan serfs, Lori, Domb, and Jatt)	1,000
Mareta (slaves)	1,000

For the ordinary Marri woman the rate is Rs. 1000, and for those of higher rank the council establishes the rate in each instance.

Pehrson (1966:30) discusses the distinction between these groups in some detail. All of these rankings are dependent upon their value to the society. The difference between the commoners and the despised groups can be readily observed in daily social intercourse. Some of the Pathans are in the Seyyal category which allows them to be viewed as unrelated equals, while other Pathans in the kamin/kum-asil category are despised. Others in this same group, the Lori, Dob, and Jatt, are essentially the same but distinguished from other groups. The Loris and Dombs are treated "as animals" and subjected to crude jesting.

C. The Persistence of Slavery

There is also a discussion of the role of slavery as an aspect of Baluchi social organization. Even though slavery may not have been too important among the Marris, it was not until 1952 that it was abolished. Among the Marri there are two origins of the slaves. There were the Hazaras who were captured during the conquest of Hazarat in the nineteenth century. They were sold to the Marris by Afghan nomads. The other group is the Mareta. According to local reports they are the descendants of prisoners taken by Mahmoud of Ghazni about 1000 A.D. and turned over to the Baluchi because of the military support given in the campaign.

The history of slavery in Baluchistan and its official abolishment is rather confusing. The Baluchs in southeastern Persia also had slaves but slavery there was abolished in 1929. However, in 1939 the Baluchi sardars of the nomads of the Bent Oasis still had two thousand slaves. (Central Asian Review, 1930:306)

Matheson (1967:43) claims that most of the slaves of the Bugti tribe descended from Mahrattas captured in battle when Emperor Humayun fought to regain Delhi. Mir Chakur, a Rind chieftain, received forty thousand prisoners for his aid to the Emperor. These prisoners were distributed as slaves to his kinsmen and warriors who were to eventually found the Bugti and Marri tribes.

There is also a reference to slaves brought by African traders to the Mekran ports of Gwadur, Pusne and Jiwani. Some of the descendants of these slaves are still in those areas and are known as Habshis. The Baluchistan Gazetteer Series (1907:188) makes reference to slaves originally imported by the Meds from the African coast. It suggests that the theory of the German scholar Glaser that the original home of the Habash was in southeast Arabia appears correct, because this portion of the population is one of the oldest in the county and in many instances their features are distinctly Negritic in character.

Matheson(1967:43) reports that the Khan of Kalat officially declared the slaves free in 1928. This decree was largely ignored except in the few towns and trading centers. In 1948, the Nawab Akbar Khan once again officially freed the slaves. Until this time they had been treated as absolute chattels. But upon their release the chieftain gave over a small portion of land received from Government, and also decreed that services performed by the Mrattas had to be paid for. As in Kalat, the Bugtis in remote areas ignored the decree.

While the slaves are now free, legislation does not erase centuries of social discrimination even though some minor modifications may occur. At Dera Bugti the Mrattas live outside the town away from the main gates in poorly built huts. They continue to live in these segregated areas because they have no skills to sell on the market place that might aid them to rise above their present status.

Even though the Mrattas are free, they are not on equal footing. For example, while they were still slaves, any Bugti male could seduce a Mratta woman and her husband dared not complain. Today, now that the Mrattas are free, in the same situation the husband is officially *supposed* to beat the Bugti male, however, he is not to kill him. He is to be without fear of reprisal and able to claim compensation. The normative structure changes when a pure-blooded Bugti seduces the wife of another Bugti. He and the wife may be killed for adultery and the offended husband receive a fine or a short term in prison.

The continuing relations between the freed slaves and their former masters still reflects the indisputable control by these masters. They will continue to be negative to these former slaves to "keep them in their place." The limited, and somewhat confused, materials available on this problem of slavery stand as a block to positive action. Systematic research needs to be done and action taken to provide these "slaves" an opportunity to improve their status in life. Even under optimum conditions it will take generations to remove the stigma of slavery.

VI. Linguistic and Religious Discrimination

A. Linguistic Discrimination

Language has been a controversial issue in the development of Pakistan. There are at least thirty-two distinct languages and a multitude of dialects. How much can be done by governmental action to alter this linguistic melange?[8] Jinnah, in 1948, speaking at a convocation at Dacca University stated, "There can be only one lingua franca... and that language should be Urdu" (Mahmood, 1970:10). This was met with disagreement by the populace of what was then the province of East Pakistan. The issue was resolved by allowing Bengali to be the official language of that Province and Urdu the official language of West Pakistan.

Urdu is the official national language of present day Pakistan even though only about ten percent of the population speak it. English, which remains from the British days, is a medium of communication for those at the higher levels of society. It is a prestige language shared by only a small portion of the population. Nevertheless, the 1962 Constitution specified that English may be used as the official language through 1972, and it continues through today.

Language differences may not break up Pakistan but they can serve as a divisive factor. While Urdu has been designated as the national language and three Provinces have made it their provincial language, such was not the case in Sind Province. On July 7, 1972 the Sind Provincial Assembly adopted Sindhi as the official language of the Province. The action of the Legislature caused a conflict between the "old" Sindhis and the "new" Sindhis. Riots and demonstrations took place in Karachi because half of its four million people spoke Urdu but not Sindhi. These were primarily the refugees from Uttar Pradesh and Bihar who felt the lack of skill in the Sindhi language would be discriminatory to them when it came to seeking occupations, particularly in the government service. The rioting spread throughout the Province and was brought to a halt when the military were brought in and a curfew imposed.

President Bhutto met with leaders of different groups including the Provincial Assembly. On July 17, 1972 an Ordinance was passed which provided that the Language Act would not be carried out in a manner detrimental to the use of Urdu. It provided twelve years for the Urdu-speaking population to learn the Sindhi language, and, during this period no person would be discriminated against in matters of appointment, promotion, and continuance in government service only on the basis that he did not know Urdu or Sindhi.

The Sindhi conflict may have contributed to the substance of Article 251 of the 1973 Constitution which states:

(1) The National language of Pakistan is Urdu, and arrangements shall be made for its being used for official and other purposes within fifteen years from the commencing day. (2) Subject to clause (1), the English language may be used for official purposes until arrangements are made for its replacement by Urdu. (3) Without prejudice to the status of the national language, a Provincial Assembly may by law prescribe measures for the teaching, promotion and use of a provincial language in addition to the national language.

Resolution of the language problem in Pakistan is not a simple task, however. Indications are that unless changes are made in the educational system such as the introduction of a common language in textbooks, dictionaries, and examinations it is unlikely that Pakistan will be linguistically unified in the near future.

B. Religious Discrimination

Believers in Islam having suffered discrimination on religious grounds became tolerant of other religions as expressed in Article 20 of the 1973 Constitution:

Subject to law, public order and morality, – (a) every citizen shall have the right to profess, practise and propagate his religion; and (b) every religious denomination and every sect thereof shall have the right to establish, maintain and manage its religious institutions.

Essentially, the same principle which appears in this Article has been part of the Constitutions of the past. Yet, in 1953 violent religious disturbances occurred in Lahore. The plundering, burning and killing lasted four days and was only halted by military action.

The 1953 riots were the result of over half a century of bitterness and resentment by orthodox Muslims. The violence was deliberately and consciously instigated by a combination of Muslim religious groups and militant political organizations. These groups had issued an ultimatum to the Prime Minister of Pakistan threatening to use force if the Qadiani Ahmadis were not declared a non-Muslim minority. The Prime Minister refused and violence followed (Vujica, 1964:16).

The Ahmadiyya movement was founded over one-hundred years ago by Mirza Ghulam Ahmad. He claimed to have received a divine revelation and viewed himself as the Promised Messiah sent to lead the Muslims as their Prophet of Allah. It is this claim to prophethood which the orthodox Muslims find so objectionable.

The Ahmadiyya movement is committed to spreading the word of Islam throughout the world.[9] It has a college located at Rabwah which trains missionaries, and a publication called *Review of Religions* which is published in a number of languages.

Religious violence involving the Ahmadis erupted once again in the summer of 1974. The violence began when nearly 200 medical students, while passing through Rabwah on a train, allegedly shouted anti-Qadiani slogans. The sect members attacked the students and the orthodox Muslims retaliated the next day. Rioting spread through the four Provinces but was most evident in the Punjab. At least 30 persons were killed, more than 200 were injured, and over 1,000 arrested. To reduce tensions Government imposed a ban on detailed reporting of the riots and restricted distribution of official information.

As a result of these riots the combined Opposition in the Parliament insisted that the Government should declare the Ahmadayyi sect a non-Muslim minority, remove all its members from influential positions, and arrest one of its leaders.

On September 7, 1974, twenty-one years after the first Anti-Ahmadiyya riots, the National Assembly and the Senate meeting separately adopted a resolution and a constitutional amendment declaring the Ahmadis to be a non-Muslim minority. This action allowed them minority representation in provincial assemblies. It also barred them from becoming President or Prime Minister. And, possibly of greatest importance, they were barred from preaching their faith in Pakistan.

It is possible that this action against the Ahmadis was not merely on religious grounds. The sect, which numbers approximately half a million in Pakistan, is a well-disciplined and close-knit community which has its own mosques and practices a rigid adherence to endogamy. The men occupy influential economic and social positions in the country far beyond the numerical strength of the sect. It is possible that economic and social reasons have also been contributing forces to the constant friction between the communities.

VII. Conclusions

Pakistan is a complex multi-ethnic society ranging from one ethnic region with a relatively high degree of urbanization to one characterized by tribal organization. The nation was founded on a firm belief in the Islamic faith, which the population assumed would serve as a cohesive force in unifying the ethnic groups. The secession of East Pakistan should demonstrate that religion alone was not sufficient. The Bengali culture, in the face of repeated discriminatory action on the part of West Pakistan, was a strong force in over-riding the presumed cohesiveness of religion. Nonetheless, in Pakistan today there are still leaders who feel that the ethnic identities can be overcome by shouting slogans of Islam. No lesson was learned from the secession of East Pakistan.

The ethnic differences in Pakistan are real and are manifest. The leadership of the society must become sensitive to this wide variety of value systems extant in the society, since it is unrealistic to assume that any one ethnic group will submit quietly to political and economic exploitation by another ethnic group for any extended length of time. Those in power must use all institutions, not simply religion, to minimize conflict within the boundaries of the state.

As contended above, there are intense intergroup problems which need to be resolved. It must be remembered that an individual's social identity is composed both of civil attributes and primordial ties. Civil attributes are earned, changeable, and perhaps impermanent like occupation and class. Primordial ties however, are the givens of the individual's social existence. These ties are the facts from his parentage: religion; language; and historical culture. When there is an unequal distribution of resources along the lines of language, religion, or cultural categories national cohesion may be at stake. Primordial ties carve up nations. The same process which split Pakistan from India also split East Pakistan from Pakistan.[10]

To resolve these problems requires not military action, as was taken in East Pakistan, but perseverance, tolerance, patience and, above all, understanding. What is needed, moreover, is a sensitivity to ethnic differences among men. Pakistan has a Constitution which recognizes the need for these attributes, but how that Constitution is implemented is another issue.

The developing countries seek aid from the Western world. Most of this aid is in the form of technological or military assistance. Very little is sought to provide better insights into the intercultural relationships within their societies. How to deal with the diverse cultures remains an area of ignorance, prejudice, and superstition in many of these countries. Leadership is needed that has the knowledge to understand cultural diversity.

This knowledge need not be acquired in a formal sense. An examination of other societies and their history of intergroup relations would be of value. The United States began with a relatively homogeneous population of English and northern European stock. It was therefore, Anglo culture which prevailed. As new immigrants came in (most of them were of similar stock)they were expected to put aside their cultural heritage and subscribe to English cultural traditions. This practice raised few questions until the time of industrialization when immigrants from all parts of the world came to the United States. They settled in enclaves in cities where they maintained their native cultures and found it difficult to become Anglicized. This was primarily because of differences in language, in religion, and in physical appearance which made assimilation a difficult process. A practice which came to be called the "melting-pot" was therefore initiated. This assumed a biological merger on the parts of the members of the various ethnic groups with the children producing a new culture. However, this also failed to work. Now, after World War II, the United States is engaging in legislation (e.g., Civil Rights of 1964) and action which supports the basic principles of cultural pluralism (i.e., an emphasis on common national values but without denial of the ethnic value system). The United States may have minimized the intergroup conflict which has been part of its history by adopting the policy of cultural pluralism. Pakistan can learn from this experience and encourage a degree of homogeneity and diversity by emphasizing the contributions from all its ethnic groups.

Earlier in the article mention was made of a policy approach for achieving national integration. One alternative might be the destruction of distinctive cultural traits of the minority community. This goal could be pursued through the control and censorship of the communications media, by the use of military action to resolve issues, or by ignoring those who have been elected by the people.

According to McCord (19:141), "The authoritarian believes that only a single party can create a sense of nationhood and unity which he regards as necessary during a period of modernization. Autonomous organizations or political parties, each articulating its own goals, are dispensable". An article in the *New York Times* (1975:E5) stated: "The Government of Pakistan has outlawed its main political opposition, the National Awami Party, ostensibly because of terrorist activity... The ban raises questions about Prime Minister Bhutto's commitment to democracy". Can the emphasis on the political institutions be any more effective in reducing discrimination and encouraging cultural pluralism than was the emphasis on the religious institution? Economic, political and social democracy is a prerequisite to the development of a condition of cultural pluralism in a nation.

References

Ahmad, S. S. (1965), *Class and Power in the Punjabi Village*, Michigan State University. (Unpublished Ph.D. dissertation.)

All-India Muslim League (n.d.), Resolutions of the all-India Muslim League from October 1937 to March 1940, Delhi, India.

Ansari, G. (1959–60), "Muslim castes in Uttar Pradesh: A study of cultural contact," *Eastern Anthropologist* 13 (December): 1–83.

Bailey, F. G. (1964), "Closed social stratification in India." *Archives européennes de Sociologie* IV: 107–124.

Baluchistan District Gazeteer Series (1907), Bombay: 7.

Barth, F. (1960), "The system of social stratification in Swat, North-West Pakistan," in E. R. Leach (ed.), *Aspects of Caste in South India, Ceylon and North-West Pakistan*, Cambridge: Cambridge University Press.

— (ed.) (1969), "Pathan identity and its maintenance," *Ethnic Groups and Boundaries*, Boston: Little, Brown & Company.

Berrman, G. D. (1960), "Caste in India and the United States," *American Journal of Sociology* 64: 120–127.

Braibanti, R. (1974), "Pakistan's experiment in political development," *Asia*, Supplement Number 1 (Fall): 25–42.

Center for International Studies, Massachusetts Institute of Technology, (1960) "Patterns and problems of political development," United States *Foreign Policy: Compilation of Studies*, Vol. II; Committee on Foreign Relations, United States Senate; Washington, D.C.: Government Printing Office; Chapter VI: 1203–1210.

Central Asian Review (1960), "The Baluchi's of Pakistan and Persia," 8: 299–309.

Das Gupta, F. (1970), *Language Conflict and National Development*, Berkeley: University of California Press.

Economist, The (1973), "The tribesmen are restless," (March 17): 42–44.

Emerson, R. (1960), "Nationalism and political development," *Journal of Politics* 22 (February): 3–28.

Evans, H. (1971), "The language problem in multi-national states: the case of India and Pakistan," *Asian Affairs* II, Part II (June): 180–189.

Griffin, K. B. (1965), "Financing development plans in Pakistan," *Pakistan Development Review* (Winter): 603–613.

Hoebel, E. (1965), "Fundamental cultural postulates in judicial lawmaking in Pakistan," *American Anthropologist* Vol. 67, No. 6, Part 2. L. Nader (ed.), *The Ethnography of Law*, 43–53.

Honigman, J. (1962), "Education and career specialization in a West Pakistan village."

Ibbetson, D. (1916), *Punjab Castes*, Lahore: Government Printing Press.

Isa, Q. M. (1973), "Baluchistan's judicial system: need for reform," *Dawn* (supplement) (August 8): I.

Karim, N. (1956), *Changing Society in India and Pakistan: A Study in Social Change and Social Stratification*, Dacca: Oxford University Press.

Khan, S. H. (1973), "Economic development: a review," *Dawn* (supplement) (August 8): III.

Levak, A. E. (1971), "Provincial conflict and nation-building in Pakistan," in W. Bell and W. Freeman (eds.), *Ethnicity and Nation-Building: Comparative, International, and Historical Perspectives*, Beverly Hills: Sage Publications, 203–221.

Levy, R. (1960), *The Social Structure of Islam*, Cambridge: Cambridge University Press.

MacEwan, A. (1971), *Development Alternatives in Pakistan*, Cambridge, Mass.: Harvard University Press.

Maddison, A. (1972), *Class Structure and Economic Growth*, New York: Norton.

Mahmood, S. (1970), "The problem of national integration," *The Pakistan Times* (May 17): 10.

Matheson, S. A. (1967), *The Tigers of Baluchistan*, London: Arthur Baker Ltd.

McCord, W. M. (1972), "The case for pluralism," in F. Tachau, *The Developing Nations: What Path to Modernization?*, New York: Dodd, Mead & Co.

New York Times (1975), "Bhutto stills opposition." (February 16): E5.

O'Malley, L. S. S. (1946), "Indian caste systems," in J. H. Hutton (ed.), *Caste in India*, Cambridge: Cambridge University Press.

The Pakistan Times (1973), "Rs. 107 Crore debt of Baluchistan and NWFP written off," (April 8).

— (1970) Editorial, (September 17, 1970): 6.

Papanek, G. F. (1967), *Pakistan's Development: Social Goals and Private Incentives*, Cambridge, Mass.: Harvard University Press.

Pehrson, R. N. (1966), *The Social Organization of the Marri Baluch* (compiled and analyzed from his notes by Fredrik Barth), Chicago: Aldine.

The Peshawar Times (1970), "Our new Provincial government," (April 11): 4.

— (1970), "Regionalism," Editorial, (September 19): 4.

— (1970), "Apostles of Hate," Editorial, (November 21): 4.

Razvi, M. (1971), *The Frontiers of Pakistan*, Karachi: National Publishing House.

Slocum, W. (1959), *Village Life in Lahore District*, Lahore: Social Sciences Research Center, Punjab University.

Syed, A. (1973), "Pakistan's security problems: a bill of constraints," Orbis. Vol. 16 No. 4 (Winter): 952–974.

Vujica, S. M. (1964), "The Ahmadiyya Movement in Islam," *Eastern World*, Vol. 15 (December): 16–20.

Weber, M. (1946), *Essays from Max Weber*, Hans Gerth and C. W. Mills (eds.), New York: Oxford University Press.

Weiner, M. (1972), "Political integration and political development," in F. Tachau, *The Developing Nations: What Path to Modernization?*, New York: Dodd, Mead & Co., 62–75.

Zafar, S. M. (1974), "Reflections on the Constitution," in Q. Ahmad, *The Constitution of the Islamic Republic of Pakistan*, Karachi: East and West Publishing Co.

NOTES

1. For a detailed historical statement of the evolution of Pakistan see Razvi, 13–44.
2. The five "constitutions" were as follows: The Indian Independence Act of 1947 as amended from time to time by the Constituent Assembly of Pakistan; the Constitution of 1956; the Constitution of 1962; the Interim Constitution of 1972; and finally, the Constitution of 1973.
3. The Six Points, in summary, are: (1) Establishment of a federation on the basis of the Lahore Resolution; (2) federal government to deal only with defence and foreign affairs; (3) two separate but freely convertible currencies, or one currency provided that effective constitutional provisions made to stop flight of capital from East to West Pakistan. Separate banking reserves and separate fiscal and monetary policy for East Pakistan; (4) denial to Central Government of the right of taxation; (5) foreign exchange handled separately by the two Wings; (6) set up of a military or para-military force by East Pakistan.
4. President Yahya Khan said he would accept and authenticate the new constitution only if it upheld the directive principles envisaged in the Order. If not, he threatened to dissolve the National Assembly and call for new elections. He felt the immediate introduction of the Six Points into the new Constitution would create chaotic conditions.
5. Sibi is a mud-walled town of 1500 population in the deserts of Baluchistan. Travel by express train from Karachi is 22 hours.
6. From December 1973 through 1974 the following occurred: The editor, publisher, and printer of the Opposition daily *Sachchai* in Quetta were arrested for disobeying the Order for pre-censorship of news and editorial comments. The paper was banned. Two other Opposition papers, the *Sangat* weekly and the *Al-Watan*, were forced to cease publication. The editor of *Jasrat* of the Opposition Jamaat-i-Islami party was arrested and the paper was banned for two months. The editor-in-chief of *Dawn*, a Karachi newspaper was removed for writing articles on India and Bangladesh. The editors of two weeklies, *Nade Sind* and *Naya Zamana*, were arrested under the emergency law. The *Frontier Journal* was closed down for two months.
7. Crore = Rs. 10,000,000.
8. For a brief but incisive discussion of the language problems in Pakistan see Das Gupta, 1970: 26–28; also see Evans, 180–189.
9. Mission centres have been established in the United States, England, France, Italy, Spain, Netherlands, Germany, Norway, Switzerland, Trinidad, Brazil, Costa Rica, Sri Lanka, Burma, the Malay States, Philippines, Iran, Iraq, Saudi Arabia, Syria, Egypt, Zanzibar, Natal, Sierra Leone, Gold Coast, Nigeria, Morocco and Mauritius.
10. Based on memo from Dr. Kenneth David, Department of Anthropology, Michigan State University, East Lansing, Michigan.

Inter-Ethnic Conflict in Africa

W. J. BREYTENBACH

Dr. W. J. BREYTENBACH is at present Senior Researcher at the Africa Institute of South Africa. Prior to joining the Africa Institute he lectured in Anthropology and Applied Anthropology at the University of Pretoria from where he obtained an M.A. degree in Anthropology. He received his doctorate in the faculty of Arts and Philosophy (African Government) from the University of South Africa. He has done research and fieldwork in a number of African states and Black Homelands, mainly in Botswana, Lesotho, Swaziland, Rhodesia, KwaZulu, Bophuthatswana, Venda, Lebowa and Kavango. His books include: *Migratory Labour Arrangements in Southern Africa*, Communications, African Institute, Pretoria, 1972; *Bantoetuislande: Verkiesings en Politieke Partye*, Mededeling, Afrika-Instituut, Pretoria, 1974; *Crocodiles and Commoners: Continuity and Change in the Rulemaking Systems of Lesotho*, Communications, Africa Institute, Pretoria, 1975; *The Black Worker of South Africa* (in collaboration with Dr. G. M. E. Leistner). Africa Institute, Pretoria, 1975. He has also written a series of articles, "Party Political and Aristocratic Rule in Swaziland," *Africanus*, Vol. 2., No. 2, 1972; "Recent Elections and the Political Parties in the Homelands," *South African Journal of African Affairs*, Vol. 4., No. 1, 1974; "Negritude and Black Consciousness in Africa," *Bulletin of the Africa Institute*, Vol. XIII, No. 9, 1973; "Recent Political Developments in Lesotho," *Ibid.*, Vol. XIII, No. 10, 1973; "Federalism in Black Africa," *Ibid.*, Vol. XIV, No. 3, 1974; "Party Politics and Traditionalism in Botswana," *Ibid.*, Vol. XV, No. 1, 1975.

Inter-Ethnic Conflict in Africa

W. J. BREYTENBACH

I. THEORETICAL FRAMEWORK

The purpose of this article is to present a general analysis of the phenomenon of "ethnicity" in Africa. Particular attention will be paid to the concepts of "social distance" and "differentiation" as manifested in occurrences of ethnically orientated power struggles, domination, suppression and the formation of diverse voluntary associations and interest aggregations in African countries.

The term ethnicity is broadly synonomous with the term tribalism. This is because no distinction is made between the terms "ethnic group" and "tribe".[1] According to Wallerstein these terms imply the existence of loyalties at various levels. These are loyalties to the family, the tribal community, the tribal government and the tribal chief.[2] There are, however, two other factors that should be stressed, namely the roles of culture and locally based interests.[3] As such the concept of ethnicity assumes certain particularistic[4] characteristics which could be prevalent in both rural and urban areas.[5]

The concept of ethnicity is of special importance in the study of contact situations in Africa, i.e. the contact between various ethnic, social and cultural groups – each with their own locally based interests. This condition basically derives from colonial times when the territorial boundaries of Africa's colonies (which became the geopolitical bases of the new states) were drawn arbitrarily regardless of Africa's indigenous divisions. Heterogeneous groups were therefore accommodated within the same political systems. This plurality in African states was furthermore enhanced by the secondary processes of culture contact and culture change as accomplished by urbanization, industrialization, migration, christianization and education. These secondary processes tended to create further (secular) divisions among Africa's modern societies. These divisions will however not be analyzed in this article, except in so far as they are relevant to ethnicity and "retribalization", "supertribalization" and "cultural revival"[6] in extra-traditional environments such as urban areas. It should be stressed that ethnic loyalties are not necessarily eroded by processes of social and cultural change. Theories of modernization based on Weberian premises assumed that ethnicity and change were mutually exclusive. This is not the case. Colin Legum, for instance, wrote that ". . . tribalism is not inherently anti-modern. Tribalism must be distinguished from traditionalism. Traditional systems may pass away while tribal affiliations remain strongly entrenched in defence of ethnocentric interests."[7] Modernization and change should therefore not be equated with detribalization. Wallerstein is correct when he asserts that "often what a writer means by detribalization is simply a decline in chiefly authority. It does not follow that an individual who is no longer loyal to his chief has rejected as well the tribe as the community to which he owes certain duties and from which he expects a certain security."[8] One could therefore anticipate the rise of *modern* ethnic nationalisms and loyalties in Africa. And this could lead to new ". . . scrambles for opportunity and the struggle to get one tribe or group into power".[9]

II. ETHNICITY AND SOCIAL DISTANCE

The concept of social distance was formed by Bogardus who defined social distance as ". . . the degree of sympathetic understanding that exists between a person and a person, a person and a group, or a group and a group".[10] Banton elaborated on this definition (which was first used by Bogardus in 1924) and stated that sympathy referred to feeling reactions of a favourable responsive type, whereas understanding involved knowledge of a person which also leads to favourably responsive behaviour.[11]

The first study of this nature ever conducted in Africa was done by Clyde Mitchell in 1956.[12] Other scientists who applied the same test were Henry Lever[13] and Melville Edelstein.[14] The results of the Lever tests will not be discussed because his research was not done on the African or Black population which is the prime concern of this article, but only on the White section of the South African population. The Edelstein study is more relevant because it makes use of a Bogardus-type social distance test. Edelstein tried to ascertain the attitudes of urban Blacks in Soweto towards 14 ethnic groups of which eight were African. The results of this study illuminated interethnic relationships and preferences.

Clyde Mitchell conducted two studies in Zambia. The first study concerned the African respondents' feeling of social distance, that is "sympathetic understanding", from a specified set of ethnic categories and the second concerned the ethnic composition of aggregates of men living together in "single-quarter" accommodation in towns situated along the north-south railway line in Zambia.[15]

The first test included questions on whether respondents would willingly admit members from arbitrarily specified ethnic groups into close kinship by marriage, a village, tribal area, work situation or allow strangers as visitors to home areas. The results of this survey enabled Mitchell (and also Epstein) to construct a distance-ordering of ethnic categories from the point of view from any particular ethnic group. The results showed *inter alia* that tribalism or ethnicity is more than mere regionalism because it is inherently characterized by cultural distinctions.[16] Mitchell refined these findings in his second survey when he determined the behaviour of single males as far as associations with those whom they identified as belonging to specific ethnic categories were concerned.

The pattern that emerged indicated that members of various tribes clustered together on account of their regional and cultural affinities, for instance, western and eastern tribes formed separate clusters due to their particularistic bonds and interests. Those results suggested that ethnic factors played significant roles in the formation of voluntary associations among heterogenous groups, and voluntary associations normally tend to be specifically organized for the pursuit of special interests, be they economic or political, etc.

These assumptions are borne out by the tests conducted by Edelstein. He tested the attitudes of African pupils in terms of eight different social distance variables, one of those being the effect of the ethnic backgrounds of respondents on social distance.[17] He reported that African respondents associated more readily with fellow Africans than they did with Indians, Afrikaners, Jews, Coloureds, and South African English. Furthermore, members of a particular ethnic group associated more readily with members of the same ethnic group, for instance, Zulu respondents associated more readily with Zulus, and Tswana with Tswana, Pedi with Pedis and Basotho with Basothos, etc.

The results of the Mitchell and Edelstein tests are important. They corresponded with the conclusions drawn by many other writers on African affairs. Kenneth Little, for instance, pointed out that tribally-orientated associations rated among the most important voluntary associations in West Africa.[18] These tribal associations ranged from small groups consisting of a few members of the same extended family or clan to much larger groups such as the Ibo State Union which is a collection of village and clan groups. The Ibo were among the first ethnic groups who established these urban associations, the purpose of which was to provide members with mutual aid, sympathy and financial assistance in the case of illness. These associations tried to keep interests alive in tribal song, history, language, etc. These tribal associations were also of practical significance since they assisted the government in the collection of taxes,[19] articulated local grievances and also formed administrative units in the towns. This was the case in former French territories in particular, where these non-traditional administrative units served as the cores of urban governmental authorities. The French term for these governmental units/authorities is *association d'originaires*.

III. ETHNICITY AND POWER STRUGGLES

It has been pointed out above that ethnicity not only plays an important role in the determining of sympathetic understanding between groups, but also exerts a significant influence on the formation of non-traditional voluntary associations. As mentioned above, voluntary associations are normally organized for the pursuit of special interests and these interest associations quite often tend to pursue conflicting goals and incompatible ends. This mainly applies to power struggles in the economical and political fields. Examples in this respect will be drawn from Nigeria, Uganda, Zambia, Ruanda, Burundi, Sudan, Mauritania, Kenya and Chad.

The segmentation of labour movements is a case in point as far as power struggles in economic fields are concerned. Trade unions are not as well developed as political parties in Africa. The most developed trade unions in Black Africa are found in Zambia, Kenya and Ghana. It is interesting to point out that trade union movements were often used as political tools during colonial times. The best examples of politicized labour unions were those in Ghana and Kenya. Labour action in those two countries went hand in hand with political aspirations, and labour discontent therefore formed inherent ingredients of the rising tides of local nationalism. The Trade Union Congress of Ghana was, for instance, taken over by Dr. Nkrumah's Convention Peoples Party.[20] As such geopolitical nationalist aspirations superseded locally-based loyalties with the consequence that ethnicity did not play important roles in the formation and direction of labour movements in these two territories.

However, in those cases where labour movements became separated from political protests, such as in Zambia, ethnic factors tended to play a much greater role in determining industrial relations. This is borne out by the strike on the Copperbelt during 1935–1940 when the reasons for striking were not political but economical,[21] and were led by a particular ethnic group, the Bembas. The trade union movement had not eliminated tribal allegiances within the industrial field, but rather tended to create further avenues for competition between various tribes, noteably the Bemba and the Lozi.[22] The Bemba tended to dominate the trade unions, whereas the Lozi tended to monopolize administrative offices. Other tribes such as the Nyakuza and Lunda resisted both Bemba and Lozi leadership on the Copperbelt and pursued secessionist policies. It is therefore clear that ethnic loyalties were of paramount importance in the ordering of labour relations in the copper mines of Zambia.

Ethnicity, however, played an even greater role in the formation of political associations and parties. This factor often threatened the stability of the social and political orders of the relevant African states.

A well-known pioneering example of the secularization of ethnic loyalties was the formation of the *Egbe Omo Odudwa*-movement (Association of the Children of Odudwa) among the Yoruba in Western Nigeria. This movement aimed at the emergence of a modern Yoruba nationalism and as such laid the foundations for the establishment of an almost exclusive Yoruba political party, the Action Group (AG), led by Chief Awolowo. In fact all the other political parties in Nigeria had their own distinctive ethnic support bases. Membership of the Northern Peoples Congress (NPC) was predominantly Hausa-Fulani and that of the National Convention for Nigerian Citizens (NCNC) predominantly Ibo.

The Action Group formed the opposition in the Federal Parliament, where the first independence government was formed by a pact between the N.P.C. and the N.C.N.C. Sir Abubakar Tafewa Balewa of the N.P.C. became executive Prime Minister and Dr. Nnamdi Azikiwe of the N.C.N.C. Governor-General (and, later, President.)

The Action Group, under Chief Obafemi Awolowo, proved to be a fairly effective opposition party and so in 1962 the Federal Government, led by Sir Abubakar, sought to crush the A.G. The strategy employed was to carve a new region – Nigeria's fourth – to be known as the Mid-West Region from the Western Region and thus split the Yoruba forces and their political strength in the A.G. The Action Group fought the establishment of the Mid-West Region tooth and nail but its forces were weakened by an internal dispute between Awolowo, the Federal leader, and Chief S.L. Akintola, the Western Regional Premier and by the fact that Awolowo and his deputy, Chief Anthony Enahoro, were jailed for treasonable felony. The opposition to the creation of the Mid-West Region collapsed and the new Region was born in July 1963. But the upheaval in the Western Region led to a noticeable increase of regional tensions and an intensification of tribal feelings. The later secession of the Ibo dominated Eastern Region which declared itself the independent Biafra was crushed militarily.

Power struggles are, however, not necessarily fought on strict tribal lines, i.e. on the basis of tribal exclusiveness, but members of various tribes may cluster and/or associate on the basis of regional and cultural bonds. Such "ethnic" alliances are formed for express political purposes. The best examples are the Kakwa, Lugbara and the Mardi tribes in Uganda who form the present ruling alliance in that country. The current President, Idi Amin, is a member of the Kakwa tribe. Their main opponents in Uganda are the Acholi and Lango tribes who were in power during Mr. Obote's regime, the former President who was ousted by General Amin. At present these two tribes are virtually excluded from any participation in central governmental processes. The explanation offered for this situation in Uganda is that, with the exception of the Baganda, there are no really powerful tribes. The Baganda, however, is not strong enough to dominate the rest. It is therefore imperative for the large number of smaller tribes to form these alliances in order to obtain power. This phenomenon can only be understood properly in terms of the Baganda's aspirations for supremacy. This has, however, changed during the times of President Milton Obote who was a member of the Lango tribe and who crushed the Baganda's aspirations. He subsequently established a ruling alliance between his own Lango and the Acholi. It would therefore seem that Uganda's ethnic alliances rotate according to the ethnic allegiance of its national leader.[23] Although Baganda aspirations have been crushed, they maintained their ethnic autonomy and still pursue separationist policies which could play an important role in the future stability and balance of power in Uganda.

Similar delicate ethnic relationships exist in Kenya where a large number of tribes have not yet reached the stage of compromise or even coalescence as far as national integration is concerned. The most powerful of these ethnic groups are undoubtedly the Kikuyu and the Luo. It is particularly noteworthy that it was the Kikuyu who took the lead in respect of the political decolonization of Kenya. Kenyan independence was therefore not the result of a comprehensive Kenyan nationalism but rather a Kikuyu nationalism. Since then, the Luo has also "awakened" politically, and the outcome of Kikuyu-Luo relationships will in all probability determine Kenya's future leadership. President Kenyatta is a Kikuyu who made overt attempts to accommodate members of other tribes in central political institutions. He tried to appease Luo and also Kalenjin demands in particular. It is interesting to note that the Kikuyu and Luo never were neighbours during traditional times. Their first contacts occurred during colonial times and this inter-ethnic contact tended to intensify their ethnic consciousness and awareness rather than lessen it. The same could be said of the Bemba and the Lozi in Zambia who had very little, if any, contact during pre-colonial times. But once they met on the Copperbelt they engaged in intense power struggles as mentioned above.

IV. Reciprocal Genocide: The Case of Ruanda and Burundi

Probably the most intense and even violent example of inter-ethnic contacts and power struggles along ethnic lines, is that of Ruanda and Burundi.

The population of this region is divided into two main sections: The Tutsi, an exceptionally tall people of Hamitic origin, cattle-owning and former ruling class comprise approx. 17 percent of the population of Ruanda and approx. 13 percent of that of Burundi; and the Hutu (ethnically Bantu) who are agriculturalists. In addition to these two ethnic groups there are a small number of the Twa, or pigmies, comprising about one percent of the population.

The racial composition of Ruanda and Burundi is one of the chief sources of social and political tension in this area. Unlike the former Belgian Congo, where tribal loyalties have produced demands for independent tribal states or for a loose form of federation, the situation in Ruanda and Burundi is one in which the traditional rule by a tribal minority, the Tutsi, over the rest of the population, the Hutu, has increasingly been opposed by the latter. The Tutsi invaded the area before the 15th century and, after subjugating the Hutu, established a highly organised political, social and economic hierarchy, at the apex of which was the Mwami or king. In this order the position of the Hutu was similar to that of the serfs in mediaeval Europe. Although the situation has changed to some extent in recent years owing to reforms introduced since 1954 traditional concepts and institutions continue to determine the behavioural patterns of a large part of the population.

The period of German rule from 1897 to 1916 was relatively peaceful, with the German authorities ruling the country through the customary institutions of the Tutsi. After the first World War the Belgians pursued a similar policy of indirect rule. However, the ruling Tutsi were deprived of a number of arbitrary powers and the hierarchical system of government was simplified.

In 1946 Ruanda-Burundi became a trust territory of the United Nations under Belgian administration. The 1950's were marked by growing opposition of the Hutu to indirect (and Tutsi) rule and the gradual introduction by the Belgians of partially democratic and non-customary institutions.

The late 1950's saw the emergence of the first political parties in Ruanda and Burundi. In the former the two more important parties were the *Parti du Mouvement de l'Emancipation Hutu* (Parmehutu), which, under its leader, Mr. Kayibanda, became the chief party of the Hutu, representing the ideals of social equality, democratisation of institutions and independence, and the *Union Nationale Ruandaise* (UNAR), which claimed to speak on behalf of all Ruandese and stood for the traditional institutions, internal self-government in 1960 and independence in 1962.

In Burundi the difference between the parties was less clearly defined. The two chief parties were the *Parti de l'Unite et du Progres National du Burundi* (UPRONA) and the *Parti Democrate Chretien du Burundi* (PDC). Both parties maintained loyalty towards the Tutsi monarchy, but while the latter favoured the democratisation of the institutions before independence, the former had called for immediate independence and was in favour of a more than nominal role for the Tutsi Mwami in the new state.

The tension between the Tutsi and the Hutu in Ruanda led to the outbreak of violence in November 1959, in the course of which some hundreds of people were killed and many thousands of huts were burnt. Many Tutsi chiefs and sub-chiefs were killed or driven from their posts, and in many cases the Belgian administration appointed Hutu to fill the vacancies until municipal elections could be held.

In Ruanda the elections resulted in a smashing victory for PARME-HUTU, which won 70 percent of the seats, and Gregoire Kayibanda (a Hutu) became Prime Minister of the provisional government. In Burundi a coalition led by PDC won the communal elections in November and December 1960 and a provisional government led by the Tutsi was formed shortly afterwards.

The Hutu therefore became the rulers in Ruanda where they are following a deliberate policy of genocide against the Tutsi. In Burundi, however, the Tutsi consolidated their position and gained modern political power through UPRONA and also engaged in large-scale massacres of Hutu tribesmen who constitute 85 percent of Burundi's population. These massacres started in 1960 and have, since then, been periodically executed.

V. Two Versions of the Arab-Negro Race War

Problems not unlike those of Ruanda and Burundi exist in the Sudan where, at least until very recently, one ethnic group has dominated and suppressed another. This is manifested in the so-called "North-South race war" which is constantly waged between the Arabs of the north and the Negroes of the south. As in the cases of Kenya and Zambia this "race war" is caused by inter-ethnic contact and consequent competition which did not prevail during pre-colonial times but resulted from recent power struggles along ethnic lines. This war began in the Equatoria Province in 1955 when a southerner, William Deng, of the Sudan African National Union rebelled against the north.

The racial war has continued unceasingly since its outbreak in 1955. In the initial rebellion, 1,000 people were massacred as civilians joined southern troops and police in an onslaught on northerners. In September 1963, there was an intensification of the war, when the northern military authorities launched an all-out campaign to suppress the rebellion. They were unsuccessful, and turned their wrath on the clergy, who were openly persecuted before eventually being deported.

Renewed rebellion broke out. Southern guerilla groups formed an interest association called Anyanya (a dialect name for a fatal poison) and began attacking government posts. Government reprisals on villagers soon followed, and this pattern has continued ever since, unaffected by the overthrow of the army regime in 1964 and the counter coup of 1969.

Southern leaders have on many occasions pressed for a constitutional solution to the problem through some kind of federal structure that would loosen the Arab stranglehold on government. With the rejection of all such suggestions many have accepted separation as the only answer. The present government led by Gaafar Nemeiri, who seized power in May 1969, has officially offered local autonomy for the south within the framework of a united "socialist" Sudan, but only after the rebellion had been crushed.[24] (After this article was completed an agreement between the contending parties was reached at Addis Ababa, providing substantial autonomy for the Southern Region as well as a general amnesty for all who had rebelled against the central government. – Editors).

The Sudanese Arab/Moslem domination of Negro peoples is not an exception. The same pattern exists in the Islamic Republic of Mauritania.

The name Mauritania derives from the old province of the Roman Empire which included all North-West Africa and, as its name implies, is a land of the Moors. The present area is much smaller with Spanish Sahara to the North and Mali and Senegal to the South. Most of it is arid desert, although the southern region in the area of the Senegal River has more fertile soil. The majority of the population are nomadic Moors but there is a Negro minority in the South.

The Moors are of Arab-Berber origin and are related to the Tuaregs of the central Sahara. Their history may be traced back to the eleventh century A.D., when the Lemtuna shepherds were brought back to the orthodox Islam faith by a chief of the Tuareg Sanhaya, Emir Yahya ibn Ibrahim.

As early as the fifteenth century, however, the Moors had contacts with the West – first the Portuguese, then the Dutch and French. France's presence in Mauritania dates back to the middle of the nineteenth century. Faidherbe and, later, Coppolani were the first French agents. France gradually achieved administrative unity while respecting the Moors' spiritual tradition and basic social structure. Mauritania was made a French protectorate in 1903, and a colony in 1920. It became independent in 1960. And the Negroes who live in the southern parts of the territory still do not share any significant political power with the Islamic rulers of the north.

Although Negro domination by Moslems is the general pattern in Africa, there are some exceptions to this rule. Chad is an example of a country in which Moslems of Arab and Berber descent are governed by people of Negro origin. The 14th century empire of Wadai was situated along the eastern frontier of the present-day Chad. However, unlike the Islam-orientated empires of Bornu and the Hausa states to the west, the Wadai empire disappeared long before the arrival of Whites in Africa. This is probably one of the reasons for the lack of a current Moslem tradition of superiority in this part of Black Africa.

The present government leader in Chad is Francois Tombalbaye, a Black Protestant from the south.

Moslems in Chad are today fighting for what they term their "national identity". They maintain that they do not want secession but merely recognition as a separate cultural group. Prior to independence, and at independence in 1960, there had been co-operation between the Moslems and Negroes and several Moslems were included in Tombalbaye's first cabinets. The Negro group (and, especially, the Sara tribe of which Tombalbaye is a member) gradually strengthened its position. But in 1963 Tombalbaye instituted a "purification" campaign during which most of the leading Moslems in government service were arrested. In 1965 another massive campaign was launched. For the northern Moslems this was the last straw. They launched a revolt, forming the *Front de liberation nationale* (FROLINAT) in June 1966.

In addition to FROLINAT, there is a smaller "liberation movement", a regional movement operating in the Biltine Province which seeks the overthrow of the Tombalbaye regime but is opposed to the centralism of FROLINAT. Relations between the two anti-Tombalbaye movements are poor. FROLINAT, in any event, claims national support for its resistance movement. Its leadership was considerably strengthened in 1967 when Dr. Abba Siddick, a one-time friend of Tombalbaye and co-founder of the governing party, joined FROLINAT and assumed the leadership.

Armed Moslem revolt began in the Ouaddai Province, the eastern province bordering on the Sudan in 1965. The FROLINAT revolt was extended in 1966. In 1966/67 a crisis developed in Sudanese-Chadian relations when Tombalbaye accused the Sudanese authorities of giving shelter to rebel fugitives. Tombalbaye closed the border with the Sudan in August 1966. The dispute between the two countries was resolved in

October through the mediation of President Hamani Diori of Niger but relations remain strained. (Relations with Libya, which supports FRO-LINAT, are also strained; in August 1971 Tombalbaye accused Libya of involvement in a plot to overthrow his government.)

The religious and specifically ethnic factor is significant in this case, because Libya is not only an Islam state, but also under the leadership of Arabs and Berbers of Hamitic origin.

In March 1968 the revolt spread to the Tibetsi area in the north. Rebel successes in May 1968 compelled President Tombalbaye to invoke a Franco/Chadian defence agreement and on August 28 formally requested French assistance in suppressing the revolt. France influenced by Chad's pro-French, pro-Western convictions, reacted favourably and sent troops to the troubled land. In 1970, there were 3,000 French troops in Chad, about half of them being members of the Foreign Legion.

The war was intensified in 1969 and, according to President Tombalbaye, 1,126 rebels were killed in the first half of that year, bringing total rebel casualties to 2,790 as opposed to an equivalent figure for Government forces of 246. The extension of the war served to highlight French involvement and this resulted in awkward questions being asked in France about the French Government's "military adventure" in Chad. Under considerable pressure, the French Government announced in 1970 that there would be a gradual withdrawal of French troops from Chad, culminating in a total withdrawal in 1971.

Early in 1970, just when there were optimistic reports that the revolt was being brought under control and that French withdrawal would not aggravate the situation, the rebellion flared up anew. This prompted Tombalbaye to express concern at the proposed French withdrawal; in France itself, however, there was increasing talk of France being involved in a "Vietnam war" in Africa. Part of the Foreign Legion contingent left Chad in July 1970 and this injected a strained element into Franco/Chadian relations. On October 11, 1970, 11 French soldiers were killed in skirmishes with rebels north-west of the Largeau area and the agitation for a French withdrawal was resumed. The French Government reiterated its intention of having all its troops out of Chad by 1971.

On the night of August 27, 1971, Tombalbaye put down an attempted coup d'etat. He accused the Libyan ruler of complicity in the attempt to overthrow his Government and relations between the two countries deteriorated.[25]

These above mentioned power struggles in the Sudan, Mauritania and Chad are, despite their overt religious natures, basically determined by ethnic animosities between Arabs and Hamites on the one hand and Negroes on the other. All these conflicts are related to the question of social distance and differentiation along ethnic lines; that is a lack of "sympathetic understanding" between ethnic groups.

The ethnic factor is also dividing the terrorists' onslaughts on the White regimes of Southern Africa, because despite some unifying (and mostly external) forces, the pattern of the nationalists' resistance is largely ethnically orientated. This will be explained in the following section.

VI. SEGMENTED STRUGGLE IN SOUTHERN AFRICA

A number of factors were responsible for internal strife among those groups engaged in revolutionary warfare against the Portuguese, Rhodesian and South African governments in Southern Africa. These factors are related to ideological socio-economic, regional and ethnic factors. The ethnic factor is particularly important as far as the participation of the masses in these revolutionary activities is concerned.

The best examples of revolutionary organizations which recruit the majority of their support from exclusive tribal groups, are the Front for National Liberation in Angola (FNLA), the Union for the Total Liberation of Angola (Unita), the South West African Peoples Organization (SWAPO), and to a lesser extent also the Zimbabwe African Peoples Union (ZAPU) and Zimbabwe African National Union (ZANU). Although the ethnic factor is not of prime importance in the case of Frelimo (in Mocambique), ethnicity nevertheless plays a significant role especially as far as support among the local population is concerned.

Terrorist activity erupted in 1961 in North Angola (the Uige and Zaire districts). There were also riots on a smaller scale in Luanda. The northern districts concerned are regarded as the traditional home of the Dembos and the various Kongo tribes, such as Sorongo, Shikongo, Zombo, Congo and Sosso.[26] The organization responsible for terrorist activity in 1961 was the *Uniao dos Populacaos de Angola* (UPA), established in 1958 under the leadership of Holden Roberto (a Shikongo-Bakongo). Roberto had been involved in 1954 in an ethnically exclusive Kongo (really Shikongo) organization (UPONA) whose objective was the restoration of the former Kongo Empire with its 12 Bakongo tribes.[27]

The former Kongo empire was divided into French, Belgian and Portuguese areas of control during the colonial period. The French Kongo area became independent as the Congo (Brazzaville) in 1960 while the Belgian Congo area also became independent in 1960 as part of Zaïre. The (Ba)Kongo revolt against the Portuguese in 1961 should accordingly be seen against this background.[28] However, there are other reasons, economic and social, that also prompted the (Ba)Kongo revolt. For these latter reasons, the Dembos who are a Mbundu tribe (Kimbundu-speaking) living south of the Shikongo-Bakongo, participated in the revolt.[29]

Originally, UPA was also augmented by some other Kongo, Ovimbundu, Mbundu and Chokwe-Lunda tribes for the reason that UPA was the only rural organization of its kind in Angola. This liaison, however, was short lived. Because of UPA's exclusively ethnic character, members of the Mbundu and Ndembu tribes broke away as early as 1961–62 and joined the MPLA, which had been formed in Luanda in 1957. In an effort to save UPA's image as a revolutionary organization in the face of internal tribal strife, UPA was renamed the *Govêrno Revolucionario de Angola no Exilo* (GRAE) in 1962. The Ovimbundu and Chokwo-Lunda eventually also broke away from GRAE, particularly in 1964. The Ovimbundu were the first to splinter in order to form an ethnically exclusive movement known as UNITA which was led by Jonas Savimbi.[30] GRAE, nevertheless, is still in existence and became known as FNLA.

As mentioned, UNITA *(Uniao National para a Independencia Total de Angola)*, was established in 1964. UNITA was formed by Jonas Savimbi (an Ovimbundu) when he resigned his portfolio of "Minister of Foreign Affairs" in Roberto's GRAE "government" in protest at (Ba)Kongo domination of GRAE.[31] The Ovimbundu are regarded as one of the most prosperous of Angola's indigenous population groups. This was due to their experience in the rubber and slave trades. In 1890, the Portuguese tried to subject the Ovimbundu and this attempt led to the Bailundu War of 1902–03. In consequence of this, the Ovimbundu lost their trading monopoly, of, in particular, rubber in 1910.[32] The Ovimbundu, the largest single ethnic group in Angola, do not have a strong nationalistic outlook, yet some of their leaders founded anti-Portuguese resistance movements as early as the 1950s.

UNITA'S headquarters is currently in Angola after originally being in Lusaka. Savimbi's guerilla activities are also directed to the Lunda-Chokwe and Ndembu tribal areas found on either side of the Angola-

Zambian border. Since the establishment of UNITA some Ndembu members have left the multi-ethnic MPLA to join UNITA. UNITA is thus grouped in the third ethnically orientated (Ovimbundu, Lunda, Ndembu and Chokwe) group. Murdock[33] points out, for instance, that the Chokwe and Ndembu were initially part of the Lunda empire of Mwata Yamwo, while the Ovimbundu on the other hand are a politically independent group maintaining good relations with the Lunda group. There has also been historical contact between the Ndembu and Ovimbundu, according to Murdock. Contact between the four tribes is thus not uncommon.

Revolutionary activities in South West Africa are organized in large measure by the South West Africa People's Organization (SWAPO). The other movement that can be mentioned is the South West Africa National Union (SWANU). SWAPO was formed in Cape Town in 1959 by Owambo migrant labourers under the name Owamboland People's Organization (OPO) and was renamed SWAPO a few years later. As will appear later, SWAPO is still an organization to which, in practice, Owambos only belong.

The birth of OPO (later SWAPO) as a nationalistic movement can be traced to certain historical events in the history of the Kuanyama-Owambo. The Kuanyama is the largest single Owambo tribe in South West Africa.[34] The tribal area of the Kuanyama stretches from deep within the Huila District of South Angola. The Kuanyama of Angola have clashed on various occasions with the Portuguese, as far back as 1904 amongst others.[35] Since 1941, the Portuguese have applied a variety of punitive measures against the Kuanyama and this has resulted in general dissatisfaction. The first nationalistic movement among the Kuanyama came, ironically enough, not in Angola, where tribesmen were under close scrutiny, but in South West Africa among westernized Kuanyamas.

The Kuanyama-orientated and exclusive membership of SWAPO is clearly to be seen in the fact, that its leaders such as Hermann Ja Toivo Jacob Kuhangwa and Sam Nujoma are all members of the Kuanyama tribe. Organizers of SWAPO have earnestly tried to camouflage the Kuanyama character of the movement by converting it into a so-called front organization. Consequently a draft programme of action was formulated in 1967 whereby membership of SWAPO was opened to members of other tribes such as the Damaras, Namas, Hereros, Basters, Tswanas and other Owambos. Interestingly enough the Bushmen were not mentioned in this programme.[36] The programme was followed up by recruitment in the Kaokoveld of South West Africa where Tjimbas, Himbas and Herero live. These attempts were, however, not very successful.

The establishment of the present nationalistic organizations in Rhodesia was preceded by the formation of the Southern Rhodesian African National Council which was created in 1957. After this organization was banned in 1959, the leaders formed the National Democratic Party which was banned, in turn, in 1961. The NDP was a multi-racial organization. After its banning Rhodesia's black nationalists formed the Zimbabwe African People's Union (ZAPU). This was banned in 1962. ZAPU, however, did not dissolve when it was banned, but went underground. The membership of ZAPU, according to observers, represented a very small percentage of Rhodesia's black population. The reason for this is that the Rhodesian Black has for many years had no specific economic and political grievances and the largest ethnic group in Rhodesia, the Shona-speaking tribes, are not aggressive by nature.[37] The Shona group comprises six ethnic units, namely the Korekore (in the north), the Zezuru (in the Salisbury area), the Nyika (in the Inyanga-Umtali area), the Ndau (in the south-east), the Karanga (in the Zimbabwe area), and the Kalanga (in the south-west). The Ndebele people of Matabeleland, which form the minority sector of the Rhodesian Black population, have an ethnic history of military and political strife – first with Chaka in Natal, then with the Voortrekkers in the Western Transvaal, then with the Tswana and Shona-Karanga and lastly with the British in Matabeleland.[38]

The leader of ZAPU is a Ndebele, namely Joshua Nkomo. In July 1963 the ZAPU movement split on ethnic as well as ideological grounds into the People's Caretaker Council (PCC) and the Zimbabwe African National Union (ZANU) with Joshua Nkomo (who enjoyed Russian support) and Ndabaningi Sithole (who enjoyed Chinese support) as leaders of the respective groups.[39] In 1964 PCC reverted to the name ZAPU.

The most important office-bearers of ZAPU are or were Joshua Nkomo, James Chikerema, Silinduka, Ndlovo and Mayo – all Ndebeles with the exception of Chikerema who is Shona-speaking.

The most important office-bearers of ZANU on the other hand are or were Ndabaningi Sithole, Peter Mtandwa, Herbert Chitepo and Takawisha. They are all Shona-speaking. Sithole is a Ndau, Chitepo a Nyika and both Mtandwa and Takawisha are of Shona origin.

ZANU members such as Everiste Africa and Edmund Ngandoro

were responsible for the first terrorist attacks in Rhodesia in 1966 when they attacked farms at Sinoia and Hartley. ZANU regards the Rhodesian/Zambian boundary northeast of the Kariba Dam wall as its area for incursions. These areas are inhabited mainly by the Korekore tribe whose socio-economic grievances are deliberately exploited in order to win local support. ZAPU is experiencing internal problems. One of these problems is that the predominant Ndebele orientation of ZAPU does not attract recruits from Shona tribesmen. Statements such as the following have appeared in the press from time to time:

"To my knowledge... they (ZAPU) will never defeat the government. There is tribalism within the party... when we were staying in the Lusaka camp, Matabeleland people were better privileged than Mashonaland people".[40]

In March 1967 there was a split in the ranks of ZAPU when one John Ndebele broke away and formed the Zimbabwe People's Democratic Party (ZPDP). It is said that this movement is given financial support by East Germany.[41]

By virtue of the clear differentiation between ZAPU and ZANU, both as far as support from the local population and outside support are concerned, it is by no means surprising that during the 1960s the two organizations planned no common offensive against the Rhodesian government.

However, in August 1967, ZAPU concluded a so-called aggressive treaty with the ANC of South Africa whereby Matabeleland and the Wankie area would be invaded from Zambia across the Zambezi River east of Mosi-oa-Toenja (Livingstone). The mutual co-operation between ZAPU and the ANC was presumably facilitated by the fact that most of the ANC members concerned were of Zulu or Xhosa origin[42] and were thus ethno-linguistically allied to the Ndebeles of ZAPU (the Ndebeles are of Zulu origin).

However, the effective counter-insurgency operations of the South African and Rhodesian security forces caused the withdrawal of this ZAPU/ANC offensive since 1970. Since then, ZANU took over the revolutionary offensive in Rhodesia with their activities in the country's north-eastern border areas.

The first outbreak of revolutionary warfare in Mocambique occurred in 1964 with the attack by 20-odd terrorists in the Cabo Delgado District (Konde tribal area). This attack was the work of the Mocambique African National Union (MANU), a "disbanded" wing of FRELIMO, under the leadership of one Lucas Fernandez. The attack was partly planned from neighbouring Tanzania, as will appear later. The MANU organization was a continuation of the Tanganyika Mocambique Makonde Union (TMMU) formed in 1954 by Mocambique Konde migrant labourers working in the then Tanganyika.[43] MANU was thus primarily a Konde-attuned organization which, however, was incorporated in FRELIMO in 1962 on the insistence of Mario de Andrade (of the CONCP in Algiers).

FRELIMO's executive comprised Dr. Eduard Mondlane (a Shangaan tribesman) as President, Uria Simango (a Beira priest) as Vice-President, David Mabunda as Secretary-General, Matthew Mmole as Treasurer, Paul Gumane as assistant Secretary-General and Leo Milas as Public Relations Officer. "FRELIMO has aspired to be a truly national party embracing the whole population of Mocambique and cutting across tribal divisions."[44] Lazarro Kavandane, a Konde chief, was appointed military leader. Bickering occurred within the organization, however, and this led to Mabunda and Gumane establishing other organisations. Gwambe fled to Uganda where he formed the UDENAMO-Monomotapa. Mmole was expelled from FRELIMO and revived MANU as a separate organization,[45] whereas Kavandane resigned his position.

After this, Gwambe and Gumane established the *Comité Revolucionério de Mocambique* (COREMO) in Lusaka in 1965.[46] MANU was also absorbed in COREMO. COREMO's declared objective was to commit sabotage and selective terrorism in Mocambique, such as the destruction of the Cabora Bassa project in the Tete district.

The activities of FRELIMO (really MANU) were confined in 1964 to the Konde tribal area. Later these were extended to the neighbouring Vila Cabral area (Nyasa District) of Mocambique, where the Yao and Nyanja tribes in particular live. A few members of the Ngoni tribes were also later recruited for guerilla training in Tanzania by FRELIMO. Most FRELIMO revolutionaries, however, came from the ranks of the Konde and Nyanja tribes, "both of whom like the Bakongo, overlap international borders. This made it easy to organize training camps in Tanzania, but the result of relying so largely on two tribes was to alienate others. This was especially the case of the Nakua, one of the largest ethnic groups in Mocambique, traditionally hostile to the Makonde."[47]

The Konde are known as a very aggressive tribe. They were, for example, the only tribe to attack David Livingstone's expeditions in 1862. Although the Konde have had contact for many centuries with

Arab traders (Konde tribal area lies across the trading route between Kilwa and Lake Malawi) they have never accepted Islam. The Konde have also clashed on numerous occasions with the Portuguese authorities, on labour regulations among other issues. In consequence, more than 10,000 Konde fled to Tanzania in the 1950s.[48] The MANU organization originated among these fugitives/migrant workers.

The Nyanja resistance to Portuguese authority can also be traced back in history. The Nyanja were, with the tacit approval of the Portuguese, the subject of Arab slave traders for many centuries.[49]

VII. Conclusion

The phenomenon of ethnicity is still of great significance in Africa. Processes of culture change did not have a particularly eroding impact on this phenomenon, which is mainly concerned with various kinds of tribal loyalties. These loyalties are manifest in both rural and urban areas and exert divisive influences on voluntary associations, political parties, labour movements, and on diverse nationalist aspirations. It therefore seems clear that contemporary African societies are currently characterized by a lack of "sympathetic understanding" between the constituent tribal groups, most of which show little loyalty to the nation which is the political aggregate of all the ethnic groups concerned, but only to their own local groups.

NOTES

1. Geertz, C., "The Integrative Revolution: Primordial Sentiments and Civil and Politics in the New States," 1963, C. Geertz, ed., *Old Societies and New States*, New York p. 177; and Klineberg, C. and Zavelloni, M., *Nationalist and Tribalism Among African Students*, Paris: Morton, 1969, p. 8.
2. Wallerstein, I., "Ethnicity and National Integration in West Africa," 1970, M. E. Doro and N. W. Stultz eds., *Governing in Black Africa*, Englewood Cliffs: Prentice Hall, p. 10.
3. Legum, C., "Tribunal Survival in the Modern African Political System," P. C. W. Gutking, ed., *The Passing of the Tribal Man in Africa*, Leiden: Brilly, *The Passing*, 1970, p.103.
4. Mair, L., *Anthropology and Social Change*, London: LSE, Athlone, 1969, p. 101.
5. Hanna, W. J. and Hanna, J. L., eds., *Urban Dynamics in Black Africa*, Chicago; Aldine, 1971, p. 105.
6. Wallerstein, I., *Africa: The Politics of Independence*, New York; Vintage, 1961, pp. 121–136.
7. Legum, C., *op. cit.*, p. 103.

8. Wallerstein, I., 1970, *op. cit.*, p. 10.
9. Lambo, Professor, Remarks made at a symposium of the Ciba Foundation held in Addis Abba 1965, In Wolstenholme, G. T. and M. O'Connor, eds., *Man and Africa*, London: London, Churchill, p. 119.
10. Bogardus, E. S., *The Development of Social Thought*, New York, 1940, p. 468.
11. Banton, N., "Social Distance: A New Appreciation," *Sociological Review* 8, 1960, p. 171.
12. Mitchell, J. C., *The Kalela Dance: Aspects of Social Relationship Among Urban Africans in Northern Russia*, 1956.
13. Lever, H., *Ethnic Attitudes of Johannesburg Youth*, Johannesburg: Wits Press, 1968.
14. Edelstein, M. L., *What do Young Africans Think?*, Johannesburg: South African Institute of Race Relations, 1972.
15. Mitchell, J. C., "Perceptions of Ethnicity and Ethics Behaviour," A. Cohen, ed., *Urban Ethnicity*, London: Tavistock, 1974, pp. 2–35.
16. *Ibid.*, p. 7.
17. Edelstein, M. L., *op. cit.*, p. 89–92.
18. Little, K., "The Role of Voluntary Associations in West African Urbanization," 1957, P. L. Van den Berghe, *Africa: Socials Problems of Change and Conflict*, San Francisco: Chandler, 1965, p. 328.
19. *Ibid.*, p. 330.
20. Henderson, I., "Wage-Earners and Political Protest in Colonial Africa," *African Affairs*, Vol. 72, No. 288, 1973, p. 290.
21. *Ibid.*, pp. 292–294.
22. Legum, C., *op. cit.*, p. 108; and Mitchell, J. C., "Tribalism and the Plural Society," 1960. Middleton, J., *Black Africa: Its Peoples and Their Cultures Today*, London: Macmillan 1970, p. 263.
23. Glentworth, G. and D. Hancock, I., "Change and Continuity in Modern Uganda Politics, pp. 237–255. *African Affairs*, Vol. 72, No. 288, 1973.
24. Kabara, J., "Sudanese Strife", *Bulletin of the African Institute*, Pretoria, Vol. XI, 1971, No. 6, pp. 261–263.
25. Chad: "Victim of a Race War." *Bulletin of the Africa Institute*, Pretoria, Vol. XII, 1972, No. 1, pp. 27–29.
26. Abshire, D. M. and Samuels, M. A., *Portuguese Africa: A. Handbook*, London: Psll Mall, 1969, p. 119.
27. Wheeler, D. L., and Pellissier, R., *Angola*, New York: Praeger, 1971, p. 169.
28. Bosgra, S. and Dijk, A., *De strijd tégen het Portugees Kolonialisme*, Amsterdam: Paris, 1968, p. 64.
29. Wheeler, D. L. and Pelissier, R., *op. cit.*, p. 166.
30. Abshire, D. M. and Samuels, M. A., *op. cit.*, p. 395.
31. Marcum, J., *The Angolan Revolution, Vol. I*, (1950–1962), M.I.T. Press, 1969, p. 138.
32. Abshire, D. M. and Samuels M. A., *op. cit.*, p. 115.
33. Murdock, G. P., *Africa, its Peoples and Their Culture History*, New York: McGraw Hill, 1958, p. 297.
34. Republic of South Africa, *Report of the Commission of Inquiry into South West African Affairs*, Pretoria, 1964, p. 34.
35. Marcum, J., *op. cit.*, p. 114.
36. *Dagbreek*, Johannesburg, 19.1.1969: It was alleged that the Khung-Bushmen of Southern Angola actively supported Portugese troops in their campaign against guerillas.
37. Welensky, R., "The Challenge of the Future", R. Shay and C. Vermaak, *The Silent War*, Salisbury: In Galaxie Press, 1971, p. 260.
38. *Ibid.*, pp. 305–306.

39. Bull, T., ed., *Rhodesia: Crisis of Color*, Chicago: Quadrangle, 1968, p. 214.
40. *The Star*, Johannesburg, 1.7.1967.
41. *The Star*, Johannesburg, 1.7.1967 and 24.7.1970.
42. *The Guardian*, Manchester, 25.8.1967.
43. Marcum, J., *op. cit.*, p. 196.
44. Abshire, D.M. and Samuels M. A., *op. cit.*, p. 403.
45. *Ibid.*, p. 402.
46. *Ibid.*
47. *Ibid.*
48. *Ibid.*, p. 123.
49. *Ibid.*, p. 125.

Ethnic Minorities in Japan

WILLIAM WETHERALL AND GEORGE A. DEVOS

WILLIAM O. WETHERALL – a doctoral candidate in Asian Studies at the University of California at Berkeley – received a B.A. in Japanese Studies in 1969 and an M.A. in Asian Studies in 1973, both from UCB. He has lived and studied in Japan for three years. He is currently engaged in dissertation research on suicide in Japan with the aid of a Fulbright grant and a Japan Foundation fellowship. His interdisciplinary interests in Japan focus on social deviancy, environmental pollution, ethnic discrimination, women's movements, popular culture in all its aspects, and mass media. Mr. Wetherall's publications include "Cybernetics and Semantics" in *California Engineer* (January, 1966); "The Educational Institution as a *Nakōdo*," *Nucleus* (Fall, 1972); and *Japan's Minorities* (Burakumin, Koreans, and Ainu), London: Minority Rights Group, 1974, a pamphlet with George DeVos. In addition to a book on suicide in Japan, he plans volumes on ethnic discrimination in Japan, stereotypes of Asians and Asian Americans in fiction by Americans of European ancestry, and sexuality in Japanese and American popular culture.

GEORGE DEVOS is a Professor of Anthropology at the University of California at Berkeley. He is author and editor, with Professor Hiroshi Wagatsuma of the University of California at Los Angeles, *Japan's Invisible Race* (Caste in Culture and Personality), University of California Press 1972. Professor DeVos is also the author of *Socialization for Achievement* (Essays on the Cultural Psychology of the Japanese), University of California Press 1973, and is author with Wagatsuma of the forthcoming *Heritage of Endurance* (Psychocultural Continuities and Delinquency in Japanese Urban Lowerclass Families).

Ethnic Minorities in Japan

WILLIAM WETHERALL AND GEORGE A. DEVOS

Japanese are often thought to be an unusually homogeneous people in two senses of the word. Not only is Japan frequently characterised as having fewer ethnic groups and minority problems than other nations, but majority Japanese are typically described as a relatively "pure" people manifesting less physical feature variation than Europeans. Japanese and foreigners alike attribute these alleged homogeneous tendencies to centuries of geographically imposed if not politically self-imposed isolation, under which circumstances Japanese racial and cultural ethnicity supposedly developed its "insular" ideosyncracies. Looking at Japanese ethnicity empirically, however, both views of homogeneity are found to be substantially incorrect. Not only does Japan have many ethnic minority groups – defined both culturally and gennetically – but the majority population continues to exhibit considerable phenotypic evidence of its heterogeneous origins.

Each of the several ethnic minority groups presently recogniseable in Japan is burdened with a distinctive history of separation and exploitation. And each of the groups has its own heritage of cultural and physical features that raise barriers of discrimination in interaction with the patterns of social acceptibility found in the majority community. Yet while surface examinations show these histories and heritages to be characteristically different for each group, deeper probings show them to be strikingly similar.

This analysis begins with an overview of ethnic minority populations in Japan with respect to who and how many. Rounded estimates drawn from figures from many sources suggest a model for ethnic composition in Japan [Table 1]. A second section looks at how Japan's minority populations are distributed in terms of co-residence not only with the majority population but with one another. A paragraph on minority occupations suggests why co-residence has probably been patterned as described.

The main body of this report is devoted to separate discussions of eleven nominally different but phenomenally overlapping minority groups in Japan. Starting with "invisible" outcastes it then proceeds to "blue-eyed" keptouts. The first several groups are taken in roughly chronological order, while the last few groups are treated without regard for their place in history. Some groups have been given more space than others, but yet all are equally important, if not numerically or politically then simply because their members face one form or another of degrading ethnic discrimination in majority Japanese society. The necessarily brief accounts even for groups receiving the most attention at best highlight the more salient historical and contemporary aspects while leaving undescribed many details that are none the less vital to a fuller understanding of minority status and discrimination in Japan.

This analysis concludes with an overview of minority status and discrimination in Japan with reference to purity myths as impediments to ethnic pluralism. Central to the patterns of systematic discrimination found in modern Japan – despite humanistic faceliftings of its legalistic structures – is a widely held belief in cultural if not racial purity. With regard to such purity myths and other aspects of discrimination discussed in this report, the writers are keenly aware that what is said about discrimination in Japan results from cultural and psychological propensities in no sense uniquely Japanese. While all cases of discrimination in Japan haven't exact analogs elsewhere, in the balance Japan is essentially like most other countries.

I. Ethnic Minority Populations

Ethnic minority groups in Japan – defined in terms of separate national origins, different genetic attributes, or a history of outcaste status based on ritual pollution – account for about four percent or 4,500,000 of Japan's 110,000,000 residents in 1974 [Table 1]. Some 2,000,000 of these minorities are indigenous burakumin [citizens of outcaste communities]. About 1,000,000 Okinawans – residing on the Ryūkū Islands – constitute the second largest minority group in Japan. An estimated 500,000 other Japanese citizens suffer minority status for allegedly genetic and cultural differences, including hibakusha [atombomb survivors], *konketsuji* [mixedblood citizens] mainly of Japanese with Korean or American parentage, *kikajin* [naturalized citizens] mainly of Korean or Japanese ancestry, *ainu* [aboriginals] on the northern island of Hokkaidō, and a number of Japanese who have returned to their homeland after once emigrating or spending formative years abroad.

In addition to 109,000,000 Japanese citizens, nearly 1,000,000 foreign nationals live in Japan. Over 600,000 of these resident aliens are Koreans, who thus form the third largest ethnic minority group in Japan – after Burakumin and Okinawans. There are much smaller though significant populations of Chinese and Americans, and still smaller but less significant numbers of other foreign nationals representing ethnic traditions around the world. Japan also boasts a fairly sizeable and socially visible flow of foreign tourists.

There are other indigenous minority groups in Japan that might be called "ethnic" in the broadest cultural sense of this word, but they are dwindling and amorphous, and are mentioned here only in passing. Included among these little-known minorities are migrant marine fisherfolk, woodworkers, hunters, ironworkers, riverine migrants, quasi-religious itinerants, and scattered other groups whose communities may be regarded as ethnic subcultures. (Norbeck 1972)

Several occupational minority groups – some of them quite extensive – are subject to considerable discrimination in Japan. Included here are miners, weavers, craftsmen working with bamboo and straw, leather products workers, truck and taxi drivers, day labourers, sewage and garbage collectors, hostesses and bartenders, waitresses and waiters, maids and servants, and models and performers. The eight general occupational classifications that include most of these low status vocations – of forty-one classifications tabulated by Japanese census takers – accounted for 9,088,700 persons in 1965, around nineteen percent of Japan's employed work force fifteen years old and older [47,633,380], or nine percent of the total resident population [98,274,961]. Moreover, the worlds of organized crime and street delinquency – not represented in these occupational statistics – attract a disproportionate number of ethnic minorities along with majorities who have fallen in status. (Sōrifu 1969[5.2]:252; sum of intermediate occupational classifications 13, 23, 24, 25, 29, 36, 37, and 40)

Roughly speaking, then, about ten percent of Japan's resident population suffers some form of social status discrimination – leaving aside women – nearly half of these for nominally cultural and genetic reasons, and the balance because of their livelihoods. While noting some overlap between these two divisions, this study is concerned primarily with ethnic groups composed of persons differentiable – apart from their vocations – on the basis of caste, race, or nationality. The number of such minorities in Japan – estimated here [Table 1] at some four percent of the total resident population – would be as high as six percent using the figures claimed by some minority movement leaders. Yet even the rate of three percent – computable from scattered published statistics – is surprisingly high when it is remembered that problems of ethnic discrimination in Japan are seldom acknowledged by Japanese or adequately assessed by foreigners in survey reports of Japanese culture and society.

But how many persons in Japan might fall into one or another socially perceived category, and why more significance may be attached to one category rather than another, should not distract attention from the people who must bear degraded status for whatever reason. While the writers might wish to enumerate, evaluate, and classify people for limited purposes of description and analysis, their ultimate concern is to fathom the psychological and social problems experienced by those who find themselves systematically discriminated against – figuratively from before the cradle to after the grave.

They have twice used the word "systematic" with implications that may require some explanation. The term is sometimes intended to describe social behaviour patterned by rather explicit rules and regulations. But they refer not so much to the overt systems of law and policy – which are found in contemporary Japan to be outwardly nondiscriminatory – as to the more covert and deeply rooted culture and personality systems. The psychocultural patterns that characterise such systems inwardly perpetuate discriminatory attitudes and behaviour among majority members of the community while concurrently they deepen propensities for low self-esteem and subnormal achievment motivation among minorities. They consider these patterns "systematic" because they are "structured" albeit subconsciously.

II. ETHNIC MINORITY CO-RESIDENCE

Burakumin are found mainly in western Japan in prefectures around the Inland Sea [Seto Naikai], where they tend to co-reside with large numbers of Koreans and Chinese, particularly in the Kinki region. The seven Kinki prefectures – including Kyōto, Ōsaka, and Hyōgo (Kōbe

City) – account for over 40 percent of all Japanese outcastes and nearly 50 and 40 percent respectively of all Korean and Chinese nationals residing in Japan, compared with less than 20 and 30 percent respectively of all residents and all other foreigners in Japan. In contrast, the seven prefectures of the Kantō region in eastern Japan – including Tōkyō and Kanagawa (Yokohama City) – account for less than 10 percent of Japan's Burakumin, with these found mainly in rural inland Kantō prefectures. But the Kantō region includes roughly 20 percent of Japan's Korean minority, some 50 and 60 percent respectively of all Chinese and all other foreign nationals in Japan (practically all of these in urban Tōkyō and Kanagawa), and nearly 30 percent of all residents in Japan.

Chinese and Euroamerican foreign nationals tend to co-reside in the high commercial activity prefectures and in the same administrative districts within the urban centres of these prefectures. Koreans are found mainly in the industrial areas that overlap these commercial regions, while Burakumin are distributed primarily in rural and industrial areas around cities and towns in western Japan where they have been distributed for the past five centuries. In most Inland Sea prefectures having large communities of outcastes and Koreans, their combined numbers account for between four and seven percent of the total prefecture population. Within these prefectures, however, Koreans tend to live in the major cities while Burakumin tend to reside outside them. In not a few prefectures in central and western Japan, Burakumin alone make up four, five, and six percent of the population, while most prefectures in northern Japan report no outcastes at all.

1959 figures show that whereas 5 and 12 percent of all Japanese employees 15 years old and older were engaged respectively as professional or sales personnel – versus 30 percent in factory production and labour – the corresponding figures for Koreans were 2 and 5 percent versus 53 percent, while for Chinese they were 8 and 28 percent versus only 9 percent. In contrast, 65 percent of all Americans in Japan in 1959 were professionals, compared with 11 percent in sales work and only 2 percent in processing of labour. Such distinct occupational differences reflect the characteristic histories of these minorities in Japan. Chinese came principally as merchants and traders, Koreans as factory workers, miners, and construction hands, while Europeans and Americans have come principally as managers, teachers, and other elites. Burakumin, on the other hand, continue to follow traditional trades, labour in the cities, or till inherited farmlands. (Hōmushō 1964:93)

Except for individuals who have recently migrated, Ainu are found mainly on Hokkaidō. There are Ainu also on the Sakhalin [Karafuto] and Kurile [Chishima] islands to the north of Hokkaidō. Although many Okinawans migrated to Japan's main islands or emigrated overseas before the Pacific War, persons recognised as Okinawans today are found principally on the Ryūkyū Islands. Indeed, all Japanese nationals residing on Okinawa during the extended American occupation there – regardless of origins within Japan – are in a sense now Okinawan minorities. Konketsuji are found primarily in the largest cities where they most readily find employment in marketing and entertainment industries.

III. Burakumin: Ritual Purity and Untouchability in Japanese History

About 2,000,000 Japanese citizens are considered genetically polluted and treated by majorities essentially like untouchables in India and as outcastes elsewhere in the world. This largest minority group in Japan is ironically the most invisible. Burakumin cannot be distinguished from other Japanese by any measureable biological attribute. The genetic differences traditionally attributed to outcastes in Japan are founded in myths of racial origin, many of them ancient but none with a convincing basis in history.

A popular but pseudo-historic view argues that Japanese outcastes are in some sense the descendants of Koreans brought to Japan as captives and slaves from peninsula campaigns during the Yamato period (300–645). Such wars are described in Japan's earliest historical documents – the *Kojiki* [Record of ancient matters] (712) and *Nihon shoki* [Chronicles of Japan] (720) – which provide also important insights into the role of ritual purity in the religious values of ancient Japanese society. It is largely through an understanding of the continuity of these values down to the present day – and the absence of biological and reliable historiographic information to the contrary – that Burakumin are most reasonably thought by contemporary anthropologists to be of indigenous rather than foreign origin. (Price 1972:12)

By the 17th century, Japanese society had become highly stratified. Below the nobility, a four-tier class system placed *samurai* [warrior] administrators above all others. Second came the farmers, third the artisans, and last the merchants. Members of these four classes were called *ryōmin* [good subjects] in contrast with *senmin* [base subjects]. The *senmin* were divided into several subclasses, the most important of which were the *eta* and *hinin*. The *hinin* [nonpeople] were typically beggars, itinerant entertainers, prostitutes, mediums, diviners, religious wanderers, fugitives from justice, or persons reduced to *hinin* status as punishment for a crime or misdemeanour. The *eta* [a word of uncertain origin commonly represented with Chinese characters meaning "much filth"] were nominally hereditary outcastes who performed tasks traditionally considered to be ritually polluting, including leatherwork and sandal making.

Polluting occupations of various kinds were recognized at the dawn of Japanese history, but *eta* origins are commonly traced to the 9th century when apparently the first outcaste communities were founded by herders of cattle and disposers of the dead, and by persons skilled in the slaughter of game – for falcons and dogs – who moved into these villages to serve as butchers and tanners, or to engage in other despised occupations. It was the rigidification of society during the Tokugawa period (1600–1868), however, that firmly established the degraded status of Japanese practicing such trades. (Price 1972:17–19)

All *senmin*, including *hinin* and *eta*, became ordinary citizens as a result of the Eta Emancipation Edict [Eta Kaihō Rei] issued by the Meiji government in 1871. The *hinin* were generally absorbed into the larger former *eta* communities, and not a few indigent majorities also found themselves living in outcaste settlements at the time of the first Meiji census. Estimates of the Burakumin population a century ago vary from 280,311 to 520,451 depending on what one wishes to consider "outcaste." Similarly, 1963 figures vary from 1,113,043 to 3,000,000 depending on whether one is a government official (low) or Burakumin movement leader (high). In the meantime, the total resident population has increased 176 percent from 34,806,000 to 96,156,000 in 1963 [Table 2].

Life for residents of the *tokushu buraku* [special communities] – as former outcaste settlements came to be called – grew physically hazardous. Non-outcaste commoners – suffering greatly themselves in the wake of social and economic upheaval, particularly in rural areas – would sometimes vent their anguish through *etagari* [*eta* hunts] organized spontaneously in mixed localities. The capacity of Japanese majorities to dissipate frustrations by indiscriminately attacking minorities has been demonstrated also against foreign nationals in Japan, both Euroamerican and Asian. Most notable was the killing of several thousand Koreans in and around Tōkyō following the Great Kantō Earthquake [Kantō Dai Shinsai] of September 1923. The destruction of outcaste lives and property early in the Meiji period (1868–1912) was not as focused as the Korean massacres, but seems to have taken a comparable toll. (Totten and Wagatsuma 1972:34–37; Mitchell 1967: 38–41)

The only way one can identify an outcaste in contemporary Japan with any degree of certainty is to know his place of residence. A Burakumin living in an area outside a known *buraku* [settlement] may attempt to hide his origins in an effort to pass into majority society –a demanding task at best in view of Japan's nationwide registration system and the continuing interest majorities have in family origins. Passing becomes most difficult when attempting marriage, which may involve a background enquiry. Added to the external social obstructions are the internal psychological barriers concerning personal integrity. Many passing Burakumin feel very guilty about passing, believing that their passing is somehow unfair to nonpassing relatives and friends, from whom they must sever all ties or risk discovery.

Most Tokugawa communities enforced proscriptions against marriage between *eta* and non-outcastes. Burakumin populations continue to be relatively endogamous, however, as non-outcaste majorities – who disfavour interethnic unions generally – continue to object most strongly to intercaste marriages. Reluctance to alter traditional patterns of prejudice towards persons of outcaste origins is reported even among Japanese emigrants overseas, some of whom came from Burakumin communities. (Wagatsuma and DeVos 1972:118–120; DeVos and Wagatsuma 1972[A]:252–256; Ito 1972)

There is a long history of political militancy among Burakumin that in recent decades has solidified into a notably proletarian orientation, as has the Korean movement in Japan. The principal Burakumin organisation today – Buraku Kaihō Dōmei [Buraku Liberation League] – serves more Burakumin community needs than all other outcaste and non-outcaste groups combined. Local governments in metropolitan areas having large populations of Burakumin have made sporadic attempts at meaningful urban redevelopment, but an off-the-tourist-beat trip through these cities will reveal still extensive neighbourhoods with conspicuously inadequate sanitation and housing. All urban outcaste neighbourhoods are not as badly off, however, and some are virtually indistinguishable from surrounding majority communities.

Agricultural Buraku in rural areas are generally much better off than their urban counterparts and are relatively self-sufficient. Outcastes during the Tokugawa period were predominately farmers, and only on occasion – if ever – did they engage in "polluting" activities. Today, few Burakumin are involved in the traditionally low status industries, although more are represented in such vocations in relation to their numbers in Japan than majorities. As late as 1920, while 52 percent of all Japanese households were engaged principally in farming activities, a comparable 49 percent of known Burakumin households also were employed in agriculture. (Price 1972:20–22; Wagatsuma and DeVos 1972:120–124)

Delinquency rates in ghettoed Buraku are often three and four times those found in majority areas. Burakumin youth tend to be left selectively unemployed as some companies prefer older Japanese, and outcastes are sometimes underpaid for similar work performed by non-outcastes. Surveys have shown also that Burakumin children may do less well in school than majority children, and may be truant more frequently. While there are notable cases of achievement and accomplishment and gradual improvements generally in Burakumin social conditions, many outcastes continue to be apathetically resigned to their depressed station in life. Finding themselves socially disadvantaged and despised by the members of majority society, Burakumin often develop hatreds towards this society. A sense of hopelessness is often passed to the next generation through parentally informed self-doubt and even self-hatred. (DeVos and Wagatsuma 1972[B]:259–269)

1971 and 1972 were years of great celebration for history conscious Burakumin. The years marked the centennial of the Emancipation Edict of 1871 and 1872, and the semicentennial of the founding in 1921 and 1922 of the predecessor to the postwar Buraku Liberation League. Recent Burakumin issues include the ongoing appeal of a death sentence handed down in the 1964 conviction of a Burakumin youth for the 1963 kidnapping, rape, and murder of a non-outcaste high school girl. Another issue involves conservative majority criticism of proletarian Burakumin literature introduced into integrated public school curricula.

Both of these issues have parallels in the United States, where minority groups rally in defence of accused or convicted fellow minorities, and fight majority objection to the introduction of ethnic studies materials into public school textbooks – materials that would impose a new point of view on majority written history. Such concern with treatment by the institutions of law and education makes it clear that Burakumin leaders today are interested in developing a profound historical sense of Burakumin ethnicity among Burakumin youth. Like other degraded minorities in Japan and throughout the world, Japan's large outcaste population generally disavows melting-pot-style assimilation. Preferring instead a salad-bowl-style pluralism, Burakumin are groping for ways to bring about a multiethnic society that would acknowledge a present continuity of differences based on past separation and exploitation.

The future of Burakumin self-esteem is perhaps best symbolized in the emergence of a Burakumin literature by Burakumin individuals having experiences in life that only they can know internally and express externally to fellow Burakumin if not sensitive majorities. The outcaste struggle has been a legitimate theme for majority artists since Shimazaki Tōson (1872–1943) published *Hakai* [Broken commandment] in 1906, his first novel and a landmark in the history of Japanese literary realism. But more significant for Japan's 2,000,000 outcastes is the recent bestseller *Hashi no nai kawa* [River without a bridge] by Burakumin authoress Sumii Sue. *Hakai* was filmed by Ichikawa Kon in 1961, while the earliest of the six volumes of *Hashi no nai kawa* published between 1961 and 1973 have been filmed in two parts by Imai Tadashi in 1969 and 1970. (McClellan 1971:79–93; *Buraku* 259 [special issue on *Hashi no nai kawa* film movement, May 1970])

IV. Ainu: Internal Expansion and the Absorption of Frontier Aborigines

The Ainu, who were possibly distributed throughout Japan two millennia ago, remained in possession of a considerable portion of northern Honshū – Japan's main island – as late as the Nara period (645–794). The Ainu were known as fierce warriors, and it took several centuries for Japanese frontiersmen to bring about their total submission. Military campaigns during the late 8th and 9th centuries settled most of northern Honshū and secured the Japanese frontier by the middle of the Heian period (794–1185). Hokkaidō was under nominal Japanese control by the Tokugawa period (1600–1868). During this period, the Matsumae feudatory occupied the southern tip of the peninsula now bearing its name directly across the Tsugaru Straits from Honshū. Assimilation efforts by the Matsumae clan were strongly resisted. When the Ainu threatened to revolt, the Japanese eased their ethnocentric policies. (Sansom 1958:12, 19, 91, 104–106; Hilger 1971:xiii)

In 1868, Ainu became subject to the laws of the new Meiji government. Hokkaidō was declared a frontierland and Japanese were encouraged to settle there. Whereas Ainu had in some sense succeeded in evading total ethnic subjugation in the past, they faced now a force of rule that was utterly determined to assimilate them into Japanese society. The Hokkaidō Former Aborigine Protection Law [Hokkaidō Kyūdojin Hogo Hō] of 1899 made Japanese education compulsory for all Ainu children, who were now to acquire a "superior" culture. (Hilger 1971:xiv)

While the Japanese population on Hokkaidō continuously increased by northward migration from 20,086 in 1701 to 151,786 in 1873, the Ainu population decreased from 23,797 reported in 1804 to 18,644 in 1873. Official censuses since 1873 have put Hokkaidō Ainu at 16,000 plus or minus 2,000 while the Japanese population has increased to a 1970 figure of over 5,000,000. (Takakura 1972: 290–291; Hokkaidō 1937:143–144)

One of the most telling changes in the Ainu population in recent centuries, but particularly since 1868, has been a phenotypic shift from nominally "pure" Ainu features to predominately "mixedblood" Ainu-Japanese if not "pure" Japanese physical characteristics. One field study found that about ten percent of those living culturally as Ainu in two sizeable Ainu communities were defineably "pure" (Suzuki 1973: 73–75). Most estimates, however, put the number of nominally "pure" Ainu at no more than one percent of the countable 16,000 Ainu population (Ainu BHTK 1969[1]:12).

Physical features that statistically distinguish Ainu from majority Japanese are many but diffuse. Ainu generally have more body hair than majority Japanese, who in turn tend to have more body hair than other Asians. Some Ainu individuals are reported to have "blue eyes," but not all observers have witnessed this attribute (Hilger 1971:x). Persons considered physically Ainu also manifest more deeply engraved facial features, including a tendency towards double-folded eyelids, and they show statistical differences in fingerprint and bloodtype indices. These population tendencies, with a notable increase in the presence of the Mongolian Birthmark (reported absent in the past) among offspring of Japanese-Ainu marriages, have led most physical anthropologists to view Ainu as caucasoid or protocaucasoid – a somewhat arbitrary classification that majority Japanese are quick to point out in their tourist guides. (Ōno 1970:22)

The present attitude of majority Japanese towards Ainu is one of condescending quaintness. Many Japanese tourists visit Ainu reservations to see professionals perform traditional dances at Bear Festivals and produce native crafts. Like Native American culture, the remnants of Ainu culture have been commercialised in the face of majority cultural oppression. Resident Ainu have reported being verbally abused by majority Japanese. Ainu are sometimes called *gaijin* [foreigners] by majority Japanese despite the legal status of Ainu as Japanese nationals. Ainu tell also of being irritated by tourists who express surprise that "Ainu speak pretty good Japanese," and wonder "How do Ainu walk?" The word for "dog" in Japanese is *inu*. Using the exclamatory "ah!", majority Japanese are known to convert *Ainu da* [It's an Ainu] to *Ah, inu da!* [Ah, a dog!]. (Asahigawa 1971:178–179, 181, 202)

Ainu youth, like Native American youth, are determined to reverse the directions of majority oppression, but as yet they face enormous social barriers in the majority community, not the least of which – as for other minorities in Japan – is the pervasive Japanese sense of racial if not cultural purity and superiority. Offspring of Japanese-Ainu intermarriages are systematically regarded as "Ainu" by Japanese majorities. A fraction of "foreign" blood in a child's veins is sufficient to deny it "pure Japanese" status in Japanese society. It becomes easier for these children to be raised as *ainu* [Ainu word meaning "man"] than as *shamo* [Ainu term for majority Japanese]. Moreover, when majority Japanese are adopted into Ainu families – a not infrequent occurence (Suzuki 1973:70) – their family register status tends to become that of a child born to the adopting family, with the result that majority adoptees become "Ainu."

V. Okinawans: National Expansion and the Assimilation of Southern Islanders

Okinawans are the people of the Ryūkyū Archipelago that extends between the main Japanese islands and Taiwan. While today Okinawans are Japanese nationals, they were not until recently subject to the strictures of Japanese society. Throughout the first millennium of Japanese history, the Ryūkyū Islands were generally independent of outside domination. The Okinawan Kingdom paid nominal tribute to

Chinese courts. The geographical location of the island chain – called Liu-ch'iu in Chinese – made it a principal trade route between China and Japan. From 1609 and throughout the Tokugawa period (1600–1868) – while main island Japanese lived under the close surveillance of Tokugawa rulers in Edo [now Tōkyō] – Okinawans endured the distant suzerainty of the Shimazu clan of Satsuma on Kyūshū. Formally, however, the islands remained tributary to China.

Okinawans first experienced the rigid legacies of Tokugawa Japanese society in 1872 when the new Meiji government – despite conflicting Chinese claims – extended its national boundaries to include the Ryūkyū Islands. The islands were annexed in 1879 and made a prefecture. Japanese sugar companies promptly infiltrated Okinawa upsetting the traditional economy and leaving thousands of farmers landless. The Okinawan people faced considerable pressure – brought to bear principally through the Ministry of Education – to assimilate into the mainstream of Japanese culture, in particular to adopt the Japanese language in place of Okinawan.

The transition to Japanese rule was not easy. Okinawans were helpless to prevent key posts in the new island government from being filled predominently by Japanese from the main islands. Newcomers by the thousands formed a new elite. Okinawans were subjected to both social and political discrimination by main islanders who enjoyed income differences and other preferential treatment in their quasi-colonial posts. Japanese businessmen and bureaucrats visiting Okinawa brought back bizarre stories of the "unsophisticated" and "strange" speech and manners that set Okinawans apart as degraded "country cousins" in the new Japanese family. (Kerr 1958:393, 399)

When Japanese main islanders were swept up in the mass hysteria that led Japan to war with America in 1941, Okinawans were reportedly unenthusiastic, having no tradition of glorifying the warrior in battle. Moreover, Okinawans clearly anticipated the inenviable role their outlying islands would undoubtedly play in a defence of the main islands. Military officers assigned to Okinawan garrisons publicly criticized the spirit and conduct of young Ryūkyū islanders, castigating their allegedly traditional easy going manners. Ethnic slurs of this kind symbolized majority Japanese distrust of Okinawan loyalty and were deeply resented. (Kerr 1958:462)

The premonitions many Okinawans had about the militarization of their homeland proved tragically justified. Some 62,489 Okinawan civilians were figuratively crushed to death between the Japanese anvil and the American hammer in the Battle of Okinawa in the spring of 1945 (Kerr 1958:472). Japanese polity held the Ryūkyū Islands to be expendible under conditions requiring their sacrifice in the national interest. The Japanese sense of ethnic identity allegedly centres on the home prefectures. Okinawa lacks territorial status for not having been a part of Japan at the dawn of Japanese history (Kerr 1958:10).

Okinawans suffered a quarter century of American military occupation, the last years of which were particularly tense and at times riotous because of Ryūkyū base involvement in the Vietnam War. But ambivalence about returning to Japanese administration in May 1972 not surprisingly became more widespread after reversion. The people of Okinawa correctly anticipated a flood of Japanese capital with inevitable inflation, monopolization, and industrial pollution. Okinawans migrating to the main islands – again as Japanese citizens – looked forward to discrimination in education, employment, and marriage.

While it is too soon to know to what extent Okinawan fears are justified, it is clear that there is not much room in Japanese government administrative philosophy for the sensitivities of 1,000,000 citizens living on the periphery of the "homeland." Most symbolic of the renewed Japanese exploitation of the southern islands is the Okinawa Development Agency established by the Japanese government at the time of reversion for the purpose of "correcting the gap with Japan proper" [*hondo nami*].

Okinawans are notably apprehensive of their future as a restored territorial arm of a revitalized Japanese industry and a possibly re-emerging military. But they are anxious also about continuing main island Japanese sensitivity to slight physical differences such as hairiness and skin colour (Maretzki 1964:104). A Japanese government source – in describing the physical characteristics of the Japanese people – observes that, "There are small local differences in skin colour: southern Japanese have darker skins than northerners, [and] the inhabitants of the Ryūkyū Archipelago [have] the darkest skin of all (Japanese UNESCO 1964:81)."

VI. CHINESE: PRELUDES OF IMPERIALISM AND REVERSALS OF HISTORICAL ROLES

Chinese in Japan symbolize two millenia of some degree of contact with continental culture. Chinese travellers to ancient Japan filed reports of their observations, a few of which survive today as the earliest external accounts of Japanese national character. Other Chinese migrating to Japan before the 8th century served as conveyors of Chinese and Indian religious thought, literature, and art. But most Chinese came to Japan as traders and merchants, sometimes through Korea but usually by way of Okinawa. Chinese have therefore always been present – however small in number – in the lively ports of Kyūshū – southwesternmost of the main Japanese islands – as well as in major cities around the Inland Sea.

From the 8th century until the beginning of the Meiji period, however, Chinese did not appear in substantial numbers as residents in Japan. After 1868 – while China continued to have difficulties coping with Euroamerican imperialism and numerous domestic problems – Japan set an example of national self-assertion that attracted not a few Chinese merchants, entrepreneurs, and political activists. When Japan humiliated China in the Sino-Japanese War of 1894–1895, even students of Japanese technology began streaming to the island empire from the continent. The Chinese student population in Japan rose from 18 in 1898 to some 13,000 by 1906, though not all succeeded in their academic studies if many even seriously tried, as reflected in the small yet impressive number of Chinese who graduated from Japanese colleges and universities. But Japan provided a less expensive source of technical training – and shelter from social unrest in China – than distant Europe or America. Not a few scientific and political terms coined by Japanese using Chinese characters found their way back to China as the ancient teacher-student roles were figuratively reversed. (Fairbank et. al. 1965:618; Miller 1967:260–261)

Many Taiwanese also came to Japan in the decades following 1895 when Formosa became part of Japanese territory. Japanese activities in Manchuria added also northern continentals to the nominally Chinese speaking population in Japan, though not in great numbers [Table 2, note 3]. Nor did very large numbers of any of these Chinese groups come to Japan as did Koreans during the Pacific War. At the end of the war, some 60,000 Chinese speaking residents in Japan – about half of these Formosans and half continentals – returned to their respective homelands. Another 30,000 Chinese – some two-thirds of these mainlanders – stayed on in Japan where they had made their lives for several decades, and where they found life – despite the occupation – more tolerable than in China where revolution brewed. Throughout the occupation, Japan's Chinese were treated along with Koreans as "third nationals" [*daisankokujin*], and like Koreans became officially foreign nationals after April 1952 when the occupation period came officially to a close (Sōrifu 1949[1]:82–83, 129)

Chinese have been treated worst by Japanese in China. The Rape of Nanking in 1937 is the best known but not the only example of Japanese brutality on the mainland. Formosans received considerably better treatment during their half-century under Japanese rule than did Koreans, who among nearby Asians have been the most consistently abused by Japanese. Chinese living in Japan also have fared much better than Korean residents, but have nevertheless been regarded as inferior people particularly after 1895 when Chinese began coming to Japan as students. (Chao 1970:132–147)

VII. Koreans: Imperialism and the Immigration
of Colonial Subjects

Over 600,000 Koreans are currently registered as resident foreign nationals with the Japanese Immigration Bureau. In addition, an unknown number of Japanese nationals risk discrimination because of real, suspect, or concealed Korean ties, including naturalized Koreans, Koreans who have managed to pass into Japanese society, Korean-Japanese mixedbloods, and miscellaneous craftsmen thought by some majorities to be culturally if not genetically in some sense the descendants of early Korean immigrants.

The prehistories of Japan and Korea are difficult to separate. Eighth century historical documents report that an early Japanese state – probably of peninsular origin – had a foothold in southern Korea and was intimately involved in conflicts between the several Korean kingdoms. By the end of the 7th century – according to an early Japanese peerage – more than one-third of the Japanese nobility claimed Chinese or Korean descent. These Chinese and Korean Japanese taught majority Japanese the philosophically and esthetically rich Indian and Chinese literary, artistic, and religious traditions to which subsequent Japanese culture was to become so heavily indebted. When their roles as teachers gradually passed, the continentals blended into the ethnic brew they had helped ferment. (Sansom 1962:44; Sansom 1963:35)

Japanese pirates, with Chinese and Korean adventurers among their tens of thousands, ravaged towns up and down the coasts of East Asia, particularly of Korea, from the 13th century until proscribed by Toyotomi Hideyoshi in the 16th century. In 1592, however, Hideyoshi himself invaded Korea on his way to conquer China, although the campaign bogged down and was finally halted when the warlord died in 1598. This seven year war – involving also Chinese – left large areas of Korean civilization in ruin and imbued Koreans with an indelible early memory of Japanese brutality. (Mitchell 1967:5–6)

Japan "opened" Korea with an unequal treaty in 1876, and fought in Korea with China in 1894 over territorial interests on the peninsula and in the Pacific. Korea was invaded by Japan in 1904 at the outbreak of the Russo-Japanese War, and was forced into an alliance with Japan as a protectorate of the Meiji state. The peninsular kingdom was annexed in 1910, fulfilling a territorial ambition traceable throughout Japanese history. Korea remained under military rule for about a decade, during which time mutual dislike between the geographical brothers further hardened.

Civil disturbances intensified by chronic unemployment and rapid population increases that continued under the civilian administration motivated hundreds of thousands of Koreans to emigrate to Japan, where life was expected to be better but seldom was. These swelling numbers of Koreans in Japan – many unemployed and politically militant – constituted a major social problem for the Japanese government before and during the Pacific War. Japanese industries, however, demanded cheap Korean labour and contracted hundreds of thousands more to replace Japanese workers as the war took Japan's labour force to the battlefields [Table 2].

Most Koreans remaining in Japan after the Pacific War were among earlier immigrants whose offspring had been born and raised in Japan. Koreans electing to stay in Japan after 1950 were divided half and half between first and second generations (Yi 1960:144), and of first generation Koreans practically all hailed from regions now part of South Korea (Hōmushō 1964:87). Postwar ambivalence regarding their legal status on the part of the Korean minority itself resulted in Koreans in Japan being officially designated aliens when the San Francisco Peace Treaty was ratified in 1952.

The status of Japan's Koreans was further complicated by the fact that Korea had divided into political divisions that did not exist when these Koreans emigrated. Both North and South Korea claimed the allegiance of Koreans in Japan despite the fact that practically all of these Koreans claimed regional origins in what was nominally South Korea. But North Korea most effectively courted the pluralistic sentiments of Japan's Korean minority in ethnic or "national" education and identification, in contrast with the South Korean government, which provided little economic and even less moral assistance and seemed at times – in alliance with the Japanese Ministry of Education – even opposed to such ethnic concerns.

Koreans in Japan became strongly divided politically. Japanese Red Cross data reports that of 563,146 Koreans in Japan in December 1964, some 122,308 or 22 percent aligned themselves with South Korea while 277,321 or 49 percent supported North Korea. The balance claimed neutrality or fell in other categories (Yi 160:112). The Japanese government, under American influence and conservative in its own right, leaned towards South Korea but was pressured from within to enter into an agreement with the North in 1959 that resulted in the repatriation of tens of thousands of Japan's Koreans to the communist sector by 1963 [Table 2, note 5].

In 1965, South Korea reached an agreement with Japan concerning the legal status and treatment of Koreans residing in Japan that made it possible for about 500,000 of Japan's Korean minorities to apply for permanent residence status and attendant social privileges. By the 16 January 1971 deadline, only 351,955 had filed for higher residence status. The Japanese Ministry of Justice reported that the majority of applicants sought permanent residency in order to be eligible for National Health Insurance benefits and compulsory education on the same basis as Japanese citizens. (*Asahi shimbun* 21 June 1971 [morning edition]:2; *Chōsen kenkyū* 106 [July 1971]:30–33)

Unification overtures from both North and South Korea have not resolved the ambivalencies Japan's divided Korean minority continues to harbour towards which Korea should lead reunity. Nor does it seem that after reunification – realistically a generation away – Japan's Koreans will be culturally if even psychologically prepared to return to a foreign "homeland" they never knew the emotional experience of leaving. By then, the Korean minority in Japan will be half second generation and half third, with fourth imminent, in all meaningful respects culturally Japanese.

In the meantime, Koreans in Japan will continue to be stereotyped much as Burakumin are, and they will continue to suffer much the same forms and consequences of social discrimination. Japan's Koreans contribute proportionally more than majority Japanese to the national crime rate – allowing for alien registration misdemeanours – and are more involved than Japanese in marginally legal professions.

At the same time, Koreans are systematically refused jobs in many Japanese companies and are denied entrance to not a few private schools – ostensibly because of their foreign citizenship – and are generally dismissed or expelled when discovered after passing as Japanese to gain employment or admission.

VIII. HIBAKUSHA: CONTAMINATION PHOBIA IN THE SHADOW OF HIROSHIMA

The bombings of Hiroshima and Nagasaki in 1945 created a new ethnic minority in Japan that shows every sign of perpetuating itself and being perpetuated into future generations. The variety of physical and psychological problems that plagued survivors in the weeks, months, and years following the two holocausts generated widespread doubts among Hibakusha and nonhibakusha alike concerning the genetic wholesomeness of Hibakusha parents and their descendants. Conspicuous atom bomb afflictions gave rise to the kinds of discrimination commonly suffered by maimed and scarred individuals around the world. But the ailment most alarmingly imputed to persons even remotely exposed to the flashes was "invisible" in the darkest recesses of the body.

Hibakusha [persons who experienced the bomb] are popularly believed to carry defective genetic mutants expected to manifest in subsequent generations. This fear of contamination has placed considerable psychological stress on Hibakusha identity and has raised difficult barriers for Hibakusha seeking marriages with nonhibakusha majorities. Atom bomb survivors are sometimes described as nuclear outcastes, and Hibakusha communities have been called *genbaku buraku* [atom bomb settlements]. The comparison with Burakumin is not without a tragic irony, however. Hiroshima has traditionally had large Burakumin communities, and not a few Hibakusha carry a double burden that tends to make them minorities within their own minority group. (Lifton 1969: 165–208)

Also among Hiroshima and Nagasaki victims and survivors were tens of thousands of Koreans and thousands of Chinese. Korean and Chinese Hibakusha returning to their homelands after the war have experienced great difficulty in securing adequate medical attention for their chronic atom bomb ailments. In 1972, however, repatriated and resident Korean atom bomb survivors publically protested on behalf of all internationally isolated Hibakusha. The movement generated considerable sympathy among sensitive newspaper columnists, and apparently among judges as well. (*Asahi shinbun* 28 July 1972 [evening edition]:11)

A landmark decision in a Fukuoka Prefecture court on 30 March 1974 held that a repatriated Korean Hibakusha convicted of illegally re-entering Japan nevertheless qualified for medical care and could not be deported. It was ruled that the standing Atom Bomb Medical Treatment Law [Genbaku Iryō Hō] – which provides for the care of atom bomb survivors regardless of citizenship – should be broadly interpreted to mean that Hibakusha be issued treatment passbooks also regardless of place of residence, and in the case of foreign national Hibakusha residing overseas, regardless of whether at the time they request treatment they are in Japan as tourists or illegal entrants. The decision was expected to invite an increase in the incidence of illegal entry across the Tsushima Straits among some 20,000 Korean Hibakusha repatriots in South Korea.

In late 1970, a popular series of children's comic magazines was criticized for featuring monsters with keloid hides as objects of disparagement (*Asahi shinbun* 30 November 1970 [morning edition]:13). A Hibakusha writer of caricature fiction counters such inadvertent media discrimination with stories that graphically express the sociopsychological problems that atom bomb survivors experience in their struggle to cope with their alleged impurity (*Shūkan asahi* 2747[13 August 1971]:126–129).

IX. KONKETSUJI: MISCEGENATION AND AMALGAMATION
IN THE WAKE OF CULTURAL CHANGE

One Japanese observer is reported to have said that there are some 200,000 mixedblood citizens in Japan today, of which 40,000 may be Black-Japanese (Takasaki 1953:194). Authorities have put the number of Eurasian Konketsuji between 10,000 and 20,000, while other estimates run as high as 50,000 (Trumbull 1967:112). Yet another writer cites guestimates from 4,000 (including 500 Black-Japanese) to 50,000 (including 10,000 Black-Japanese), then states matter-of-factly that some 20,000 mixedbloods have been fathered by American servicemen in Japan since the end of the Pacific War, that most of these Konketsuji have been illegitimate, and that 2,000 of them are offspring of Black-Japanese unions (Thompson 1967:44, 46).

The Japanese Ministry of Welfare reported that by February 1953 as many as 3,289 mixedblood children had been abandoned in Japan by their foreign fathers. By December 1952, when a new immigration law went into effect simplifying legal registration of international marriages, American consulates had granted passports to some 2,585 other children born of interracial unions in Japan. Passage of the Refugee Relief Act in 1953 – which permitted up to 4,000 orphans under ten years of age to be admitted to the United States if adopted by an American citizen with a spouse – and later legislation providing children of American-Alien parentage with nonquota visas, had allowed some 700 Black-Japanese to be adopted by Black families in America by the middle of the Sixties. (Graham 1954:330; Thompson 1967:49)

In contrast with nominal American encouragement for mixedblood immigration, the Japanese government made Brazil the target of an emigration movement whereby Black-Japanese in particular would be resettled in South America. In the summer of 1967, the director of the Elizabeth Saunders Home – the best known institution for the care of Konketsuji orphans in Japan – was turned down after several attempts to resettle Black mixedbloods. The Brazilian government wanted to know why it should assume the responsibility for children neglected by the countries of their parents and virtually made outcast by the country of their birth. (Thompson 1967:54)

Offspring of interracial unions are not new to Japan, which has witnessed considerable gentic mixing throughout its history. Particularly around its periphery – where contact with nominally nonjapanese was common – the Japanese archipelago has experienced a "mixedblood problem" [konketsuji mondai] for thousands of years. While Eurasians born after the Pacific War are statistically the most visible Japanese-Nonasian Konketsuji in Japan, they are not the first historically, nor were postwar Black-Japanese the first mixedbloods to face banishment.

Decrees throughout the 17th century dealt with the Eurasian progeny of hairy foreigners and Japanese women, often requiring that such children be deported to the Asian colonies of European homelands of their fathers. The Tokugawa hegemony was bent on sweeping Japanese soil clean of all subjects who had Red-haired-men [kōmōjin] from Holland, or Southern Barbarians [nanbanjin] from Portugal or Spain for fathers or grandfathers – until edicts in the early 18th century forbade Konketsuji from leaving Japan. Foreigners on the island of Deshima in Nagasaki Bay during the Tokugawa period (1600–1868) had access to special pleasure quarters in the city, but these districts were prohibited to the few Blacks who had begun to appear in the port. Women found entertaining Blacks were harshly punished, as were their patrons. (Takasaki 1953:195–197)

Japan's largest encyclopedia reports that a July 1959 Ministry of Education survey found 2,401 Konketsuji registered in elementary and junior high schools throughout Japan, with 2,115 or 88 percent of these children in the first six grades. A peak of three to four hundred mixed-blood students completed their compulsory nine-year education between 1964 and 1965 and were looking for jobs if not going on to high school. A study of the world of Japanese entertainment leaves a clear impression that a large proportion of Japan's postwar mixedbloods have found careers as models, singers, and dancers, if not as hostesses and waiters in drinking establishments. (Heibonsha 1964–1968[9]: 199–200)

Postwar Konketsuji seem the most commonly featured minority group in Japanese popular arts. Some caricature fiction and derivative films centre around mixedblood characters often depicted in deviant roles. Japan's mixedbloods are stereotyped as fast, loose, and confused. Many Konketsuji invite these images through the lifestyles they inadvertently choose, but no evidence whatsoever supports the popular notion that such social tendencies are attributable to genetic mixing rather than widespread discrimination.

Pulp magazines in Japan give direct insights into Japanese interethnic sexuality. Eurasian if not caucasian models regularly frequent nude gravure sections in not a few popular periodicals. Mixedblood models are almost always White-Japanese, as Black-Japanese usually manifest physical features most despised by majorities. But despite superficial inclinations of the indigenous paradigm of physical beauty towards nominally Eurasian features – witnessed particularly in contemporary fashions and cosmetics – Japanese continue to be essentially proud of their own attributes. (Wagatsuma 1967)

X. KIKAJIN: ACCOMMODATING FOREIGNERS LOOKING FOR A HOME

Some 28,579 persons holding foreign citizenship were naturalized in Japan in the first decade of postwar independence from 1952 to 1962 [Table 2]. 3,928 or 14 percent of these "new" Japanese citizens were former Japanese nationals. 1,152 or 4 percent were spouses of Japanese nationals; 3,105 or 11 percent were offspring of Japanese nationals;

8,828 or 31 percent were offspring of former Japanese nationals; 11,108 or 39 percent were "pure" Koreans; and 386 or one percent were "pure" aliens other than Koreans, while another 72 naturalized persons were adoptees. (These and following figures computed on basis of data in Hōmushō 1964:91–92)

Ninety percent or 25,723 of these 28,579 Kikajin had nominally been Korean nationals before naturalization as Japanese citizens. Of these Korean nationals, some 2,867 or 11 percent were women twenty years old or older who had once been Japanese citizens. 885 or 3 percent were spouses of Japanese (606 or 68 percent of these spouses were women). Offspring of parents one or both of whom were Japanese citizens accounted for 2,310 or 9 percent (1,756 or 76 percent of these offspring were of Japanese women who had married Koreans). In contrast, the progeny of parents one or both of whom formerly held Japanese citizenship accounted for some 8,411 or 33 percent (all of 8,387 or 99.7 percent of these offspring were of women who had married Koreans and surrendered their Japanese nationality). 11,108 or 43 percent of all Korean nationals who acquired Japanese citizenship during this ten year period were "pure" Koreans, that is, Korean nationals claiming no relationship – genetic or marital – with Japanese. The remaining 142 or one percent were divided between adoptees and former Japanese citizens in other categories.

Stated yet another way – some 14,473 or 56 percent of all Korean nationals who acquired Japanese citizenship during this ten year period did so on the basis of genetic or marital relationships with majority Japanese. Some 10,143 or 70 percent of these 56 percent – or 39 percent of all Korean citizens who took Japanese citizenship – were the children of mothers who held or formerly held Japanese citizenship. In short – not only were most [56 percent] of the Korean nationals who gained Japanese citizenship during this decade categorically not "pure" Koreans, but most [70 percent] of those who were not "pure" Koreans were apparently the offspring of Japanese women who had married Korean men.

These multidimensional figures suggest that naturalization is not an inviting pathway of identification except for Korean nationals who already have a genetic foot in the door of Japanese ethnicity. The notably low naturalization rate among ethnic minorities not having genetic ties with Japanese or former Japanese majorities may reflect the ambivalent attitude these minorities have towards becoming nominal members of Japanese society. Indeed, it is not difficult to conclude from these figures that motivation seeking naturalization in Japan is primarily familial in nature and involves the satisfaction of needs to cultivate formal relations with the Japanese half of one's biethnic self.

Legal status as a Japanese citizen does not reduce the visceral disdain that Japanese majorities are capable of displaying towards Kikajin, particularly those naturalized from other Asian nations. Foreign nationals who take Japanese citizenship continue to be regarded as *gaijin* [outsiders, i.e., foreigners] by many majority Japanese, but endure the process in the hope that their offspring may find a greater nominal kinship with their contemporaries. This becomes particularly important for ethnically mixed citizens who have majority relatives. Bicultural and biracial children must covet close ties with their Japanese majority kin if they are to have full access to the nurturing benefits of family belonging enjoyed by most majority citizens.

While naturalization in Japan does not necessarily benefit directly those who change citizenship, it would be wrong to leave the impression that Japan is peculiar in this regard. It is sometimes pointed out that Japanese sharply distinguish between themselves and "foreigners," while residents in the United States supposedly believe that everyone living in America is more or less "American" regardless of whether native born or how long they have resided in the country if an immigrant.

While there is a notable tendency for Japanese to believe that only the descendents of Japanese can be truly Japanese, America is not the haven it is often made out to be for ethnic pluralism. The spiritual "americanness" of several noneuropean minority groups in the United States is widely doubted by European ancestry majorities. A case in point is the treatment of Asian Americans generally and Japanese Americans particularly. The widespread "suspicion" that popularly supported the unconstitutional encampment of Japanese Americans during the Pacific War has not yet diminished. Not a few European ancestry majorities continue to regard Japanese Americans with a sense of disbelief that "they" are really "true" countrymen like "we" are.

XI. Kaigai no Nihonjin:
The Defiling Experiences of Japanese Citizens Overseas

Some 267,246 Japanese citizens were reported living abroad in 1970, of which 144,853 or 54 percent were in Brazil while 173,382 or 65 percent were in all of South America. Another 52,161 or 20 percent were in North America, all but 4,172 of these in the United States. Among these overseas Japanese – along with resident immigrants – are diplomats, professionals, businessmen, students, and other Japanese who intend to return to Japan within two or three years. (Asahi 1973[1]:532)

Not a few Japanese are hesitant about company or government assignments in foreign countries, afraid that time and experience overseas may retard their advancement in highly inbred, permanent employment systems back home. Japanese graduates of American universities and Japanese professors who spend too many years abroad as visiting faculty may find upon their return to Japan a sense of distance if not hostility directed towards their *batakusai* [butter reeking] professional attitudes and career expectations. Young Japanese women who work or study in North America or Europe may experience some difficulty in courting a husband in Japan if it is suspect that their experiences abroad have corrupted their views of sexual status in Japan, or worse, included companionships with foreign men.

Japanese nationals returning "home" after living overseas for long periods of time have traditionally been viewed with mixed feelings by their stay-at-home countrymen. Japanese [*nihonjin*] abroad [*kaiga*] are thought to lose their "japaneseness" in proportion to their exposure to foreign cultures. In the 17th century, the Tokugawa government curtailed the spirited migration trend that had begun in the 16th century and resulted in Japanese settlements in countries throughout Southeast Asia and the East Indies. Decrees issued between 1633 and 1639 excluded Japanese who had been outside Japan for more than five years, and provided death penalties for those who attempted to return to Japan from a foreign territory – and for those who tried to leave Japan – without a valid license. (Sansom 1963:36)

The decrees re-enforced antichristian orders issued between 1611 and 1614, and coincided with the Shimabara uprisings from 1637 to 1638 that resulted in the massacre of tens of thousands of Japanese Christians on Kyūshū, and the systematic persecution of Christians scattered throughout Japan. A 1636 decree dealt with the children and grandchildren of foreign fathers and Japanese mothers, while another order posted that year required all foreigners in Japan to move to the island of Deshima in Nagasaki Bay. The seclusion edict that nominally closed Japan until the 19th century was promulgated in 1639.

Japanese who continued to take an interest in foreign culture during the centuries of relative isolation were regarded much like Japanese translators are today. Conservative Japanese have usually recognized the need to maintain some semblance of a relationship with the outside world, but a countryman assuming the role of cultural intermediary has traditionally been considered somehow soiled by his expertise in foreign customs and languages. Approaching the Meiji period (1868–1912), when enormous emotions were rallied against the reopening of Japan to foreigners, not a few Japanese advocates of *sonnō jōi* [revere the emperor and repel the barbarian] advised against Euroamerican relations because such contact "polluted" Japan in the indigenous Shintō sense of the term. Japanese who traded with the Dutch were considered fools, and those who pursued Dutch learning [*rangaku*] were viewed with great fear and contempt. (Harootunian 1970:265)

Japanese women who became mistresses to foreign dignitaries in Shimoda and other ports opened to barbarian traffic in the middle of the 19th century were scorned by their contemporaries, but were later depicted as citizens who had been sacrificed for their country (Statler 1971:560–561). The theme of martyrdom is found also in the occupation period prostitute, and in Japanese women who continue to act as repositories of the contaminants of American military personnel still in Japan under the auspices of the Mutual Security Agreement. Even businessmen and government officials who reluctantly live abroad for their companies and country evoke strong sentiments of martyrdom.

But Japanese are not the only people who tend to view intimate foreign experiences with ambivalency and suspicion. Nor are Japanese companies and government bureaucracies unique among business and official organs elsewhere in their concern with the loyalty of their employees and officers assigned abroad. The United States Department of State, for example, is known to have discouraged if not prohibited marriages between its career foreign service officers and nonamericans, it being popularly feared that international marriages divide national loyalties. Even the assignment of area specialists to foreign countries is rare, as the specialist may genuinely respect if not covet the values of his adopted culture, and be that less able to serve the ethnocentric interests of his home office. Provinciality – if this is the appropriate word – is in no sense a Japanese specialty, nor viewed objectively is there much evidence that Japanese are better at ethnocentricity than other peoples around the world. Americans in Tokyo and Hong Kong enclave as much as Asians in San Francisco.

XII. Nikkeijin: Status Deprivation of Foreigners of Japanese Ancestry

By October 1969, there were reported to be 1,062,293 foreigners of Japanese ancestry in countries throughout the world. Some 491,418 or 46 percent of these Nikkeijin were settled in Brazil, while 554,744 or 52 percent were found in all of South America. In contrast, the United States was listed as having 464,587 or 44 percent of all foreign citizens thought to be of Japanese ancestry. All of North America accounted for some 494,153 or 47 percent of the world total. Of an estimated 1,329,539 Japanese citizens or persons of Japanese ancestry living abroad in 1969 and 1970, a total of 728,126 or 55 percent were in South America with most of the balance in the United States. The statistical presence of Japanese or persons of Japanese ancestry in Brazil – about one percent of that nation's 1970 population of 95,305,000 persons – is similar to the numerical presence of Korean nationals in Japan. (Asahi 1973[1]:191, 532)

The Japanese treatment of Nikkeijin [persons of Japanese ancestry] by Japanese citizens living temporarily abroad – and the treatment in Japan of Nikkeijin who have come to sightsee, study, work, or visit relatives – is not unlike the treatment in Japan of citizens returning from extended stays overseas. The two groups are sharply distinguished, however, in several important respects. Not only are Nikkeijin citizens of foreign countries, but they exemplify patterns of behaviour compatible with those of their majority countrymen. Many second generation [nisei] and most third generation [sansei] Nikkeijin have no competency in the Japanese language, and are kinesthetically identifiable among Japanese.

While many Japanese sense a physical affinity with their surname and phenotypic kinfolk abroad – and while some Nikkeijin view Japanese with a similar feeling of cultural and racial brotherhood – this common attraction affords Nikkeijin little more than a tentatively closer initial relationship with Japanese in Japan and overseas. The fact that Nikkeijin are ultimately not Japanese in the Japanese view, however, gives rise to discrimination as in marriage with Japanese, and may lead to conflicts of interest as when Japanese financiers and developers overseas take advantage of superficial affinities in the process of establishing often exploitive footholds in Nikkeijin communities.

Japanese visiting Japanese language classes in North America are sometimes surprised at the number of "Japanese" studying "their" language. Although themselves foreigners when outside Japan, Japanese living abroad often continue to refer to Euroamerican majorities as *gaijin* [aliens] while excluding Nikkeijin from this category. Thus Japanese are known to use the word *amerikajin* [American] with reference to White Americans if not also Blacks, but tend to distinguish Japanese Americans as Nikkeijin if not *nihonjin* [Japanese].

Nor do Japanese regard Nikkeijin coming to Japan from America or Canada, for example, as "real" Americans or Canadians. Japanese tend to feel that being a "real" foreigner has genetic requisites that categorically disqualify Nikkeijin from being genuinely foreign. Japanese Americans and Japanese Canadians are sometimes denied teaching positions in fly-by-night English schools in Japan because of popular beliefs that their English – though natively acquired and college polished – is somehow not "pure" enough to serve as a model for Japanese students. Yet Japanese employers at such schools will accept Europeans whose command of English in no sense qualifies them as teachers of the language much less as models of usage, but whose faces satisfy the often more important requirement that teachers of a foreign language ought to "look" foreign. (Okimoto 1971:178–179)

To make matters more difficult for Nikkeijin living in Japan, among some English schools willing to hire them have been known to require that "Japanese looking foreigners" dress and behave like "proper" Japanese – at least less radically or conspicuously "foreign" than their European ancestry faculty fellows. Some school administrators have felt that their students will identify more with Nikkeijin teachers and aspire towards Nikkeijin lifestyles, whereas "genuine" foreigners would be viewed more like carnival performers observed by spectators mainly for amusement.

Foreigners of Japanese ancestry are criticized by majority countrymen when they fail to reflect fully the values of the majority community, and they are criticized by Japanese for not retaining the values of their surname relatives in Japan. Nikkeijin are not as overtly discriminated against in Japan as Burakumin, Koreans, and Konketsuji, nor are they as covertly degraded as Euroamerican caucasians. But they are subjected to a more subtle and for this reason painful form of discrimination whereby status is deprived in a crossfire of conflicting expectations. If not considered ethnically "impure" by Japanese, Nikkeijin are often looked down upon as offspring of immigrants sometimes thought to have shown their disloyalty by leaving Japan. Yet

Japanese citizens venturing overseas do not often plan to stay permanently. Years simply become decades as the point of easy return recedes from their horizon. Before they are fully aware of it, they find themselves *issei* [first generation] immigrants replenishing the roots of Nikkeijin ethnicity.

XIII. Aoime no Gaijin: Special Ttreatment of Euroamericans as a Form of Degradation

The term "aoime no gaijin" [blue-eyed foreigner] and its abbreviated form "aoime" [blue eyes] are used rather freely by Japanese to designate caucasian foreigners. It makes little difference that individuals called "aoime" may not have blue eyes. Japan's Emigration and Immigration Control Law defines *gaikokujin* [alien] as "a person who does not possess Japanese citizenship" (*Koria hyōron* 122[May 1971]:50). But one must note that such legal definitions seldom regulate the emotional meanings that words assume in vernacular usage. An observant Japanese psychologist has pointed out that *gaikokujin* [literally "outside country person" or "foreigner"] and its more common colloquial abbreviation *gaijin* [outsider] conjure up caucasian [*hakujin*] images for many Japanese users (Minami 1971:106–107).

The basis for this often restricted meaning of *gaikokujin* and *gaijin* is attributed to a tendency for Japanese to fantasize their world in terms of a "heterogeneous and global" yet somehow "monolithic west" [Europe and America] versus a "homogeneous and insular Japan" [ambivalently "east" yet "unique in Asia and the rest of the world" for many Japanese and foreigners]. Japanese are often said to harbour a sense of inferiority towards Euroamerican caucasians while they feel superior to other Asians and Third World peoples. Japanese themselves have described their alleged "middlority complex" as a case of "worshipping whites and despising blacks (Suzuki 1973:57)."

The presence in Japan of European ancestry foreigners from Europe and North America has always been a source of anxiety for Japanese, who have sometimes blamed national or local calamities on resident aliens. The "blue-eyed" co-pilot of the ill-fated Japanese YS11 "Bandai" that crashed in Hokkaidō in early July 1971 was rumoured in the mass media – within hours after the plane had been reported missing, before the wreckage had been discovered much less an investigation made – to have been the "cause" of the accident (*Asahi shinbun* 5 July

1971 [morning edition]:2). Foreign pilots in Japan reacted promptly, charging that such premature conclusions exemplified the nativistic attitude with which they felt Japanese regard their homeland, and the racist attitude they felt pervaded the way Japanese may suspect the ability of nonjapanese to navigate Japanese in *fūdo* [(mystical) climate and (mystical) terrain] (*Japan Times* 18 July 1971; 5 August 1971:14).

Not a few foreigners who have seriously attempted to assimilate into Japanese society – including some who have naturalized and taken Japanese names – describe a great reluctance on the part of Japanese majorities generally to recognize the "visible" foreigner as a genuine member of the national much less local community. But totally committed aspirants are not well represented in Japan. They are usually minorities among an enclaved, protective foreign population that every bit deserves the skepticism Japanese tend to harbour about the ability of Euroamericans to get inside Japanese culture. (Bell 1973)

One of the most common complaints voiced by Euroamerican residents in Japan concerns the word *gaijin* as used by Japanese, particularly children. The English language press in Japan abounds with letters-to-the-editor decrying repeated experiences of being pointed at and called *gaijin* by Japanese children if not adults. The word is considered somewhat offensive by resident foreigners marginally familiar enough with the Japanese language to detect the word when used towards them. Less known to the paranoid foreigner is the occasional rendering of *gaijin* as *jingai* to give it both a more secretive and pejorative character. (*Japan Times* 31 October 1971; *Japan Times* 16, 22, 28 November 1971; *Japan Times* 6, 9, 21 December 1971; *Shūkan yomiuri* 1203[2 February 1972]:148–149)

Caucasian foreigners are frequently used in television and magazine advertising in Japan. *Aiome no gaijin* appeared in a recent promotion of a kind of soup that excites Japanese nostalgia for their national cuisine. Despite the tendency for Japanese to feel that Euroamerican foreigners are not quite capable of appreciating the textures of Japanese foods, the television viewer was shown that this brand is the one to buy because even *gaijin* ask for second helpings. Yet one could not help but feel – after several viewings of the appeal – that what was intended to sell the soup was not the explicit message, but its humorous implications that the reason the soup is excellent is because foreigners have so conspicuously pretended to like it.

The Euroamerican foreigner in Japan is expected to like what is classic and refined about Japanese culture while at the same time it is popularly felt that he really doesn't know what he is involved in. The mossy and subdued are supposed to have an exotic appeal even for nonjapanese. But relishing what is considered the core of Japanese ethnicity is not seen as tantamount to understanding Japan. It is viewed instead as superficial appreciation. What the foreigner often dislikes most about Japan – its material "westernization" and industrial clamour – is what most Japanese take for granted as part of their daily lives. The Japanese marketing technician who suggested that *gaijin* be used to sell Japanese what is felt to be most Japanese well understood the reverse psychology that allows foreigners to regard as an "honour" what is really a form of degradation.

XIV. Purity Myths and Ethnocentricity

Japanese tend to view "racial discrimination" [*jinshuteki sabetsu*] as a problem of other countries in the world but not of Japan. The term *jinshu* [variety of man] is ordinarily used to designate groups like "mongoloid" or "caucasoid" or "negroid," but is seldom applied to groups like "Japanese" or "Korean" or "Chinese," which are commonly referred to as *minzoku* [group of people]. Accordingly, Japanese society is seen as supporting "racial discrimination" [*jinshu sabetsu*] only in reference to the Ainu, who are often characterized as "caucasoid." Treatment of Koreans in Japan is viewed as a case of "ethnic discrimination" [*minzoku sabetsu*], whereas discrimination towards Burakumin is thought to play essentially the same "objective role" [*kyakkanteki yakuwari*] as "racial discrimination" in the broader sense of the term in countries throughout the world. (Suzuki 1973:56–57)

The term "racism" in English, used vernacularly in reference to contemporary social issues, has much broader applications. The American quite easily labels this "racist" and that "sexist" without many technical restrictions, and tends to believe that such "isms" are universal even though not always recognized in racist and sexist societies. In Japan, the word *sabetsu* [discrimination] is used widely with a number of prefixes but not often vernacularly with *jinshu* [race] by majorities in reference to minority problems in Japan. Nor is *minzoku sabetsu* [ethnic discrimination] commonly used except with broader implications of race than culture.

It must be noted, however, that the word *minzoku* is often translated "race" and with some justification. The colloquial, emotional use of the term – as it is frequently suffixed to nationalities – is very definitely with a sense that a *minzoku* has very special attributes that distinguish it from all other *minzoku*. This is not altogether unlike our use of the term "ethnicity" throughout this report, with the exception of the emphasis that *minzoku* seems often to place on the physical or genetic dimension of ethnicity. The words *nippon minzoku* are particularly used with such a visceral sense of absolute ethnic distinction with a genetic flavour that one cannot easily avoid "Japanese Race" as a suitably emotional English equivalent. In not a few English language publications by the Japanese government, the word "race" is used with reference to the "Japanese people" both consistently enough, and in sufficiently "racist" contexts, to warrant our suspicion that – technical definitions aside – Japanese are as "racist" as Europeans and Americans but simply will not describe themselves as such.

A number of surveys have shown, however, that Japanese have specific preconceptions [*sennyūkan*] concerning physical types, and specific biases [*henken*], including worshipping Whites and despising Blacks. Koreans are also despised, according to other surveys, second only to Blacks (Mitchell 1967:133; *Shūkan bunshun* 760 [20 January 1974]:25). The prevailing view, however, is that discrimination against other Asians does not result from "racial" preconceptions or biases, but from "objective" *minzoku* distinctions [*minzoku sabetsu* being here also *jinshu sabetsu* in the wider vernacular senses of both terms]. A prominent Japanese social scientist, in commenting on these nominalistic problems, has stated emphatically that, "In Japan, racial discrimination as a *social system* does not exist" (Suzuki 1973:57; italics in original).

The differential treatment of Black versus White mixedbloods in Japan suggests that Japanese are somehow sensitive to "colour" if not "race." Yet the director of the Elizabeth Saunders Home is quoted to have said that the prejudice Japanese harbour towards Konketsuji, especially Black-Japanese, "is a moral judgement rather than colour consciousness (Trumbull 1967:113)." Supposedly, if mixedbloods in Japan are discriminated against, it is because some were the offspring of foreign soldiers and Japanese prostitutes, or because some were born out of wedlock if not also abandoned by their parents. The tendency for Japanese to deny a colour or racial basis for Konketsuji discrimination is widespread. Japan's largest encyclopedia specifically observes that, "Japan's mixedblood child problem [*konketsuji mondai*] is not an allegedly race problem (Heibonsha 1964–1968[9]:200)."

Yet the overwhelming evidence is that Japanese have been colour sensitive throughout their history, in which light if not white skin has been highly regarded and cultivated, while dark if not black skin has been viewed negatively and avoided (Wagatsuma 1967:407–415). A deep pigmentation of the skin is sometimes imagined to be an identifying characteristic of Burakumin and Koreans, and is also said to stigmatize atom bomb survivors (Lifton 1969:170–172). The contemporary emphasis in Japanese cosmetics on literally snowy or milky skin, smooth as silk and clear as honey, continues a tradition traceable from the court literature of the Heian Period (794–1185) to the pulp fiction of the present era.

The concept of Japanese "racial" uniqueness cannot be understood except as part of a larger concept of "cultural" uniqueness, and neither of these notions of "uniqueness" can be adequately articulated without recourse to a "purity "myth. The belief that Japanese racial and cultural ethnicity is not simply "different" from other national ethnicities but is "uniquely different" and somehow "pure" pervades the Japanese language press on Japan. An official publication intended for foreigners reading English – compiled by the Japanese National Commission for UNESCO under the auspices of the Japanese Ministry of Education – assumes such a thesis of "unique difference" and "purity" throughout its thousand pages. The opening lines set the tone for the entire volume and are worth repeating here at some length (Japanese UNESCO 1964:5; italics ours).

The Japanese race has formed a unique mode of living and shaped an extremely individual pattern of culture in the past two thousand years. The Japanese have maintained individual characteristics through every stage of their history – ever since the first ancient state emerged and established its domain over central and western Japan, between the third and sixth centuries. These traditions lived through the social and political reformations patterned after the states of the West in the Meiji Era, which spanned the last half of the nineteenth century.

It cannot be denied that every race has its own hallmark imprinted in the process of cultural evolution in its particular geographical surroundings. But the Japanese race has, besides the characteristics derived from the geographical and climatic peculiarities of the land, a very individual mode of living and a *special* pattern of culture which *almost defies comparison* with those of other peoples. ...

What is truly remarkable in the cultural attitude of the Japanese race is the coexistence of exclusive and receptive tendencies. The former is best exemplified by the fact that the Japanese people have chosen to retain a mode of living and a pattern of culture that are *purely and peculiarly* Japanese in character. The peculiar mode and pattern of this way of life *cannot* be

(continued on page 370)

TABLE 1: *Estimated Ethnic Minority Populations in Japan (Circa 1974)*

Japanese Population Categories	1. All Residents	2. Japanese Nationals 1.—6.	3. Japanese Majorities 2.—5.	4. All Minorities 5.+6.	5. National Minorities 5. below	6. Foreign Minorities 6. below
Estimated % of 1.	110,000,000[1] 100%	109,000,000 99%	105,500,000 96%	4,500,000 4%	3,500,000 3%	1,000,000[2] 1%
Published % of 1.	104,665,171[4] 100%	103,956,713 99%	101,513,981 97%	3,151,191 3%	2,442,733 2%	708,458[3] 1%

5. National Minorities	5a. Burakumin	5b. Okinawans	5c. Hibakusha	5d. Kikajin	5e. Konketsuji	5f. Ainu
Estimated % of 5.	2,000,000[5] 57%	1,000,000 29%	300,000 9%	100,000[6] 3%	80,000[7] 2%	20,000 1%
Published % of 5.	1,113,043[8] 46%	945,111[9] 39%	290,000[10] 12%	28,579[11] 1%	50,000[12] 2%	16,000[13] 1%

6. Foreign Minorities	6a. Koreans	6b. Chinese	6c. Americans	6d. English	6e. Others	6f. Tourists
Estimated % of 6.	670,000 67%	70,000 7%	25,000 3%	5,000 1%	30,000 3%	200,000[14] 20%
Published % of 6.	614,202[16] 87%	51,481[16] 7%	19,045[16] 3%	3,001[16] 0%	20,729[16] 3%	660,715[15] omitted

1. Estimation based on 1970 census and 1975 projection, plus Okinawa. Sōrifu 1973 [22]: 10, 14.
2. Sum of 6a. through 6f. estimated. Includes steady state tourist population. See note [14] below.
3. Sum of 6a. through 6e. published. Excludes tourist population. See note [15].
4. Sōrifu 1973 [22]: 14. 1970 census. Includes Okinawans and foreign nationals but excludes American military personnel and dependents in Japan under Status of Forces Agreement.
5. See section on Burakumin for explanation of this estimate.
6. Estimate based on 35 year extrapolation of 10 year rate. See note [11] below.
7. Estimate intended to reflect probable increases through 1960's and to include the emerging second generation of mixedblood offspring (the children of marriages involving first generation mixedblood citizens). See section on Konketsuji.
8. *Buraku* 228 [special issue March 1968]: 50. 1963 government figures.
9. Sōrifu 1973 [22]: 14. 1970 census.
10. Lifton 1967: 191. 1950's estimate. See section on Hibakusha.
11. Hōmushō 1964: 91. 1952–1962 total. See section on Kikajin.
12. Trumbull 1967: 112. 1950's estimate. See section on Konketsuji.
13. Ainu BHTK 1969 [1]: 12. 1960's estimate. See section on Ainu.
14. Estimate of steady state tourist flow (the population of tourists in Japan at any given time of year).
15. Asahi 1973 [1]: 512. 1971 foreign national entry rate. Excludes residents. Includes sightseeing, business, and transit status foreign nationals only.
16. Asahi 1973 [1]: 532. 1970 Immigration Bureau, Ministry of Justice figures. Excludes foreign nationals on Okinawa, and excludes American military personnel or dependents in Japan or on Okinawa under Status of Forces Agreement.

TABLE 2: Selected Minority Populations in Japan (1872–1970)

Year	1. All Residents[1]	5a. Burakumin[2]	% of 1.	6. Foreign Minorities	% of 1.	6a. Koreans	% of 6.	% of 1.	6b. Chinese	% of 6.	% of 1.	6c. Americans	% of 6.	% of 1.
1876[3]	35,555,000	*520,451	*1.495%	4,348	.012%	0		0	2,371	55%	.007%	132	3%	.000%
*1872	*34,806,000													
1890[3]	39,902,000			9,707	.024%	9	0%	.000%	5,498	57%	.014%	972	10%	.002%
1910[3]	49,184,000	*799,430	*1.686%	17,474	.036%	2,577	15%	.005%	8,420	48%	.017%	1,633	9%	.003%
*1907	47,416,000													
1920[3]	55,963,000	829,773	1.483%	78,027	.139%	40,755	52%	.073%	24,130	31%	.043%	3,966	5%	.007%
1930[3]	64,450,000			478,940	.742%	419,009	88%	.650%	44,051	9%	.068%	3,638	1%	.006%
*1935	*69,254,000	*999,687	*1.444%											
1940[3]	71,933,000			1,303,051	1.811%	1,241,315	95%	1.726%	45,739	4%	.064%	4,755	0%	.007%
1950[4]	83,200,000			598,696	.720%	544,903	91%	.655%	40,481	7%	.049%	4,962	1%	.006%
*1958	*91,767,000	*1,220,157	1.330%											
1960[4]	93,419,000			650,566	.696%	581,257	89%	.622%	45,535	7%	.049%	11,594	1%	.012%
*1963	*96,156,000	*1,113,043	1.158%											
1970[5]	103,720,000			708,458	.683%	614,202	87%	.592%	51,481	7%	.050%	19,045	3%	.018%

1. Total resident population excluding Okinawa after 1940. All figures have been rounded to nearest thousand. Sōrifu 1973 [22]: 9–14.

2. Ōtsuki 1968: 50. Source gives simply "beginning of Meiji period" for 1872, and 1921 for 1920. We have taken 1872 for convenience (earliest national census) and 1920 as year ordinarily given for 829,773 population. See section on Burakumin for alternate early and recent estimates.

3. Hōmushō 1964: 10. Chinese population for 1920 includes 1,703 Taiwanese. 1930 Chinese population includes 4,611 Taiwanese, while the population for 1940 includes 22,499 Taiwanese and 3,787 Manchurian Chinese. There were 186 Germans, 190 French, and 1,025 British in Japan in 1876. The British population rose to a peak of 4,188 in 1920, fell to a low of 1,115 in 1950, and was 3,001 in 1970. The German national population in Japan fell to third place among Euroamerican foreigners by 1890 but rose to second place with 2,713 residents in Japan by 1940.

4. Asahi 1962: 245. Figures do not include military personnel or their dependents in Japan under Status of Forces Agreement. By 1945 there were 2,365,263 Koreans in Japan according to Chōng 1970: 110, which was 3.278 percent of the 1945 Japanese population [72,147,000]. Rapid repatriation immediately following the Pacific War resulted in a reduction to 593,914 registered Koreans in Japan by 1947 [census that year gives 508,905 Koreans] (Sōrifu 1949 [1]: 80–83).

5. Asahi 1973 [1]: 532. Leveling of Korean population during the early 1960's is partly accounted for by the repatriation of 80,843 Korean nationals to North Korea between 14 December 1959 and 15 December 1963 (Mitchell 1967: 158–159). Leveling during the 1950's and 1960's seems also attributable to naturalization (Hōmushō 1964: 91–92). See section on Kikajin.

adapted to those of other communities. For instance, most of the fundamental habits retained by the Japanese in the matter of clothing, food and housing – though, no doubt, subject to influences from the West and the world at large – may hardly be adopted as they are by the peoples of Western countries and even other Asian countries. . . .

Language is perhaps more emotionally involved in a people's self-perception of their ethnicity than any other cultural artifact, and thus how Japanese regard their language is instructive as to how they tend to feel about their total ethnicity. The Japanese UNESCO volume quoted above describes the Japanese language in highly hyperbolic terms as having "little relation with any [other language] of the world." The Japanese language is said to occupy "a unique position among the languages of the modern world [because] the Japanese people were confined to their native islands and never suffered the invasion of foreign races. . . The confusion of the Japanese language [is therefore] *fundamentally different* from that of foreign languages in that it is not the result of *intermingling*. . . Chinese [and] other foreign languages. . . have influenced the vocabulary of the Japanese language but *never* its structure (Japanese UNESCO 1964:89; italics ours)."

The structure of the Japanese language – whatever is meant by this – has allegedly never been "contaminated" by the structural incursions of foreign tongues. Like all else that is perceived as most Japanese about Japanese culture, the Japanese language is described here as being "pure" in a world of languages steeped in amalgamation. Nor do many Japanese consider the pristine structure of their language-culture susceptible of change.

Indigenous Shintō beliefs embodied concepts of ritual pollution and avoidance, and required offerings of propitiation to cleanse one of contamination associated with blood and death. Persons involved in the handling of the dead – as well as those who assisted in childbirth, and even itinerants who occupied abandoned parturition shelters – were considered unclean. Acts of murder, wounding, incest, and bestiality were also regarded as polluting, and required expiation by ceremonial cleansing sometimes accompanied by purgatory fines or other punishments. (Price 1972:17; Sansom 1958:31)

Dissections in 18th century Japan were performed by *eta* outcastes under the direction of majority anatomists who were psychologically and socially incapable of handling dead flesh (Keene 1969:21–22). Majority Japanese medical students in the early 20th century are known to have stood back and let fellow Chinese students handle cadavers and make the incisions required for laboratory exercises: the Japanese students would wash their hands after touching exposed parts with forceps (Chao 1970:141).

The Japanese concern for ritual cleanliness is given considerable scriptual authority in the myths of origin found in the *Kojiki* and *Nihon shoki* [called also the *Nihongi*]. The god Izanagi and the goddess Izanami are the heavenly parents of the Japanese islands, to which Izanami gave birth. Izanami gives birth also to a number of gods, but when the god of fire emerges from her body she is burned and dies. Izanagi finds his mate a mass of putrefaction in the polluted Land of Darkness [*yomi no kuni*], and he purifies himself upon returning to Japan by bathing in a stream. Throwing down his garments on the bank of the stream, Izanagi produces the sun goddess Amaterasu and the storm god Susanowo. There is frequent discord between the sister who reigns in heaven, and the brother who stays on earth. In one dramatic quarrel, when Susanowo pollutes his sister's palace, she withdraws into a cave and the light of the world goes out. (Sansom 1958:30–32)

References

Ainu Bunka Hozon Taisaku Kyōgikai [Ainu Culture Preservation Policy Council]
> 1969 *Ainu minzoku shi* [Document of Ainu ethnology], Tokyo: Daiichi Hōki Shuppan. 2 vols.

Asahi shinbun [Asahi news]. Asahi Shinbun Sha [Asahi Newspaper Company].

Asahi Shinbun Sha [Asahi Newspaper Company]
> 1962 *Asahi nenkan 1962* [Asahi yearbook 1962], Tokyo: Asahi Shinbun Sha.
> 1973 *Asahi nenkan 1973* [Asahi yearbook 1973], Tokyo: Asahi Shinbun Sha. 2 vols.

Asahigawa Jinken Yōgo Iin Rengōkai [Asahigawa Coalition for Protecting Human Rights]
> 1971 *Kotan no konseki* (Ainu jinken sho no ichidanmen) [Vestiges of the village (A phase in the history of Ainu human rights)], Sapporo: Asahigawa Jinken Yōgo Iin Rengōkai.

Aston, W. G.
> 1972 *Nihongi* (Chronicles of Japan from the Earliest Times to A.D. 697), Tokyo: Charles E. Tuttle Company. Reprint of 1924 edition.

Bell, Ronald (ed.)
> 1973 *The Japan Experience*, Tokyo: John Weatherhill, Inc.

Buraku [Community]. Buraku Mondai Kenkyūjo [Buraku Problems Research Center].

Chao, Buwei Yang
    1970 *Autobiography of a Chinese Woman,* Westport (Connecticut): Greenwood Press. Reprint of 1947 edition.
Chŏng Ch'ol.
    1970 *Zainichi Kankokujin no minzoku undō* [Ethnic movements of Koreans in Japan], Tokyo: Yoyosha.
Chōsen kenkyū [Korean studies]. Nihon Chōsen Kenkyūjo [Japan Korean Studies Center].
DeVos, George, and Hiroshi Wagatsuma
    1972 *Japan's Invisible Race,* Berkeley: University of California Press. Revision of 1966 edition.
    1972A "Group Solidarity and Individual Mobility", in: DeVos and Wagatsuma 1972: 241–257.
    1972B "Minority Status and Attitudes Toward Authority", in: DeVos and Wagatsuma 1972: 258–272.
Fairbank, John K.; Edwin O. Reischauer; and Albert M. Craig
    1965 *East Asia: The Modern Transformation,* Tokyo: Charles E. Tuttle Company.
Gekkan ekonomisuto [Monthly economist]. Mainichi Shinbun Sha [Mainichi Newspaper Company].
Graham, Lloyd B.
    1954 "Those G. I.'s in Japan", in: *The Christian Century,* Vol. 71, No. 11 [17 March 1954]: 330–332.
Harootunian, H. D.
    1970 *Toward Restoration* (The Growth of Political Consciousness in Tokugawa Japan), Berkeley: University of California Press.
Heibonsha [Heibonsha Publishing Company]
    1964–1968 *Sekai dai hyakka jiten* [World encyclopedia], Tokyo: Heibonsha. 26 vols.
Hilger, M. Inez
    1971 *Together with the Ainu* (A Vanishing People), Norman: University of Oklahoma Press.
Hokkaidō Chō [Hokkaidō Prefecture]
    1937 *Shinsen Hokkaidō shi* [New Hokkaidō history]. [Sapporo?]: Hokkaidō Chō. 7 vols.
Hōmushō Nyūkoku Kanri Kyoku [Ministry of Justice Immigration Bureau]
    1964 *Shutsunyūkoku kanri to sono jittai* (Showa 39 nen) [Emigration and immigration control and its disposition (1964)], Tokyo: Ōkurashō Insatsu Kyoku [Ministry of Finance Printing Bureau].
Ito, Hiroshi
    1972 "Japan's Outcastes in the United States", in: DeVos and Wagatsuma 1972: 200–221.
Japan Times. The Japan Times.
Japanese National Commission for UNESCO (comp.), Ministry of Education
    1964 *Japan: Its Land, People, and Culture,* Tokyo: Ministry of Finance Printing Bureau. Revision of 1958 edition. English translation of *Gaikokujin no tame no Nihon jiten* [A Japan encyclopedia for foreigners].

Keene, Donald
1969 *The Japanese Discovery of Europe, 1720–1830,* Stanford: Stanford University Press. Revision of 1952 edition.
Kerr, George H.
1958 Okinawa (The History of an Island People), Tokyo: Charles E. Tuttle Company.
Koria hyōron [Korea review]. Minzoku Mondai Kenkyūjo [Ethnic Problems Research Center].
Lifton, Robert Jay
1969 *Death in Life* (Survivors of Hiroshima), New York: Vintage Books.
Maretzki, Thomas W.
1964 "Personality in Rural Okinawa", in: Smith 1964: 99–111.
McClellan, Edwin
1971 *Two Japanese Novelists: Sōseki and Tōson,* Tokyo: Charles E. Tuttle Company.
Miller, Roy Andrew
1967 *The Japanese Language,* Chicago: The University of Chicago Press.
Minami Hiroshi
1971 "Naze Gaikokujin ni henken o motsu ka (Yurusenu Ajiajin besshi)" [Why are we prejudiced toward sforeigners? (Our unforgivable disdain for Asians], in:*Gekkan ekonomisuto,* Vol. 2, No. 1 Ser. 9 [January 1971]: 106–111.
Mitchell, Richard H.
1967 *The Korean Minority in Japan,* Berkeley: University of California Press.
Norbeck, Edward
1972 "Little-Known Minority Groups of Japan", in: DeVos and Wagatsuma 1972: 183–199.
Okimoto, Daniel I.
1971 *American in Disguise,* Tokyo: John Weatherhill, Inc.
Ōno Susumu
1970 *The Origin of the Japanese Language,* Tokyo: Kokusai Bunka Shinkōkai [Japan Cultural Society]. English translation of *Nihongo no kigen* (Tokyo: Iwanami Shoten, 1957).
Ōtsuki Makoto
1968 "Buraku no genjō" [The present situation in outcaste communities], in: *Buraku,* Vol. 20, No. 4, Ser. 228 [special issue March 1968]: 49–58.
Philippi, Donald L. (trans.)
1969 *Kojiki,* Princeton: Princeton University Press.
Price, John
1972 "A History of the Outcaste: Untouchability in Japan", in: DeVos and Wagatsuma 1972: 6–30.
Reischauer, Edwin O., and John K. Fairbank
1962 *East Asia: The Great Tradition,* Tokyo: Charles E. Tuttle Company.
Sansom, George
1958 *A History of Japan to 1334,* Stanford: Stanford University Press.
1961 *A History of Japan, 1334–1615,* Stanford: Stanford University Press.

1962 *Japan: A Short Cultural History*, New York: Applieton-Century-Crofts. Revision of 1943 edition.

1963 *A History of Japan, 1615–1867*, Stanford: Stanford University Press.

Shūkan asahi [Weekly asahi]. Asahi Shinbun Sha [Asahi Newspaper Company].

Shūkan bunshun [Weekly bunshun]. Bungei Shunjū [Bungei Shunjū Publishers].

Shūkan yomiuri [Weekly yomiuri]. Yomiuri Shinbun Sha [Yomiuri Newspaper Company].

Smith, Allan H. (ed.)

1964 *Ryukyuan Culture and Society*, Honolulu: University of Hawaii Press.

Sōrifu Tōkeikyoku [Office of the Prime Minister, Bureau of Statistics]

1949 *Nihon tōkei nenkan* [Japan Statistical Yearbook], Tokyo: Nihon Tokei
(1) Kyokai [Japan Statistics Association].

1969 *Shōwa 40 nen kokusei chōsa hōkoku; Dai 5 kan: 20% chūshutsu shūkei kekka*
(5.1) *zenkokuhen; Sono 1: Nenrei, shussei no tsuki, haigū kankei, kokuseki, setai, junsetai jinin, jūkyo no jōtai* [1965 Population Census of Japan; Vol. 5: Twenty Percent Sample Tabulation Results for Whole Japan; Part 1: Age, Month of Birth, Marital Status, Legal Nationality, Households, Quasi-Household Members, and Housing Conditions], Tokyo: Nihon Tōkei Kyōkai [Japan Statistics Association].

1969 *Shōwa 40 nen kokusei chōsa hōkoku; Dai 5 kan: 20% chūshutsu shūkei kekka*
(5.2) *zenkokuhen; Sono 2: Sangyō to shokugyō* [1965 Population Census of Japan; Vol. 5: Twenty Percent Sample Tabulation Results for Whole Japan; Part 2: Industry and Occupation], Tokyo: Nihon Tōkei Kyōkai [Japan Statistics Association].

1973 *Nihon tōkei nenkan* [Japan Statistical Yearbook], Tokyo: Ōkurashō
(22) Insatsukyoku [Ministry of Finance, Printing Bureau].

Statler, Oliver

1971 *Shimoda Story*, Tokyo: Charles E. Tuttle Company.

Suzuki Jirō

1973 *Shiro, Kuro, Kiiro* (Sabetsu to henken no kōzō) [White, Black, and Yellow (The anatomy of discrimination and prejudice)], Tokyo: Otowa Shobō.

Takakura Shin'ichirō

1972 *Ainu seisaku shi* [History of Ainu policy], Tokyo: San'ichi Shobō. Revision of 1942 edition.

Takasaki Setsuko

1953 *Konketsuji* [Mixed Blood Children], Tokyo: Isobe Shobō.

Thompson, Era Bell

1967 "Japan's Rejected", in: *Ebony*, Vol. 22, No. 11 [September 1967]: 42–54.

Totten, George O., and Hiroshi Wagatsuma

1972 "Emancipation: Growth and Transformation of a Political Movement" in: DeVos and Wagatsuma 1972: 33–67.

Trumbull, Robert

1967 "Amerasians", in: *New York Times Magazine*, 30 April 1967: 112–114.

Wagatsuma, Hiroshi
  1967 "The Social Perception of Skin Color in Japan" in: *Daedalus*, Spring
    1967 [Color and Race]: 407–443.
Wagatsuma, Hiroshi, and George DeVos
  1972 "The Ecology of Special Buraku", in: DeVos and Wagatsuma 1972:
    113–128.
Yi Yu-hwan
  1960 *Zainichi Kankokujin no gojūnen shi* (Hasseiin ni okeru rekishiteki haikei
    to kaihōgo ni okeru dōkō) [A fifty year history of Koreans in Japan
    (The historical background of generating factors and tendencies in
    the wake of liberation)], Tokyo: Shinju Bussan.

Human Rights in Communist Ruled East-Central Europe

WALTER DUSHNYCK

WALTER DUSHNYCK was born in Western Ukraine, where he graduated from a classical *Gymnasium*. After earning a Bachelor of Arts degree at the University of Louvain, he emigrated in 1935 to the United States. He attended Columbia University in New York City (1937–1940) and received his Master of Arts in Government. Later, he earned his Ph.D., awarded in 1965, at the Ukrainian Free University in Munich (which has German university accreditation). He served in the United States Armed Forces (1942–1945) and took part in the campaigns for Saipan, The Philippines, Okinawa, and Japan. He served as interpreter in General Mac-Arthur's headquarters in Manila and Tokyo.

A lifelong student of Soviet and East European affairs, Dr. Dushnyck is the author of *The Russian Provisional Government and the Ukrainian Central Rada*; *Death and Devastation on the Curzon Line*; *Martyrdom in Ukraine*; *In Quest of Freedom*; and *The Ukrainian-Rite Catholic Church at the Ecumenical Council, 1962–1965*. He was an associate editor of *Encyclopedia Slavonica* (1949) and of *Ukraine: A Concise Encyclopaedia* (1963 and 1971).

Since 1957, Dr. Dushnyck has been editor-in-chief of *The Ukrainian Quarterly*, an English-language review founded in 1944 and published in New York City.

Human Rights in Communist Ruled East-Central Europe

WALTER DUSHNYCK

In 1945, as World War II was entering its decisive phase, the USSR, as inheritor of the Russian imperial domain, was presented with an unexpected and centuries-dreamt-of opportunity to extend its political power into Eastern and Central Europe and into the Balkans. With the benign approval of the Western powers, the USSR not only retained its 1939–40 territorial gains, but added to them East Prussia and the Ukrainian land of Carpatho-Ukraine and, in the Far East, the southern part of the Sakhalin and Kurile Islands.

Against the background of the collapse of Hitler's empire and the acute differences between British and American policies, Soviet expansion rolled unchecked into Central Europe. In January, 1944, the Soviet army overran the Polish territory and, in April, the Rumanian. The Soviet-Rumanian armistice of September 12, 1944 sanctioned Russia's "temporary" occupation of Rumania; three days later, on September 15, following a Soviet-Bulgarian war that lasted twenty-four hours, Soviet tanks rumbled into Sofia. On September 6, the Red Army entered Yugoslavia, and on September 23, Hungary. By the end of April, 1945, Soviet troops had reached the outskirts of Berlin. On May 9, they marched into Prague. It was by agreement with the Allied Command that the Russians were allowed to occupy both Prague and Berlin.[1]

The basic factor in establishing Soviet domination over East-Central Europe and the Balkans was military occupation. This allowed the process of implanting "People's Democracies" that was to follow. In all the occupied countries the pattern was the same. A small core of local Communists, trained in Moscow, would be brought in to form the nucleus of a new government. Under the "protection" of Soviet troops and large contingents of security and police forces, the non-Communist political structures would be destroyed, and all the non-Communist elements disposed of in a matter of weeks. All the elaborate arrangements that had been worked out at Teheran, Yalta and Potsdam collapsed, with the result that between 1945 and 1948 the nations of Eastern and Central Europe were squeezed into the confining mold of Russian Communism; once proud and sovereign nations now became subservient puppet governments of the Kremlin.

The rule in the Communist-dominated countries of Central and Eastern Europe and the Balkans can be divided roughly into two phases: a) the Stalinist rule with all its terroristic features (from 1945 to 1953); b) the post-Stalinist rule, marked by trends toward "liberalization" and "national Communism." The difference between the two is undoubtedly substantial, but the basic objectives of Soviet policies remained unchanged: political, social, economic and cultural integration; strict adherence to the Party line and, above all, complete subordination of the policies and development of the satellite countries to Soviet foreign policy goals.

After the death of Stalin and the introduction of "liberalization" policies in the USSR by Khrushchev, there were widespread manifestations of dissent and attempts to get rid of the Communist rule: the Hungarian rebellion in the fall of 1956; the Poznan riots in Poland, also in the fall of 1956, and the "liberalization course" in Czechoslovakia under Alexander Dubcek in 1968. Sooner or later, all these attempts were crushed by Soviet troops.

Yugoslavia and Albania succeeded in breaking away from Moscow's tutelage because of the happy circumstance that their territories are not adjacent to the mass-land of the USSR.

The post-Stalinist rule in the satellite countries eased up for a number of reasons, the main one being the USSR's fear of getting involved in a new European conflict by trying to implement the "Brezhnev Doctrine," say, in Yugoslavia, or Rumania. But political control in the satellite countries is maintained firmly by Moscow through a number of covert devices: a Soviet representative is in the Ministry of Interior (secret police) in every satellite country; all the general staffs of the satellite armies are directed by Soviet military leaders; all the Communist Parties in these countries are directed by the Politburo of Moscow and all the economic and financial institutions are subordinated to the overall plans of the USSR, as are the educational policies as well.[2]

Like the USSR, all the satellite governments signed not only the Universal Declaration of Human Rights, adopted on December 10, 1948, but also three other important documents dealing with all aspects of human rights and human freedoms. These are the International Covenant on Economic, Social and Cultural Rights; the International Covenant on Civil and Political Rights and the Optional Protocol to the International Covenant on Civil and Political Rights, which were all adopted at the U.N. General Assembly on December 16, 1966.[3]

These four basic documents all deal with such problems as the right of self-determination; prevention of discrimination; war crimes and crimes against humanity, including genocide; slavery, servitude, forced labour and similar institutio nsand practices; nationality, statelessness, asylum and refugees; freedom of information; freedom of association; employment policy; political rights of women; marriage and the family, childhood and youth; social welfare, progress and development, and the right to enjoy culture, and international cultural development and cooperation.

Yet, despite these covenants and the U.N. Universal Declaration of Human Rights itself, the satellite governments practice discrimination in every field of man's endeavour and get away with it. In the countries under study there is no other but the white race, and there is no "racial problem" as such. But it exists in various forms, notably, in the form of national or ethnic discrimination against the *national minorities*. Every Communist-dominated country of Central Eastern Europe possesses national minorities which suffer political and cultural persecution. As a rule, the minorities are under-represented at local councils; their languages are discriminated against, and the minorities are barred from high government positions. In Poland, there exists a sizable Ukrainian minority (over 500,000 persons), as well as German, Jewish and Byelorussian minorities. In Czechoslovakia, where the Slovaks, the largest ethnic entity after the Czechs, have attained a degree of "autonomy," there are the Hungarian, German, Ukrainian and Polish minorities whose treatment is far from just or equitable. Rumania's national minorities comprise sizable German, Hungarian and Ukrainian groups. Substantial numbers of Germans, Slovaks and Rumanians are also to be found in Hungary. Bulgaria has the Macedonian and Greek minorities whose problems as ethnic entities date back for centuries.

Although Yugoslavia is a federation consisting of six quasi "independent" republics, nationalist frictions are very strong, especially between the Serbs on the one hand and the Croatians and Slovenes on the other; such minorities as the Albanians and Macedonians are subjected to systematic ethnic discrimination and persecution.

As a rule, all the Communist governments in the satellite states deny political rights to their citizens, such as freedom of expression and as non-conformist political parties or groups (detailed in the description of individual countries).

There also exists vast discrimination as regards emigration of the citizenry to foreign countries; moreover, severe restrictions are also imposed upon internal emigration within the country. There is no freedom of choice with respect to jobs, residence and, sometimes, even marriage.

The Communist legal system is likewise wholly discriminatory as far as the individual is concerned, partly because penalties are severe and unreasonable and partly because appeal procedures are inoperative; defendants have no real capability of defending themselves, as they have to prove their innocence rather than the government proving their guilt.

Likewise, there exists a wide area of economic discrimination against certain strata of the population in every satellite country. It is probably true, as the Communists claim, that women are not discriminated against in the sense that they have access to a number of professions from which they were excluded or barred in earlier times, but it's also probably true that women are paid less than men, which is the case in most "capitalist" countries. Nevertheless, the equality of the sexes is often abused, as women are used in employment requiring superior physical strength, such as street cleaning, employment in mines, road paving, laying railway tracks, and so forth.

Children are discriminated against on an extensive scale, despite specific provisions against the abuse of children written into the basic documents on human rights by the U.N. Usually, child labour exploitation and abuse is veiled in the euphemistic phrase of "voluntary" service to the state.

Such unpaid labour is standard policy in all Communist countries. In most cases work of that kind is done after regular school hours. But larger tasks, including the beautification of parks and other projects, are undertaken on weekends. This practice has been confirmed by a report from Warsaw, the capital of Poland:

Polish parents are complaining that their children are being compelled to perform increasingly longer periods of "voluntary" unpaid labour for the state.

There have been no official statements on the subject, but children in Warsaw were reportedly told recently that they would be expected to perform 80 hours of unpaid labour a year from now on.

There have been reports that some children have been used to help bring in the potato crop... Rumours are circulating that some school time will be cancelled to enable students to work on public service projects...

Unpaid labour by office employees and others is especially encouraged on Sundays. Though Saturdays are still working days, the country is supposedly moving toward free Saturdays... but in many areas it seems to be making up for the free Saturdays with "voluntary" labour on Sundays...[4]

The peasant situation is deplorable in all Communist lands, not only in the USSR. Although the idea of general collectivization of agriculture in the satellite countries has been abandoned and peasants are allowed to hold limited individual homesteads, the peasantry as a class is discriminated against brutally. This discrimination is carried out through exhorbitant taxation and interest rates, inflationary wage and price manipulations, and buying-selling regulations by the state (buy low, sell high, especially with respect to food). Under such conditions, peasants cannot live on "officially planned" incomes. In order to have additional income, they are forced to "moonlight," thus risking arrest and trial as "economic saboteurs."

The ruling class in every Communist state consists of Communist officials and members of the Communist Party and Communist Youth Organizations. They receive preferential treatment in every phase of life: salary, housing, education, social benefits (better pensions), privilege to travel abroad, especially on various "cultural exchange programmes" to the Western countries, etc. An exception to the rule is Poland, which allows thousands of its citizens to travel abroad, especially to the United States and Canada, to visit relatives. The Communist government of Poland also allows a limited number of its citizens to emigrate abroad.

The ironclad rules against private property, emanating from the concept of a "classless" and "propertyless" society is also a flagrant violation of the U.N. Universal Declaration of Human Rights and the four other covenants concerning human rights and human freedoms. The state-manipulated regulations actually deprive an individual of the product of his or her labour, thus weakening the family foundations and, consequently, contributing to the progressive decline of the population.

In conclusion, the Communist regimes of Central and Eastern Europe are *anti-people* governments *par excellence*. Politically and militarily, East Germany, Poland, Czechoslovakia, Hungary, Rumania and Bulgaria are in fact colonies of the Soviet Union, forming the outer periphery of the Soviet Russian empire. The Soviet rulers are ever mindful of the theory of Sir Halford J. Mackinder, who postulated:

> Who rules East Europe commands the Heartland;
> Who rules the Heartland commands the World Island;
> Who rules the World Island commands the World.[5]

I. ALBANIA

A. A Plaything in Aggressive Power Politics

Situated on the west coast of the Balkan peninsula, separated by less than 100 miles of Adriatic Sea from the heel of Italy, Albania is unique in being geographically isolated from the Soviet block. With impunity, Albania has become an ideological outpost of Chinese Communism in Europe. On the north and east it is bordered by Yugoslavia, on the south by Greece and on the west by the Adriatic. It has 10,757 square miles of territory and from 1,840,000 to 2 million people of Thraco-Illyrian ethnic origin, divided into two principal linguistic groups, the *Gheghs* in the north and the *Tosks* in the south. Seventy percent of the population is Moslem twenty percent belong to the Orthodox Church, while the remaining ten percent are Roman Catholic.[6]

Albania was long dominated by the Roman and Byzantine empires. In the XVth century the Albanian people were united under the leadership of a native prince, George Kastrioti (Skanderbeg), who fought for a quarter of a century against the Turks. Incorporated into the Ottoman Empire, Albania succeeded in proclaiming its independence only in 1912, and remained free until 1939, when it was taken over by Fascist Italy.

Despite the fact that the people of Albania have usually held aloof from the affairs of their neighbours, they found themselves engulfed by the events of World War II. Soon after Albania's occupation by Italian troops, the Albanian people rose against the foreign occupiers and developed a large-scale armed resistance movement. The Albanian Communist Party, supported by the Yugoslav Communists led by Joseph Broz, later to be known as Marshal Tito, did not fight against the Axis troops until after Germany invaded the USSR in June, 1941.[7] Later on, when the Allied forces set foot in Italy, the Allied Command supplied the Albanian Communists with necessary war materiel, thus enabling them to win the local war against true Albanian patriots and nationalists.

An obscure schoolteacher who appointed himself a General, Enver Hoxha, succeeded in organizing a "National Liberation Front," and as the Italians and Germans withdrew in 1944, Hoxha and his Communist-dominated "National Front" proceeded to liquidate most of the Albanian anti-Communist organizations, such as the "Nationalist Front" *(Balli Kombetar)* and the Legality *(Legaliteri)* movements, as well as all partisans and supporters of King Zog.[8]

B. *Transition of Power and Unbridled Terror*

The transition to Communism and its "People's Republic of Albania" regime began early in November, 1944, with the unleashing of unbridled terror by the Communist government of Hoxha. The policies of terror continued when the Hoxha regime was under the influence of Tito (until 1948), and then also under the direct dictation of Moscow, especially after Tito broke away from Stalin's control in 1948.

According to non-Communist Albanian sources some 20,000 persons were executed without trial; hundreds perished in concentration camps, which had been swiftly established by Hoxha, and at least 50,000 were thrown into labour camps, where they were kept for years.[9]

Following the Khrushchev-Tito reconciliation, the Kremlin plotted the ouster of the Hoxha group in favour of a group more acceptable to Yugoslavia and less harsh on the Albanian people. But the plot was discovered and Hoxha reacted violently. He openly identified himself with Stalinism and sought political shelter and protection under the wings of the by then openly schismatic and rebellious Peking. Lacking direct territorial access, Moscow was unable to take effective action against Hoxha, especially since Tito's Yugoslavia would not cooperate against its avowed enemy. While opposed to any punitive steps on the part of Moscow, Tito was equally opposed to any Western intervention. As a result, Albania today is firmly lodged in the Peking camp, and is a police state *par excellence*.[10]

C. *Abuse of Political Power*

Despite the fact that the Albanian Communists could claim only 4 or 5 percent of the population as supporters, the spurious "democratic front" received 93.2 percent of the votes cast to elect a constituent assembly in January, 1946. Undoubtedly, all methods of coercion were used to achieve this "resounding" success. In 1948, the Communist Party of Albania, now euphemistically known as the "Albanian Labour Party," adopted an organizational structure based on the principles of "democratic centralism." The highest organ is the party congress, made up of delegates nominally elected by district, regional and city conferences, whose principal functions include rubber-stamping of reports submitted by the Central Committee and other organs.[11]

The constitution of 1950 recognized the privileged and controlling position of the Albanian Labour Party, which also controls the government and the 264-member "People's Assembly." The latter is purportedly the highest and the most important organ in the government structure. Members of the Assembly are "elected" for four years on a single-slate ballot from candidates put up by the party. Thus, the vast majority of the people have no voice in the government, not even in the so-called "People's Assembly."

The overall situation of Albania after the Communist takeover is described in a statement of the Free Albania Committee:

Once the Communist regime had consolidated its control over the country, an extremely savage process of transforming Albania into a "socialist" state was initiated which, at first, was carried out with the help of Yugoslav advisers and technicians, then with those from Soviet Russia, and at present from Communist China.

The first stage in this direction was the nationalization of all banks, mines and oil industries and the collectivization of agriculture, farms and other industrial and agricultural enterprises, all without any compensation to the owners. This was followed by the confiscation of all private property. Private land, industrial plants, factories, warehouses, commercial shops of every kind, restaurants, hotels and even private homes were seized without any compensation, often by use of harsh forms of intimidation and terror... Schools, theatres, music organizations, writers and artist groups, the press and other mass media, as well as all youth and physical education organizations, were transformed into strict, obedient servants of the so-called "Workers" (Communist) Party and its regime.

In spite of the fact that the so-called "People's Republic of Albania" has since 1965 been a member of the United Nations, and is one of the signatories of its Charter and the Universal Declaration of Human Rights, the Tirana regime has remained the most rigid and despotic among the Communist-ruled countries of Mid-Eastern Europe. All human rights and fundamental freedoms enumerated in the above-mentioned U.N. Charter and Universal Declaration, have been systematically and grossly violated and ignored.

By these means the regime achieved the following objectives; it took under its control the entire national wealth and resources, the nation's activity in all levels of life; it forced the Albanian people to live in an extreme state of poverty, strictly isolated from the outside world and to be totally dependent upon the Party and its regime.[12]

D. Religious Repression

Until Stalin's death in 1953, hundreds of people in Albania were executed without benefit of trial or jury. In the past two decades some 200 church leaders of the three religious denominations either were killed outright or perished from torture in the country's 14 concentration camps.[13]

At the beginning of its rule, the regime devised various methods of depriving the three churches of their income, curbing their influence and outlawing any form of religious instruction. Expropriated were church buildings, schools, monasteries, libraries, printing presses and so forth. The appointment of clergy had to be approved by the government, and all pastoral letters, sermons, etc., became subject to censorship. The education of the youth became a matter for the government exclusively.

The fatal blow to religious life in Albania was delivered in 1967, the year Albania was officially declared the "first atheistic state in the world." The three churches were ordered to turn over to the government some 3,000 church buildings, mosques and monasteries, as well as other religious institutions; many buildings were either destroyed or turned into movie and theatre houses and dance halls. On November 22, 1967, the *Official Gazette* in Tirana published decree No. 4337, which annulled all the laws on the state-church relationship; this act delivered a legal *coup de grace* to any formal religious life in Albania.[14]

In March, 1973, a priest, Rev. Shtjefn Kurti, was murdered by the secret police for secretly baptizing a child on the request of its mother.

II. BULGARIA

Wedged into Europe's southeastern corner and with a coast on the Black Sea, Bulgaria is separated from the Aegean by a thin strip of Greek territory. It borders only with one bloc neighbour, Rumania, to the north, with two non-bloc states, Turkey and Greece, to the south, and with Yugoslavia to the west. It encompasses an area of 42,808 square miles and as of 1969 had a population of 8,500,000, of whom about 91 percent are ethnic Bulgarians, a Slavic people, with the remainder consisting of Turks, Greeks and Macedonians. The national language is Slavic with some Turkish and Greek linguistic influence. 84.4 percent of the population adhere to the Orthodox faith, 13.5 percent are Moslem and 2.1 percent belong to other religious denominations.

A constant rival to the ancient Byzantine empire, Bulgaria fell under Byzantine rule in 1014, and remained under it until 1186. After some 200 years of independence, the country was overrun by the Turks in 1393, and was held by the Ottoman Empire for the next five centuries. Actually, the Bulgarian people were under a dual yoke: that of Greek spiritual control and Turkish political rule.[15]

After the Russo-Turkish war of 1877–78 Bulgaria again emerged as an independent state mainly as a result of Russia's "war for the deliverance of brother' Christians in the Balkans from the Moslem oppressors."

In World War I Bulgaria became a member of the Central Powers, and with their defeat in 1918 it lost a part of Western Thrace to Greece and some territory to Yugoslavia.[16]

When World War II broke out Bulgaria hoped to regain its lost territories from Greece and Yugoslavia. When it became apparent that Nazi Germany was going to lose the war, a moderate Bulgarian regime sent emissaries to Cairo at the British and American headquarters to negotiate a separate peace. Not only did they get nowhere, but also the Western Allies saw fit to inform Stalin about the Bulgarian move.[17] Although not at war with Bulgaria, the USSR promptly declared war on the country and proceeded to occupy it.

A. Soviet Occupation and Terror

In September, under the bayonets of the Soviet Army, a new government was formed in Sofia, with the four most important ministries occupied by Communists. Ironically, Winston Churchill wrote in his memoirs, "I never felt that our relations with Rumania and Bulgaria in the past called for any special sacrifices from us...," but also admitted that "Communism raised its head behind the thundering Russian battlefront. Russia was the Deliverer, and Communism the gospel she brought..."[18]

As soon as the country was firmly in Russian hands, the Communists, headed by George Dimitrov, let loose unbridled terror: according to official Bulgarian sources, 2,000 persons were executed within a period of two months in the fall of 1944; but the actual number is probably higher.[19]

The "Sovietization" of the country was carried out by Bulgarian repatriates who had spent years in the USSR and then came back on the heels of the Soviet armies. Amongst them were Premier Vulko Chervenkov; General Ivan Mihailov, deputy prime minister; Peter Pauchevski, defence minister; Raiko Damianov and Karlo Lukanov, deputy prime ministers; Dimo Dichev, head of the State Control Commission; Ferdinand Kozovski, president of the National Assembly (*Sobranje*); Georgi Damianov, chairman of the Presidium; Kiril Lazarov, finance minister, and many others.[20]

These men proceeded to remake Bulgaria on the Soviet model, including the government, the armed forces, the legislative branch, political parties, as well as youth and social organizations and educational institutions.

B. Government, Party and Political Repressions

In September, 1946, on "popular request," an all-national plebiscite was held to repudiate the old "Tirnovo Constitution" (1878) and to introduce the "Dimitrov Constitution," patterned after the 1936 Soviet constitution. It was adopted in 1947. In establishing a "People's Republic of Bulgaria," the new constitution ordained collective ownership of the means of production and the right of the state to nationalize all facets of the national economy.

The unicameral National Assembly is elected from candidates of the Bulgarian Communist Party, which theoretically elects its presidium, appoints the judges to the Supreme Courts and the members of the Council of Ministers, amends the Constitution and performs a number of other legislative functions.[21]

The 52-man Council of Ministers is defined by the constitution as the "supreme executive and administrative organ of the state." The present premier, Todor Zhivkov, is also the first secretary of the Bulgarian Communist Party, a member of the party Central Committee, a deputy of the National Assembly and chairman of the constitutional drafting committee.[22]

The Bulgarian Communist Party and its facade, the "Fatherland Front," is the real power in Bulgaria, and since its membership has never been larger than 672,000 (as of 1970), the great majority of the Bulgarian people have no voice in their government.

During the rule of Georgi Dimitrov (who died in 1949) and that of his successor, Vulko Chervenkov, the Bulgarian Communists were busy "liquidating the enemies of the people." An official statement issued in March, 1945, admitted to 2,138 executions and 1,940 prison sentences of twenty years and 1,689 of ten to fifteen years.[23]

Purges within the party have been so thoroughgoing that they have left few leaders of stature willing or able to offer any effective opposition to Moscow dictation. Even prominent Peasant Party leader Nikola Petkov was executed on September 23, 1947, as a "traitor."

The current party leadership under Zhivkov demonstrates a tendency toward political unity, but the mediocre leadership is completely devoted to the policies of Moscow. There seem to be no groups or individuals of sufficient importance and strength (in contrast to a degree of opposition seen in Poland and Rumania) to plump for a more independent course of action.

In 1965, the KGB allegedly uncovered an "anti-state plot" to overthrow the government of Sofia; one of the defendants committed suicide, while nine others were condemned to various prison terms.[24]

C. Religious Curbs and Repressions

The Bulgarian Communist government has been able to gain substantial control over the religious life of the country. In 1945, in order to support a Soviet design to extend Soviet political influence in the Middle East by means of the Orthodox Church, the Bulgarian Orthodox Church was allowed to elect an Exarch on January 21, 1945, and to effect a reconciliation with the Patriarch of Constantinople on February 22 of the same year. This was important because the majority of people belong to the Bulgarian Orthodox Church. The other faiths – Roman Catholic, Moslem, Protestant as well as the Jewish – were each separately handled with great effectiveness. As in the USSR, the Orthodox Church was the most important because of its traditional role in Bulgarian national life; the Communist government fully capitalized upon this attachment:

The Communists have patronized the Church as the traditional national church of Bulgaria, not only to obtain support from Church devotees, but also to unify national Orthodox churches under the aegis of the Soviet controlled Russian Orthodox Church. [The new Patriarch] Kiril clearly demonstrated his attitude. . . when he thanked the regime for the re-establishment of the Bulgarian Patriarchate and called on all the faithful to support the government in its policies. . . [25]

In 1949 a special statute was issued whereby all religious denominations were required to register with the Committee for Religious Affairs and to obtain approval for their operations; all church leadership was made "responsible to the state," and religious functionaries could not "take office or be dismissed or transferred without approval of the government. . ."[26]

Church schools could be maintained, if government permission was secured, but no church was allowed the right of secular education. These restrictions have curtailed virtually all religious freedoms and have reduced religious leaders to mere spokesmen of the Communist regime.

There are no Catholic churches open in Bulgaria today. In 1952 some forty leading Catholics were tried on charges of "spying" and were executed. There still are some 56,000 persons of the Roman Catholic faith, but no church organization.

During 1950–1951 some 150,000 Moslems were forcibly expelled to Turkey, and some 700,000 have been organized into Moslem religious communities. Their leader, the Grand Mufti, makes docile favourable statements concerning the regime.

The Protestants have suffered a fate similar to that of the Catholics but their denominations are active if on a limited scale. Several Protestant leaders were tried on "espionage" charges in March, 1949, and a number of them received severe prison terms.

Some 45,000 Jews emigrated to Israel in 1949–1950, and the remainder of some 5,500 still maintain 14 synagogues; their Grand Rabbi, however, is wholly subordinate to the Communist regime.

Discrimination exists in every field of endeavour, especially in the field of education, where enrollment into higher institution of learning is based on exclusive privileges. The status of women in Bulgaria is very much the same as in the USSR. Few women hold responsible positions in government, industry or science or educational and diplomatic posts. Those few who do are, as a rule, "window-dressing" to display the "equality of sexes."

III. Czechoslovakia

Czechoslovakia (Czecho-Slovakia according to Slovak usage) was founded in 1918 at the close of World War I. For the ensuing two decades it was the most prosperous and (except in relations with ethnic minorities) democratic government in Central-Eastern Europe, based on a Western-style constitution adopted in 1920.

The present-day population of Czechoslovakia is 14,467,000, comprising 9,389,000 Czechs (64.9%), 4,210,000 Slovaks (29.1%), 564,000 Hungarians (3.9%), 130,000 Germans (0.9%), 73,000 Poles (0.5%), 58,000 Ukrainians (0.4%) and 43,000 others (0.3%).[27]

About 77 percent of the population are Roman Catholics, 8 percent belong to the Czechoslovak Church, 7 percent are Protestant, 0.5 Jewish, 0.5 percent Orthodox and 7 percent are without religious affiliation.[28]

The ancestors of the Czechs and Slovaks once created a powerful Moravian empire. After its disintegration the Czechs founded their own Kingdom of Bohemia, including Moravia and Silesia, which grew into one of the most powerful kingdoms in Europe of the time. Despite its flourishing culture the Czech state then declined, falling under the domination of the Hapsburgs, while the Slovaks were annexed by the Hungarian kingdom. This forcible separation lasted for centuries.

The modern Czechoslovak Republic was established on October 28, 1918. Subcarpathian Ruthenia (Carpathe-Ukraine) was joined to it later. Between 1918 and 1938 Czechoslovakia developed advanced social and economic institutions. With the Munich agreement in 1938, it became a victim of Nazi aggression, lost a large part of its territory, its democratic system and its very liberty and independence.

A. Communist "Coup d'État" and Terror

After World War II Czechoslovakia was re-established as a free independent republic under the presidency of Eduard Beneš. But in February, 1948, the Communist Party of Czechoslovakia carried out a *coup d'état*, whereby it terminated democracy and established the dictatorship of the Communist Party. The February 1948 takeover was violent but bloodless, and it did occur without direct military intervention on the part of the USSR.

The new regime, based on the swiftly-adopted constitution of May, 1948, was an exact replica of the Moscow-controlled governments in other satellite countries, combining Marxism with some features of the "capitalist" regime. The constitution cleverly masked the true nature of Communist rule by providing a facade of democratic respectability. The existence of a coalition government and the temporary continuance of President Beneš in office were also factors, because many of the provisions in the constitution had been formulated before the Communist takeover.[29]

The Communist rule in Czechoslovakia, especially the form of "democratic elections," was aptly characterized in an official U.S. Congress document:

1. There was always a privileged list of candidates the so-called government bloc. The bloc (identified by different names in different countries) was always controlled by the Communists. It always included groups masquerading under the names of traditional, democratic parties. This was done in order to create the appearance of a "coalition of democratic elements,";

2. Thousands of democratic leaders were arrested during such election campaigns on trumped-up charges. Some were released within 48 hours only to be re-arrested a few hours later;

3. Opposition lists were prevented whenever possible from being filed on time by last-minute changes of rules and forms;

4. The secret police exercised "pressure" on the candidates from opposition lists to withdraw their candidacies;

5. Opposition candidates' names were often eliminated from the lists on unfounded charges of collaboration with the Nazis or "unfriendliness" toward the Soviet Union;

6. Large numbers of people were disfranchised for a variety of flimsy reasons;

7. The chairmen of the electoral commissions were overt or secret agents of the Communists. There was no secret ballot although it was provided by the law;

8. Counting of the ballots was done by the Communist chairmen of the electoral commissions and their stooges. The results were always assured;

9. The Communists introduced the technique of "mass voting." Factories, offices, institutions and, in some cases, whole villages were required to vote *en masse*. Such practices were introduced in order to frighten those who, if the secret ballot were allowed, would have voted against the government;

10. An atmosphere of general terror was created in connection with every organized election...[30]

"Show trials" soon followed in which charges of treason were manufactured against the democratic leaders. Most of them were executed or imprisoned, some managing to escape abroad. Thus one-party rule became firmly established and Communist domination over the lives of citizens under its control became absolute.

B. Church-State Relations and Religious Repression

Roman Catholicism has been the dominant force in Czechoslovak religious history, as close to 77 percent of the population are of the Catholic faith. The Communist regime has not dared to annihilate the churches, yet it has spared no effort to bring them under the complete control of the government, thereby demonstrating the Kremlin's thesis that the religious issue has to be handled the Communist way.[31]

As early as 1948, articles published in the Communist press and speeches of Communist leaders criticized the Catholic hierarchy for opposing Communist policies. On June 10, 1949, an organization of opportunistic clergymen and laymen, called "Catholic Action," was founded as an instrument for bringing the Roman Catholic Church under the complete control of the atheistic government. When Archbishop Joseph Beran ordered the clergy to boycott the spurious organization, he was placed under house arrest and his closest collaborators were carted off to concentration camps. In March, 1951, Archbishop Beran was committed to the "concentration monastery" in Nova Rise, Moravia. He was released in 1963 upon the direct intervention of the late Pope John XXIII and allowed to come to Rome, where he was made a cardinal; he died a few years ago.

The "Church Bill," which made its appearance on October 14, 1949, and was enacted into law on November 1, 1949, reduced the Roman Catholic Church to an adjunct of the Communist state. All Catholic schools, religious orders and publications were abolished, and by 1952 relations with the Vatican had been severed.

All Catholic priests were forced to take the following oath:

I promise on my honour and conscience that I will be faithful to the Czechoslovak Republic and its people's democratic order and that I will not do anything against its interests, security or integrity. I will, as a citizen of the people's democratic state, fulfill honestly the duties which result from my position and I will try with all my might to support constructive efforts exerted for the welfare of the people...[32]

In 1963, when the USSR moved to improve relations with the Vatican, the Prague authorities released a number of jailed Catholic clergy, but it was only during the short-lived regime of Alexander Dubček that the rights of the Roman Catholic Church were restored, with Bishop František Tomašek as apostolic administrator. However, by 1970, new restrictions had been imposed on religious activities. A regime-sponsored association of Catholic clergymen, *Pacem in Terris*, has since been circumventing the bishops and dealing directly with the government.[33]

But the fate of the Ukrainian Greek Catholics was far worse than that of the Roman Catholics. After the annexation of Carpatho-Ukraine to the Ukrainian SSR on June 29, 1945, some 300,000 Ukrainian Catholics of the Eastern Rite remained within the boundaries of postwar Czechoslovakia, comprising groups in the Diocese of Priashiv (Prešov) and the Mukachevo Diocese and some 68,000 Ukrainian Catholics living in Bohemia and Moravia, who were organized in a separate ecclesiastical body.[34]

On April 28, 1950, when the Communist power was firmly entrenched in Czechoslovakia, the Communists convoked a synod in Priashiv, abrogated the Act of Union of 1946 (union with the Vatican) and imprisoned two Ukrainian Catholic Bishops, Bishop Paul Goydych, OSBM (who died in prison in 1960), and Bishop Basil Hopko. The clergy were given a choice – to join the Orthodox Church or suffer being placed in a concentration camp.[35]

The year of 1968 ushered in the liberalization of the Communist regime of Czechoslovakia under the leadership of Alexander Dubček, first secretary of the Communist Party. The new regime initiated steps toward relaxing tension between State and Church and agreed to restore the Greek Catholic Church. But in the process of rehabilitating the Greek Catholic Church, the Slovak extremist clergy, supported by the Slovak Communists, removed Bishop Basil Hopko as head of the Church and appointed a Slovak priest, Jan Hirka, to administer it. Now, with new restrictions heaped down from Prague, the Ukrainian Catholic Church in Eastern Slovakia is undergoing a ruthless process of Slovakization.[36]

C. Deportations and Treatment of National Minorities

The Sudeten Germans are the largest group of Germans, living outside the 1937 frontiers of the *Reich*, to be expelled from their homeland after 1945. They inhabited the areas of Bohemia and Moravia for many centuries, and although they had been bound together by a common fate in the old Austrian monarchy, their political consciousness up to 1918 was much more an Austrian than a German one.[37]

The Sudeten Germans attempted to join the Republic of German Austria and were included in the Czechoslovak Republic without their consent; they were not represented in the assembly that drew up its 1920 Constitution. In the ensuing struggle against the principle of the centralized national state the Germans of Bohemia and Moravia-Silesia awakened to their national and political identity. In 1933 a new phase began with the rise of Nazism and Hitler's policy of expansion. The Sudeten German demand for autonomy, which arose as part of their own political tradition and for which they had long struggled, found a great "protector" in Hitler, who, however, not only wanted to annex the Sudetenland to the *Grossdeutsche Reich* but to destroy Czechoslovakia as an independent state as well.

According to the 1930 population census, the Sudeten Germans constituted 3,365,341 people out of the total of 14,729,536 in all Czechoslovakia.[38] In the parliamentary election of 1935, two-thirds of the Sudeten Germans voted for Konrad Henlein's Sudeten German Party (SdP). By the spring of 1938 the Sudeten German leadership had been taken over by Hitler, who in the Munich agreement of September 29, 1938, forced the cession of the Sudetenland to the *Reich*.

Undoubtedly, hundreds of thousands of the Sudeten Germans took part in Hitler's war machine and helped to build what they believed to be their future in union with the Greater *Reich*.

When the greater part of Slovakia was occupied by Soviet troops in the spring of 1945, a provisional government of the "National Front of Czechs and Slovaks" was established in Košice. A coalition government in union with the Communists oriented towards Moscow, it was under the chairmanship of Zdenek Fierlinger, Czech ambassador to the USSR, and was approved by President Beneš, who had returned to Slovakia from London via Moscow. At its first meeting on April 5, 1945, the new government set forth a comprehensive programme for the rebuilding of the republic. Paragraphs VIII and XI of the programme statement dealt with the treatment of the German and Hungarian minorities. Three groups of these peoples were differentiated as follows:

1) Those Germans and Hungarians who, as "anti-Nazis or anti-Fascists," had actively fought for the preservation of the Republic before the Munich Agreement or who had been persecuted because of their resistance against the Nazi regime after the integration of the Sudetenland into the German

Reich and the creation of the *Protektorat*, or who had taken part, as refugees in exile, in the fight for the re-creation of Czechoslovakia. All these were to have their Czechoslovak citizenship reaffirmed and would be assured of their eventual return to the Republic.

2) The "other" citizens of German and Hungarian nationality: their citizenship was to be annulled. But they were to be permitted a renewed option for Czechoslovakia, which would be acted upon in each individual case by the authorities of the Republic.

3) Those Germans and Hungarians who, because of their crimes against the Republic, were liable to be convicted. Those not sentenced to death would be expelled.[39]

According to Karel Lisicky, a Czech diplomat who served with the Czechoslovak government-in-exile in London, the Beneš government made the following analysis:

Out of the 3,200,000 Germans counted in the census taken in 1930 in Czechoslovakia, 250,000 would have to be written off as war casualties and 500,000 adherents of the Henlein movement would flee the country. Of the rest – nearly 2,500,000 Sudeten Germans – more than 1,600,000 would be expelled by organized transfer, while some 800,000 would be permitted to remain in the country.[40]

Actually, the number of expelled Sudeten Germans was much higher: according to official Bonn government sources a total of 2,921,000 Germans were ejected from Czechoslovakia.[41]

Up to 1968, that is, until the advent to power of Alexander Dubček, all the national minorities in Czechoslovakia had had a hard time. The new Czech constitution of 1968 officially recognized four national minorities; the Hungarian, German, Ukrainian and Polish. By West German count, there are today only 140,000 Germans in Czechoslovakia. There are some 68 German organizations, but no German-language schools; also there are 25 German Catholic parishes, but no German Catholic bishop. They muster two deputies – one Communist, another from the Labour Union – both controlled by the Communist Party.

The long years of intimidation prevented the Sudeten Germans from producing an articulate leadership amongst the German minority. The situation of the Ukrainians was rather better. According to Vladimir V. Kusin:

Unlike the German minority, which has not produced an intelligentsia after the war, the Ukrainians in Eastern Slovakia became well known beyond the Czechoslovak border precisely because of their scholarship and literary merit. The Ukrainian language and literary sections of the two Prešov faculties (of philosophy and education) of the P. J. Šafarik University in Košice, the Ukrainian section of the Czechoslovak (Slovak) Writers' Union, the four Ukrainian-language newspapers and periodicals with their supplements and the Radio Prešov Ukrainian broadcasting have all been actively promoting the Ukrainian cause and have given wholehearted support to the Prague Spring...[42]

But with the takeover of the Czechoslovak government by Gustav Husak a reaction set in, and the rights of the Ukrainian minority, as well as those of other minorities, have been considerably curtailed. A number of Ukrainian cultural leaders have been arrested and jailed for supporting Dubček's "liberalization" programme of Czechoslovakia.

The general oppression and the discrimination exercised against all the intellectual forces in Czechoslovakia in the post-Dubček era have also been stressed by a Czechoslovak scholarly association in the United States:

"Woe betide the nations whose literature is interrupted by force; this is not simply a violation of 'freedom of the press.' It is the incarceration of the nation's heart, the amputation of the nation's memory. The nation can no longer remember itself, the nation is deprived of its spiritual unity; and despite what seems like their common language, the members of that nation suddenly cease to understand one another..."

These words were prepared by Alexander Solzhenitsyn for his Nobel Prize lecture. They describe fittingly the situation now prevailing in Czechoslovakia, which is a reflection of the circumstances existing in the Soviet Union. Should this current development be permitted to continue unchecked, the outcome could only be as predicted by Solzhenitsyn.

The members of the Czechoslovak Society of Arts and Sciences in America, gathered in the City of Toronto, Ontario, on the occasion of their annual convention, voiced the most solemn protest against the new wave of harassment directed not only against the Czech and Slovak writers, but against all artists, scientists, scholars and intellectuals in general.

The meeting of the CSASA was held on November 17, 1973, the calendar day on which the Nazi authorities in the year 1939 closed the Czech universities of Bohemia and Moravia, deported the students to concentration camps and executed their leaders.

The similarity of what happened then and what is happening now stresses the urgent need for all people of good will to do their utmost in order to enable the arts and sciences in Czechoslovakia to flourish in freedom again.[43]

D. The Slovaks

The Czechoslovak Socialist Republic is a quasi-federal state, composed of two constituent republics and governed under the provisions of a constitution promulgated for the first time in 1969. The supreme organ of state power is the Federal Assembly, a bicameral legislature, elected in general elections. It is composed of the House of People, to which 200 deputies are elected in proportion to the number of Czechs and Slovaks of the country's population, and the House of Nations, whose 150 members are divided equally between representatives of the Czech Socialist Republic and the Slovak Socialist Republic; it also includes representatives of the national minorities.

The Federal Assembly is elected for a term of four years and is empowered to pass laws and to supervise the activities of the executive and judicial branches of government. The Constitution recognizes the Communist Party of Czechoslovakia as "the guiding force in society and in the State..."[44]

Slovakia's federal status suffers from the "normalization" process since the Soviet occupation in 1968 and from the new "consolidation" trends. The party has moved toward the recentralization of controls in most areas of activity.[45] The "December 1970 amendment to the 1968 law on federation restored the control of planning and economic management to the federal government, reintroduced a single Czechoslovak citizenship, and transferred the entire administration of state security to the Federal Ministry of the Interior."[46]

Denial of Human Rights

The Constitution declares that it will protect the dignity and freedom of citizens, regardless of their class status, religion or national origin, but the reality is something quite different. Through constant purges or so-called "verifications" (investigations), the regime has excluded 70% of former functionaries from trade unions, student associations and organizations of agricultural workers. It has removed 40% from the ranks of journalists and economic experts; over 1200 professors and teachers were transferred to other jobs, mostly to subalternate manual work. The regime has regimented the judiciary, so as to be able to carry out the so-called "socialist legality" through newly-appointed judges who are totally subservient to the regime. No one can say how many hundreds of persons were sentenced to prison terms for "subversive activities and incitement" against the socialist order. Thousands of people have lost their jobs without having been sentenced by the courts, and without recourse to courts, writers have been deprived by the regime of the right to publish, artists to create, actors to act, and young people to study. Former members of the Academy of Sciences or university professors are often seen selling lottery tickets, or loading or unloading construction material.[47]

Religious Life

With administrative-police methods the regime in Slovakia is trying to paralyse all activities of the churches and religious institutions. It has ordered the forced retirement of all clergymen who reach the age of 60, regardless of the fact that there are few members of the clergy. It has revoked the right of many young clergymen to carry out their priestly assignments. Likewise, the Communist government has reduced even further the small number of theological students by expelling scores of them from the seminaries. Nuns are forbidden to participate in social and charitable work, and in order to separate them from the rest of the population, they are confined in special concentration camps.[48] By a ruling of the Secretariat for Religious Affairs in Slovakia, all religious orders were dissolved, their property confiscated and the monks dispersed; many of them were compelled to give up their civilian jobs and many were transferred to homes for the handicapped and retarded, to take care of them.[49]

According to the official publication of the Catholic Church of Slovakia, the number of priests decreased from 2075 secular priests and 742 monastic priests in 1948–49 to a total of 1858 priests, including all retired priests, in November, 1971.[50]

The regime also forbids the importation of religious literature from abroad, and is restricting more and more the publication of religious periodicals and literature at home. It has eliminated the teaching of religion in schools of all levels and has expelled teachers who did not display a willingness to give up their religious convictions. Atheistic education proceeds step by step: in 1969–70 62% of the children attended religious classes: in 1970–71 – 53%; in 1971–72 – 48%, and in 1972–73 – only 30%. In one year alone 14 theological students were expelled from the Theological Seminary in Bratislava by a decision of the Secretariat for Religious Affairs.

The Sovietization of Czechoslovakia

All fields of human creativity are patterned after the Soviet example; official admiration of Soviet achievements is a fee for a career and survival. The Russification of Slovak life is accentuated in the arts, literature, recreation and in all areas. The submissiveness of the regime to Moscow is best characterized by the new law on the "National Security Corps" in Czechoslovakia. One article of the law reflects the "Brezhnev Doctrine" fully in stressing that one of the duties of the Security is to "protect friendly socialist countries and to safeguard the security of the whole Socialist World." The legalization of the so-called "Auxiliary Guard" of the Public Security that numbers 90,000 members of the "Elite Guard," underscores the terroristic persecution of non-Communist citizens.[51] In the words of George Moldau, "... we have a leadership without leaders or leadership without policies, and thus it is that Czechoslovakia and its tragedy seems so ridiculous and unreal. It is only a poor caricature of Soviet politics..."[52]

IV. EAST GERMANY

Germany, following her defeat in World War II, was divided into four zones of occupation – one each for the United States, the Soviet Union, the United Kingdom and France, with Berlin as seat of the Allied Control Commission. This arrangement was agreed upon and confirmed at Yalta on February 4–11, 1945. The occupation was to last until a general peace treaty between Germany and the Allies could be concluded.

On August 12, 1970, the Brandt Government signed a treaty with the Soviet Union in Moscow whereby the division of Germany was officially recognized by the Federal Republic. Both Foreign Minister Walter Scheel and State Secretary Egon Bahr stated that the treaty would; a) safeguard the preservation of the national interests of the German people; b) help to anchor peace in Europe; c) set the future relations of the Federal Republic with the Soviet Union; d) bring the German people more security.[53]

Although the West German constitution states that "the entire German people are called upon to achieve in free self-determination the unity and freedom of Germany," State Secretary Bahr did not hesitate to mention that "if peace in Europe is to be made more secure and co-operation is to be intensified, then the existing frontiers – whether this course please us or not – must be respected and regarded as inviolable..."[54]

A. Establishment of the Communist State: Looting and Sovietization

East Germany, even before its recognition by the Federal Republic of Germany, had been functioning as a quasi-independent state ever since the so-called "Pankow regime" had been established by the Soviet occupation authorities in 1949.

The German Democratic Republic (GDR) occupies 41,390 square miles, or about one-fourth the area of the Germany of 1937. In 1949, the year it was created, it had a population of 18,320,259. Now the population has fallen below 17 million: between 1949 and 1961, the year the Berlin "Wall" was erected, some 3 million Germans escaped to West Germany.

The entry of the Red Army into German territory was accompanied by wide-scale raping of German women, mass murdering of Germans and wholesale looting. Official German reports stated:

The raping of German women and children by Soviet officers and men was systematic in the truest sense of the word... This is proved by the fact that official searches were made for women, that many women were repeatedly raped by a series of men, and that the rapings very often took place quite publicly. Even aged women and children were not spared, and this had a particularly repelling and terrifying effect upon the German population.

In soldiers' newspapers, circulars and broadcasting, for instance, by author Ilya Ehrenburg, the Soviet troops were incited with brutal candour... Soviet officers and men, particularly the confirmed Stalinists, were influenced by this hate propaganda of Ilya Ehrenburg and other Soviet journalists to regard the violation of German women as an act of revenge against the Germans. The Soviet soldiers were not satisfied with merely raping women, but afterwards often killed them and in some cases even mutilated them in a sadistic manner...[55]

During the first two years (1945–1946) of occupation, the Russians looked forward to the eventual reunification of Germany, and the extension of their brand of "socialism" over the whole country, and therefore proceeded with the process of Sovietization of East Germany under an umbrella of "democratizing and anti-fascist activity," even allowing for the existence of "socialist" parties. But first among these was the German Communist Party (KPD), which soon forced a merger with the Socialist Party to form the Socialist Unity Party (SED). A decade after the fusion, however, only some three of the 40 original Socialists remained on the SED Executive Committee.[56]

The GDR promulgated two "constitutions," one in 1949 and the other in 1968, both of which were modelled strictly on the Soviet constitution, in which preference is given to representatives of the ruling party, that is, the Communist Party (SED). One of the most characteristic acts of the totalitarian regime of East Germany was its abolition of the autonomous provinces *(Länder)* in 1952 even though the action was unconstitutional. The dissolved provinces were replaced by fourteen administrative districts *(Bezirke)*, each containing fifteen or more counties *(Kreise)*. This move completely eliminated local government as a source of power and any eventual opposition to the central regime. Thus the overwhelming majority of the people of the GDR are not represented in the government, inasmuch as they are not members of SED.

For all intents and purposes, the GDR is a police state, masked by a democratic façade. Discrimination is rampant: there is no freedom of assembly, unless it be a Communist gathering; there is neither freedom of the press nor freedom of religion. The secret police as well as the regular "people's police," or *Volkspolizei* (VOPO), are resolutely keeping the German people in total submission.

B. Economic Strangulation

From the very inception of the Soviet military occupation of East Germany the Kremlin saw fit to strip East Germany of all worthwhile assets. Under the pretext of "de-Nazification" and "de-militarization," the Russians revelled in a conqueror's license to take anything they wanted. Thus they dismantled 13 airports, all radio stations; they tore up 2,500 kilometers of railroad trackage, while reducing another 11,680 kilometers of railway to a single line. The Kremlin also dismantled 1,667 factories and parts of 254 others; 213 of the largest concerns were confiscated or converted into Soviet corporations. It is estimated that at least 50 percent of all industrial facilities were dismantled and carted off to the USSR.[57]

The Russians exacted these enormous reparations with a total disregard for the needs of the populace. An American review stated:

... The Soviets actually bled the East German economy for almost ten years after the war. Factories were dismantled and their equipment shipped off to the Soviet Union. Expert technicians were sent for long-term duty in Russian laboratories and plants. Huge reparations were exacted by the Soviets. Estimates range from $16 to $27 million.[58]

Under the pretext of the Allied-approved de-Nazification programme, the Russians dispossessed large farms, killing or deporting in the process. Some 7,112 farms of over 250 acres, and 4,248 farms under 250 acres, were confiscated and handed to the "proletariat."

In 1952 collectivization was initiated (it terminated in 1960), causing a sharp decline in production. The Pankow government then allowed peasants to own small plots on which to raise additional foodstuffs. But the barns of the former large farms were empty; the state's elimination of private property resulted in a shrinking of rural communities.

A tourist who visited East Germany after two decades of absence made the following observations:

... one could see abandoned houses with broken window panes, lop-sided hanging shutters, decayed stucco, missing roof tiles... Similar conditions could be observed in smaller factories... closed by the government and left vacant and unused because the production there had been transferred to a large *kombinat* in another locality...[59]

The fact remains that East Germany's living standard is the highest in the Communist world. Yet it is still a low standard in comparison to the Western. The April 23, 1973 issue of *U.S. News & World Report* presented these facts:

Of 6 million homes in East Germany, only 2.5 million have private toilets: fewer than 600,000 have central heating. Private baths? Luxuries; 40 percent of East German homes have none. A million have no running water.

The Communist economy has failed to produce the worker's paradise, merely scarcity and deprivation.

The GDR suffers from another Soviet Russian exploitation as well. After dismantling and expropriations ceased, price bleeding began. Moscow tells the GDR what to produce and what to sell to and buy from the Soviet Union. And it puts the price of what it sells to the GDR so high and on what it buys so low that in effect it exacts continued reparations. On December 3, 1965, Dr. Erich Apel, GDR minister, committed suicide in protest against the Soviet economic exploitation.

C. Persecution and Destruction of Human Rights

Terrorization and persecution of the people of East Germany is part and parcel of the Pankow regime. The Soviet secret police, as the sole master in East Germany, has had a free hand in liquidating whomever they wished. In the first year alone of the Soviet occupation of East Germany, 80 percent of all the judges and 72 percent of all the teachers were arrested and tried, with most being sent either to prison or concentration camps in the USSR. Some 520,000 persons were arrested for belonging to the Nazi Party, of whom 150,000 went to concentration camps, 80,000 of these perishing there.

Despite ruthless indoctrination and terror, the East Germans refused to accept the role of robots of the Communist state. They fled to the West in great numbers, and expressed themselves in mass uprisings in East Berlin after the death of Stalin in 1953. In order to check the loss to the West of so many people, the Pankow regime, upon instigation of the Kremlin, erected the "Berlin Wall" on August 13, 1961. From Hof to Luebeck, a distance of some 836 kilometers, the Pankow government stretched barbed wire, built 152 bunkers and 741 observation towers and laid 1,700 personnel mines. Along a critical stretch of 79 kilometers they installed automatic weapons. Around West Berlin they have 243 observation towers and 144 bunkers. And between West and East Berlin there is the "Berlin Wall," measuring some 28 miles in length. As an admission of sociopolitical failure, this artificial remonstrance has no parallel in Communist futility. Yet despite all these barbed wire entanglements, in 1972 alone 6,782 persons risked their lives to escape the "Worker's Paradise," but it is not known how many of them were killed or captured.[60]

Most East Germans apparently have an undying yearning to unite with their brothers and sisters on the other side of the barbed wire and become part of the West. Late Soviet Premier Nikita S. Khrushchev, speaking in East Berlin, told the party leaders that they must produce more consumer goods for the people if they wished to quench these yearnings for the West. But man does not live by bread alone, a principle which even the Communist rulers must recognize here.

Religion has not been abolished in East Germany; both the Roman Catholic and Lutheran Churches exist. On February 6, 1966, the late East German leader, Walter Ulbricht, wrote to the United Nations that the GDR subscribes to Art. 18 of the U.N. *Universal Declaration of*

Human Rights, namely, "Every man is entitled to freedom of thought, conscience and religion." And indeed, around 1963 the Pankow government came to realize "that to endure as a national regime an accommodation with the church is necessary."[61] However, it ingeniously smothers religion at every other step. Clergymen may not proselytize for new members; they may not preach on topics which touch on any atheistic system: seminarians must take "compulsory courses in Marxism-Leninism"; religious publications are censured, and even when a publication is approved, only enough paper for a small edition is available.[62]

D. Intellectual Restlessness and Ferment

The new leadership of the SED under Erich Honecker, who succeeded the forbidding Walter Ulbricht, has not altered the stance of the great majority of the German people toward Communist rule. The general trend of dissent in the satellite countries and the USSR has also had its effect upon the East German intellectuals. One of the most vocal dissidents has been Robert Havemann, former professor of chemistry at Humboldt University in East Berlin. A lifelong Communist, he nevertheless criticized the SED leadership as too "dogmatic" and assailed the Party for not allowing public discussion on the "mistakes and shortcomings" of the Party. He was ousted from the university, the Party and the East German Academy of Sciences. He was not executed and was not even arrested, something that allows speculation as to a possible official relaxation, particularly when all dissension among East Germans was suppressed after the invasion of Czechoslovakia in August, 1968. Two of Prof. Havemann's sons subsequently were arrested and tried for "anti-State incitement."[63]

E. The National Minorities

There is a general concensus of opinion that the East German population of 17,000,000 people is homogenous, except for two small national minorities, namely Jews and Wends. The latter are also called *Sorbs*. The *Oxford English Dictionary* defines a *Wend* as "a member of the Slavonic race now inhabiting Lusatia in the east of Saxony, but formerly extending over Northern Germany; a Sorb." It also defines *Sorb* as a variant of Serb, "a Slavonic race inhabiting Lusatia, in the east of Saxony; a Wend." Lusatia, according to *Webster's Geographical Dictionary*, is the German Lausitz, a region "between the Elbe and Oder (Modern Silesia)."

The *Area Handbook for East Germany* (Keefe and Bernier, 328 pages, 1972), in conformity with all other sources consulted, explains:

... [the] only Slavs with distinct identification surviving on German territory as a group are the Sorbs, sometimes known as Wends...

They constitute pockets in the Cottbus and Dresden districts, near the Czech and Polish borders. They were estimated between 38,000 and 70,000 in the late 1960's, the lower estimate being probably closer to their actual number.

According to Arthur M. Handhardt, Jr., there were some 62,000 "Sorbs in the German Reich" in 1925; he further stated that "during the Nazi period they were persecuted," but they had persisted in their culture. He writes:

In 1948 the *Land* government of Saxony passed the "Law for the Protection of the Rights of the Sorbic Population," guaranteeing Sorbic elementary and secondary schools, making Sorbic the official language of the local bureaucracy and establishing the Office for Sorbic Culture...[64]

The Pankow regime, emulating the Soviet "nationality policy," encourages the development of Sorbic culture and the use of the Sorbic language in schools for children of Sorbic parents. In 1952 the Ulbricht regime was even considering a plan to clear the Sorbic areas of Germans to achieve a purely Sorbic population. "But neither Sorbs nor Germans... were enthusiastic about this plan and the idea was dropped..."[65]

The only other recognized non-German minority in the German Democratic Republic is the Jewish minority. It is very small, perhaps 10% of that of the Sorbs, but one that has considerable international significance. Handhardt writes of the Jewish minority:

... It is unlikely that the Jewish community has grown since 1945. A western source gives the number of Jews in the Soviet Zone and East Berlin in 1946 as 3,100. In 1952 this figure declined to 2,600 and in 1962 to 1,800. In 1961 there were nine Jewish communities in the GDR...[66]

Another writer, Welles Hangen, writes of some 1,500 Jews in the GDR, mostly older people, and contends that the Jews in East Germany are slowly dying out, not because of immigration (which is prohibited), but of old age and because of the low natural increase among them. The synagogues in the larger cities have been rebuilt with government help; some Jewish cemeteries and community houses have also been reopened, and "anti-Semitic propaganda is officially forbidden in the GDR." Hangen also contends that as of 1966 Jews did not play an important part in the industrial or social life of East Germany. The only ones in important posts, he found, were Otto Winzer, foreign minister, Albert Norden, the propaganda expert of the Politburo, and the late Dr. Gerhart Eisler, chairman of the radio and television department.[67]

It would seem that the national minorities in East Germany, the Sorbs and Jews, do not represent any internal threat to the Communist regime of the Pankow government and, therefore, they are left relatively unmolested.

The position of the Jewish minority in East Germany has improved even more with the establishment of diplomatic relations between the United States and the GDR on September 4, 1974. In granting diplomatic recognition *(Staatsvertrag)* to the German Democratic Republic the U.S. government prevailed upon the Pankow regime to make reparations to East German Jews who suffered from Nazi persecution.[68]

In conclusion, mention must be made of several thousands of foreign workers *(Fremdarbeiter)*, mostly Hungarians, Poles, Yugoslavs and perhaps Czechs who work on a contractual basis in industry and agriculture at salaries higher than domestic labour. There also are a great number of foreign students in East Germany, particularly those from Asia and Africa, as well as those from various "people's democracies" in Central and Eastern Europe.

The East German government is adamant in not allowing its citizens to emigrate abroad, particularly those persons whose kin escaped to West Berlin or West Germany in general.

V. HUNGARY

Hungary, situated in southeastern Europe in the Danubian Plain, is bordered by Czechoslovakia, Austria, Yugoslavia, Rumania and Ukraine. Encompassing an area of 35,912 square miles, it has a population of 10,280,000, of whom 97 percent are Hungarians, the remainder being German, Slovak and Rumanian minorities. About 67 percent of the Hungarians are Roman Catholics, 22.8 percent Calvinists, 3.3 percent Lutherans and 6.9 percent of other faiths, including Greek Orthodox and Jewish.

The Hungarian people stem from Finno-Ugric stock who broke off from their main ethnic family and after long years of migration settled the valley south of the Carpathian Mountains in 895. By the XVIth century Hungary was a leading power in Europe, challenging the Turkish penetration of Europe for several decades. Its defeat by Turkey in 1526 resulted in the Turkish occupation of the central part of the country until 1686, with the western portion going under Hapsburg rule.

Periodic national uprisings and revolutions occurred in the XVIIth and XIXth centuries, one of which was led by Lajos Kossuth in 1848–49 and which was finally put down by Russian troops called in by the Hapsburg rulers. The defeat of Austria by Prussia in 1866 led to the establishment of the Austro-Hungarian monarchy upon whose collapse in 1918 an independent state of Hungary was established under a Regency that lasted until World War II.[69]

Rear Admiral Count Stephen Horthy, who was supported by Hungarian nationalists and members of the old Magyar aristocracy, was able to form a coalition government which dealt a crushing defeat to the Hungarian Communists under the leadership of Bela Kun in what the Communists describe as the "White Terror" campaign, in which some 300 Communists were massacred.[70]

In accepting the provisions of the Treaty of Trianon (1920), the Horthy regime was forced to give up some three-fourths of its lands and thenceforth had to walk a tightrope thanks to an impoverished economy. With the expectation of regaining what it considered her lost territories, Hungary aligned itself with Nazi Germany and Fascist Italy. In 1939, during Hitler's final assault on Czechoslovakia, Hitler allowed the Budapest government to invade Carpatho-Ukraine, a tiny Ukrainian land which on the day before, March 15, 1939, had proclaimed its own independence.

Hungary entered the war as an ally of Germany and Italy only in mid-1941, and sent several divisions to the Eastern front. Horthy stated in his *Memoirs* that Hungary was an "unwilling satellite" of Germany, and tried, with "meagre means" to "defend herself against two encroaching forces: against the Soviets with all their available arms: against the Nazi ideology with all her diplomatic powers..."[71]

A. Communist Takeover and Terror

The Soviet armies marched into Hungary on October 6, 1944, bringing along a well-prepared plan for the future of the country. A nucleus of trained Hungarian Communists, led by Matyas Rakosi, Zoltan Vas and Mihaly Farkas, formed a Communist government which had the indispensable support of the Red Army.[72] The "armistice" agreement with the USSR compelled Hungary to deliver up reparations amounting to the 1938 equivalent of $200,000,000. Moscow ruthlessly exploited this arrangement, virtually bankrupting the country through so-called Soviet-Hungarian joint stock companies under Soviet control and management.[73]

The new government, in cooperation with the NKVD (Soviet secret police) and Soviet Army security forces, arrested thousands of Hungarians on the pretext of their Nazi collaboration and membership in Hungarian fascist organizations. Hungarian exile sources estimated that by 1960 some 60,000 Hungarian prisoners of war were still in the USSR, with at least 120,000 civilian deportees also being held in the Soviet Union.[74]

B. The "People's Government"

Although universal suffrage was adopted and the November, 1945, election for a regular government was held in the presence of the Soviet army, the Communists reaped only 17 percent of the vote, electing only 70 deputies as against 245 gained by the Smallholder Party. This was not only embarrassing to the Rakosi government but to the USSR as well. Despite the fact that by every parliamentary rule the Smallholders should have been called upon to form a government, they were not. On the contrary, the police arrested Bela Kovacs, Secretary general of the Smallholder Party, and many Social Democratic leaders; new pro-Communist parties were formed to counter-balance the Smallholders. Backed by Moscow the Communists forced the liquidation of the opposition parties, arrested and killed their leaders, and compelled other parties to join the Marxist ranks.

On August 18, 1949, a new constitution was proclaimed establishing a "People's Republic" of Hungary after the Soviet Communist pattern, a formula which expresses special recognition and thanks to the USSR in making possible a development toward socialism. The preamble of the constitution states:

The armed forces of the great Soviet Union liberated our country from the yoke of the German fascists, crushed the power of the great landowners and capitalists who were ever hostile to the people, and opened the road of democratic progress to our people... supported by the Soviet Union, our people began to lay down the foundations of socialism and now our country is advancing towards socialism along the road of people's democracy...[75]

Hungary possesses a unicameral system, with the parliament designated as the highest organ of state authority. It is charged with the responsibility of enacting laws, determining the state budget, formulating national economic plans, electing the Presidential Council and the Council of Ministers, controlling ministers, declaring war and concluding peace, and exercising the prerogative of amnesty.[76]

The Presidential Council and the Council of Ministers are the second and third highest organs of state authority, and are supposed to be the highest organs of state administration.

The country is divided into counties, districts, towns and boroughs; in each of these the power is exercised by councils elected for a period of four years.

The degree of discrimination in political matters against the Hungarian people is indicated by the fact that all candidates for election to national and local organs of government hail from the Hungarian Socialist Workers' Party, which is the euphemistic name for the Communist Party; participation of nonparty members is purportedly provided through the so-called People's Patriotic Front, which some people in the West think of as an opposition party but which in fact it has never been.

For more than a decade Hungary was ruled by the "Muscovites," i.e., Matyas Rakosi, Erno Gero, Imre Nagy and Mihaly Farkas, all of whom had spent long years in the USSR and then had been brought back to Hungary by the Soviet troops. It was an iron rule characterized by various forms of oppression and discrimination; by arrests and trials of and severe sentences meted out to all real and imagined opponents of the Communist regime; persecution of the church and intellectuals, and so forth.

C. The 1956 Uprising

No discussion of Hungary could be deemed complete without a reference to the traumatic Hungarian uprising in the autumn of 1956, which was an open challenge to Soviet domination. Although Rakosi was removed from power by Moscow in 1953–1955 and his successor, Imre Nagy, initiated softer policies, the trend did not continue, as many Hungarians hoped and thought it would. In February, 1955, the line hardened as Rakosi returned to full power and Nagy was expelled from the Party. As Rakosi waxed more and more tyrannical in his rule of the country, the various dissident and opposition elements, notably the students and the intelligentsia, began contemplating a return to democracy through open revolt. With Khrushchev's' "de-Stalinization" policies in effect in the USSR, and the general trend of "relaxation" in the Communist-dominated countries, many Hungarian leaders were optimistic as to a revolt, and believed that neither the Kremlin nor the U.N. would interfere with Hungary's return to democratic rule. But they sadly erred on both counts. Neither the U.N. nor any of the Western powers did anything despite the desperate appeal the Hungarian freedom fighters made as the Soviet tanks crushed them.

The Kremlin could not allow one of its subservient satellites to escape from its control and become neutral, let alone become Western-oriented, because of the effect this might have on the other satellite countries. The uprising of the Hungarian people was thus doomed to failure.

Janos Kadar, chosen by the USSR as both the new premier of Hungary and first secretary of the Party, initiated bloody "purges," carried out by the AVH (Hungarian secret police). Details of this massive repression of the Hungarian people have been brought to the fore by outstanding Hungarian leaders who took part in the revolt, such as Joseph Kovago, former Mayor of Budapest, Gen. Bela Kiraly and Imre Kovacs.

Thousands of Hungarian men and women were arrested, tried and either executed or sent to concentration camps in the USSR. Dr. Kurt Rabl, an international law expert, reported:

As we know, the Red Army deported a large number – just how many we shall never know – of Hungarian freedom fighters to the Soviet Union. This action was unanimously condemned by international public opinion and especially by the investigating committees of the U.N. on Hungary; these deportations were cited as a "disregard of human rights and basic freedoms..." [Yet] as for Soviet Russia, it never called into question this juridical qualification. The deportations were denied, and it was said that no deportations had taken place and if so, they were "blunders of subordinate officers..."[77]

It was reported, in the aftermath of the revolt, some 63,000 Hungarians were deported to Siberia. According to the same source, by January, 1965, 463 participants still languished in the Central Prison in Budapest; and 143 who had been under eighteen years of age in 1956 were subsequently executed.[78]

D. Religious Oppression and Persecution

Organized religion, especially the Roman Catholic Church in Hungary suffered the same fate as in the other Communist-dominated countries. During 1944–1947, a comparatively tolerant attitude toward religion was adopted by the rulers. But in January, 1948, the most violent phase of the anti-religious campaign was initiated. Denominational schools were nationalized; by 1950 all religious instruction was controlled and regimented; religious orders abolished; convents and monasteries closed, and priests, monks and nuns arrested and deported. The trial in 1949 of Archbishop Joseph Mindszenty shocked the Western world. His dramatic release during the Hungarian uprising in the fall of 1956 and his subsequent refuge in the United States Legation in Budapest only heightened and symbolized the religious persecution in Hungary.[79]

By late 1963, following the softening of Moscow's policies towards the Vatican, the Hungarian government also relaxed its attitude toward the Catholic Church. Five bishops and an apostolic administrator were allowed to attend the Vatican Council in Rome. On September 19, 1964, an agreement was signed in Budapest between the Vatican, represented by Msgr. Agostino Casaroli, and the Hungarian government, represented by Joseph Prantner, chief of the State Office of Religious Affairs. The accord restored the Church's right to form a hierarchy and to communicate with the Vatican. But neither the case of Joseph Cardinal Mindszenty nor that of the Archbishopric of Esztergom was resolved. The agreement, on the other hand, dealt with such matters as the appointment of bishops, the "citizens' allegiance oath" to be taken by priests and the Papal Hungarian Institute in Rome.[80]

Cardinal Mindszenty took refuge in the U.S. Legation in Budapest until September 28, 1971, at which time he consented to leave Hungary under an agreement between the Vatican and Budapest. For a short time he was a guest of the Pope in the Vatican, but later on he went to live in a Hungarian church institution in Vienna.

On February 5, 1974, Pope Paul VI relieved Cardinal Mindszenty of the jurisdiction he had still nominally maintained in Hungary and deprived him of his honorary function as the Roman Catholic primate of Hungary, a move that has been widely interpreted as a gesture on the part of the Vatican towards improving its relations with Communist Hungary. The Vatican itself said that it had acted after "prolonged, mature and grave reflection." For his part, Cardinal Mindszenty denied that he had retired "voluntarily" or that he had "abdicated" his offices.[81]

It should be pointed out that the Vatican's hopes for improved relations with the Communist regime of Hungary had been shattered immediately after the signing of the 1964 agreement by a new wave of arrests and trials of the Hungarian Roman Catholic clergy. Rev. Laszlo Emodi and Rev. Laszlo Rozsa were tried and sentenced to five and eight years hard labour, respectively. Several unidentified priests were sentenced in a "conspiracy" case, while others were apprehended for organizing "discussions" among youth groups.[82]

E. Pro-Soviet Education and Intellectual Dissent

The building of a "new Socialist man" in Hungary is accompanied by an extensive indoctrination in Russian history, literature and arts and in the Russian language. Concomitantly there proceeds an eradication of the Hungarian cultural heritage, which has always been anti-Russian and pro-Western. (The Kremlin cannot follow here a "pan-Slavic" approach as it does in such Slavic countries as Poland, Czechoslovakia, Bulgaria and Yugoslavia.) Stress is put on the "outstanding achievements" of the Russian people and the Soviet leadership.

Aside from institutional manipulation, Russification has been fostered through governmental measures, monopoly of the press and control of all mass communication media. In Hungarian universities, Russian is the main language taught; in order to receive a degree an individual must pass state examinations designed to show a command of the Russian language.

Many students receive scholarships and other forms of financial aid which enable them to travel to the USSR and to study at Russian universities. Instrumental in pressing the Russification policy have been the school clubs and youth organizations, particularly the latter. In all schools, literature, history and geography are taught with a strong Russian slant; literature courses for instance, deal mainly with Russian authors, characters and events.[83]

Despite the severity of the Communist regime, however, opposition among Hungarian intellectuals is growing and making inroads in Hungarian society as a whole.

Charging "subversion," the Communist regime indicted in 1972 Miklos Haraszti, a 28-year-old author, for privately circulating a work he had written on human problems in connection with industrial work. The writer had been reprimanded by the Party a few years ago and expelled from the university for writing a poem in which he declared that "those who use force, will be destroyed by force." His new work grew out of his labour in a factory. Tried on January 10, 1973, he was given a suspended 8-month sentence.[84]

The former prime minister who was ousted as the result of the Hungarian revolution in 1956, sociologist Andras Hegedus, along with philosophers Mihaly Vajda and Janos Kis, was expelled from the Communist Party for his critical and dissident views (he was called an "intellectual renegade" by Gregory Aczel, "cultural commissar" of the Budapest regime). Hegedus, who in 1956 had called on the Russians to suppress the Hungarian revolt, was converted to liberal and reformist views because of Soviet brutality, and joined other social scientists at an international conference held in 1968 in Yugoslavia in condemning the Soviet invasion of Czechoslovakia.

To cope with ever-growing dissension, the government has introduced in all schools a course entitled "Civic Knowledge," and for two hours each week students are drilled in "socialist patriotism."

Open challenge to the regime comes from the youth, particularly the university youth. On March 15, 1972, and March 15, 1973, in commemorating the Kossuth Revolution of 1848, students demonstrated against government speakers, calling for "freedom" and human rights. At least 250 students were arrested, with many of them being convicted of "anti-state" crimes.[85]

VI. POLAND

Poland, situated in northeastern Europe, is one of the most important of the Communist-dominated nations because of its strategic position and its population. It lies between the Baltic Sea and the western Carpathians and is bordered by Germany to the west, Ukraine and Byelorussia to the east and Czechoslovakia to the southwest. It embraces an area of 120,664 sq. miles and has a population of 33,070,000, of whom 95 percent are Roman Catholic, the remaining 5 percent of other denominations.

The national traditions of the Polish state date back to 966, the year Poland introduced Christianity; it was a powerful state until the beginning of the XVIIIth century. In three successive partitions (1772, 1793 and 1795), Poland was apportioned among Prussia, Russia and Austria.

On November 11, 1918, Poland was resurrected as an independent state and, on June 28, 1919, it was recognized by the Treaty of Versailles.

One of the major weaknesses suffered by Poland, especially after its conquests in its 1920 war with Russia, was the presence of a great number of national minorities – 6 million Ukrainians, 2 million Byelorussians, 1.5 million Germans and about 2 million Jews, plus hundreds of thousands of Lithuanians. Despite the fact that the Polish government had obligated itself to grant national autonomy to the Ukrainians, its semidictatorial government, especially that under Marshal Josef Pilsudski (1926–1935), treated the Ukrainians and other minority peoples harshly, trying to Polonize them and depriving them of their national rights which the Warsaw government had guaranteed under its obligations to the League of Nations. Poland also had endless territorial conflicts with Lithuania, Czechoslovakia and Germany, which rendered the new Polish state weak and vulnerable to outside pressure.

On August 23, 1939, Hitler and Stalin, on the basis of the Ribbentrop-Molotov agreement, divided Poland and thus precipitated the outbreak of World War II.[86]

A. Communist Takeover and Establishment of a Totalitarian Regime

The shape and future of postwar Poland was determined at the Yalta Conference, February 4–11, 1945, by Franklin D. Roosevelt, Winston

Churchill and Joseph Stalin, who collectively agreed that the Polish people would choose their form of government in "free and unfettered elections as soon as possible on the basis of universal suffrage and secret ballot."[87]

New Poland was also assigned approximately 40,000 sq. miles of German territory east of the Oder-Neisse line comprising Silesia, Pomerania, West Prussia and a part of East Prussia, as a "compensation" for the loss of Western Ukraine (Galicia and Volhynia) and of a part of Western Byelorussia to the Ukrainian SSR and the Byelorussian SSR, respectively.

During the war the Polish government-in-exile in London had co-operated well with the Allies. A strong and effective Polish army was organized and it took part in the war against German forces in North Africa and Italy. Immediately after the Nazi attack on the USSR, the Kremlin proceeded to organize the so-called "Lublin Committee," which they later began championing as a true "free Polish government," in opposition to the Polish government-in-exile in London.

After the war, while the Western statesmen believed that they had assured a democratic government for Poland, the Russians proceeded with their own plan for Poland, a plan which they had had all along. To them a "free" or liberated government meant one that would act in blind obedience to the Kremlin. Therefore, Russian agents and Soviet-trained Poles quickly exploited the early postwar situation and pushed ahead with typical Trojan horse tactics and various "United Fronts."

With the Communist takeover of Poland accomplished, the Warsaw regime asked the people of Poland to decide on the composition of the government through so-called "free elections," which had been agreed upon by the USSR and the Western Allies at Yalta. But in the immediate period after the war, the Kremlin and its Polish political puppets endeavoured to eliminate all possible opposition by means of violence and deception and by falsifying the results of the "elections." In the first "election," held in 1947, the Communist campaign was directed especially against the Polish Peasant Party and the Roman Catholic Church, as all other prewar Polish political parties had been banned outright as "enemies of the people." Regardless of the widespread dissatisfaction and opposition on the part of the Polish people, the ruthlessly waged campaign brought the Communists a resounding success.

The results of the "elections" were wholly falsified. On January 25, 1947, six days after the election, the Department of State in Washington issued a note to the Polish government to point out that "the electoral procedure was in complete contradiction of the Yalta Agreement." The note was ignored and the Communist terror, abuses, murders and violence continued.[88] The travesty and tragedy is vividly described by an eyewitness, Stanislaw Mikolajczyk, the leader of the Polish Peasant Party, who barely escaped with his life.[89]

In the subsequent elections of 1952, 1957, 1961, 1965 and 1969, the Communists were in full control of the electorate; new statutes were enacted and promulgated which have made impossible the election of any candidate of the opposition.

The ruling political party of Poland is the Polish United Workers' Party, known by its Polish initials PZPR *(Polska Zjednoczona Partia Robotnicza)*, which numbers only about 2.3 million members out of the total 33 million population of Poland. One may wonder how it may be possible for a few men to maintain total control and regimentation over a great nation. This can be explained by the party structure, which includes several levels: local, district, provincial and national.

Although the statutes of the party call for "democratic centralism" and although all the ruling authorities from the lowest to the highest echelon are "elected" in a democratic manner, the only real authority resides with the first secretary of the Party, who since 1970 has been Edward Gierek (replacing Wladyslaw Gomulka). It is not known to what degree Moscow controls the Polish first secretary, but that Moscow exercises a strong influence cannot be doubted.

In October, 1956, before the "Poznan Revolt," it is a known fact that during the eighth plenary session of the Polish Party's Central Committee Nikita S. Khrushchev, Vyacheslav Molotov, Lazar Kaganovich and Anastas Mikoyan arrived unexpectedly and uninvited from Moscow after learning that "standard operating procedures were not being observed in Poland."[90] This "visit" was timed to coincide with Soviet troop movements in the direction of Warsaw.

Also, for world consumption, a few "independent" parties are allowed to put up their own candidates, but in the final outcome it is always the candidates of the Communist Party who win.

The legislative powers are theoretically vested in the parliament, or *Sejm*, which formally elects the State Council. This latter body, which has replaced the former presidency, calls general elections, interprets laws, appoints and recalls diplomatic representatives, and so forth. The Council of Ministers is the highest executive and administrative organ of state authority. The local "State Councils" are the equivalent of local *soviets* in the USSR.

These instruments of State authority, including the Supreme Court, and the provincial and county courts as well – all are but a legal facade for the party organization, which, like the party in the USSR, is the sole source of political power.

B. Restraint of Religious Freedom

In prewar Poland the Catholic population constituted over three-fourths of the entire population, including some 5 to 6 million Ukrainian Catholics of Eastern Rite; the remainder were some 3 million Jews, over 1.5 million Germans (mostly Lutherans) and some Lithuanian Catholics and Byelorussians (both Catholics and Orthodox). After the war, most of the Ukrainians were forcibly transferred to the Ukrainian SSR. Close to 200,000 Ukrainians were resettled in the "recovered territories," that is, the former German territories in the west and northwest. Almost all the Germans were expelled to West Germany, while the Byelorussians were sent to the USSR. Thus from the religious viewpoint, postwar Poland was almost totally (95 percent) a Catholic country, with a devout and nationalist people.

In 1945, the new Communist-dominated Council of Ministers in Warsaw abrogated the twenty-year-old concordat between Warsaw and the Vatican, which had regulated the activities of the Church in Poland. The reason for the annulment of the agreement allegedly was the Church's "favouring" of Germany during the war. From that time on a systematic campaign beset the church, beginning in the press and on the radio. Subsequently, religious instruction was gradually curtailed.

When in January, 1949, Archbishop Stefan Wyszynski became the primate of Poland, he initiated a series of negotiations with the regime in order to "normalize" state-church relations. Almost at the same time the Vatican issued a decree which would excommunicate all Catholics who actively supported Communism. The Polish regime reacted by declaring that priests enforcing this law would be tried under Polish law on the ground that they were taking orders from a "foreign power."

Eventually, in April, 1950, an agreement between the government and the Polish episcopate was signed: the church agreed to abstain from all political activities and keep the clergy from opposing the Communist regime; the government, in turn, pledged to guarantee freedom of worship, religious education in public schools and noninterference with the Catholic press. But the agreement was weighted down with restrictive clauses which were used by the regime to circumvent the spirit of the agreement.[91]

In subsequent years the persecution of the Catholic Church mounted in intensity. In September, 1953, Cardinal Wyszynski was arrested and forbidden to discharge the functions of his office; the arrests of clergy followed as well, and all Catholic organizations and the Catholic press were systematically hounded.

In 1956, after the "Poznan Revolt" and the assumption of power by Wladyslaw Gomulka, Cardinal Wyszynski and other arrested clergymen were released, and a new precarious "truce" between the church and the state was made.

As in a number of other countries, the Communists organized a rival Catholic layman organization, "Znak" (Omen), in opposition to the legitimate Catholic layman organization, "Pax." The "Znak" group is composed of Catholic laymen who are staunch supporters of the Communist regime. One of them, Stanislaw Stomma, a member of the *Sejm*, declared:

The question of peace today is a single one – it embraces the whole world. Declarations are not enough... The policy of Peoples Poland has stood on a firm foundation, and has chosen an alliance consistent with the development prospects of our nation. This is the alliance with the Soviet Union.[92]

Outwardly, the regime tries to creat the impression of a free church; it granted Cardinal Wyszynski and 25 Polish bishops permission to go to Rome to participate in the Ecumenical Council, but 54 other Polish Catholic bishops were refused such permission. In the fall of 1965 and early in 1966 Cardinal Wyszynski was singled out for attack because of a letter from the Polish episcopate to the German Catholic bishops inviting them to take part in the observances of the Christian millenium in Poland in May, 1966. The Warsaw regime reacted violently by accusing the church hierarchy of encroaching upon foreign affairs. Also, on December 15, 1969, Cardinal Wyszynski submitted to Pope Paul VI a memorandum asking for the appointment of regular bishops in the Oder-Neisse areas; as a result, the church and the Vatican were accused by the government of a "pro-German attitude." Eventually, new bishops were appointed, and on December 7, 1970, the Bonn-Warsaw treaty was signed recognizing the *status quo*, that is, the German-Polish border on the Oder and Neisse Rivers.

The future of the church in Poland does not seem to be a bright one, but the Catholic population is very strong and continues to be an important political factor, one which the Communist regime cannot afford to ignore or underestimate.

C. Deportations and Treatment of National Minorities

In accordance with the Yalta Agreement of February, 1945, it was deemed necessary to expel all Germans east of the Oder-Neisse line, as was similarly decided in the case of the Sudeten Germans in Czechoslovakia. The pro-Soviet "Polish Liberation Committee," which on December 31, 1944 constituted itself as the "Provisional Government of the Polish Republic," consisted exclusively of Polish and Russified Communists, and the programme of expulsion was implemented by the new Polish government in cooperation with the Soviet Union.

A total of 9,758,000 Germans were involved in this massive expulsion. According to West German government statistics these Germans lived, as of 1937, in East Prussia (2,519,000), in East Pomerania (1,861,000), in East Brandenburg (660,000) and in Silesia (4,717,000).[93]

This operation was accompanied by feelings of bitterness, hatred and a spirit of retribution resulting from the inhuman and barbaric treatment suffered by the Polish people at the hands of the Nazi regime. Thus there were excesses: mass rape of women, murder of German civilians, plundering and incendiarism.[94]

Another ethnic community which suffered untold harshness and persecution on the part of the new provisional Communist government of Poland was the Ukrainian one.

When the new Polish-Soviet frontier was definitely agreed upon in 1945 following the Yalta agreement, there were some 1,200,000 Ukrainians living west of the new boundary. They inhabited the western parts of the province of Peremyshl (Lemko Land in the south) and the areas of Lublin, Kholm, Polisia and Pidliasia in the north. The Soviet government, in conjunction with the Warsaw regime, tried every trick of propaganda to induce these Ukrainians to move "voluntarily" to Soviet Ukraine, but to no avail. The dreadful memories of the executions and the mass deportations in Eastern Galicia were far too vivid. So, the Ukrainians en masse refused to heed the Soviet bid.

As a result, the Polish government decided to resettle them in the newly-acquired German territories by force. This drove thousands of able-bodied young Ukrainian men and women into joining the ranks of the anti-Soviet Ukrainian Insurgent Army (UPA) to wage a fanatical underground struggle against the Polish Communist forces and the Soviet security and regular army troops for several years on both sides of the Curzon Line.[95]

For almost four years (1945–1948), the Polish government unsuccessfully deployed thousands of troops and police against the insurgents. In 1947, a joint Soviet-Polish-Czechoslovak agreement was signed calling for combined operations on the part of several divisions of the three states to end the insurgency of the UPA. Khrushchev wrote in his latest book:

There were quite a few uprisings in the eastern region of Poland along the Ukrainian border. However, these flare-ups, which sometimes amounted to war, were instigated by Ukrainian nationalists, not Poles; and the flames were fueled by the Americans, who parachuted arms, machinery, communication equipment, and other supplies to the insurgents. We sympathized with the troubles the new Polish government was having in its eastern territories, for on our own side of the border, in the Western Ukraine, we were engaged in a cruel struggle against the OUN and the followers of Stepan Bandera. We also had to contend with strong resistance to the new system among the kulaks and an armed insurrection in Lithuania.

The Ukrainian nationalist activity within Poland became so serious that the Polish armed forces had to conduct full-scale military operations in the frontier areas of their republic near the Carpathian Mountains...[96]

The military operations by the Polish government against the Ukrainian underground were accompanied by the forcible "resettlement" of the Ukrainian population in the "recovered territories." Taking part in the operations were the *Fifth* and *Ninth* Divisions of the *WP* (*Wojsko Polskie* – the Polish Army), known as the "punitive divisions," several detachments of *O.R.M.O.* (Voluntary Citizens' Militia Reserve), units of *KBW* (Corps of Internal Security) as well as several units of *UB* (Office of Security) – all in all about 100,000 armed men against some 1,200,000 Ukrainians.

In the course of this "punitive action" hundreds of Ukrainians were slaughtered without any trial or investigation. For instance, the 34th Infantry Regiment of the Polish Army attacked the Ukrainian village of Zavadka Morochivska, slaying 54 persons in cold blood.[97]

Other atrocities by the Polish troops were committed in the village of Valva (80 persons killed); Hnatkovychi (40 killed, including children); Richytsi (35 killed) and Sosnytsi (25 killed).[98]

By February 19, 1947, the Polish government announced that at least 97,935 Ukrainians had been sent to the USSR in accordance with the Soviet-Polish pact and some 700,000 had been "resettled" in former German lands.

Today, Poland's national minorities include some 500,000 to 600,000 Ukrainians; the total membership of the German, Jewish and Byelorussian communities accounts for a similar figure. Up to the advent to power of Wladyslaw Gomulka in 1956, these minorities were in reality outside the law: they were discriminated against as a foreign ethnic element; their languages were unrecognized; they were, as a rule, discriminated against in obtaining state employment, unless they chose to change their nationality, and in the case of Ukrainians, they were penalized because of religion and even their family names.

After the 1956 upheaval in Poland some improvements were made in the status of the national minorities, especially the Ukrainians. They were allowed by the government to establish a national Ukrainian Cultural Association and to open a number of Ukrainian-language schools on the elementary and secondary levels; also, a chair of Ukrainian literature and language was created at the University of Warsaw. A weekly newspaper and books are published in Ukrainian. But the pleas by the Ukrainian Catholic clergy to reopen Ukrainian Catholic churches – the overwhelming majority of Ukrainians in Poland are Catholics of the Eastern Rite – are continually rejected by both the Polish Catholic hierarchy and the Polish Communist government.

The German minority, however, fares much worse. There is no German-organized life, so to speak; German-language schools are forbidden as are German associations and press, despite the fact that the German language is taught in a number of secondary schools in Western Poland.

D. Progressive Suppression of Freedom

In 1970, a few days before Christmas, riots and protests by Polish workers flared up in the Baltic port of Danzig because of the sudden increase in prices of food and fuel. But the price increases were not the only reason behind the riots. There was also the new wage structure designed to stimulate productivity which would have resulted in a drop in the earnings of hundreds of thousands of workers. But behind these economic grievances were deeper feelings: the hate of the workers for the system as a whole lent violence to their anger. The crisis had been growing for years, the economy was stagnating. A prominent Polish scholar assessed the situation in an interview in a Warsaw publication:

The excessive centralization of decisions and the monstrously overgrown bureaucracy ruined many potential sources of energy, inventiveness, creativity. The morale of the working masses has been undermined, human activity drowned in a sea of paper. Technical programmes have been halted because the most false criteria were selected by which to judge – and not merely economic accomplishments... This grotesque evaluation of human activities was applied not only in the economy. Schools were judged, for instance, by the number of good marks the pupils were getting. Therefore, schools graduated under-educated people in order to acquire a good reputation...[99]

The most important result of the 1970 riots was the emergence in Poland of labour power. The assertiveness of workers, standing alone and not associated with any intellectual forces, was a new phenomenon in the Communist world.

But after the events in Czechoslovakia in 1969, the Soviet course in the Polish 1970 crisis was less drastic if no less firm. Moscow made it known in Warsaw that it would abandon any ally overnight – no matter how loyal, "internationalist" and "Marxist" – if such would try to "betray" the "friendship" with the USSR. So Gomulka's successor, Edward Gierek, beat a strategic retreat.

Thus far, the new Polish regime has proved to be as subservient to Moscow in foreign policy as was that of Gomulka. The complete jamming of "Radio Free Europe" has been resumed. There has been a big build-up of the secret police and the militia, and there has been an ominous return in Poland to secret trials of political dissidents. As many as three such trials were held in Lodz and Warsaw, at which 17 defendants, members of a new underground organization *Ruch* (Movement), received sentences of up to 10 years.

Thus Poland, despite some recent changes, is slipping back into the mold of the total police state with all its attendant traits of discrimination against the Polish people in all their strata and all their rights.

VII. RUMANIA

Rumania is one of the easternmost nations under the satellite Communist rule, bordering on Ukraine in the north and east, the Black Sea in the east, Bulgaria to the south, Yugoslavia to the southwest and Hungary to the west. Its population is estimated to be now about 20,770,000, of whom 87 percent are Rumanians, while the remainder consists of minorities, namely 1,680,000 Hungarians, 400,000 Germans and 360,000 others, including 150,000 Ukrainians, 100,000 Jews and some Greeks and Bulgarians.[100]

The Rumanian nation traces its origins back to the union between the Roman conquerors and the native Dac-Getae peoples. The first unification of the three major principalities that constitute Rumania today occurred in 1599, when Michael the Brave united the Principalities of Wallachia, Moldavia and Transylvania. Subsequently, the union was dissolved, but in 1859 Wallachia and Moldavia were united again, forming the nucleus of the modern Rumanian state which proclaimed its independence in 1877. In World War I it was defeated by Austria-Hungary (1914–1915). Later Rumania rejoined the Allies and emerged in 1919 with a territory almost doubled in extent: it received Bukovina, part of Bessarabia, Transylvania and Banat, and over 4 million national minorities, including 1,500,000 Hungarians (in Transylvania), 1,000,000 Ukrainians (in Bukovina and Bessarabia), 750,000 Germans, 750,000 Jews and 250,000 Bulgarians (in southern Dobrudja).[101] King Carol II made himself dictator in 1938, but he abdicated in 1940 in favour of his son Michael I.

In 1940, Marshal Ion Antonescu, leader of the militarist movement, came to power and aligned himself with the Axis Powers. Caught between the millstones of the Ribbentrop-Molotov Pact, Rumania's vast territorial gains were lost: in June, 1940, the USSR demanded both Bessarabia and Northern Bukovina, the Bulgarians asked for and got southern Dobrudja and the Hungarians regained much of Transylvania.[102] King Carol, unable to withstand such blows, fled into exile, and although young King Michael became the nominal successor, the real power fell to Marshal Antonescu. With the German attack on the USSR, Rumania entered the Second World War and occupied a substantial part of Ukraine, including the port of Odessa; Bucharest also thought of regaining at least a part of Transylvania from Hungary.

But the fortunes of war now turned against Rumania: the German debacle at Stalingrad and the American bombing of the Rumanian oil fields at Ploesti contributed to an about-face of Rumanian policy. Dropping Antonescu, King Michael assumed power and immediately sought an armistice with the USSR. Such an agreement was signed on September 12, 1944, restoring Transylvania to Rumania but indemnifying the USSR to the tune of $300,000,000 plus the entire cost of the "liberation."

A. *"Liberation" and Communist Takeover*

The Communist takeover in Rumania was depicted authoritatively in a U.S. Congress document:

> ... In January, 1945, Gheorghiu-Dej and Ana Pauker, leaders of the Communist Party, were called to Moscow and given orders to proceed with the overthrow of the legal government and to seize power. Accordingly, the Communists prepared their coup for February 24, 1945. Large groups of Communist Party shock detachments were massed in the center of Bucharest and an attempt was made to take over key positions in the capital and assassinate Prime Minister General Radescu. Both attempts failed and the same evening the Prime Minister announced to an indignant country the Communist plot to overthrow the government. This defeat of Moscow's puppets forced the Kremlin to discard all pretense and to take overt action. Vishinsky and Malinovsky were sent to Bucharest and through the famous two-hour ultimatum given the King by Vishinsky, combined with an impressive display of military force, the Kremlin imposed the Soviet-controlled Groza government [Dr. Peter Groza].
> ... After a period of terror and denial of political freedom... elections were held. The declaration by the Department of State, of November 26, 1946, regarding these elections states... "the United States Government cannot regard those elections as a compliance by the Rumanian Government with assurances it gave in implementation of the Moscow decision..."[103]

The "war crimes" trials of Rumanian leaders responsible for support of the Axis Powers helped to immobilize the opposition and prevented the formation of a coalition of liberal and peasant parties to offset the Communists. Opposition parties were withered by the use or threat of violence, and in December, 1947, King Michael was forced to abdicate. From a group of fewer than a thousand members in 1944 the Communist Workers' Party grew up to 217,000 by September, 1945, and soon mustered the manpower to staff the new regime.[104]

All the organs of power in Rumania, such as the Grand National Assembly (parliament), the State Council, the Council of Ministers and the local (provincial) organs merely constitute a facade for the Communist Party, which is, as in other satellite countries, the single source of power.

Stalin's death in 1953, the 1956 upheavals in Hungary and Poland, the Moscow-Peking rift, increasing internal pressures and economic difficulties – all led to a somewhat less rigid relationship between Moscow and Bucharest. Some "concessions" to national feelings were made in Rumania, but they did not appreciably alter the strong political, military and economic ties binding Rumania to the USSR.

B. The Ceausescu "Independent Course"

Apropos the general "relaxation" atmosphere in the satellite countries, Rumania received a good deal of publicity to the effect that the Rumanian Communist leadership under Nikolae Ceausescu was conducting an "independent foreign policy" bordering on defiance of the Soviet Union, especially as regards Rumania's opposition to the Soviet economic integration plans embracing all the Communist countries. But Rumania ultimately went along in joining CEMA (Council of Economic Mutual Aid). On January 1, 1972, Rumania also became a founder of the Communist-sponsored International Investment Bank, designed to further economic integration in the Communist bloc. Cordial relations were also established by Rumania with the Arab nations, particularly with Egypt, Algeria and Iraq, while her establishment of diplomatic relations with Israel is largely interpreted as the need of a "listening post" for the Communist countries in Tel Aviv.

The consonance of Rumanian foreign policy with that of the USSR had been confirmed by Rumania's stand on Vietnam, the European Security Conference and a number of other international issues.

A recent document circulated by Rumanian exiled leaders in the United States discloses a number of new repressive measures undertaken by the Ceausescu regime towards curtailment of the freedom and human rights of the citizens of Rumania:

1) A new internal trade law (Art. 75, b.) provides for prison terms up to five years for those guilty of selling items received in "gift packages" from abroad (a severe measure since relief packages from abroad provide a means of livelihood for the impoverished population).

2) On December 17, 1971, the National Assembly approved a "Law on the Organization of Contribution in Cash or Labour for Public Works," which stipulates that all able-bodied men between 18–55 and women between 18–50 must contribute six days of their free time for public works or make cash contributions in lieu of work. For peasants who are always lacking in money, this is tantamount to a new form of slave labour.

3) Also in December, 1971, the National Assembly voted a new "Law on

Citizenship," whereby the acquisition of foreign citizenship by Rumanian citizens is not recognized (Art. 3), except for "sound reasons," and accompanied by a declaration by the individual renouncing his Rumanian citizenship that "he will not commit any act prejudicial to the interests of the Rumanian state and people."

4) Likewise, in December, 1971, a new "Law on Preservation of State Secrets" was enacted, by which it is forbidden "to issue, distribute or publish abroad works or writings of any kind which could be harmful to the interests of the Rumanian state." Art. 14 states that Rumanian citizens cannot maintain relations with persons abroad which would be harmful to the interests of the Rumanian state. Further, Art. 5 provides that a Rumanian citizen can grant interviews for foreign media, at home or abroad, only with the prior approval of the central or local higher authority...[105]

Although the Ceausescu government has been widely publicized by the Western press as a regime engaged in a process of "liberalization," the fact is, according to Rumanian observers abroad, that the present government of Rumania is "the most Stalinist in Eastern Europe." The Bucharest regime had never shown the permissiveness writers and artists enjoyed at least for short periods in Czechoslovakia, Hungary and Poland. The only freedom conceded to Rumanian writers was that of choosing non-political subjects. But even this limited concession was withdrawn, with *socialist realism* being reimposed with vigour. Speaking before the Central Committee of the Rumanian Communist Party on November 5, 1971, Ceausescu declared:

It is necessary to strengthen the role played by the Party in science, industry, education, culture and the arts... Our writers must render in art the great socialist transformation of the country...[106]

One of the immediate effects of this new literary policy of *socialist realism* has been the replacement of practically all editors of literary and cultural periodicals with reliable party hacks. Numerous works scheduled for publication were stopped, while several books were withdrawn from the bookstores and several plays dropped from the repertory of theatres for alleged "anti-government" tendencies.

C. Plight of the Church and Religious Persecution

The independent Rumanian Orthodox Church embraces 79 percent of the population; Catholics of the Eastern Rite (Uniates) number 8 per-

cent, the Roman Catholics and Protestants 6 percent each and the Jews
1 percent.[107]

Today in Rumania, as in other Communist-dominated countries, the
church is allowed to exist as a *malum necessarium*, a "necessary evil." The
government takes every opportunity to use the church as a tool while at
the same time it oppresses and hampers church life at every step.

The Byzantine (Eastern) Rite Catholic Church, as in neighbouring
Western Ukraine, has been subjected to a savage police terror. By a
decree (Decree No. 318 of December 1, 1948), its faithful were forced
to join the Orthodox Church. More than 600 priests were arrested and
sent to jail and concentration camps – among them six bishops and 75
prelates. Five of the bishops – Ion Suciu, Basil Aftenie, Valeriu Frent-
ziu, Ion Balan and Alexander Rusu – died in prison.

The religious situation in Rumania is extensively described in a
"Top Secret" Memorandum sent by reliable religious leaders in Rumania to the Rumanian National Committee in the United States.[108]

Briefly, the religious development in Rumania is as follows:

1. The most harshly persecuted religious group is the Catholic because of its strong anti-Communist stance. Those Roman Catholic
bishops who informed the Vatican about the deplorable church situation in Rumania were thrust into prisons as "spies."

2. The Greek Catholic (Uniate) Church was liquidated and put outside the law.

3. The Protestant Churches (Reformed Christians, German Lutherans, Hungarian Lutherans, Baptists, Seventh-Day Adventists, Pen-
tacostals, "Open Brethren" and Unitarians) are persecuted because of
their spirited activities and because of their relationship with the Protestant Churches abroad.

4. A special target of the Bucharest government is a layman's organization known as *Oastea Domnului* ("The Army of the Lord" – which is
not to be confused with the Salvation Army) that claims a membership
of 300,000. It was founded by an Orthodox priest as a revival movement within the Orthodox Church. The official Rumanian Orthodox
hierarchy, which collaborates with the Communist regime, is opposed
to the organization (its founder has been excommunicated).

5. No Bibles may be printed in Rumania. Religious periodicals are
tolerated, provided that they wholeheartedly support the Communist
regime and abstain from criticizing it. Sunday schools are formally forbidden, and the youth especially are persecuted for participation in any
religious activity.

VIII. Yugoslavia

The Socialist Federated Republic of Yugoslavia, like the USSR, is a Communist state embracing a conglomerate of states that possess diverse ethnic, cultural, linguistic and religious characteristics; it is one of the most complex Communist structures. According to the 1970 population census, Yugoslavia contains 20,400,000 people of whom 8,527,200 (42 percent) are Serbs, followed by 4,692,000 Croats, 1,876,800 Slovenes, 1,142,400 Macedonians, 1,060,800 Moslems and 550,000 Montenegrins. The remaining 2.5 million are Albanians (Shiptars), Hungarians, Turks, Slovaks, Rumanians, Bulgarians, Italians, Jews, Ukrainians and others.[109]

Religion is one of the most important determinants underlying the ethnic-religious diversity of the Communist state of Yugoslavia. Some 42 percent of the population is Serbian, espousing the Orthodox faith; almost 32 percent belong to the Roman Catholic Church and 12 percent are communicants of the Islamic religion, with the remainder belonging to other faiths or having none. The Orthodox Church is strongly identified with the Serbian nationality, the Roman Catholic Church is linked with the Croats and Slovenes, a division which dates back to the Middle Ages. There is no Yugoslav "nation" *per se*, nor is there a Yugoslav language. The country has two basic alphabets; the Latin variety is used by the Croats and Slovenes, the Cyrillic by the Serbs and Macedonians.

A. World War II and Communist Takeover

In 1941, Yugoslavia, then a kingdom under King Peter II, was attacked by the German armies and overrun in a matter of days. The Nazi government immediately threw their support to the Croats to establish an independent state of Croatia under the premiership of Ante Pavelich, head of the "Ustashi" movement, which, like the Macedonian revolutionary terrorist organization, IMRO, engaged in acts of terrorism and assassination, including the assassination in Marseilles in 1934 of King Alexander of Yugoslavia.

The first anti-Nazi guerilla operations in Yugoslavia were conducted by General Draja Mihajlovich, leader of the Serbian nationalist *Chetnik* movement; he fought not only the Nazis but the "Ustashi" forces as well. The Italian and Nazi occupation is a tale of horrors. The Bulgarians, Hungarians and Italians under benevolent Nazi supervision contributed their share to the virtual civil war waged between the Serbs and Croats.

Significantly, the Yugoslav Communists were "neutral" until the German attack on the USSR in June, 1941, whereupon the Communists were ordered by Moscow to start an anti-Nazi warfare under the leadership of Joseph Broz-Tito, an obscure and somewhat mysterious figure who was hardly known in Yugoslavia. In 1943 Tito formed a *de facto* government, with the "Anti-Fascist Council for National Liberation" proclaiming itself a "supreme representative of the peoples and the state of Yugoslavia."[110] The Council also declared the Yugoslav government-in-exile "illegal," forbade King Peter II to return to Yugoslavia and adopted a "federalist system."

There is no question today that the Western Allies favoured Tito and his "Partisans" over the *Chetniks*. The result in 1945 was that the monarchy was abolished and a new Soviet-type constitution proclaimed; a year later, Gen. Mihajlovich was tried and executed as a "traitor." Thus the Tito resistance movement after the end of World War II became a "legitimate and constitutional" regime without the direct support of the USSR and without the approval of the great majority of the peoples of Yugoslavia.

The "constituent states" of the Yugoslav federation are Serbia, Croatia, Slovenia, Bosnia-Herzegovina, Macedonia and Montenegro, the setup approved by the 1946 constitution. There are also two autonomous regions, Voivodina and Kosovo, located in Serbia. Although Yugoslavia is not a genuine federation in the Western sense, it is certainly more of a federation than, for instance, the Soviet Union. Nonetheless, all the Yugoslav "federated" republics are subordinated in vital and most important matters to the central government. There are at least twenty-four areas of jurisdiction in which the Central Belgrade government exercises sole and exclusive authority.[111]

The totalitarian and monolithic character of the Yugoslav "federation" is denoted by its complete nationalization of industry, centralized economic planning, rule by a single party and suppression of all opposition.

A new constitution, adopted on April 7, 1963, brought some changes: the "People's Republic" became the "Socialist Federal Republic of Yugoslavia" and the Federal Assembly consists now of six chambers, instead of the former two. In the course of the last decade a total of 42 amendments to the constitution were passed, in 1967, 1968 and 1971, giving the six constituent republics and various nationalities a greater measure of autonomy.

Executive power is vested in the office of the president, who is elected for four years and can be re-elected only once. But Tito has been elected president for life, for his "historic merits."

On January 30 and 31, 1974, a "combined session of all Chambers of the Federal Assembly was held at which the new Constitution of the Socialist Federal Republic of Yugoslavia was adopted, whereafter it was promulgated on February 21, 1974, by the Chambers of Nationalities."[112]

The reason for the promulgation of the new constitution was stated in an "Introductory Note," which said:

This phase of constitutional reform meant a partial revision of relations between the Federation and the Republics. These relations were placed on new foundations, more in line with the level of socio-economic and political development which had by then been attained. In addition to defining the new status, role and powers of the Federation and introducing corresponding changes in its organization, the amendments, especially those referred to as the "workers' amendments," (XXI to XXIII), laid down certain new principles and forms for self-management socio-political relations. In the course of implementing these partial constitutional changes it was realized that a fuller and more precise regulation of some basic questions concerning the socio-economic system of self-management, the communal system, the assembly system and the political system in general, were needed. For this reason the Federal Assembly decided that immediately after the adoption of Constitutional Amendments XX–XLII preparations should be started for the second phase of constitutional reform (i.e. the drafting of a new constitution) which was seen as an organic continuation of the first phase.[113]

From the above statement two things are abundantly clear: the new 1974 constitution was necessitated by growing unrest and dissatisfaction among the "Republics," and by economic pressures, and a new constitutional reform is forecast.

Although the Albanians in Yugoslavia constitute 38 percent of all Albanians, they do not have their own "Socialist Republic," but rate only the status of "Socialist Province," that of Kosovo, where they form 67.7% of the population. In the other "Socialist Province" of Voivodina the Serbs constitute 54.9 percent, while the rest are various other national groups,[114] with the Hungarians the most numerous. No national autonomy is provided for the Greek, Turkish, Ukrainian, Italian and Bulgarian minorities.

The ruling party is the Communist Party, which in 1952 adopted the name "League of Communists of Yugoslavia" (SKJ), and which is directed by a 52-member Presidium, with the following ethnic composition: Serbs – 17; Croats – 8; Slovenes – 8; Macedonians – 6; Montenegrins – 6; Albanians – 3; Moslems – 3; Hungarians – 1.[115] The Presidium is thus dominated by the Serbs, who make up 51.5 percent of the Communist Party membership.

The so-called "general elections" are conducted in the form of a plebiscite, wherein voters simply cast their ballots for candidates designated by the party.

B. Bigotry and Persecution of Religion

The Yugoslav government and the Yugoslav constitution guarantee, on paper, full freedom of conscience and religion, that all churches are equal but have to conform to the constitution and the laws of the republic and that religious instruction can be given only in churches or buildings belonging to religious denominations.

In reality, however, the Communist government of Yugoslavia immediately displayed a hostile attitude toward religion, especially toward Catholicism, which was the official religion of the independent, pro-Nazi government in Croatia; the "Ustashi massacres of Croatian Serbs as often as not were carried out in the name of Catholicism."[116]

The war on Catholicism was unlimited in inhumanity and lawlessness:

All denominational schools, particularly numerous in Slovenia and Croatia, were closed or nationalized. The teaching of catechism to children was forbidden even in the church building. School children were incited to immorality in the youth groups, and school textbooks told them that they incurred no moral guilt in sexual intercourse. The bishops were compelled to liquidate all Catholic associations and organizations... The nuns were expelled from charitable institutions...

Nearly all church property, used to maintain the clergy and provide for divine service and charitable works, was confiscated... State subventions and ecclesiastical taxes were eliminated...

In addition to these legal and administrative restrictions, a constant campaign of terror was maintained. By summer, 1946, out of a total of 2,700 Croat priests, more than 400 had been killed, 200 fled abroad, and several hundred were in jail...[117]

The culmination of the anti-Catholic campaign was the trial in 1946 of Archbishop Aloysius Stepinac of Zagreb, who was sentenced to 16 years hard labour for "collaboration" and "opposition" to the regime.

The Protestant denominations, which numbered about 280,000 people, the German Lutherans (175,000) and the Hungarian Calvinists (60,000), as well as the numerous Moslem communities – all were decimated or dispersed; by 1950 only small numbers of them were still in existence.

The Jewish community was similarly subjected to pressure and persecution; all Zionist organizations were disbanded and the religious life harshly curtailed, and subsequently thousands of Jews emigrated from the country.

In 1970 the Belgrade government and the Vatican resumed diplomatic relations after an 18-year break. But even before this renewed contact with the Vatican, the Yugoslav regime had relaxed somewhat its hold on the religious life; overt persecution had disappeared, but government regulations, especially in the field of religious education, hampered the normal development of the religious life of every nationality, a process which continues today.

C. Nationalist Tensions and Dissent

Yugoslavia's foreign and domestic policies have vacillated depending on the relations of Belgrade with Moscow at a given time. Many people believed that after the Stalin-Tito rift in 1948 Yugoslavia would "liberalize" its rule in order to receive some economic help from the West. But these hopes did not materialize. In 1953 came the reconciliation between Tito and Khrushchev, and in 1956 Tito unabashedly supported the USSR in its violent suppression of the 1956 Hungarian revolt. In 1958, however, Moscow assailed Tito's "revisionism," and Tito again turned to the West. His policy since has been one of vacillating between Moscow and Peking. Yet on balance Yugoslavia has supported major Soviet foreign policy moves to the present day.

There is no question that Tito's primary objective has been to preserve Yugoslavia as a "unified" state in the face of incurable and bedeviling nationalist and religious problems which historically have plagued Yugoslavia.

Antagonism and armed conflict between Serbs and Croats, and, to a lesser degree, between Serbs and Slovenes, are centuries-old; and there is a strong Moslem minority which is now rediscovering its own ethnic and self-administering a birth of Moslem nationalism in Yugoslavia. Furthermore, the Macedonians have very strong centrifugal tendencies while the Albanians have been gravitating to Albania for several decades. To complicate matters, there exist severe rifts among and in these minority "nations" as well.

Religion plays a powerful role also in the interrelations existing among the various nationalities, especially between the Catholic Croats and Slovenes and the Orthodox Serbs. Here the Serbs, as the strongest and most numerous ethnic element, are undoubtedly guilty of discrimination as regards the weaker nationalities. But all nationalities, including the Serbs, suffer equally from the stern and rigid rule of Communism. No criticism of it, nor of the Soviet Union as a whole, is permitted in Yugoslavia; a number of intellectuals have been penalized for daring to raise a dissenting voice.

One such voice is that of Milovan Djilas, former member of the Politburo of the Communist Party of Yugoslavia and Vice President until he was expelled from the party in 1954. He became a *cause célèbre* for writing two outstanding books, *The New Class* (1957), which was a devastating commentary on the Communist system, and *Conversations with Stalin* (1962), which resulted in his imprisonment for allegedly disclosing "official secrets."[118] He served prison terms from 1956 to 1961 and from 1962 to 1966. In 1968 he was allowed to come to the United States for a short visit.

Another belongs to Mihajlo Mihajlov, a young writer twice imprisoned since 1965, the first time for writing an account of Soviet concentration camps, and then for writings critical of the Yugoslav government and for having a letter printed in *The New York Times*.

In March, 1974, both Djilas and Mihajlov were invited to take part in a symposium: "The Social Context of Inquiry: Problems of Forbidden and Discouraged Knowledge," sponsored by the American Association of Advancement of Science in San Francisco, Calif. Both were denied visas by the Belgrade government.

In a message to the symposium's sponsors, Mr. Djilas caustically stated:

My greetings are sent in the name of writers and scholars who are forced to remain quiet or to say what they do not believe; in the name of those imprisoned for making a joke or for answering criticism, which only yesterday was still allowed; in the name of Marxist humanists who just now are being chased out of Yugoslav universities... [119]

Mr. Mihajlov's terse message read:

Since I live in a country which does not honor the United Nations Declaration on the free movement from and into one's own country, I do not have the chance to address you in person...

He added his wishes for the success of the conference "in exposing" all the spiritual and ideological limitations which represent constraint on scientific research."[120]

On October 7, 1974 Mihajlo Mihajlov was arrested again in his home in Novy Sad, charged with "disseminating hostile propaganda and associating with foreign emigré organizations." The evidence against him consisted of five essays he was commissioned to write by Western publications in 1974 when the ban on his publishing expired and he was free to write. Mr. Mihajlov has already served a four-year sentence for "disseminating foreign propaganda."[121]

D. The Macedonians

Macedonia is a region in the heart of the Balkan Peninsula with a dominant strategic position. Possession of Macedonia has been the common objective of the nationalistic and strategic ambition of three Balkan powers: Greece, Serbia (Yugoslavia) and Bulgaria. In fact, Macedonia has been a political problem rather than a geographical unity.[122]

Historically speaking, Macedonia became prominent under Philip of Macedon and Alexander the Great (IVth c. B.C.), although a publication of the Yugoslav government flatly stated the contrary:

The present inhabitants of Macedonia have no direct connection with the people who populated the region at the time of Alexander of Macedonia.[113]

It had been under the domination of several foreign powers which tried to denationalize the population and seize its territory. The modern Macedonian question began to take political significance in the XIXth century with the development of the Balkan nations which tried to extend their nationalistic ambitions to Macedonia. Another important factor in Macedonia's turbulent history was the Internal Macedonian Revolutionary Organization (IMRO), founded in 1893 in Salonika, which aimed to establish an autonomous Macedonia; the IMRO waged bloody terroristic campaigns, often supported by Bulgaria, against Greece and Serbia. After the Second Balkan War (1913), Macedonia was partitioned between Serbia and Greece, while Bulgaria obtained only a small portion of its territory. Under the Treaty of Versailles (1919), the Macedonian territory was divided again, with only a little part of it going to Bulgaria. With the German-Italian conquest of Greece in 1941, Greek Macedonia was occupied by Bulgaria, which it lost again after the collapse of the Axis powers in 1945.

After the First World War the population of Macedonia was distributed as follows: Greece – 1,412,477 (1928); Yugoslavia – 818,377 (1921) and Bulgaria – 429,744 (1926).[124] According to official Greek figures, the population of Greek Macedonia on December 31, 1938 was 1,686,479 of whom 77,000 were "Slavophones" and 93,000 others, such as Jews, Armenians, gypsies, etc.

According to the Macedonian Patriotic Organization of the U.S.A. and Canada, which in general follows the policies of the IMRO, the area known as Macedonia is about 25,000 square miles and had a total population of about 3,000,000, by 1962. In its *Golden Book*, published in 1972, the same organization stated that as a "result of the Balkan Wars and the First and Second World Wars, 51 percent of the Macedonian territory became part of Greece, 39 percent of Yugoslavia and only 10 percent of Bulgaria."

Through the transfer of populations after World War I and through the policy of "hellenization," much of the population of Greek Macedonia has simply disappeared.

During World War II, according to official Yugoslav sources, the Macedonians "rose in large numbers against the enemy and fought side by side with the other Yugoslav nations until the final victory."[125]

Macedonia is inhabited by five different races: Slavs, Turks, Greeks, Albanians and Wallachians (Vlakhs). Most of the population is Slavic, with a historic development different from that of the Serbs or Bulgarians. Hence they are somewhat distinct from both Serbs and Bulgarians, speaking a Slav dialect, or dialects of the Slavic tongue, which is understood by Serbs and Bulgarians, both of whom claim Macedonia for their own.

The majority of Macedonians are of the Orthodox faith, but among the Macedonians in Yugoslavia there are some 40,000 Moslems (Torbesi).

It is to be stressed that only in Yugoslavia is there a possibility for the development of Macedonian culture, even though under Communist rule. In 1945 a university was founded in Skopje, the nominal capital of

Macedonia, which has departments of Macedonian history, language and culture, something that is non-existent for the Macedonian minorities in Bulgaria and Greece.

No matter what degree of suppression the Tito regime may bring to bear upon the population, Communism in Yugoslavia, as Djilas charged, has betrayed the people by denying them the most rudimentary human rights: the corollary of the rule of a small minority whose power is assumed without the consent of the people as a whole. His prediction that the world will eventually move in the direction of greater unity, progress and peace with freedom and justice, cannot be taken lightly by the ruling Communists of Yugoslavia nor, for that matter, by any other Communist government in Communist-dominated Europe.

IX. CONCLUSIONS

To sum up the status of human rights, discrimination and abuse of power throughout the Communist-dominated countries of Central Europe:

1. In each of the eight countries the overwhelming majority of people are forcibly deprived of any meaningful possibility for participating in the rule of their country.

2. In each exist crass violations of human rights and the constitutions which were proclaimed by the Communist governments themselves.

3. Extensive official discrimination is aimed against religion in each of the eight countries, its extent and severity depending on the stance of the population and also on the relations at a given time with the Western nations.

4. Oppression, in one form or another, of national minorities, is undeniable. But more sinister are the efforts to create a soulless robot-like man, the so-called "Soviet man," a reduction of humanity which the Soviet government and the Communist Party of the Soviet Union have been endeavouring to effect for the past three decades.

5. Intellectual dissent and dissatisfaction persists in every Communist country.

6. The influence of Soviet policies and the ideological control over the Communist-dominated countries of Europe remain rigid and uncompromising, despite much-publicized "liberalization" policies which are more of a pipe-dream of Western statesmen and policy-makers than a reality in the countries concerned.

NOTES

1. Joseph S. Roucek and Kenneth V. Lottich, *Behind the Iron Curtain:* The Soviet Satellite States – East European Nationalisms and Education, Caldwell, Idaho: The Caxton Printers, Ltd., 1964, pp. 22–23.
2. *Ibid.*, pp. 25–29.
3. *United Nations: Human Rights:* A Compilation of International Instruments of the United Nations, New York: United Nations, 1973, p. 106.
4. "Poles are Vexed by Unpaid Labor," *The New York Times,* October 20 1974.
5. Sir Halford J. Mackinder, *Democratic Ideals and Reality*, New York: Holt, Rinehart and Winston, Inc., 1942, p. 150.
6. *Communist Oppression of Central and Eastern Europe:* Deeds and Practices of Russian Imperialism and Colonialism in the Heart of Civilized Europe, New York: Conference of Americans of Central and Eastern European Descent (CACEED), 1966, p. 7.
7. Dr. Rexhep Krasniqi, "Albania Under Communist Rule" Dr. Krasniqi is President of *Komiteti Shqiperia e Lirë*, Free Albania Committee. His statement was issued on February 20, 1974 in New York.
8. Richard F. Staar, *The Communist Regimes in Eastern Europe*, second rev. ed., Stanford, Calif.: Hoover Institution Press, 1971, pp. 3–4.
9. Krasniqi, *op. cit.* also See U.S., Congress House, Select Committee on Communist Aggression, *Communist Takeover and Occupation of Albania*, Report No. 13, Printing Office, 1954.
10. CACEED, *Communist Oppression*, p. 8.
11. Staar, *op. cit.* p. 5.
12. Krasniqi, *op. cit.* pp. 1–2.
13. U.S. Congress, House, Committee on Foreign Affairs, *Anti-Religious Activities in the Soviet Union and in Eastern Europe*, Hearings before the Subcommittee on Europe, 89th Cong., 1st Sess., Washington, D.C.: U.S. Government Printing Office, 1965.
14. Krasniqi, *op. cit.*, p. 3.
15. CACEED, *Communist Oppression*, p. 9.
16. Staar, *op. cit.* p. 28.
17. Roucek and Lottich, *Behind the Iron Curtain*, p. 375.
18. Winston S. Churchill, *Triumph and Tragedy*, New York: Houghton Mifflin 1953, p. 208.
19. Roucek and Lottich, *op. cit.* p. 376.
20. *Ibid.*, p. 404.
21. Staar, *op. cit.* pp. 30–31.
22. U.S. Department of State, *Directory of Bulgarian Officials*, Washington, D.C.: U.S. Government Printing Office, 1969, pp. 1, 43, 51, 54, 76.
23. Stanley G. Evans, *A Short History of Bulgaria*, London, 1960, p. 189.
24. Bulgarian Telegraphic Agency dispatch, June 19, 1965, cited in RFE report, "Sentences of Bulgarians in April Conspiracy," June 24, 1965, pp. 2–4.
25. L. A. D. Dellin, ed., *Bulgaria*, New York: 1957, p. 187.
26. Staar *op. cit.* p. 43.
27. Czechoslovakia, Statni Statisticky Urad (State Statistical Office), *Statisticka Rocenka ČSSR* (Statistical Yearbook of the CSSR), Prague: 1969, p. 87. Radio Prague, July 28, 1970.
28. CACEED, *Communist Oppression*, p. 12.
29. Staar, *op. cit.* pp. 51–52.

30. U.S., Congress, House, *Second Interim Report of the Select Committee on Communist Aggression*, Report of the Select Committee to Investigate Communist Aggression Against Poland, Hungary, Czechoslovakia, Bulgaria, Rumania, Lithuania, Latvia, Estonia, East Germany, Russia and the Non-Russian Nations of the U.S.S.R., 83rd Cong., 2nd Sess., Under Authority of H. Res. 346 and H. Res. 438, Washington, D.C.: U.S. Government Printing Office, 1954, p. 8.

31. Roucek and Lottich, *Behind the Iron Curtain*, pp. 255–56.

32. *Ibid.*, p. 257.

33. *Christ und Welt* (The Christian and the World), May 1, 1970: statement by Catholic clergy in *Lidova demokracie* (People's Democracy), May 14, 1970, p. 1; on the Greek Catholic Church, see *Pravda*, Bratislava, September 1, 1970 – all cited in Staar, *The Communist Regimes*, p. 78.

34. *Schematismus Dioecesos Presovensis pro. A.D.* 1948 (Directory of the Diocese of Priashiv in 1948), Priashiv: 1948, pp. 148, 206.

35. Rev Athanasius B. Pekář. OSBM, "Restoration of the Greek Catholic Church in Czechoslovakia," *The Ukrainian Quarterly*, XXIX, No. 3 Autumn, 1973, pp. 283–84.

36. Pekář, *op. cit.*, pp. 288–96; also see *The Tragedy of the Greek Catholic Church in Czechoslovakia*, New York: Carpathian Alliance, Inc., 1971, pp. 40–43.

37. Federal Republic of Germany, the Federal Ministry for Expellees, Refugee and War Victims, *Documents on the Expulsion of the Germans from Eastern-Central-Europe.* (hereinafter cited as *FRG Documents*) Vol. IV: *The Expulsion of the German Population from Czechoslovakia*, Bonn: The Federal Ministry for Expellees, Refugees and War Victims. 1960, p. 3.

38. *Ibid.*, pp. 7–15.

39. *Ibid.*, p. 36.

40. *Ibid.*, p. 45.

41. Federal Republic of Germany, The Federal Ministry for Expellees, Refugees and War Victims, *FACTS Concerning the Problem of the German Expellees and Refugees*, Bonn: The Federal Ministry for Expellees, Refugees and War Victims, 1967, Table 4.

42. Vladimir V. Kusin, *Political Groupings in the Czech Reform Movement*, New York, Columbia University Press, 1972, pp. 156–57.

43. Statement of the Czechoslovak Society of Arts and Sciences in America, Inc., November 17, 1973, New York.

44. *Clemens Encyclopedia of World Governments*, 1973, p. 64.

45. G. M. Zainovich and D. A. Brown, "Political Integration in Czechoslovakia: The Implications of the Prague Spring and Soviet Intervention," *Journal of International Affairs*, Vol. 27, No. 1, 1973, pp. 66–69.

46. Robert W. Dean. "Czechoslovakia: Consolidation and Beyond," *Survey*, Vol. 17, No. 3, Summer, 1971, p. 106.

47. Dr. Martin Kvetko, "Report to the Assembly of Captive European Nations, Doc. No. 495 (XIX), September 19, 1972, New York, N.Y.

48. *Ibid.*

49. P. Skoda, "Tchecoslovaquie 1972: le peuple de Dieu dans la 'normalisation,'" *Documentation sur l'Europe Centrale*, Vol. 10, No. 2, 1972, pp. 95–113.

50. *Schematizmus slov. kat. diocez. Trnava*, 1971.

51. *Documentation sur l'Europe Centrale*, Vol. 12. No. 2, 1974, pp. 153–154.

52. George Moldau, "A Letter from Prague," *Survey*, Vol. 19, No. 2 (1973), 247.

53. Federal Republic of Germany, Press and Information Office, *The Bulletin*, Vol. 18, No. 28, Bonn: August 13, 1970.

54. *Ibid.*

55. FRG Documents, Vol. I: *The Expulsion of the German Population from the Territories East of the Oder-Neisse-Line*, 1960, p. 49.
56. Staar, *The Communist Regimes*, p. 84.
57. Manfred Rieder, "DDR: Russisches Regime – Deutsche Leistung" (The GDR: The Russian Regime – The German Performance), *Deutsche Annalen*, Druffel Verlag, 1972, p. 106.
58. *U.S. News and World Report*, March 14, 1966.
59. "A Visit Behind the Iron Curtain," *Der Deutsch-Amerikaner*, Chicago: November, 1973.
60. See General Reinhard Gehlen's article in *Zeichen Der Zeit* (Signs of the Times), Mainz: 1973, pp. 110–12.
61. Jean Smith, ed., *Germany Beyond the Wall*, New York, Little, Brown & Co., 1969, p. 138.
62. *Ibid.*, p. 39.
63. Staar, *op. cit.* p. 92.
64. Arthur M. Handhardt, *The German Democratic Republic*, Baltimore: Johns Hopkins University Press, 1968, p. 10.
65. *Ibid.*, p. 11.
66. *Ibid.*, p. 44.
67. Welles Hangen, *Der Unbequeme Nachbar*, Munich, 1967, p. 30.
68. *The New York Times*, September 4, 1974.
69. CACEED, *Communist Oppression*, pp. 17–18.
70. Roucek and Lottich, *Behind the Iron Curtain*, p. 279.
71. William L. Shirer, *The Rise and Fall of the Third Reich*, New York: Simon and Schuster, 1960, pp. 387–88, 449–50.
72. U.S. Congress Senate, Committee on Foreign Relations, *Tensions Within the Soviet Captive Nations*, Part 7: Hungary, 83rd Cong., 2nd Sess., Washington, D.C.: U.S. Government Printing Office, 1954.
73. Howard J. Hilton, *Hungary: A Case of Soviet Economic Imperialism*, Washington, D.C.: U.S. Department of State, Office of Public Affairs, 1951, pp. 323–27.
74. Joseph S. Roucek, "The Forced Labor Camps in the Soviet Orbit," *Prologue*, Vol. IV, Nos. 1–2, Spring-Summer, 1960, pp. 64–65.
75. Amos J. Peaslee, ed., *Constitutions of Nations*, The Hague: 1956, II, p. 185.
76. Radio Free Europe (RFE), *Situation Report*, New York, June 28, 1966, pp. 1–2.
77. Kurt Rabl, "Mass Expulsions and International Law," *The Sudeten Bulletin*, VIII, Nos. 8–9, August–September, 1960, p. 210.
78. Bela Fabian and Imre Kovacs, "Kadar's Hungary," a letter to *The New York Times Magazine*, January 24, 1965, p. 6.
79. Roucek and Lottich, *op. cit.* pp. 312–13.
80. CACEED, *Communist Oppression*, pp. 19–20.
81. "The Mindszenty Case: A Religious Detente," *America*, Philadelphia, February 28, 1974.
82. *L'Osservatore Romano*, July 9, 1965.
83. Roucek and Lottich, *Behind the Iron Curtain*, pp. 313–26.
84. Imre Kovacs, "Hungary Adjusts Herself to Detente," *The Rising Tide*, Washington, D.C.: Coordinating Committee of Hungarian Organizations in the U.S.A., November 19, 1973,
85. *Ibid.*
86. CACEED, *Communist Oppression*, p. 26.
87. U.S., Congress, Senate, Committee on Foreign Relations, *A Decade of American Foreign Policy*, Washington, D.C.: U.S. Government Printing Office, 1950, p. 30.
88. CACEED, *Communist Oppression*, p. 26.

89. Stanislaw Mikolajczyk, *The Rape of Poland: Pattern of Soviet Aggression*, New York: 1948, pp. 180–202.
90. Staar, *The Communist Regimes*, pp. 138–39.
91. *Ibid.*, pp. 149–50.
92. *Życie Warszawy* (Warsaw Life), as cited in Joseph Mackiewicz, *In the Shadow of the Cross*, trans. by Wiktor Moszynski, New York: Contra Publishing, 1973, p. 61.
93. *FRG Documents*, Vol. I (Expulsions East of Oder-Neisse Line), p. 4.
94. *Ibid.*, p. 104.
95. For a detailed account, see Yuriy Tys-Krokhmaliuk, *UPA Warfare in Ukraine*, trans. by Walter Dushnyck, New York: Society of Veterans of Ukrainian Insurgent Army, 1972, Chapter Twenty-Four, "Operations of UPA in the Curzon Line Territory," pp. 375–88.
96. Strobe Talbott, ed. and trans., *Khrushchev Remembers: The Last Testament,*with a foreword by Edward Crankshaw and an introduction by Jerrold L. Schecter, Boston-Toronto: Little, Brown and Company, 1974, p. 175.
97. Walter Dushnyck, *Death and Devastation on the Curzon Line:* The Story of Deportations from Ukraine, New York: Committee Against Mass Expulsion and Ukrainian Congress Committee of America, 1948, pp. 15–32.
98. *Ibid.*, pp. 22–23.
99. See article by Prof. Edward Lipinski in *Kulisy* (The Curtains), Warsaw, May, 1971.
100. Staar, *op. cit.*, p. 169; Roucek and Lottich, *op. cit.*, pp. 331–33.
101. Roucek and Lottich, *op. cit.* p. 332.
102. *Ibid.*, p. 336.
103. U.S., Congress, House, Select Committee on Communist Aggression, *Communist Takeover and Occupation of Rumania*, Special Report No. 11, 83rd Cong., 2nd Sess., Washington, D.C.: Government Printing Office, 1954.
104. Staar, *op. cit.* p. 155.
105. See "Statement on Recent Developments in East Central Europe," a research document prepared on the basis of official data by Prof. Brutus Coste of Fairleigh Dickinson University, March, 1974.
106. *Ibid.*, pp. 7–8.
107. CACEED, *Communist Oppression*, p. 29.
108. See "Religious Situation in Rumania," a research document prepared on the basis of official data by Prof. Brutus Coste of Fairleigh Dickinson University, March, 1974.
109. Yugoslavia, *Statisticki godisnjak Jugoslavije* 1970 (Statistical Yearbook of Yugoslavia 1970), Belgrade: 1970, p. 330.
110. Dragoljub Durovic, ed., *Narodna vlast i socijalisticka demokratija*, 1943–1963 (People's Power and Socialist Democracy, 1943–1963) Belgrade: 1964, pp. 22–23.
111. Staar, *op. cit.* p. 192.
112. The Constitution of the Socialist Federal Republic of Yugoslavia, Belgrade: 1974, p. 5.
113. *Ibid.*, p. 6.
114. Dr. Koca Joncić, *Relations Among the Peoples and National Minorities in Yugoslavia*, Belgrade: 1969, pp. 7–8 for the Ukrainian minority in Yugoslavia see, "The Rusyny: A Ukrainian Ethnic Element in Yugoslavia," by Leonard J. Manko, *The Ukrainian Quarterly*, Vol. XXX, No. 3, Autumn, 1974, pp. (249–274), New York, N.Y.).
115. Radio Free Europe Reports, by Slobodan Stanković, "Yugoslav Party Con-

gress Round-Up," I, March, 1969, 8 pp., and "Analysis of Yugoslav Party Presidium Meeting," April 27, 1970, 9 pp.

116. Charles P. McVicker, *Titoism*, New York: St. Martin's Press, 1957, p. 48.
117. Gary Mac Eoin, *The Communist War on Religion*, New York: The Devin-Adair Company, 1951, pp. 148–49.
118. Staar, *The Communist Regimes*, pp. 211–12.
119. Walter Sullivan, "2 Yugoslavs Told They Can't Travel," *The New York Times*, March 2, 1974.
120. *Ibid.*
121. "Mihajlov, Yugoslav Writer, Is Arrested Again," *The New York Times*, October 10, 1974.
122. *Slavonic Encyclopaedia*, ed. by Joseph S. Roucek, Ph. D., University of Bridgeport, Philosophical Library, New York, 1959, see: "Macedonia," pp. 734–736.
123. *Facts About Yugoslavia*, published by *Yugoslav Review*, Belgrade: (undated, p. 17.
124. *Slavonic Encyclopaedia*, p. 736.
125. *Facts About Yugoslavia*, p. 18.

Pessimism in Australian Race Relations

COLIN TATZ

Colin Martin Tatz, a native of South Africa, graduated in political science and public administration from the University of Natal, where he obtained his B. A. (Honours) in 1955. For the next several years he taught English, geography, history and business subjects in secondary schools, whilst completing his M. A. in public and native administration, which he received from the University of Natal in 1960. The following year he was granted a research scholarship to the Australian National University, where he completed his Ph.D. in political science and public administration in 1964. He has since become an Australian citizen.

From 1964 to 1970, Dr. Tatz served at Monash University, as Lecturer in Politics, Senior Lecturer in Sociology, and Director of the Centre for Research into Aboriginal Affairs. He also became a member of the Aborigines Welfare Board of Victoria and chairman of several of its committees. Since 1971 he has been Foundation Professor of Politics and head of the department at the University of New England, Armidale, N.S.W. He is a member of the Council and the Executive of the Australian Institute of Aboriginal Studies. During a visiting professorship at Queen's University, Kingston, Ontario, Professor Tatz engaged in research on problems of Canadian Indians.

From Professor Tatz's long list of publications, the following pertinent to this study may be mentioned: "The Relationship Between Aboriginal Health and Employment, Wages and Training," in Ian Sharp and Colin Tatz, eds., *Aborigines in the Economy* (1966); "Aborigines: Law and Political Development," *The Australian Quarterly*, Vol. 42, No. 4 (December, 1970), pp. 14ff.; and (as editor) *Black Viewpoints: The Aboriginal Experience* (Sydney: Australia and New Zealand Book Company, 1975).

Pessimism in Australian Race Relations

COLIN TATZ

I. The White Problem

Race relations in Australia can only get worse. Some optimists will counter this view with evidence of changes in political philosophies on race, the enactment of human rights bills, abolition of discriminatory laws, appointment of royal commissions, creation of new administrative structures and the infusion of more money. My pessimism rests on the seeming insolubility of four very basic – yet barely conceded – problems.

Firstly, there is the psychological inability of most whites to stop talking *about* blacks rather than *with* them, to cease being their proctors and curators and allow them to act on their own behalf. Secondly, there is a growing trend among bureaucrats and liberal humanitarians to analyse the black problem in terms of the deficiencies of the victims. The avowed racists have always done it: but the new paradox is the spectacle of the kind and the concerned becoming increasingly adept at what William Ryan calls *Blaming the Victim*.[1] Thirdly – and this perhaps explains victim-blaming – there is the cultural impossibility, for most whites, of evincing empathy rather than sympathy for Aboriginal viewpoints on black consciousness and identity, on their frustration, alienation and deprivation. Finally, there is the improbability of whites ever comprehending, let alone conceding, that a major avenue towards blacks' survival and progress is *their* rejection of white society, its values and the programmes mounted for their benefit – especially those which aim at changing the victim.

This chapter is a revised and updated version of the paper "Aborigines: Political Options and Strategies," read to the Australian Institute of Aboriginal Studies Conference, Social and Cultural Change symposium, Canberra, 23–24 May 1974.

To recognize a group as a problem has the merit of implying the need for action towards its solution. Professor Charles Rowley has portrayed the action of neglect, and occasionally the action of concern, until the 1960's.[2] Theses, symposia, anthologies and articles have expounded the variety of welfare and social change programmes this past decade. Social history shows one feature common to both the eras of neglect and concern: that white society unilaterally defines the problems, prescribes the policy dicta, enacts the laws, creates the administrative machinery and determines the nature, content, personnel and flavour of remedial programmes.

At the national policy level, the Federal-State conferences of Ministers and Officials – in 1937, 1951, 1961, 1963, 1965 and those following decided on philosophies of limited segregation, or assimilation, or integration, or cultural choice, or illogical mixtures of them all. Until the 1960's, legislatures elected solely by whites enacted what was believed to be in the Aborigines' best interests. The Aboriginal franchise has not affected that situation. Governmental and mission agencies were, and are, so structured as to leave no doubt about who are the administrators and who the administered.

Under the Federal Labor government – which came to power in December 1972 – the Aboriginal Land Rights Commission[3] (Justice Woodward) has been a white-centred operation. It is true that land is for the whites to give, but it would not have been calamitous to grant persistent Aboriginal demands for one or two black deputy commissioners. In 1965, New Zealand had the integrity, or the public relations foresight, to include Hemi Tono Waetford as a junior partner to Justice Ivor Prichard in their inquiry into Maori land laws and land courts.[4] The National Aboriginal Consultative Committee (now the National Aboriginal Congress) was the concept and creation of the Minister for Aboriginal Affairs, Gordon Bryant. Even the laudable "new" policy that "community self-determination should be the over-riding principle of government policy" was decided by an all-white seminar of officials and a few missionaries at Batchelor (N.T.) in February 1973.[5]

Innumerable examples show that Aboriginal affairs have always been, and still remain, a white activity. This tradition has become a cultural norm, as deeply ingrained as our acceptance of an elaborate private school system or private enterprise in social welfare activity. How does one break this psychological yoke? Some Ministers and officials state vaguely that there has to be a new order of things, namely, the cessation of proctorship and guardianship, the need to cede power, authority, responsibility and accountability to the "problem" people. But no one really *believes* it, or in it. Hence the unending token mechanisms of reserve councils, advisory councils, national consultative committees, specially convened seminars, and some Aboriginal membership in such bodies as the Institute of Aboriginal Studies and the Council for the Arts.

One could argue that in some countries the ceding of economic, legal and political autonomy to black people could destroy the present order of social control and social cohesion. Such a cession to between one and two per cent of the population – especially to (Rowley's) *Remote Aborigines* – will assuredly not crumble the walls of the white Australian Jericho. The conundrum remains: how does white society – the defining group – come to recognize and to *accept* the reverse situation, namely, that by its words, promises, policy slogans, attitudes and actions, *it* is the very essence of the problem?

The conscious or even unconscious game of victim-blaming is simple: one defines a "problem" group on the criterion that "they" – the sick, the poor, the jobless, the black – unwittingly cause their own troubles. Thus, the sick are sick because as victims they have poor motivation and lack health information! The failure of the miseducated or "culturally deprived" child is *his* failure. The problem of the Negro family is just that: the Negro family. The Aboriginal problem arises from *their* differences: their Aboriginality is a social stigma located within themselves, inside their skins. The Different Ones, says Ryan, are seen as less competent, less skilled, less knowing – in short, less human.

In 1972, (then) Prime Minister McMahon[6] admitted what had been known for a long time, namely, the high incidence of child mortality. But, he explained, *"the semi-nomadic life of some of the Aborigines,* which has aspects not compatible with normal standards of health, *is a contributory factor."* The infant mortality rate – in the region of 120 to 150 per 1000 – occurs in fact among the Aborigines living permanently on settlements and missions. In March 1975 the general secretary of the West Australian Police Union inserted an open statement in a Perth newspaper,[7] entitled "Laverton – The Facts," arising out of events there in December 1974 and January 1975. Fact one was that Laverton Aborigines drank heavily. The consequent "facts" were: Aborigines "stood over the local police"; "confronted them in huge numbers"; began a "running fight"; "attacked police with bottles and stones"; began "running wild"; engaged in "violent behaviour"; "stole food and drink"; "ganged up on two policemen"; "frightened and alarmed" the publican. An experienced inspector then stopped the "riotous behaviour" and arrested many "among the drunken natives milling around in the streets." The writer denied all police baton-usage, burning Aborigines with cigarettes and refusal to feed prisoners. All Aboriginal injuries – and there were many such – "*in fact* were the results of altercations with other Aborigines"! (Italics added.)

There are two logical outcomes of analysing social problems in terms of the deficiencies of the victim. One is to take refuge in the hopelessness of their helplessness and derive some sort of comfort from this paralysis of mind and action. The other is to develop programmes aimed at correcting their deficiencies. In short, change the victim. Australians have laboured mightily to change him – by exterminating, pacifying, segregating, integrating or assimilating him. What we cannot, or will not, see is that *our* perception of their differences is not necessarily *their* perception of them. There is an unbridgeable difference between our seeing "them" as different and their seeing it.

The third fundamental problem is therefore even less likely of solution: that a mainstream society can ever have empathy with a depressed minority, the group at the receiving end (of white activity and white analysis), the group which wishes to identify itself in its own way, its different though doubtless to many less comfortable way. One can feel and display sympathy, that is, have compassion, share *intellectually* another person's emotion or sensation or condition. But can one – never having *experienced* the same kind of frustration, alienation and discrimination – feel empathy, that is, project one's personality into and so fully comprehend another's experiences? I think not.

W. E. du Bois' concept of "double consciousness" is more than a concept: it is the reality of being forever aware that one is black (or yellow, or female, or Mormon, or Jewish) in a WASP society and aware that the WASP's are aware of one's difference. Those of "single consciousness" can intellectualize du Bois but they cannot, by definition, experience it or fully comprehend it. Hence the predilection of so many administrators and academics for prescribing a variety of remedies for Aboriginal ills – all of which demand that the price for better health status, or equality before the law, or equal wages, is the impossible surrender of cultural attitudes, values, beliefs – and being.

Dr. Peter Moodie has stated two important propositions.[8] First:

If Aboriginal (or part-Aboriginal) cultural differences are tolerated fully...
then health disadvantages within certain parameters must be tolerated also.

If Aborigines must have the same health status as other Australians, then there is no alternative to complete social and economic assimilation, with complete loss of Aboriginal identity.

Professor John Cawte[9] says much the same thing: "The truth is that there is no return to an aboriginal culture possible for Aborigines, and if health and growth is the aim, there can be no standing still in the present marginal society." Secondly, in the context of guilt being a bad motivator, Dr. Moodie "deplores the tendency by many Australians – including some Aboriginal opinion-leaders – to harp upon the past misdeeds of citizens and governments in the treatment of Aborigines."

These points may be disputed. First, there is no watertight theoretical logic or irrefutable empirical proof that good health is *contingent* on cultural surrender and consequent assimilation. There is only the smugness of the present-day biocrat whose technological scientism "tells" him that this is so. One can assuredly be an Herero tribesman and healthy, a Sicilian peasant and healthy, Jewish and healthy. Health in the sense of physical, mental and social well-being is not a state reserved and preserved by definition for white, western, Anglo Protestants and Catholics. Secondly, there *is* a return to Aboriginal culture – even if awkward, gross, artificial, diluted or symbolic – on the part of those whom *we* decree have lost it. Thirdly, these men (and others) reveal an ethnocentricity – and if not quite that, then a lack of empathy with the reality that the deeds and misdeeds of the Aboriginal past are an intrinsic part of their folklore and folkways. Massacre, persecution, violent clash and culture clash are often the social cements that hold an identity together. Identity is rarely an issue for a secure member of a WASP-ish mainstream. The let's-turn-over-a-new-leaf-and-start-from-scratch philosophy is simply not possible.

This contention about empathy has moved some social anthropologists, doctors and lawyers to ask whether I am saying that only "ethnics with experience of discrimination" should work – as academics or officials – among Aborigines. The answer is no – in such areas of research and administration as archaeology, blood group genetics and running the Aboriginal Capital Enterprises Fund. But those who have it are much more likely to succeed when designing houses to suit Aboriginal family culture, establishing special medical and nursing services, and giving sensitive legal aid and counsel.

It is highly unlikely that white Australian society can swallow the proposition that black progress is, in part, contingent on black rejection of white society. The reasoning that follows here is not this writer's but that of the "black consciousness" movement generally. The basic precept is that blacks want to know, and must know, more about who they were and who they are if they are seriously concerned about *whom they intend to become*. Black consciousness is an attitude of mind, a way of life. A basic tenet is that the black man must reject all value systems that seek to make him a foreigner in his own country and which reduce his human dignity. He must build up his own value systems, and see himself as self-defined and not defined by others. As Charles Perkins put it: "However favourably a white person, or any other racial or national group in this country mingles with Aborigines or looks upon the Aboriginal question, he is just not part of that definition."[10]

Group cohesion and solidarity therefore become much more important than ever before. Thus, in order to join the open society on anything like equal terms, black people *should first close their ranks*; not as evidence of anti-whitism, but as an exclusion of whites for the time it will take to realise their immediate aspirations of black consciousness. Rejection is not hatred.[11] Integration is the ultimate goal: but one that cannot be achieved in a suspicious, hostile and distrustful atmosphere. It does not mean assimilation of blacks into an already established set of norms drawn up and motivated by white society. Integration means *voluntary separatism* first: the development of group cohesion, an awareness of political and economic strength, a feeling of power arising out of knowing who and what they are; followed then by free participation in a given society, *on their own terms*. This is not the kind of integration espoused by some political parties, advancement organizations or academic critics. Their efforts, say blacks, are directed merely at relaxing certain oppressive measures and practices and *allowing* blacks into a white type of society.

On this emergent philosophy, two South African writers have made interesting comments. Nadine Gordimer sees black consciousness as a revival of the philosophy of *négritude* propounded by Aime Cesaire and Leopold Senghor, and as an essential step towards liberation. She views the black rejection of whites as a sign of "healthy *négritude*."[12] Alan Paton states that this philosophy wants to change the order of things: but the order cannot be changed without power. "How long will the young zealots be satisfied with a mush of culture, mysticism, lyricism and going around saying 'haven't I a lovely skin'?" This philosophy could lead, he speculates, to a "refusal to believe, on principle, that any white man speaks the truth, and might end up being a twin of white

nationalism."[13] Paton is doubtless the more pragmatic, in terms of ultimate change and the need for power to effect it. But, I believe, Jewish
Nadine has a greater empathy and comprehension of the black viewpoint than has Protestant Alan. Her perception – which I share – is
that the "mush" (if that's what it is) is a pre-requisite to establishing
the power needed for change.

II. POLITICAL PARTICIPANTS?

Where does this leave Aborigines in a white-dominated, decision-
making, uncomprehending society, one which twitches neurotically at
the mention of separatism for fear of being equated with South Africa?
What *political* strategies or options are available to Aborigines, or can
be devised by them, which can produce any modicum of change in
"The System"? Our political institutions offer little of comfort to the
poor, the disorganized, the uneducated, the religious and racial minorities. Do Aborigines have access to political institutions and do they influence, let alone participate in, decision-making that affects them? The
commonplace, moronic response is that by virtue of their franchise they
are political participants.

Just what does political equality – in the sense of having a vote –
mean for such a group? Representation theory in political science is a
tortuous concept. There are four sub-theories.[14] "Authorisation" theory is that a community consciously sets aside one or more men to act on
behalf of the whole. Once the representative is so authorised, the principals (constituents) are bound by his actions. This transmission of authority from constituents to representative is sometimes held to embrace
a "doctrine of mandate." On the other hand, the paradoxical "accountability" theory is that representatives are *themselves* responsible for
their actions to their principals. The underlying notion is that the representative should be kept both responsive and responsible to his electorate.

"Microcosmic" theory reflects the view that in a modern society all
cannot meet to decide on policies best for them. Therefore, elected
representatives should meet on their behalf. Thus an assembly of representatives should reflect the composition of the nation, or the region, at
large. John Adams, an early U.S. President, once said the legislature
"should be an exact portrait, in miniature, of the people at large, as it
should think, feel, reason and act like them." In most societies such a

legislature could only begin to be achieved by adoption, at the very least, of a proportional system of representation. Finally, there is the "acting for" theory: the notion that we elect representatives to do our politicking for us. The MP is the specialist, the one who acts efficiently to advance our particular objectives.

If these concepts have any practical validity for us – which I very much doubt – what meaning have they for Aborigines? Many Aboriginal populations in north and western Australia are to be found in Country Party-controlled Federal and State electorates. Aborigines there have assuredly not given that rurally-based, ultra-conservative Party a mandate to block attempts at granting land rights, to peg wages on pastoral properties, to create legions of "slow workers," to inhibit and prohibit the provision of accommodation in crop-picking districts, and so on.

In the absence of proportional representation, the Adams concept is simply not available for Aborigines (and many other groups). There is no way that white representatives (or officials) – of whatever ilk – can think, feel, reason and act like them. This reality doubtless led Bruce McGuinness and Elizabeth Hoffman to stand for the "Aboriginal Independent Party" in the May 1974 Senate election, the Senate being the one political institution that has a vague tinge of proportional representation about it. But unlike Senator Neville Bonner – and more recently Eric Deeral – Bruce McGuinness has yet to learn that our system works for the party machines, not for independent, or even symbolically representative, individuals.[15]

Given the over-riding consideration in our system of keeping the party machine oiled and efficient, there is little room for the individual representative to politick for us. The party symbol men – the largely unknown but party pre-selected men – do it, in an arena all their own, with an agenda all of their own. That Gordon Bryant – co-founder of the Aborigines Advancement League in Victoria and the Federal Council for the Advancement of Aborigines and Torres Strait Islanders (F.C.A.A.T.S.I.) – became very temporarily Minister for Aboriginal Affairs in 1973 was simply a short-lived, happy coincidence.

Ironies abound when white society deals with dark people. One particular twist comes to mind. The recent past has heard the cry: "if only there was an emergent Aboriginal leadership!" But once it emerges, the retreat and defence tactic is to deny the "representativeness" of those who criticise. When confronted with Aboriginal spokesmen from pressure groups or civil services, the first reaction is to demand of each:

"how representative are *you* of Aborigines?" Unless they can show –
and clearly they cannot – that each of them has been given a unani-
mous mandate, that each has been authorised, is accountable for his
actions, is truly symbolic of and each is acting for the total "grass-
roots" community, then they are written off as non-leaders and non-
spokesmen. In 1973 Charles Perkins claimed – before 220 teacher
trainees in Armidale – that some departmental colleagues had told him
to stop insisting on being consulted – because by virtue of his B.A. degree he was no longer an Aborigine.

The system of reserve, mission and settlement councils can hardly be
accepted as political structures providing participatory decision-making. These councils operate not in a normal, open milieu but in what
Erving Goffman defines as "total institutions":[16]

...place(s) of residence and work where a large number of like-situated individuals, cut off from the wider society for an appreciable period of time,
together lead an enclosed, formally administered round of life. Prisons serve
as a clear example, providing we appreciate that what is prison-like about
prisons is found in institutions whose members have broken no laws.

Where councils exist, permissively, in legislation, and more so where
they are created by administrative fiat, they are contained and constrained in terms of effective authority and power.[17] The spending of
canteen profits, the purchase of football equipment, the punishment of
those who offend the white system or the tribal system – these are not
real powers. In Queensland, especially, there is always an overriding
authority in the form of the managership or Directorship.

Now, after decades of total institutionalisation and white-imposed
programmes, the federal government has announced a policy of "self-
determination" for reserve communities – at least in its Aboriginal jurisdictions of the Northern Territory, South Australia, Victoria, New
South Wales, and Western Australia. There is an official resolve to get
out of the way of Aboriginal decision-making, to opt for "non-directive" rather than directed programmes. The February 1973 gathering
at Batchelor (N.T.), albeit a wholly white affair, decided that henceforth the fulcrum of policy will be "self-determination," defined as
"Aboriginal communities deciding the pace and nature of their future
development." Government will have to create the necessary climate
and conditions therefor. "Unwilling communities" should not have
other government goals – such as on health, education and housing –
imposed on them. Land ownership is "a fundamental pre-requisite"

and "communities should be free to evolve their own forms of organization." Further, "no decision-making structure should be imposed upon them and they should be free from an externally imposed time scale." If communities waste money initially, this should be tolerated. Finally, where Aborigines seek funds based on their own view of their needs, "the merits of the case should not necessarily be judged in European terms."

There is significance in all this. But what, precisely? First, this wording and thinking is not Ministerial or political party gloss, but radical attitude change by *senior* officials in the field. Secondly, as a wave of "freedom" it embraces the large black population in the Territory and most States. Thirdly, several points of substance have been publicly pinned to the masthead, yardsticks for staff to follow and for others to measure, prod, praise or criticise governmental action.

However, several doubts come to mind. As of 1975, reserve communities in Queensland[18] will continue under the existing constraints – and they constitute about 23 per cent of the total Aboriginal population. The new dictates apply to very specific communities and say or do nothing of meaning for urban and peri-urban "non-reserve" Aborigines. Officials believe they can inspire and imbue general field staff with this new ethos. This writer does not. His scepticism stems from being an "administration-watcher" from way back: observing the same people, or the same kind of people, administering a succession of segregationist, assimilationist, welfare, wardship and social welfare tenets. The slogans come and go, but the attitudes and behaviour of the men and women who deal daily with Aborigines remain. "Non-directive" approaches cannot become reality overnight following a seminar at Batchelor. Quite often they can emerge from three or four years of "directive" socialisation in social work degree programmes. Assuredly the field staffs in these jurisdictions will not be going back to school. Finally, one cannot avoid the cynical comment that it has taken officialdom in Australia 55 years to catch up with President Woodrow Wilson's liberation slogan at Versailles for the oppressed and depressed in Europe.

Writing in 1970, Charles Rowley could say that "the Aboriginal voice is now continuous; and this in itself is indicative of a growing confidence, of a potential for leadership."[19] But, he added, "as yet these voices are not backed by effective *Aboriginal* pressure groups and organization." It is regrettable that there has been no serious research into the origins, aims, personnel, tactics and effectiveness of Aboriginal pressure groups, that is, those earlier dominated by whites, and later, those taken over by blacks. Such research might well conclude that pressure tactics had the effect of highlighting, but not directly preventing or ending, the removal of the Mapoon people, the Woomera rocket range people, the nightmarish drinking laws epitomised by Namatjira's case, and police treatment of Aboriginal "offenders". More successful has been pressure on the Yirrkala and other mining situations, on land rights, abolition of the Victorian Aborigines Welfare Board, on the retention of Lake Tyers reserve and on federal responsibility for Aboriginal affairs.

A significant feature of this activity, however, has been the consistent white tactic of asserting that advancement organizations have been manipulated, or infiltrated, by the Communists and their fellow-travellers. Communism as a menace is now *passé*: it has been replaced, notably in the eyes and utterances of Victorian and Queensland officials and Ministers, by academic critics or more loosely, "do-gooders". On a television programme in 1974, dealing with Aboriginal stoning of white cars in Alice Springs,[20] the president of the Northern Territory Legislative Council said that he knew blacks very well: "They have no organization, they can't organize themselves out of a paper bag; therefore the violence involved must be being organized by someone else."

Group pressure, has in my view, gained something for Aborigines. But as a strategy it has serious weaknesses: decision-makers can and do accuse the organizations of being puppets for "other" mysterious and dark forces; they can and do claim that these bodies are not truly representative or that their views conflict with other Aboriginal views; and there is the inability of these groups to pose any real threat to the security of the corporate men in the political or bureaucratic power structure. There is no hard (or even softish) evidence of a cattleman being deprived of his labour for his misuse of it; or of a policeman being dismissed for his ill-treatment; or of an official being suspended, demoted or punished for his graft, or embezzlement of Aboriginal funds, or his gross misuse of such funds, or abuse of his statutory obligations.

The emergence of young, articulate and militant Aborigines, supported by student action groups, civil rights movements and above all, the media, finally made Aboriginal affairs a real political issue in the 1970's. Aborigines were espoused by the socially and politically concerned – together with female equality, abortion and homosexual law reform, drugs, the poor and the environment. The 1972 federal election was notable for Labor's priority cry that "we will solve the Aboriginal problem and we will grant land rights." Gough Whitlam's classic policy speech read, in part:[21]

Let us never forget this: Australia's real test as far as the rest of the world, and particularly our own region is concerned, is the role we create for our own Aborigines... Australia's treatment of her Aboriginal people will be the thing upon which the rest of the world will judge Australia and Australians – not just now, but in the greater perspective of history.

The following seventeen months saw almost daily headlines as to what was awry in Aboriginal affairs. Charles Perkins, a first assistant secretary in the Department of Aboriginal Affairs, consistently criticized his Department in the media. His Minister and some colleagues reacted strongly but no formal disciplinary action was taken. Aborigines expressed bitterness about the Woodward Land Rights Commission. The Auditor-General produced a blistering criticism of the Department of Aboriginal Affairs for its abuse of Treasury regulations on financial disbursements. The popular Gordon Bryant was dismissed as the first ever Federal Minister to hold a full and solo portfolio of Aboriginal Affairs. His replacement by Senator J. Cavanagh was much resented by Aborigines. The Department's considerable financial investment in a turtle farm project in the Torres Strait Islands became a *cause célébre*. The formation of "Rights for Whites" groups in the Northern Territory made a gloomy end to what had been a euphoric beginning by Labor. The abnormally early Federal election in May 1974 was notable for some 40 sentences *in toto* on Aborigines. The Opposition said little, though it did produce a policy manifesto giving credence to the concept of cultural diversity – from which, presumably, there will not be any turning back when they come to govern. Labor said it agreed *in principle* with the Woodward Commission recommendations. The Prime Minister – dealing with social policies in a TV speech – promised remedies for the poor, the sick, the old, the retarded and the Aborigines. Mr. Whitlam seemed to be classifying them all as a generic group, the way in which the well-to-do and the mentally well treat the totality of those they consider as the "poor" and the "retarded" – the victims of themselves.

There is some surprise in all this. Given that it took so long for Aborigines to become a major issue in a federal election, one can be forgiven for expecting it to remain one for a reasonable time. One could have surmised that as more discontent and less "solution" became manifest under Labor, so the Opposition would have retained

"Aborigines" as an election or criticism issue. They didn't. The conclusion for Aborigines will doubtless be that the party political system has nothing to offer them in the way of real change.

III. FROM N.A.C.C. TO N.A.C.

Charles Rowley's prescription[22] is that the Aboriginal future rests on the willingness of governments to *negotiate* with leaders "willing to operate within the political system or face others who have rejected it." *Negotiation*, he says, is preferable to *reconciliation*, because that would involve "the admission of injustices by government in the past." Negotiation requires representative spokesmen: "and these in turn depend on accepted ways of making decisions which are binding on the group and on recognition of authority held by some individuals to express views based on consensus." What must emerge then is "*a structure of authority* within Aboriginal groups." This in turn "requires that the limits of the groups be defined, probably on the basis of location and common interests." Finally, new Aboriginal institutions need to be established "which can be entrusted with property, either in perpetuity... or in the form of loans, user rights, or on a rental basis": "they may also be entrusted to administer government property which is devoted to the requirements of Aborigines."

Almost as if he had used Rowley's prescription, and as an answer to Aboriginal frustration within our political system, Gordon Bryant announced in January 1973 the creation of a National Aboriginal Consultative Committee. What follows here is not intended as a definitive history and analysis of the N.A.C.C. Rather, the tangle of promises, expectations and misunderstandings arising from its creation is set down as illustration and evidence of what has been contended above.

Albeit a white-conceived political structure, we have in the N.A.C.C. a reasonable appearance of what Rowley says is needed: 41 elected Aborigines having most of the characteristics Hanna Pitkin deems necessary for true representatives (duly authorised, accountable, reflecting the composition of the community and standing or acting for them). There were certainly defects in compiling the electoral roll, and it is possible that some 20,000 potential voters were missed. Nevertheless, an analysis of the voting returns shows that the majority of those elected have a Pitkin-type "mandate" or a Rowley-type "authority to express views based on consensus". On the face of it, we have in the N.A.C.C. a *structure of authority* based on reasonably distinct locations of populations and common interests of such groups. The electoral divisions are open to criticism, but represent a brave first attempt at groupings.

The poll result, announced on 13 February 1973, was as follows:

State	No. of Contestants	No. Elected	Total Voters Roll	Total Votes Cast	Votes Cast as % of Roll
N.S.W.	38	8	5810	4368	75.18
VIC.	8	3	1044	846	81.03
QLD.	55	9	11614	10398	89.52
S.A.	14	4	2376	1929	81.18
TAS.	3	1	354	300	84.75
W.A.	35	8	7859	4964	(63.16 or 76.71*)
N.T.	39	8	7281	5434	74.63
Totals:	192	41	36338	28239	(80.05 or 77.71*)

* One W.A. candidate, in an electorate of 1065 voters, was returned unopposed. If those electors are omitted, the percentage vote is 63.16% for W.A. and 77.71% for Australia. The Department of Aboriginal Affairs has weighted the unopposed seat in some way, thereby arriving at figures of 76.17% and 80.05% respectively. The overall informal vote was 2.54%. (Figures from *Aboriginal News*, vol. 1, no. 5, February 1974, a Department of Aboriginal Affairs publication.)

The N.A.C.C. appears to be an authority structure from within Aboriginal groups, a new political institution allowing for governmental negotiation with reasonably authoritative black leadership. If this is the appearance, what is the reality? Reality, it is suggested, hinges on the reconciliation of Aboriginal and governmental perceptions of the purpose, role, powers and function of the new body.

Gordon Bryant's January, 1973, announcement of the N.A.C.C. indicated that body would "*advise* me directly on Aboriginal problems."[23] Departmental material sent to Aboriginal communities declared the new Committee would "*advise* the Government in matters pertinent to Aboriginal citizens."[24] Senator Cavanagh, having replaced Bryant as Minister, told the first N.A.C.C. meeting: "if your proposals are wise and logical the Government would reject them only at its own peril... we want the Committee to be a forum for the expression of Aboriginal people."[25] The governmental perception is clear: a forum, an advisory role and function, a consultation mechanism, a black "think-tank" whose views will be listened to – provided the white mind sees them as both wise and logical.

It may have been unhappy coincidence, but the Prime Minister's statement to the N.A.C.C. on official polling day was: "our most important objective now is to restore to Aborigines the power to make decisions about their own way of life."[26] The N.A.C.C. would, he added, also take part in the process of transferring responsibility for community affairs from government superintendents and managers to Aborigines. The minutes of the first N.A.C.C. meeting reflect the black interpretation, or misinterpretation, namely, that the N.A.C.C. "will take part in the machinery of policy-making on all matters affecting Aborigines." If Mr. Whitlam was in fact announcing the "de-institutionalisation" policy for reserve communities, he chose a very peculiar time, place and context in which to do it. The Aboriginal perception is, or was, clear then: a representative body invested with power to participate in decisions on all Aboriginal matters. The extent and strength of this view became clearer early in 1974.

The problems began early in the new year. On Monday, 4 February 1974, the N.A.C.C. voted to change its name to the National Aboriginal Congress – with all the black nationalistic connotations of the word "Congress" – and announced its aim of making the Department of Aboriginal Affairs its "secretariat".[27] On the 6th, Senator Cavanagh said he would not accept these proposals. They were, he said, spending too much time worrying about their own powers and not enough on voicing the demands of their constituents: "Whatever they call themselves they are only a consultative committee. Their power is only to advise the Government."[28] To which the Chairman, Bruce McGuinness, replied that the Minister's statement was "reactionary": "We are setting ourselves up as a separate body that is not going to be dictated to by the Government and Senator Cavanagh."

On the 8th, the Congress censured the Senator, accusing him of being "ignorant of the plight of the Aboriginal people of Australia" and calling on the Prime Minister to take over the portfolio.[29] The Minister's response was to ask his officers to meet with Treasury: to discuss whether delegates could be paid their salaries now that the N.A.C.C. "had adopted different functions from those described in the Cabinet minute establishing that body":

I may have no power to pay them in view of the fact that they have changed their name, and become, not a body to consult with me, but a directive body. If my belief is correct, they have put themselves off the payroll.

The *impasse* was resolved by a publicly announced compromise: the Congress would revert to its original name of N.A.C.C. and its members would continue to be paid. Further, the N.A.C.C. would continue its advisory role until the government considered a new constitution for a National Aboriginal Congress. However, said Bruce McGuinness, "we must have a dual role: we cannot operate just as a tool of the government."[30]

In July 1974 Senator Cavanagh met with eight N.A.C.C. members. The following day he announced[31] that upon adoption of the new proposed constitution, the body "shall be known as the National Aboriginal Congress":

The desire of members is to have an independent congress with permanency, with or without government support. It is, and shall remain an advisory body to the Government while possessing certain executive powers of its own.

The Minister said that objections to previous constitutional amendments – which vested "excessive power" in the executive – had been met: there were now "safeguards" which would make the constitution acceptable, on this point, to Cabinet. The executive would "head" committees dealing with health, housing, education and so on. It would be known as the Executive Council, and its members as Councillors – not "Ministers" as was previously proposed.

The constitution would provide for the "setting up of local community groups, which shall select, or appoint, at least one representative to a regional assembly in which the local community group is situated." These regional assemblies will meet regularly "to advise and assist the local Congress member, make recommendations and communicate information to the Aboriginal people within the region." These bodies "will establish consultation with the Aboriginal people at grass roots level and will better look after the special needs and situations" of all Aborigines. At this time of writing, Cabinet has yet to agree to the new constitution and the level of expenditure for the N.A.C.'s operations.

What is the nub of all this? What factors have led to a rigid Ministerial decree that they are "only a consultative committee", with power "only to advise" and, ominously, "with or without government support"? What has led to a Congress "attempting vaingloriously to arrogate to themselves a quasi-parliamentary right of running their own affairs"?[32] The *Australian* suggested that both Gordon Bryant and the Aborigines "were obsessed by the euphoria of their own idealism and the result was that they over-interpreted the liberalism of a new government which, in fact, is basically as conservative as its predecessors."[33]

Early in 1973 euphoria was rampant: the quick elevation and expansion of the new Department of Aboriginal Affairs; its "victory" after years of defeat by the hardline, Country Party controlled Department of the Interior; the full-circle swing by which a number of key advancement league people – like Bryant – became *the* decision-makers; the flushness of a vastly greater budget; the Labor tenet of not whether land rights but how land rights would be granted, and so on.

There is, however, a real sense in which the psychological climate of Aboriginal communities has run way ahead of Departmental and Ministerial thinking. Consultation ideas are white ideas. Congress notions and power-sharing are black ones. The Department has invested its eggs in the "self-determination" basket, a policy essentially for northern reserve-dwelling communities. It envisages a new set of black organizations working out their own destiny, at their own pace, in their own terms – new structures which could possibly give greater autonomy and responsibility to local communities than they could obtain from an N.A.C. The notion of local community groups appointing representatives to regional assemblies – to "assist" the N.A.C. member – is part of the same kind of thinking. As with the South African Bantustan philosophy, it provides for a separate development ethos for reserve blacks. Unlike urban Africans, urban Aborigines cannot be fobbed off with the response that if they want autonomy and power-sharing on their own terms they must return to their "homelands". Urban Aborigines cannot be faulted for seeing the N.A.C. as one political vehicle available to them – and most resent the dilution they see implicit in a series of local and regional assemblies.

A core group of Aborigines – led by Charles Perkins – has already rejected the N.A.C. instrument. That body clearly failed in its efforts to become the decision-makers and the controllers of the $117 million: level budget of 1973–74. Perkins has been consistent in calling for an Aboriginal Affairs Commission – a statutory body, totally independent – in place of the Department. The F.C.A.A.T.S.I. national conference in March 1975, which Perkins convened, resolved to form a National Bureau of Aboriginal Affairs to replace the Department.[34] Such a Bureau would "demand the right to accept and hold all funds that become available to Aborigines each year." The rationale, said the conference, was simple: the Department of Aboriginal Affairs had failed; further, the Prime Minister, in opening the conference, had urged Aborigines "to do more for themselves."

It seems obvious that an important factor in these confrontations is the desire by the present Minister and some of his officials to assert *their* control over Aboriginal policy. It is a desire for Aboriginal approval of what they decide is in their best interests, for co-operative endorsement, nor for hectic criticism – let alone a role reversal by which the Aborigines become the policy-makers and the Department *their* civil service! The vastly expanded Department of Aboriginal Affairs now has a bureaucratic rhythm, an almost nationwide staff, "territory" – and a pretty grand budget. Like all other Aboriginal administrations before it, it believes it has to expand before it can disappear – "disappearance" being the age-old cry of them all. Within Labor's first 1000 days, it is not about to surrender its control, its power, its internal status considerations, and its newfound muscles because a new black princess – the N.A.C. – has been born. The Department is now hell-bent on its third or fourth re-organization. While it will introduce procedures that may stop financial abuses, the changes will assuredly mean less access, "openness" and response to Aborigines.

Quite a deal of the Department's first 1000 days under Labor has been spent defending itself against trenchant criticism from the Auditor-General; appearing before the Joint Parliamentary Committee of Public Accounts;[35] and submitting evidence in defence of itself and against former Minister Bryant to the Royal Commission on Australian Government Administration.[36] The linen has been washed in public – and it remains, sadly and painfully, very grey. Questions about Ministerial accountability, the rights of a permanent head, the right of a Minister to fashion his own (new) department, the role and rights of Ministerial advisors and consultants outside of the department: all this and more is fascinating reading for the constitutional lawyers and political scientists. But where does this leave Aborigines? Their own disaffection with the Department is now supplemented by "The System's" castigations. Whether the Department can ever retrieve credibility is doubtful; but while trying, the Aboriginal communities move further and further away from any faith, any belief in our political and public service institutions.

The cry for an alternative political mechanism, in the form of an independent, statutory Bureau or Commission, is doomed to be yet another frustration along the Aboriginal road to alienation, withdrawal and violence. No parliament will create an independent agency, generously funded, autonomous in policy and practice, answerable to no one. Even ASIO, our one exception, has lost its "freedom".

The conclusion then is that the existing political institutions and options available offer Aborigines no hope of real change, and that present institutions (including the new N.A.C.) are incapable of yielding to their legitimate grievances. It is not simply that the mechanisms of the structures cannot yield; rather it is the states of mind of the white organization men – discussed earlier – that prohibit the cessions and concessions being demanded.

IV. Convivial Law or Racial Violence?

What other options are there? This writer has discussed elsewhere the possibility of new political and politico-legal mechanisms.[37] Briefly, it is worth considering the New Zealand concept of *Maori Incorporations* under the *Maori Social and Economic Advancement Act*. These legal incorporations, usually but not always based on land and its economic use, provide an umbrella of legal rights, making far more use of lawyers as advisers than welfare officers as guides. The system shows, as it does in Canada with statutorily derived and empowered Band Councils, that decision-makers give greater credence to the "incorporated" viewpoints of Aborigines than has ever been given to the claims or complaints of the "naked" individual, no matter how articulate. The present government intends legislating for Aboriginal corporations, and supports several communities already incorporated in this way. It will be of interest to see how soon these incorporations are "entrusted to administer government property" – as Rowley suggests they should be.

Secondly, the participation-consultation mechanisms that exist are based very much on western notions of representativeness and representative structures. Our idea of committees, councils, chairmen, treasurers, agendas and rules of procedure have been foisted upon Aborigines – as if they were the only valid forms or vehicles for decision-making and spokesmanship. One of the reasons for the failure of these mechanisms to "produce" is that they are intrinsically alien to the cultural configurations of the indigenous societies. With the vast anthropological research material available, it seems a not too difficult exercise to interpret the indigenous system of decision-making and to create structures that have strands of thought and practice substitutive for our equivalent cultural assumptions. The "self-determination" policy fully recognizes this point when it states that communities "should be free to evolve their own forms of organization." One can only speculate about the use officials will make of political science-anthropological expertise in these exercises, or will allow Aborigines to make use of it.

There are a number of legal strategies that may prove valuable in the short and long term. These options may not ignite a new or better set of race relations and they may not obviate or eliminate the four fundamental problems posed at the beginning. But a political scientist may look more optimistically at the legal framework than at the political one. Ivan Illich talks about the need to "recover legal procedure":[38]

> Most of the present laws and present legislators, most of the present courts and their decisions, most of the claimants and their demands are deeply corrupted by an overarching industrial consensus: that more is better, and that corporations serve the public interest better than men. But this entrenched consensus does not invalidate my thesis that any revolution which neglects the use of formal legal and political procedures will fail. Only an active majority in which all individuals and groups insist for their own reasons on their own rights, and whose members share the same convivial procedure, can recover the rights of men against corporations.
>
> The use of procedure for the purpose of hampering, stopping, and inverting our major institutions will appear to their managers and addicts as a misuse of the law and as subversion of the only order which they recognize. The use of due convivial procedure appears corrupt and criminal to the bureaucrat, even one who calls himself a judge.

In short, law or legal procedure can be viewed as a convivial tool of social change – as opposed to political strategies that are, or become, a violent tool of social revolution. Can Aborigines recover their rights against the corporations, be they Swiss aluminium companies, or Federal and State bureaucracies, or mission societies?

Political battles can be won in the legal arena. Often failing, expensive and tortuous, resort to civil law processes has nevertheless won considerable concessions for blacks and Indians in the United States and in Canada. The Aboriginal legal aid programmes have to date shown that their only concern is providing ambulance-type, last minute representation for Aborigines appearing in criminal courts.[39] This function should not be denigrated, but its natural concomitant is counteraction in the civil courts. There is a crying need for this kind of Nader-type or Illich-type function. Aborigines are currently suffering legal and political wrongs for which white society (and other minority groups abroad) has found resolution through actions in contract theory, intentional and negligent torts, the laws of descent and distribution. Recourse has been had to the developing body of international conventions on human rights and the notion of an official tort in the adminisstration of governmental programmes. These areas of action need to be explored by and for Aborigines. The threat of expensive, time-consuming, and very public civil litigation, or the actual pursuit of that litigation, has the remarkable effect of causing institutional men to think twice before acting and above all, to feel a modicum of insecurity for the first time in their lives. Nader has shown just how very effective such civil action can be. The litigation instituted by a group of Methodist ex-missionaries on behalf of the Yirrkala clans in *Milirrpum* resulted in the loss of that particular case. But in the end it won the *principle* of land rights. The *reality* of land rights could well come about through civil cases, if there is no political action by government.

Finally, there is one other option – one far less theoretical and far more realistic each day: the threat of, or actual violence. In another essay the present writer has dealt at some length with the prospects of civil violence in Australia.[40]

Civil violence occurs when expectations about rights and status are continually frustrated, and peaceful efforts to press these claims yield inadequate results. Whether frustration or protest erupts in violence depends mainly on the degree and consistency of social control and the degree to which social and political institutions afford peaceful alternatives for the redress of group grievances. One can postulate this syndrome: frustration, then alienation, then withdrawal from the larger society, then violence, or the threat of it.

In Australia we are said to have an "open" political system. "Open" implies greater flexibility of the existing channels for redress of grievances – and the creation of new institutions for that purpose. It has been shown that our system offers few, if any, peaceful alternatives for Aborigines, and that the new institution, the N.A.C., is tailor-made for the creation rather than the diminution of new frustration. Aborigines, in Rowley's words, have always been met with the whip, the lash and the gun, or the threat of them. There are decrees of "advisory" roles only, and threats of cutting off salaries. For some, there is in turn a recognition that threats of rocks and bottles, or actual rocks and bottles, call more attention to their grievances than any other single tactic.

The erection of the Aboriginal tent "Embassy" on the lawns outside Parliament House in Canberra on 26 January 1972 was indicative of Aboriginal frustration, alienation and feeling of foreignness in their own country. It was a piece of political genius, a peaceful confrontation. But not for long. As Stewart Harris puts it: "On 20 July the Government's patience, which had much to do with the growing public sympathy with the Aborigines, finally snapped."[41] The police moved in, producing and provoking, or provoking and producing, considerable violence. A week later, when the tent was re-erected, Bobbi Sykes shouted at supporters protecting the embassy from police:

> "What do we want?"
> "Land Rights!" they shouted back.
> "When do we get them?!"
> "Now!"
> "And what have we got?"
> "Fuck all!"

The Harris and other accounts of the Embassy – voluntarily dismantled by Aborigines in 1975 as having served its purpose – illustrate only too well the point being made: that in the end, confrontation and eyeball to eyeball tactics produce hastier – though not always better – results. Much of Labor's rush to formulate new policies stemmed from these events. Certainly it has no wish for more world TV coverage of another round between those armed with weapons and those only with tongues.

Our white philosophy in a democratic system is that effective democracy is equated with stable democracy, that is, that violence is pathological to the system of governance. This philosophy, as expounded in political science textbooks, is not reality. Violence is, in many ways, an integral facet of western societies, especially frontier societies such as ours. Violence may be abhorrent to those socialised as we are, but we cannot demand that it be abhorrent to, or abstained from, those people historically at the receiving end of it. That black people see violence as a legitimate political weapon is a reality. One can deplore it, attack it, and wish it were not there, but it is there and we have to learn to comprehend it, even appreciate it. If we fail to do so, we will unleash a shocked, surprised and vengeful violence in return, to the irretrievable detriment of race relations in Australia.

NOTES

1. *Blaming the Victim*, New York: Vintage paperback, 1972.
2. Charles Rowley, trilogy: *The Destruction of Aboriginal Society, The Remote Aborigines*, and *Outcasts in White Australia*, Harmondsworth: Pelican Books, 1972.
3. *Aboriginal Land Rights Commission*, Second Report, April 1974, Justice A. E. Woodward, Canberra: Australian Government Publishing Service.

4. *Committee of Inquiry into Laws Affecting Maori Land and Powers of the Maori Land Court*, New Zealand: Government Printer, duplicated, 15 December 1965.
5. Press Release: "Statement by the Minister for Aboriginal Affairs, Mr. Gordon Bryant," Canberra and Darwin, 29 March 1973.
6. *Australian Aborigines: Commonwealth Policy and Achievements*, Rt. Hon. William McMahon, C.H., M.P., 26 January, 1972, Canberra: Government Printer, p. 15 (italics added).
7. *West Australian*, 18 March 1975.
8. See F. S. Stevens, ed., *Racism: The Australian Experience*, Vol. 2, *Black Versus White*, Sydney: ANZ Book Co., 1972. Chapter: "The Health Disadvantages of Aborigines," pp. 235–242, at pp. 236–237 and at p. 240.
9. *Ibid.*, chapter: "Racial Prejudice and Aboriginal Adjustment: The Social Psychiatric View," pp. 43–62, at p. 56.
10. In Garth Nettheim, ed., *Aborigines, Human Rights and the Law*, Sydney 1974: ANZ Book Co., p. 9. This volume is a report of the proceedings of a conference with the above title, sponsored by the international Commission of Jurists.
11. It certainly is, at least for this particular phase of black consciousness in Australia. How long it will last I cannot say. The vehemence with which *invited* whites at assorted conferences are greeted by "what right has this fucking white got to address us?" is explicable, but hardly acceptable to the guest on the receiving end. Some Aborigines delight in the newly found reverse insult situation, especially on camera. Others, of course, really do hate.
12. *Reality*, a South African journal, November 1972.
13. *Reality*, March 1972.
14. Drawn from Hanna F. Pitkin, *The Concept of Representation*, Berkeley and Los Angeles: University of California Press, 1972.
15. Bruce McGuinness (4183) and Elizabeth Hoffman (434) polled a total of 4617 primary votes for the Aboriginal Independent Party. (A fair number of non-Aborigines voted for them: we know that the National Aboriginal Congress electoral roll in Victoria has 1044 voters.) On the other hand, Senator Neville Bonner gained 17,824 primary votes as number three on the Queensland "Liberal National Party" ticket. Eric Deeral was elected on preferences as a National Party candidate in the Queensland "Deep North" seat of Cook at the December 1974 State elections. He defeated the Labor man by 3459 votes to 3115. (The new National Party is a renaming of the conservative Liberal- and Country Party coalition.)
16. *Asylums*, Harmondsworth: Penguin, 1961, at p. 11.
17. See, *inter alia*, J. P. M. Long's *Aboriginal Settlements: A Survey of Institutional Communities in Eastern Australia*, Canberra; A.N.U. Press (Aborigines in Australian Society series), 1970; and Garth Nettheim, *Out Lawed*, Sydney: ANZ Book Co., 1973.
18. All other States have placed their Aboriginal administrations under the Federal Department of Aboriginal Affairs. Given the frenetic anticentralism of the non-Labor States, it is surprising that they should cede authority to Canberra. Perhaps the seriousness, or insolubility, of the situation has finally come home to them.
19. Rowley, *Outcasts in White Australia*, pp. 384–385.
20. Relations between the races are at their lowest point ever in Alice Springs. Whites are buying guns, talking of vigilante groups and pressing for the reintroduction of "vagrancy" as a criminal offence. Aboriginal drinking, one sexual assault, and white backlash at the extent of Federal spending on Aboriginal affairs have triggered this activity. There is no policial or urban guerilla activity among local Aborigines.

21. *Canberra Times*, 5 January 1973.
22. Rowley, *op. cit.*, p. 422.
23. *Australian Government Digest*, vol. 1, no. 1, p. 63.
24. Circular from Charles Perkins to the chairmen of various Aboriginal Councils (no date).
25. *Sydney Morning Herald*, 14 December 1973.
26. *Australian*, 24 November 1973.
27. *Australian*, 7 February 1974.
28. *Ibid.*
29. *Australian*, 11 February 1974.
30. *Sydney Morning Herald*, 13 February 1974.
31. Media Release, Department of Aboriginal Affairs, (Australia) JC/72, 3 July 1974.
32. *Australian*, 12 February 1974.
33. *Ibid.*
34. *Australian*, 31 March 1975.
35. *Joint Committee of Public Accounts: Inquiry into the Financial Administration of the Department of Aboriginal Affairs*, submission no. 1, 26 March 1974. See also the two-part *Submission by the Hon. Gordon Munro Bryant, Minister for the Capital Territory and Former Minister for Aboriginal Affairs to the Joint Committee* (no date).
36. See Department of Aboriginal Affairs, *Submission to the Royal Commission on Australian Government Administration*, duplicated, no date. The subchapters include: (1) Problems of Establishing a New Department; (2) Problems of Establishing a New Unit within an Existing Department; (3) Aboriginal Involvement; (4) Redeployment of Certain Officers; (5) The Relationship between the Minister and the Permanent Head.
37. Colin Tatz, "Aborigines: Law and Political Development," in Stevens, ed., *Racism: The Australian Experience*, *op. cit.*, at pp. 97–109.
38. Ivan Illich, *Tools for Conviviality*, New York, Harper and Row, 1974, pp. 98–99.
39. See Charles Potter's critique: "The Aboriginal Legal Service in New South Wales," *Sydney Law Review*, vol. 7, no. 2, September, 1974, pp. 237–256.
40. Colin Tatz, "Is Law to Help or Hinder Aboriginal Advancement?" and "Aborigines – the Struggle for Law," in Nettheim, ed., *Aborigines, Human Rights and the Law*, pp. 127–134 and 174–189 respectively.
41. *This Our Land*, Canberra: A.N.U. Press, 1972, pp. viii-x.

The Mongolian Nation within The People's Republic of China

PAUL V. HYER

PAUL V. HYER, a native of Utah, served in the Pacific Theatre in World War II, and graduated in history from Brigham Young University. He earned an M.A. in Asian history from the University of California at Berkeley and a Ph.D. in Asian history and social institutions from the same university. He has been on the faculty of BYU since 1957 and is now Professor of History, but he has also had extensive experience as a Japanese language teacher and has mastered the Chinese and Mongolian languages as well. Professor Hyer has travelled extensively in Asian countries, and has worked as a visiting researcher and teacher in Taiwan and Japan. He has served in a number of academic groups concerned with Asian affairs, and has been a consultant on Manchuria for the Civil Affairs Group, United States Army. Among Professor Hyer's many publications, the following may be mentioned as pertinent to the present study: "Ulanfu and Inner Mongolian Autonomy Under the Chinese People's Republic," monograph in *Mongolia Society Bulletin,* Vol. 8, 1969, pp. 24–62; "The China-Mongol Frontier: The Cultural Revolution and After," in *Collected Documents of the First Sino-American Conference on Mainland China* (Taipei, Institute of International Relations, Republic of China, 1971), pp. 635–50; and "Mongolia's Measured Steps," *Problems of Communism,* Vol. XXI, No. 6 (November–December, 1972), pp. 28–37.

The Mongolian Nation within The People's Republic of China

PAUL. V. HYER

I. Introduction and Background[1]

In considering the status of the Mongolian minority within the People's Republic of China, and evaluating conflicts and injustices as well as the allocation of roles, rights and rewards, the fact must be kept in mind that Inner Mongolia was only recently integrated into China. The political and economic subordination of the region, first to the Republic of China and now to the People's Republic, is resented by the Mongols but reluctantly accepted as a necessary expedient because mass resistance would be suicide. The Mongols were integrated by force and only after many long decades of struggle. They negotiated and received the support and protection of the Japanese against Chinese pressure during the 1930's up until 1945. From the end of the war they obtained the sympathetic support of Americans familiar with their cause, but since 1950 they have been subject to the turbulent winds of fate in Chinese politics.

Ancient Imperial China enjoyed one of the highest levels of culture and the most sophisticated institutional system in world history; it achieved these whilst Europe was still in the Middle Ages. Never had such a large number of people over such a wide geographic range been ruled with as much stability as was the Chinese Empire over two millennia. A little known tragedy of her historic development, however, which has brought and continues to bring untold anguish to minority nationalities, is China's general rejection of cultural pluralism except for certain cases of expediency.

Within its various systems certain contradictions created great problems as China moved into the modern period. Some of these contradictions were resolved with the Republican Revolution (1911) and the events following. Others were resolved with the Communist take-over in 1949 and the turbulent revolutionary changes that followed. Still others have not been resolved at all, but only aggravated by political and ideological turmoil.

One factor that has not been changed by the successive revolutions is the notion of the Chinese that their cultural or institutional model is not just one of many in the world, but rather *the* pre-eminent model, all others being inferior. Chinese, in their particularistic closed society, have always had a very precise image as to what a model society should be. Part of their mental set is a rejection of cultural plurality for a compulsion towards uniformity and unity, a thrust towards a monolithic, orthodox social structure. Obviously this creates great problems for minority nationalities whose society and culture do not conform to the Chinese pattern. The problems of minority peoples were compounded with the rise of nationalism, modern transportation, mass media and an authoritarian society, with its emphasis on unification and defence against foreign threats. Mao Tse-tung unified the Chinese and launched great social and economic engineering programmes, based on models compounded of Marxist and traditional Chinese views. Mao's projects are applied with great power on the grass-roots level, even in such outer regions as Mongolia and Tibet.

Before the Communist take-over peripheral nations such as the Mongol and Tibetan were beyond the Chinese pale. They are now ruled by a Chinese administrative structure in which policy is often executed by militant idealists and radical activist cadres. Theoretical policy statements to the contrary, Communist China, like Imperial China before it, basically rejects in practice cultural or institutional plurality and consequently traumatic changes are incumbent upon the Mongols and other minority people. The tight integration of Mongol lands into the Chinese administrative system is justified on grounds of defence against foreign imperialism – more recently "socialist (Soviet) imperialism". It is also argued that the people need to be liberated from "feudal" or "bourgeois" oppression.

Historically, the "Han" or Chinese and the Mongols have always been two distinct peoples, with a clear-cut political, economic and cultural demarcation guarded by the Great Wall, built over two thousand years ago. It is important in this connection to note that historically "Chineseness", by definition, was not based on ethnic criteria or nationality, but rather was determined by culture. There is no ideal type, in ethnic terms, of a traditional Chinese; the question was whether a person or group accepted Chinese culture – the "Confucian" value system, expressing a particular family structure as the core group of society, the use of the Chinese script and to a lesser extent, such factors as dress, diet, and life-style.

It is hard to imagine two societies or cultures more different than those of the sedentary Chinese agriculturalists and the pastoral nomadic Mongols. The long historical separation of these two peoples was by mutual rejection as the nomads placed great value upon free movement in the steppes, reaped the benefits of the power of their mobility, and had a strong aversion to living in towns or villages and farming the land like the Chinese. Conversely, the Chinese had only contempt for the nomads, to whom they invariably referred as "barbarians". The Great Wall stands as a symbol of the rejection by the Chinese of any hope of symbiosis or amalgamation with or acculturation of the nomadic peoples beyond. Chinese dynasties espoused a policy of an embargo against trade or cultural exchange with the nomads beyond the pale. Any exception to this pattern was invariably based upon expediency, when a nomadic power such as the empire of Chingis Khan imposed its presence upon the Chinese or when, as during the Ming period, the Chinese conceded to the nomads the opening of market sites along the Great Wall to supply such necessities as grain or cloth which the nomadic peoples could not produce themselves, but which, nevertheless, were very important to them. The separation was maintained until very recent times, and the merger with the Chinese has meant, from the Mongol point of view, a cultural tragedy and an impending end as an ethnic group.

The expansion of Chinese culture from its cradle around the Yellow River began during the great Han and T'ang dynasties, and was essentially complete centuries ago. As it moved primarily southward, Mongol society north of the Wall remained largely undisturbed. An important change in the situation of the Mongols, however, took place with the rise of a non-Chinese, Manchu-Ch'ing dynasty (1644–1911). While this brought Mongols into a political structure centered in Peking, the non-Chinese Manchu banned Chinese migration beyond the Wall into Manchuria and Mongolia and they furthermore, banned inter-marriage and the study of the Chinese language by the Mongols. Still, the main factor inhibiting the imposition of Chinese culture upon the Mongols was not primarily these statuatory limitations but rather the geography of Mongolia, which is inhospitable to the customary pattern of life most familiar to the Chinese. The Mongols were linked to the Chinese politically and economically through their Manchu alliance but were not integrated into China. This situation only changed significantly at the end of the 19th century and somewhat more during the first three decades of this century. In addition to foreign pressures, there were those resulting from a tripling of Chinese population within less than two centuries. Malthusian population problems were compounded by floods, famines, droughts, and other natural disasters. The result was a continual migration of Chinese beyond the Wall into Mongol lands.

The expansion of Russian influence into Siberia, and from there into Mongolia and Manchuria, provoked further Chinese migration into the border lands as a defence measure. Such settlement, for defence, to relieve population pressure, or to gain wealth, was facilitated by the cultural assimilation of the Manchu minority by their Chinese subjects and by the shift from a Manchu policy of forbidding migration beyond the Wall to a Chinese policy of promoting it. This policy was stepped up after 1900; Chinese invasion across the traditional border took place on an unprecedented scale, aided by the building of railways and the introduction of modern firearms. Especially important was the greatly reduced mobility of the Mongols in a steppe region, which now lost its role as a pivotal refuge and became a pawn among competing Asian colonial powers. Areas which had been Mongol pastures for ages past, now became the objects of Chinese land-grabbers and Japanese or Russians interested in railway development and minerals. The tide of Chinese settlers was certain to dominate. On the Mongol side banner land traditionally held in common by the Mongol banners was now sold by princes as their private land, often over the objections of their people, in order to maintain an aristocratic standard of living at the expense of the common people. One theme in modern Mongolian development followed the Chinese traditional pattern of centuries past – Chinese political power, economic domination and cultural assimilation pentrated Mongol areas wherever Chinese settled. Chinese *hsien* or district administration came to take precedence over Mongol banner rule and the Mongols lost in virtually every phase of life. As the number of Chinese immigrants grew they generally formed enclaves surrounded by the nomadic minority population, some of which in time settled as a rural segment. In successive stages Chinese military and administrative personnel came as reinforcements and they now dominate the entire scene. Chinese colonization of the Mongol lands created a plural or stratified social structure, the injustices and conflicts of which are the main subject of this chapter.

Any bi-cultural symbiosis was precluded by the Chinese thrust for a monolithic socio-political structure; the Chinese had the numbers and power to reject co-existence or alternative social and political approaches. Consequently, the standard of living of the Mongols declined with the loss of their lands and the economy came to be dominated by Chinese. Next, Mongolian political autonomy was lost and the very ethnic survival of the Mongols as a people was threatened. The response of the Mongols was frequently violent as local patriots struck back at Chinese officials and settlers and at Mongol princes who were a party to selling Mongol lands to Chinese migrants.

The most dramatic event was the declaration of independence in 1911 from the Manchu court by the Khalkha or Outer Mongolian princes. This movement, after some difficulties, was successful in establishing the Outer Mongolian Peoples Republic in 1924, although not without assistance from the Russians. The Republic was able to remain independent from China because of its distance from the centre of Chinese power – with the Gobi Desert in between. General support for this separatist movement in Inner Mongolia was frustrated by the military, economic, geographical and political situation, which favoured the Chinese. But an upsurge of Mongol education and national awareness strengthened desires for self-determination and self-rule. They were fostered by world currents of nationalism and the wide-spread influence of such liberal ideas as those of Woodrow Wilson. The thrust for Mongol self-determination reached a peak in the 1930's with a movement for autonomy – an important conference was held at Batkhalagsume (Pailingmiao – 1933), the beginning of a series of events culminating in the establishment of a Mongol government in Kalgan (1939–1945). This period, a bright spot in the struggle for a Mongolia ruled by Mongols, was made possible only by the reluctant acceptance of Japanese patronage to counter the overwhelming Chinese pressure which was forcing the expulsion or assimilation of the Mongolian people. This course of action was forced upon the Mongol leaders when the Central Government demonstrated that it was unwilling or unable to restrain such warlords as Fu-Tso-yi from exploiting Mongolia. The collapse of Japanese expansion left Inner Mongolia open to "liberation" by the Chinese Communist Party, a new strong Chinese government, and an efficient "People's Liberation Army" (PLA).

The Mongols have never ceased to stress their independent and unique historical development, at once demonstrated by Chingis Khan's empire, when Mongols ruled China rather than the other way round. They still emphasize their separate language, culture, religion and life-style totally unrelated to those of the Chinese. Chinese leaders from Sun Yat-sen to Mao Tse-tung have made great promises to the Mongols for autonomy and self-determination but these fine sentiments have always been forgotten when the time comes for their realization.

Mongol leaders who have been forced to work within a Chinese context over the decades speak of their people as a "minority nationality" *(Hsiao-shu min-tsu)*, in distinction from the Chinese who use the pejorative term "local nationalism" *(ti-fang min-tsu chu-i)* to minimize the significance of the Mongol movement.

The Chinese, in addressing themselves to the problem of minority nationalism, now make use of Marxist-Leninist theory and maintain that class, not nationality, is the fundamental cleavage in modern development. They see the efforts of such minorities as the Mongols to maintain their traditional culture and separate identity as an undesirable aberration to be tolerated only rarely where expedient and ultimately to be eradicated.

II. Mongol Nationality and the Mongol-Chinese Plural Structure

The Mongol minority concentrated in the Inner Mongolian Autonomous Region is a numerical one against which injustice is directed in a socio-political sense, primarily because of its unique cultural characteristics. Mongols are singled out for differential treatment and seem without any known exception to regard themselves as the objects of coercion to adapt to a Chinese-dictated, collectivized way of life. They are being removed from political power, fragmented as a nationality, and forced towards ultimate assimilation by the Chinese majority.

Mongols within the Mongolian People's Republic (M.P.R./Outer Mongolia) are a nationality in terms of citizenship as defined by the United Nations and international law, but the Mongols within the People's Republic of China in no sense constitute a sovereign state. They are, rather, a nationality: a people united by ties of culture, political community, a sense of common history and destiny. These factors are reinforced by a homogeneity of race, language and religion, plus geographical contiguity. Many attempts have been made over the centuries, and still continues to integrate the Mongols into China. Some progress has been made in a broad economic and political sense. But since the Mongols have not been given adequate political expression, territorial or ethnic autonomy, the whole undertaking has fallen far short of success and innumerable rebellions and periodic separatist movements have occurred.

The Mongols are not a nationality group like the Indians and Blacks of the United States, who are too scattered to think of sovereign states. They are a full-fledged nation, traditionally separate from China, and united in sovereign states at different times. The Mongols created highly developed economic and political institutions, which are now weakened or obliterated by Chinese migrations and Chinese colonization of Mongol territory. The current dislocation is compounded by a Chinese Communist administration, which has carried out radical changes in the Mongolian political, economic and cultural structure. The latter have made it largely impossible for the Mongol nation within the present state of China to function as a separate, unified entity.

It is difficult to obtain precise information regarding the Mongolian nationality living in China, in part because the Chinese criteria for determining ethnic affiliation are rather vague and arbitrary. It is doubtful that Peking itself has been able to amass a data-bank of a very satisfactory size and quality. Reporting on such information as ethnicity, for example, has been left up to the discretion of the head of the household in certain important surveys.[2] Nevertheless, some population figures are available from the following authoritative sources, giving some idea of the number of Mongols involved:

1953 Census	Hsin ming-tz'u tz'u-tien (Dictionary of New Terms) (1953 Shanghai)	Jen-min shou-ts'e (People's Handbook) 1957	Nationalities in China (Peking, 1961)
1,462,956	1,700,000	1,460,000	1.645,695

The key demographic factor, however, is not the number of Mongols, but rather the ratio of Chinese to Mongols. The latter have now been displaced by demographic dilution to become a minority even in pastoral areas once ruled by Mongols.[3] The changing population ratio between the Chinese and the Mongols, is a result of Peking's impacting of Mongol lands with Chinese settlers, Production-Construction Corps, People's Liberation Army personnel and others, often under coercion. It reflects various aspects of government policy: from defence and development to the demonstration that so-called "Autonomous Regions" are but a hollow facade without concrete provisions for self-determination or self-rule. When the Communists took over what is now the "Inner Mongolian Autonomous Region" in 1947, the population ratio of Chinese to Mongols was approximately four to one. The population of the Region as announced in November 1954 was 6,100,104 – about 1,460,000 being Mongols and the remainder Chinese; the ratio had risen to about six to one in favour of the Chinese. By May 1957 the total population of Inner Mongolia was officially given as 8,700,000 (a 7 to 1 ratio) and in 1960, with a 10,000,000 total, the ratio was up again: 9 to 1. In May 1967, the *Jen-min jih-pao* (People's Daily) set the total population at 13,000,000 and numbered the Mongols as some 1,518,005. The trend has continued and the ratio is now estimated at from twelve to fourteen Chinese for every one Mongol. This has greatly aggravated an already strong Mongol nationalism and is the basis for most problems.

Among the various alternative factors which integrate and define nations, the Mongols reflect the ethnic-cultural-linguistic concept – somewhat like the idea of the *Volk* as developed by J. G. Herder. Their case differs from Western Europe, where rationalistic factors of dynasty and territory are stressed. The Mongols are distressed to find their nationality nullified by the way the Chinese now redefine them as Chinese. This requires a word of explanation. Historically the concept of a nation-state was quite foreign to the thinking of the Chinese. They thought rather in terms of *t'ien hsia* the "universe" of the "realm". "Chineseness" was defined in cultural rather than ethnic terms and one result was the retarded development of nationalism. In the modern period, as Chinese began anew to redefine their identity, iconoclastic movements arose, striving to change the culture and to shift the loyalty of all to the nation. In this process, since the Chinese do not have an ethnic homogeneity, the term *"Chung-kuo-jen"* (Chinese) has been expanded in a politically shrewd, semantic change as a catch-all to include all ethnic groups in the old Manchu empire. The term *"Han"* (also Chinese), previously used interchangeably with the term *Chung-juo-jen*, is now used in a more narrow sense to refer to those people of original, ancient or root Chinese stock while other peoples, formerly referred to as "barbarians", are now called *Chung-kuo-jen*.

The basic dichotomy separating Chinese from Mongol society has been the contrasting economic base: Chinese sedentary, intensive agriculture and concentrated settlement versus the extensive, pastoral nomadism of the Mongols. But the structure and dynamics of Chinese society form another differentiating and antagonizing factor.[4] The negative image the Chinese hold of Mongol society is one of the reasons for tensions and for Chinese pressures to reform, assimilate and/or eradicate the Mongols. Families are the cores or social atoms of both Chinese and Mongolian society, but their structure and dynamics are very different. Both have their own unique patterns of exogamous marriage and patterns of propriety in interpersonal relationships. In Mongol society more value is traditionally placed upon youth than in Chinese society and the status and freedom of Mongolian women is significantly greater than that of Chinese. The situational ethic of Chinese society stresses a unique approach to contracting marriage, stricter limits on exogamous marriage, non-marriage of widows, peculiar burial rites, great emphasis on ancestral veneration and many other such factors, all of which inclines Chinese to be very critical of peoples with different patterns. This has been particularly true of the Chinese assessment of Mongol society and results in the stereotype of Mongols as barbarians and in various forms of discrimination. Family, lineage and clan are basic units in both societies. But while China maintained an age-oriented male dominant society, Mongols place comparatively higher value on and grant more prerogatives to young people and females. In many social situations, there is considerably more freedom amongst the Mongols than in Chinese society. Economic necessity dictates close co-operation in both societies. But in Chinese society, there has been virtually no place for a person outside his traditional family; the customary ideal of the extended family is five generations under one roof if at all possible. Fragmentation of clan and family is much easier in Mongol areas, where there are economically viable alternatives open to newly married couples, and where patri-local marriage and intensively authoritarian family patterns were by no means so strict as in China. Chinese agrarian society and culture has been intensely "familial" in that all attitudes, behaviour and values centre in or stem from the family and its welfare. In contrast, the nature of Mongol life on the steppes, associated with pastural migrations, gave rise to a very different life-style, but one which is idealized and romanticized by the Mongols and which they have been very reluctant to give up.

In defining the Mongol nationality, or rather in contrasting it with the Chinese, it may be noted that while China as a state contains a certain plurality of nations or nationality groups the Chinese ("Han") have long played the role of a *Staatsvolk*. They occupy about ninety-four percent of the People's Republic and, despite formal equality, a balance is impossible because the Chinese operate the state as their own exclusive political instrument. In ethnic or linguistic terms the political expression on the macro-level is entirely Chinese and politically "Han" is the preferred ethnicity.

III. Differential Treatment of Chinese and Mongols

The historical cleavage between the Chinese and the Mongols came to be more accentuated rather than less with a rising level of Mongolian literacy, education and national consciousness. For centuries the Mongols had but a dim awareness of their ancient imperial glory in the days of Chingis Khan and the Mongol empire. But in the first decades of this century they became increasingly more aware of their heritage, both as a cause and as a result of nationalist thinking. This alone was a factor perpetuating tension and situations conducive to discrimination. The nomadic life-style, so very different from that of the Chinese in diet, dress, daily routine and livelihood to say nothing of language, family or social structure, became a value to be defended at all costs.

During the medieval Mongol Empire (1260–1368), the Mongols ruled China but were not really of China. Many nomadic groups had invaded the Middle Kingdom but had been assimilated and lost their identity. The Mongols, more than any other group, resisted amalgamation, acculturation or assimilation and consequently they, more than any other group, have received a rather bad press in Chinese historical treatments; Mongols are traditionally the barbarians par-excellence. These attitudes are not easily changed, and continue to poison the relationship between the two peoples. Since the Chinese have incomparably greater numbers and power, as conflict has arisen Chinese "pacification" has often, though not always, tended to take the form of suppression by force.

Ethnicity in the sense of biological heritage is not an important factor in perpetuating tension between Chinese and Mongols and thus differential treatment. There is no distinct Chinese type, nor are there any marked physical differences between Chinese and Mongols. Mongols engulfed by Chinese settlers are often wholly assimilated in a generation or two and lose their identity. Similarly, nationality, in the sense of citizenship, has not been a problem in the relationship between the two peoples as in the case of the multistate systems in Europe. Passports, proof of citizenship and other such factors have not been required in this century for people to move from Mongol regions into China or vice-versa. Earlier the Manchu-Ch'ing did employ certain checks to restrict Chinese movement north of the Wall and thus place Mongol areas out of bounds. Indeed, if there had been a more precise definition of nationality in this century and stronger controls on the flow of Chinese into Mongol areas it is quite possible that many economic and political problems could have been averted or more easily resolved.

A cause of conflict and oppression alluded to earlier is the monolithic political structure of modern China, which disallows regional autonomy and self-rule by minority nationalities. Mongols generally assert that had the Chinese adopted and carefully observed the separation, checks and balances of a federal system following the Chinese Republican Revolution (1911), Mongol areas would quite likely have made a transition and found a place within the Chinese state system. As things happened, Outer Mongolia separated from the Manchu Empire, including China; its greatest fear is being annexed again by Peking.[5] Many Mongols supported Mongolian independence in 1911 but were soon disillusioned by the young country's growing pains. They returned to Inner Mongolia and a closer association with China upon hearing that there was going to be a "new deal" for minority nationalities within the new Republic – symbolized by the slogan of "Union and Harmony of the Five Peoples" (Mongol, Manchu, Moslem, Tibetan and "Han" Chinese). This policy was represented for a time by a five coloured flag symbolizing the five peoples and the hoped for union. Some Chinese leaders sincerely espoused this ideal but for others it was merely a propaganda tactic, a facade to allay alarming separatist tendencies among peoples brought together by the Manchu – tendencies that found their most notable expression in the Mongolian and Tibetan declarations of independence. In the event, fine slogans not withstanding, the ideals of the Founding Father Sun Yat-sen were soon set aside and in actual practice warlords encouraged Chinese settlement on Mongol lands. Frequently colonization was carried out by force; troops were used to remove Mongols from their pasture lands in order to settle Chinese migrants in the areas.

Ironically, religion was once a factor binding together the Mongols and the Manchu rulers of China. But with the fall of the Manchu and modern reforms, religion became another point of contention on the Mongol border. The strong element of Lamaist Buddhism in Mongolia, with its obvious tendency to perpetuate "Mongolness," came under pressure and criticism by the Chinese, whether justified or not, and thus emerged as an element perpetuating a discrimination differential in Mongol-Chinese relations. Chinese and Manchu officials took a sympathetic view of Lamaist Buddhism during the Ch'ing period, were particularly solicitous of the Lamas as a group and contributed greatly to the building of temples and the perpetuation of the exalted status of the higher ecclesiastical dignitaries of the Buddhist church.[6] The situation, however, changed significantly with the onset of reform and revolution. Chinese and Mongols were drawn together by transportation, migration and political change after the turn of the century. At the same time there was a marked movement towards secularization as an inevitable part of modernization amongst Chinese political leaders, particularly the intellectual elite. This change was facilitated by the fact that the traditional Chinese elite had always maintained a rather agnostic view of religion, which was strengthened as a so-called "scientific" view of life developed.

Chinese officials began to place more restrictions on Mongolian Buddhism and took steps to reform the church and to curb its power. There is a broad concensus that Lamaist Buddhism in Mongolia had fallen to an exceptionally low level; the illiteracy and immorality of the clergy were scandalous by any criterion. Rascals escaped, moved into the monasteries and many men became a non-productive class of society. The monastic segment of the male population reached fifty percent in some regions and the incidences of social disease among this "celibate" group reached as high as ninety percent. Be this as it may and the need for reform notwithstanding, the stereotype of the Mongols as a superstitious, backward people was used as justification for many actions of persecution and discrimination. Naked force was often the instrument: Chinese troops campaigning in Mongolia killed lamas and destroyed temples or quartered their troops in them. Changes in Mongolian Buddhism directed by communist cadres were brutal in the late 1940's; there followed a comparatively moderate course during the early 1950's and a return to militancy and radicalism with the Great Leap Forward and Commune movements beginning in 1957.

Caste, class and sex now play virtually no real role in the Mongol minority problem in China. The factor of class in both a social and economic sense, however, has been seized upon by Chinese Communist leaders and cadres as a sort of smoke screen or rationalization for radical policies in Mongolia. The Government line asserts that the real problem of the Mongols is a class problem of exploitation by their own hereditary nobility (an argument not without truth in earlier times) and that race or ethnicity apart from socio-cultural factors are not the real cause of poor relations between the Mongols and the Chinese. The strategy is to blame the old Mongol upper-class for the misery of the common people, not the loss of their land to Chinese Colonists.[7]

Socio-cultural norms and the images associated with them are, as has been observed, the main factors setting Mongols off from Chinese and perpetuating antagonism between the two peoples. The life-style of the Mongols in diet, dress, family organization, daily activity for a livelihood and so forth have stereotyped an image of the Mongols as barbarians in the minds of the Chinese. This negative stereotype has its roots in a Sino-centric "culturalism," in counter-distinction to the generally accepted concept of nationalism.[8] This syndrome and its implications require some attention here, because the problem of differential perception and treatment could more easily be resolved were it not for the present lack of a strong commitment on the part of the communist Chinese to a pluralism in socio-cultural matters.

The traditional Confucian society of China was one of the most highly organized and stable in the world and was extremely resistant to change. This caused the long persistence of a folk-society, the retarded development of modernization, the late emergence of nationalism. One consequence is the radical nature of social and economic change as it has finally come to the Chinese country-side. It has been impossible for Mongolia to isolate itself from this social revolution in China. As the old stability was lost, Chinese banditry spilled over into Mongolia, followed by Chinese warlordism, the Sino-Japanese struggle for control of Manchuria, the Chinese civil war and currently the social and economic radicalism of Peking's approach to national defence and development.

The conflict is exacerbated by the fact that Mongolian steppe society is itself highly resistent to change. The general stereotype projects a simplistic view of the dynamics of nomadic society, and most people fail to realize that it, in the words of Owen Lattimore, "requires tight organization, technically skilled division of labour to cover a wide spread of activities, close gradation of responsibility and authority, precise legal concepts of territory..."[9] Mongolian nomadic life is a constant challenge and families or groups realize the importance of maintaining their integrity and of approaching life in terms of mutual aid and collective security. Mongols are, or were until the Chinese Communist take-over, self-reliant, and place great value on mobility and freedom in the steppe.

One Chinese stereotype and contempt or condescension for Mongols is seen in a common household phrase according to which an incompetent, inept, bungling person is often referred to by the pejorative term "Mengku taifu" (literally, Mongolian doctor). In North China it was customary in earlier days to discipline children by warning them that the Mongols, ie. the "bogey man" or goblins, would get them if they misbehaved. On the other hand, Mongol antipathy for the Chinese who "dig in the dirt" is seen in one of their common swear words, which is to call another Mongol a *khara khitat* ("Black Chinese"). In at least one case in a Mongol prison, this epithet sparked a riot and death.[10]

IV. How Discrimination is Manifested

Many imponderables arise in considering the problem of injustices and discrimination against the Mongols under Chinese rule. On the national level, for example, policy is not explicit and it is difficult to determine whether discrimination is intentional. Maltreatment does not appear to be premeditated and dictated from Peking but rather is unavoidable when high level policies are carried out, whether through confiscation of property in a socialist economy or drastic measures taken to secure the Mongol frontier against a Russian threat, real or imagined. The problem is a lack of protection for the interests of the Mongol minority who cannot compete with the Chinese majority.

On the local level, the crux of the matter is one of perception; conflicting values and the demands of national defence versus the continuity of Mongol identity and self-determination. As data are lacking regarding the interpersonal relationships between Chinese and Mongols the extent to which locally determined anti-Mongolian Chinese policy and action are explicitly intended is problematical. The reports available seem to indicate that discrimination is most often circumstantial. The recruitment of administrators or officials, for example, in many cases is based on conformity to cultural norms and an ideology which the Mongolian minority do not share, namely, an eclectic dogma combining traditional Chinese thought with a mixture of Marxism-Leninism as filtered through the dominant element of "Mao Think" *(Mao ssu-shiang)*.

Before the Cultural Revolution (1966–1969), and particularly before the Great Leap Forward (1957), there was a considerable preference for recruiting young Mongols as cadres and awarding them the privileges and advantages accruing to that elite group. This seems, however, to have been a matter of expediency – a policy adopted with the ulterior motive of instituting special programmes to integrate selected young Mongol leaders into the Chinese Communist system. They in turn were to help carry out programmes designed for the larger Mongolian minority.

Chinese pressure on Mongolian society is complex and many-faceted and its form tends to differ according to time and place. In the premodern period the object of rulers in Peking was to protect China from the raids and plunder of nomadic incursions across the wall from Mongolia. Under the Manchu (1644–1911), formal restrictions were placed on nomadic movements, and traditional Mongol political movements were fragmented through a policy of "divide and rule." Also, by certain formal rules, institutionalized in statutes, restrictions were placed on the flow of Chinese language and learning and commercial activity into Mongol areas, not as a form of discrimination but in an attempt to protect the Mongols and to maintain their separateness and distinct culture. These laws and policies had become a dead letter by the turn of this century and there was growing Chinese pressure on Mongolian land and society. This was spontaneous at first but by the second and third decade the expulsion of Mongols from their traditional lands – the greatest single form of injustice or discrimination – was actively encouraged by warlords for their own aggrandizement. The land-grabbing was condoned by Peking as a necessary measure to counter Russian and Japanese movements into an area which constituted a power vacuum. Only under the Communists did the confiscation of Mongol land for Chinese settlement become highly institutionalized and systematic.

The incidence of slavery on the Chinese-Mongol frontier has not been carefully studied. It seems to have been negligible and at no time to have created a critical problem. There were instances of poor Chinese peasants selling their children in time of famine and depression to Mongol areas having a low birth rate and a high infant mortality. Cases of Chinese capture, enslavement and sale of Mongols into Chinese society are unknown. Rather, Mongols were despised and until this century Chinese interaction with them was not institutionalized but was more spontaneous and sporadic. Dynastic histories contain evidence of a certain inclination among the Chinese toward Mongol genocide, but this seldom became explicit and was carried out only in rare cases by border commanders. It was more of a popular notion akin to that regarding the Indian on the American frontier, a feeling that the "best Mongol is a dead Mongol." Institutional and cultural assimilation of minorities by coercion has really only reached its peak on a broad scale under Chinese communist rule.

Under the new system of the People's Republic the traditional Mongol elite has been liquidated, disfranchised or socially ostracised, and certain persons, depending on individual cases, have been harshly circumscribed in their activities and at times subject to house arrest, as in the case of Prince Demchugdongrub. In their place a newly differentiated, functional elite has developed, namely the party cadre. Earlier in this century there was considerable discontent among the young Mongolian intellectuals against the traditionally privileged segment of Mongolian society, including the nobles. This, however, could hardly fall within the usual purview of Marxist class struggle because many princes, like De Wang, were extremely popular leaders with great charisma, against whom there was virtually no antipathy.

V. INJUSTICE, CONFLICT AND VIOLENCE

One of the most serious problems of the Mongols has been economic exploitation by Chinese, who penetrated the Mongol border areas to sell tea, cloth and a wide variety of manufactured items unavailable in a nomadic economy. Mongol indebtedness rose as Chinese merchants manipulated credit and made loans at usurious rates. The average Mongolian nomad had little understanding of the value of either Chinese manufactured goods or their own livestock in the Chinese market, and it was easy for the Chinese to take advantage of him. The central government in Peking was either unwilling or unable to place restrictions on merchant exploitation of the nomads. Many princes among the Mongol nobility and large numbers of the common people became virtually bankrupt or totally indebted to the Chinese merchants. The Chinese Communists in the last two decades have eradicated this problem of merchant exploitation, but now the old problem has mutated into many new ones of an entirely different nature and much greater magnitude.

One overwhelming problem which the Mongol finds great difficulty in coping with is that of the controlled consumption of goods and services managed by a spartan elite, namely the party cadre, in the name of "the people" and the nation. For the Mongols, the "nation" is China, perceived as hostile to the interest of their own people. It is doubtful whether there is overt discrimination against the Mongols, particularly in food rations, clothing and other essentials, but refugees escaping Chinese control and crossing into the Mongolian People's Republic complain about the dire need of the common people in Inner Mongolia for cloth, as but one example. They claim that the clothing of many Mongols is in tatters and that in some places there have been no cloth rations for as long as a year.[11]

More serious is the confiscation of all private herds and property of the Mongol people in the name of the People's Republic of China, which the Mongols feel does not represent their welfare. They have had strong adverse conditioning through the confiscatory requisitions of Chinese warlords and border officials, and it is their experience that the problem has now grown worse in view of the strength and penetrating power of the Peking administration, and the fact that the Chinese element is the single greatest factor in the Communist system confronting them. The Mongols are strongly individualistic in the customary ownership of their herds. They have very little of the cultural conditioning that predisposes the Chinese toward collectivism. Thus the entire new system, the approach to ownership and so forth, are distressing factors to the Mongols.

One of the greatest problems of the Mongol nation is the socialist collectivization of their lands. Land had traditionally been tribal or banner land held in common, but the National Government in Nanking claimed administrative sovereignty over Mongolia in 1928, holding that all land belongs to the nation and that its disposition is to be determined by the Central Government. It was asserted that Mongol banners could no longer lay claim to land, that leases and rentals from Mongols by Chinese tenants were no longer legally valid and that title to the public domain was vested in the Government. This meant that the Mongol banners no longer had legal jurisdiction over their traditional grazing fields and that the Chinese government could survey and register the land and sell it to Chinese settlers in a systematic manner. An early pioneer in this field was the controversial figure I Ku, a high official who represented Peking in land development in the Mongol regions of Ulanchab, Tumed and the Ordos or Yeke-juu league in Western Mongolia (Suiyuan) in the first decade of this century. He was eventually arrested for one of the biggest land scandals in Modern China.

One of the greatest recent changes in Mongolia is in the area of organization. In the past, Chinese warlords, the Nationalist government of China and the Japanese militarists were never really ever able to penetrate to the grass roots on the local level. Now a tight network of relationships and organization has been woven, which may be characterized as close to totalitarian. With this has come a modicum of security and the people are guaranteed a minimal livelihood and the rudiments of human dignity, but the greatest loss for the Mongols is their former freedom. This has been a traumatic experience for them and there is no way to measure the mental anguish or damage to humanity that has accompanied this final overwhelming subordination to the Chinese majority in this border area.

While the trend is toward the inexorable assimilation of their people, the process has not been an unmixed curse for the Mongols. There are frequent reports of efforts made by the Peking government to improve the health and sanitation facilities in Mongolia, to build schools and improve housing. Some of the most ambitious programmes have been in digging wells to water the herds, to build protective stockades to protect them in the winter, and other such measures. Another impressive undertaking is reforestation to reclaim areas into which the Gobi Desert has encroached.[12] Changes like these relieve some of the tensions of the Mongols, but their benefits are offset by the frustation of the over-riding fact that the changes are largely imposed on the Mongols by outside policy; the improvements are often designed to make way for more and more Chinese. During the Great Leap Forward and the forced socialization of the late 1950's, the Mongols were particularly unsettled by pressure to discard their traditional dress, to reform their customs, to learn the Chinese language and accept greater communization at the local level.

One area of tension arises out of the Mongol accusation that the Chinese are prejudicial in personal relationships. The Chinese are criticized as being arrogant and overbearing, unwilling to learn the Mongolian language, for failing to observe customary etiquette and for generally exhibiting what the party itself calls "Great Han Chauvinism." The Mongols, like other minorities, accuse the Chinese, in the words of the party itself, of not allowing them to be "masters in their own house."[13]

The Great Leap Forward (1957–58) was a critical period when the power structure in Mongolia and the decision-making process carefully developed over the previous decade came under intense pressure. Chinese party activists directed a deluge of criticism against Mongol leaders and cadres who had been trying to protect their people and perpetuate their own culture. The main sin was Mongolian "local nationalism," and "bourgeois nationalism." Many Mongol leaders were in a precarious position during the "anti-Rightist campaign" – the drive to increase production, eliminate bureaucracy and simplify administration. Many Mongol leaders became the targets of the concurrent *Cheng-feng* movement – a rectification campaign to reform attitudes and behaviour unacceptable to the party.

One index used to monitor the stress in a place such as Inner Mongolia under Chinese rule is the incidence and movement of refugees. In the 1930s when Soviet-inspired purges and oppression were at a peak in the Mongolian People's Republic the overwhelming movement was from north to south out of the MPR and into Inner Mongolia to escape the liquidations, confiscation of herds and other property, forced secularization of Lamas and so forth. One report estimated that 19,800 persons fled to take refuge in the area controlled by the Kalgan Mongol Government under Japan's patronage.[14] This trend ended in 1945 and for the next several years tens of thousands moved from Inner Mongolia north. At first this was motivated more out of a Pan-Mongol urge for unification, but after 1947 it was to escape from the radical movements within the Inner Mongolian Autonomous Region – the Great Leap Forward, the Commune Movement, the Cultural Revolution. No figures have been released by either Peking or Ulan Bator (MPR) but certainly the numbers of people involved do not approach those who fled from Hungary, East Germany, North Korea, North Vietnam or even Tibet. Some have been interviewed and speak of how tight the border is and how difficult it is to gain the collusion of the imported Chinese herdsmen that is necessary to make good an escape to the Mongolian People's Republic.

The most notable recent case of conflict and overt violence between the Mongols and the Chinese took place during the Cultural Revolution.[15] For almost two decades after the takeover in 1947, Peking followed a United Front approach in Mongolia. This meant, essentially, working through veteran minority leaders – that is men who had risen from the people through their own career – rather than young cadres who had been recruited and trained by the Chinese Communist Party. The key figure on the Mongol side is Ulanfu who, a decade before the Cultural Revolution (1966–69), had arisen to the peak of his career, serving simultaneously as First Party Secretary of the Inner Mongolian Branch of the Chinese Communist Party, Political Commander of the Inner Mongolian Military Region and head of the government as Chairman of the Inner Mongolian Autonomous Region.

Ulanfu had followed a "no struggle" moderate policy, which protected the Mongol nation beyond the Wall from a flood of radical Chinese communizing and assimilating forces. The latter could have been much more disruptive to Mongolian institutions than those experienced up to 1966. But "Maoist" forces felt that a restrained approach in Mongolia favouring the minority nationality: (1) frustrated Chinese needs for unity in the face of a keenly felt Russian threat; (2) retarded socialization and development of the nation on a strategic front; (3) was in fact a form of revisionism – perpetuating bureaucratism, privilege and other evils; and (4) contributed to the delinquency of "local" or "bourgeois" nationalists.

To many "hard-line" cadres and officials, Ulanfu symbolized situations and elements in Mongolia to which they were diametrically opposed. He, like Li-shou-shin, commander of the Mongol army of the earlier Kalgan Government, and some other Sinicized Tumet and Kharachin Mongols, went through a process of "role transference," and became a confirmed Mongol nationalist. This may be explained as follows. Before Ulanfu became head of the Inner Mongolian Autonomous Region he had spent most of his adult life in some type of academic situation in Peking, Moscow or Yenan, and had been only peripherally associated with the Mongol masses and the real problems of the Mongol nation. He had been thoroughly immersed in Communist thought and techniques, but had also maintained a strain of latent nationalism. Blood being thicker than ideology, Ulanfu became increasingly sympathetic with the cause of the Mongols when he found himself their top leader. Living with them, he became aware of their day to day problems and the threat to their very existence under the Chinese. In time he learned the Mongolian language and his earlier opportunism gave way to increased courage to resist Chinese policies which he felt were unwise and unjust. Ulanfu's nationalism and concern about the fate of Mongolia are amply confirmed (1) by interviews of the writer with Mongols who know him, (2) by his activities in the late 1920's in the Mongolian People's Republic, (3) to a lesser degree by his guerrilla activities during Japan's occupation of Mongolia, but (4) particularly by his actions to protect Mongol areas until his purge by the "Maoists." Many "Red Guard" accusations against Ulanfu are exaggerated but many have a basis in fact. Exaggerated, for example, is the idea that he was trying to detach Inner Mongolia from China. This was quite possibly the ideal of many Mongols but after 1947 it was totally unrealistic and Ulanfu knew this. Such a move had failed in all Inner

Mongolia in 1911; it failed in the Barga and Hulunbuir area of Mongolia in 1929; it failed in the Kalgan Mongolian Government Movement in the 1930's and again in the concerted attempt made following the war in 1945–46.[16] The most the Mongols could hope for was an equitable place within China with enlightened protection for Mongol rights under Chinese rule. The arrangement dividing Mongolia was made by Peking and Moscow and the Mongols are keenly aware of this.

VI. THE CHINESE POWER STRUCTURE AND DISCRIMINATION

Since political power determines the establishment of priorities, the assignment or allocation of roles and the distribution of goods and services, the shape of the political structure in Inner Mongolia is really the decisive problem from which most others stem. The Mongols were able, more or less, to hold on to political power at the grass roots level through the continuity of their traditional league and banner system until very recently. But, because the latter perpetuated certain "feudal" tendencies, it came under fire from young Mongol activists in the early period of the Chinese Republic. Still the system remained largely unchanged and, even during the period of Japanese occupation outside interference was strongly resisted. With the Communist takeover, however, Chinese political power was able to penetrate into the most basic units of the Mongolian power structure and from that time the decision-making process rapidly changed.

A. Shifts in Leadership Prior to the Cultural Revolution

As soon as Japanese power collapsed, there was a spontaneous emergence of independent or autonomous local or regional governments, some of them with Outer Mongolian Communist support. Most of these spontaneous governments were headed by natural, popular leaders who had long previous experience, who were influential during World War II and who now assumed direction during the confusion following the war. Over a period of several years the Communists, largely through their own Mongol cadre led by Ulanfu, gained the submission or subversion of these local "governments," "committees" or "councils." In this process the Mongols lost a whole generation of their best leaders. Some were liquidated, as were Manibadara and Buyandali; others merely disappeared, as did Hadanbatar, still others were forced to flee, as was Demchugdongrub, who fought to the end against both Chinese and Communist domination and became widely idolized as a heroic defender of his people. He made his last stand in Alashan in the far western deserts of Mongolia, finally, when the cause was lost, fleeing for his life into Outer Mongolia. He was later extradited from the Mongolian People's Republic to Peking upon the demand of the Chinese Communists, during the honeymoon of the Sino-Soviet alliance, and placed under house arrest. Several hundred leaders fled Mongolia knowing that their lives were threatened. Other leaders who remained and who had struggled long against all outsiders, whether Chinese or Japanese, were in time stigmatized by Peking as "narrow nationalists" and liquidated or purged. This removed many minor leaders.

A second generation of leaders under the Chinese Communists may be discerned largely as men who had been educated in Russia or the Mongolian People's Republic, initially under the auspices of the Inner Mongolian People's Revolutionary Party, organized by Pai-yun-ti and Merse. Most of them had "laid low" during the period of Japanese occupation when leftists or liberally-inclined Mongols were under strict surveillance and when life was quite precarious. Examples are Temurbagan and Uljiochir, both of whom were successful in hiding their Communist affiliations during the war, and purposely remained behind the scenes carrying on covert activities. An exception is Haffenga, a progressive leader sympathetic to the (Outer) Mongolian People's Republic under the patronage of the USSR, who came under Japanese suspicion. He was threatened with death but was saved when Japanese intelligence officers found it expedient not to kill him because he had strong popular support amongst the Mongolian people. The leaders in this generation were given posts in the new political administration under the Chinese, partly in response to pressures of Mongol nationalism, and partly as an attempt to transmit communism to the Mongolians through a cadre from their own people, so it would not be rejected out of hand. Another factor was that when Mao Tse-tung took over in 1949, his policy was to solve nationality problems by working through leaders drawn from the minority nationalities themselves.[17]

For almost the first decade the Communist Chinese government apparatus in the Mongol areas of Inner Mongolia was formed to a large degree by Mongol Nationalists, who in some cases had Communist leanings, but only as a secondary inclination. This situation was not unique to Mongolia, for in many minority areas the Chinese were constrained to use "politically doubtful material." during a transitional period.[18] Most of these leaders have been purged since the Sino-Soviet split, largely, it would seem, because of their old international Communist connections from the 1920s and 30s including long standing contacts with the Mongolian People's Republic and, not least, their latent Mongolian nationalism. A third group, and the only one which has persisted until the present time, are Mongolian party members who have a long history of co-operation with the Chinese extending back to the Yenan period. Most outstanding is Ulanfu, who from beginning to end has been the single most prominent leader of the Mongols, or for that matter of any minority, under the Chinese Communists. He studied at the Sun Yat-sen University in Moscow from 1925 to 1930 and established firm relations with leading Chinese Communists from 1941 to 1944 as Principal of the Nationalities Institute in Yenan. Wang To, another Mongol colleague of Ulanfu's, also began his party career in Yenan while serving there as Secretary General of the Nationalities Institute.[19]

By the middle 1960s most old Mongol leaders had been eliminated in favour of young Mongol cadres who had been trained and indoctrinated in the minority institutes in China and who worked on the grassroots level. As noted, the only senior Mongol leaders still holding high positions in the administration were politically reliable persons long associated with the Chinese Communist Party. The general trend in Mongolia was the gradual consolidation of Chinese control through such minority leaders as Ulanfu, who were both "red" and "expert." There followed, however, an important and persistent transition which gradually made Mongol cadres expendable and permitted their replacement with activist Chinese cadres – particularly in such "forward" movements as the Great Leap Forward, the Commune Movement and the Cultural Revolution. This was a trend that has been termed "demographic dilution".

Work in the Chinese Communist system is done in some form of a group – work team, committee or council – often affiliated with the People's Liberation Army. These units are manipulated by the Chinese to make it appear that the minorities are really involved in decision making, but while a Chinese would "generally occupy a position nominally secondary to a member of the minority nationality of the area, he was nearly always a Han Chinese and the real power..."[20] From the foregoing it may be seen that during the first decade of Communist Chinese rule in Mongolia there was a "go-slow" policy and a concerted attempt to gain popular, voluntary acceptance of party leadership. It also appears that the traditional elite maintained much of its prestige and property.[21]

Towards the end of the 1950s, however, as the balance of actual power and actual decision making came increasingly into the hands of the Chinese cadre, greater tensions arose. Many frustrations of the Mongols, as well as others, were made known during the Hundred Flowers Movement (1957-1958). Local leaders complained that their Chinese comrades ignored them or failed to take their advice.[22] Another complaint that came out in the "speak bitterness meetings" was that the so-called self-government, granted by Peking, had been as useful as a "deaf ear", that the people were simply messenger boys for the Chinese. There were demands for a rectification of the situation described as "many rights in theory, few rights in practice."[23]

Some of the typical sentiments or criticisms which surfaced during this period and which were evident symptoms of widespread discontent are listed by Roderick MacFarquhar as follows:

"We want independence even if that means we have to forego socialism."
"If Han settlers continue to come, we will close the borders."
"So long as Han cadres remain [in autonomous areas] the national minorities will not be able to exercise their power. All Han cadres should be evacuated from national minorities areas."
"Party members are a privileged class and party leadership is a rule by great Han nationalism."

Some Mongol leaders advocated the partition of Inner Mongolia into a "purely Chinese area" and a "purely Mongol area", so as to separate the nationalities, even if this meant losing such important industrial complexes as the Paotow Steel Industry.[24]

Fortunately for the Mongol leaders the problems of the Great Leap soon became obvious to more moderate Peking leaders and the pressure was released before too much damage could be done. The strong "forward policy" of the Chinese party activists was re-assessed and the general feelings by late 1958 seemed to be that "at present there are still very few Communists hard-core elements who are capable of political leadership among the minority nationalities."[25] The course of action was again a more pragmatic approach to minority problems, but, as we can see now, it was, consciously or unconsciously, to be a "two steps forward, one step back" policy. During the early 1960s key Chinese leaders working with minority nationalities had continued what has been termed a "United Front" approach. This meant a policy of compromise rather than struggle, of peaceful negotiations and the avoidance of nationality and religious strife. Hardline officials felt that minority leaders participating in administrative decision-making were often more of a hindrance than a help, but moderate leaders saw it as bowing to the inevitable. The leftist faction of hard-core party militants, bided their time only to re-emerge later with the Cultural Revolution. The direction of Peking's policy during the early 1960's may be seen from a leading article in *Min-tsu t'uan-chieh*, the main party organ devoted to nationalities work. Lu Chien-jen, a member of the Nationalities Affairs Commission of the State Council projected the party line in his article entitled "Continue Good United Front Work with Nationality Religious and Upper Classes".[26]

The power structure during the 1960's remained largely unchanged, and the Mongolian input into decisions affecting their people continued at a low level – matters were not as good as pre-communist days, but also not as bad as during the periods when hardline Maoist types controlled policy. A less obvious trend, seen by June Dreyer, who has made a close examination of the fate of minority personnel within the Party-government structure, is that while old veterans like Ulanfu maintained top positions, minority cadres trained by the party after liberation were relegated to low levels. "Whatever their backgrounds, [they] had little prospect of success outside of the Party-government minorities structure, or indeed outside their own minority area," and "one must conclude that the minorities' chances for advancement were highly circumscribed by geography and that their integration into the CPR had taken place in only a limited sense."[27] Actually, while integration was the Chinese policy, the Mongols felt it inimical to their interests and they resisted it with any means at their disposal. Ulanfu, and no doubt others, took what subtle actions they could to protect their people from the various inroads of the Chinese, but high policy and its application in Mongolia ultimately lay with Peking. It became increasingly more difficult to restrain injustices in the political and economic system.

B. The Cultural Revolution and Liquidation of Mongol Leadership

Ulanfu was one of the first persons attacked in the campaigns and turbulence of the Cultural Revolution. It came in May, 1966, and Mongolia was soon thoroughly involved in the movement.[28] It would be out of place here to undertake a general discussion of the Cultural Revolution and its various interpretations. The focus here is the fact that the movement brought great new stresses to Mongolia and eventually swept out many Mongolian leaders and cadres who had so far generally maintained moderate policies. These moderates had held at a distance the rapacious policies of militant cadres who could have wreaked havoc in Mongolia. One reason Ulanfu and his associates came under fire was their association in China's power structure with moderates who themselves were attacked as enemies of Mao and the "people," particularly Liu Shao-chi (stigmatized as "China's Khruschev"). Ulangfu and other Mongol leaders, responding with a strategy of fighting fire with fire, concocted their own "Great Socialist Cultural Revolution." This move also need not be detailed here, but it should be noted that there was a spontaneous rising of organizations and peoples to defend the interests of Mongolia and Ulanfu.

A few examples here will serve to give some idea of the types of tensions, conflict and rhetoric involved in the Mongolian struggle. The attack on Ulanfu, a February 1967 report claims, "exerts great pressure on Mongolian poor and lower middle peasants. The destiny of us Mongolian people is in the hands of Ulanfu."[29] Then, during the autumn of 1966, the Vice-Chancellor of the Mongolian University in Huhehot and a number of the faculty were purged for opposing "class struggle" and for promoting autonomy in Mongolia.[30] Other Mongols were set upon for such remarks as "Han (Chinese) oppress Mongols; they give us new rice bowls and then make us beg for rice!"[31]

Many Chinese militants condemned, in particular, the priority placed on the Mongolian language in Inner Mongolia. Cultural criticism along these lines brought tensions to a new peak and by February, 1967, factional strife between moderates and radicals and between Chinese and Mongols had reached such a point that the capital, Huhehot, was in a state of anarchy. Immediately the People's Liberation Army forces were brought in and the area was placed under martial law. Chou En-lai called representatives of the factions to come to Peking for mediation and for several weeks there was a cessation of rioting, demonstrating and armed parades. But soon the truce was broken and when violence erupted again, Peking moved the 21st Army of the People's Liberation Army from Shensi into Inner Mongolia. A week later, April 13, 1967, the Central Committee of the Chinese Communist Party issued a directive purging Ulanfu, reorganizing the Inner Mongolian military region, establishing a Revolutionary Committee for the region, and initiating a struggle against Mongolian leaders and "anti-party elements".[32] Open clashes broke out between Chinese troops and the supporters of Ulanfu. The commander, T'eng Hai-ch'ing, determined that of over 5,000 top-level cadres of the region, only 500 would be retained by the new Revolutionary Committees.[33] Many key departments and bureaus in Inner Mongolia were overhauled to weed out those elements who had been "poisoned" by Ulanfu and a soft Mongolia line. In one department 230 cadres were replaced by 40 army cadres and in another Huhehot bureau, 80 percent of the cadres were purged. In many cases this meant Mongols.[34]

In a move to consolidate their position, the Chinese Army and the new Revolutionary Committee undertook an unprecedented propaganda campaign against Ulanfu, his supporters and the values they symbolized. The following extracts from monitored broadcasts in Mongolia give some idea of the standard line pursued: Ulanfu (and by implication most Mongol leaders) "practices national separatism, undermines the unity of the Motherland... plots to separate this region from the big family of nationalities... is an arch traitor to the Chinese people and to the Mongol nationality."[35] Mongol leaders were further condemned for "carrying out a peaceful transition in the pastoral areas, realizing peaceful co-existence with the reactionary superstructure..." and for "attempting to substitute the nationality question for class struggle." Particularly condemned was an attempt "to remould Inner Mongolia with bourgeois nationalism and nationalist splitting activities designed to separate Mongolia from the big community of the Fatherland."[36]

Much of the rhetoric and radio broadcasts of the period were propaganda, and while they cannot be taken literally, they are an accurate indication of what was in the minds of the Chinese and the Mongols. Mongol leaders were condemned for promoting the publication of books about traditional Mongol culture and about Chingis Khan. Ulanfu, for example, is quoted as having said even before the Chinese People's Republic was established, "Sons of Chingis Khan unite!" He was condemned for projecting himself as "the Chingis Khan of the Twentieth Century," and for criticizing Chairman Mao while speaking to his people in the Mongolian language. Singing Mongol songs about Chingis Khan, composed during the Japanese period, was condemned as were such lines as the following, claimed to be found in Mongolian language publications: "The ancestors of the Mongol nation are in

Ulan Bator and the capital of China is Nanking," and "Inner and Outer Mongolia unite, hail Ulanfu!"[37] To defend their interests, a number of the Mongol students organized an "East is Red Revolutionary Rebel Joint Association", from which came "Chingis Khan combat squads" to do battle with Peking's Red Guards.[38]

The Revolutionary Committee in Mongolia reversed many of the earlier policies favouring the Mongolian language and cultural development and moved again to the "fusion" approach of the Great Leap period. Earlier plans for development of Mongolian art, drama and other cultural programmes were condemned as "revisionist". A reformation of Mongolian opera was undertaken to base it on "the creative application of Mao Tse-tung Thought on the cultural front."[39]

The period of the Cultural Revolution and that following saw a phenomenal upsurge in Chinese colonization of Mongol lands. The Institute for Chinese Communist Problems in Taiwan estimates that while the rate of population increase in China as a whole was around 10.1 percent from 1960 to 1969, it reached in Inner Mongolia, 41.3 percent. There were scattered reports of Mongols resisting the Chinese colonizers and of troops being used to quell disturbances.[40] One of Ulanfu's alleged sins was the assertion that his native Tumet Banner belonged to the Mongols and had been seized by the Chinese to develop agriculture in the Mongol grass lands. Efforts to retrieve this area from the Chinese were condemned as were attempts to form "groups of pure Mongol nationals" and "groups of pure Han nationals" in a diversion of the programme of collective labour.[41]

A particular feature of new trends after the Cultural Revolution was the massive influx of Production and Construction Corps of the People's Liberation Army. One feature of this movement was the relocation of hundreds of thousands of former Red Guards from urban areas to the rural or frontier areas of Inner Mongolia and other places. Another important measure was the dissolution of the Mongolian military district with the integration of the western section into the district under the command of Peking, while the northeast section was placed under the jurisdiction of Shen-yang.

The policy of the last several years seems to be one of benign assimilation. It lacks the militant coercion of the "fusion" approach of the Great Leap, but nevertheless, the situation seems rather bleak for Mongol life and culture. While some allowance for bias and propaganda must be allowed, judging from the Soviet-sponsored "Radio Peace and Progress" and reports emanating from Ulan Bator, there are distressing indications of restrictions placed on Mongols in detention camps, forced teaching of the Chinese language, restrictions on the Mongolian language, attempts at forcing intermarriage as well as the suppression of Mongolian holidays, festivals and music.[42]

VII. PERPETUATION OF CHINESE-MONGOL PLURALISM

The ideal of the Mongols has long been a unified nation enjoying self-determination. They abhor the perpetuation and regularization of national partition with the southern or Inner Mongolian half subordinated under China. Their situation has some similarities and also some unique aspects when compared to that of the Germans, Koreans and Vietnamese. The plural or stratified society which has been forced upon the Mongols, with its attendant injustices, is very unlikely to change without some form of self-rule and self-determination. Movements in this direction in this century have been frustrated by the Chinese and at different times by the Russians and Japanese. The political integration of Mongolia into China was initiated by the Manchu, in some cases by treaty and mutual alliance, in others through coercion and violence depending on the particular tribe or area. While northern Mongolia, Khalkha, regained its independence with the collapse of Manchu rule, the forced integration of Inner Mongolia into China continued through the rule of Yuan Shih-k'ai in the early Republic and under other warlords and through the establishment of the National Government at Nanking (1928). With the collapse of Japan's empire in 1945 it appeared for a short time that the Mongols would realize their hope of unification under the independent state of the Mongolian People's Republic. This possibility, however, was frustrated by the agreement of the great powers, China and Russia, which opted for a division of Mongolia along the de facto lines set by the Chinese Revolution, leaving an overlapping plural society of Mongols and Chinese in southern Mongolia along the wall.

Political integration of Mongolia into China after World War II was accomplished in part by violence and military force and in part by political activity. The coercion and force began with the takeover of Mongolian territories from the Japanese by either Nationalist or Communist forces. In some cases the Chinese took over Mongol areas from Russian and Mongol (MPR) troops who had moved south from the Mongolian People's Republic with the collapse of the Japanese. Drawing on their experience of the 1930's and 1940's, the Chinese Communists realized that a policy of bare force was ill-advised and consequently it was veiled with more subtle political means to gain the acquiescence of Mongol leaders in the political integration of Mongolia into the People's Republic of China. The agent of these political manipulations was Ulanfu and a group of Mongol cadre trained by the Chinese during the war at Yenan.

The New China took special pains to avoid offending the nationalist sensitivities of the Mongols, setting up an elaborate facade of autonomy and making many promises to allay the fears of the people. However, successive mass movements and waves of radicalism have eroded the original idealism of the party and government and the protection of minority peoples has now been lost by default. With the trend towards formalization of Peking's control in Mongolia there have been fewer efforts to treat it as a special area and less concern about maintaining a pretence of autonomy. This diminished concern for minority nationalities is reflected in (1) the State Constitution of 1954, (2) the Party Constitution of 1956, and (3) the Party Constitution of 1969. In the first document of 1954, seven articles out of one hundred and six were concerned directly with the ratification or confirmation of minority self-government and related matters. This basic law held, for example, that:

The People's Republic of China is a single multi-national state. All the nationalities are equal. Discrimination against, or oppression of any nationality, and acts which undermine the unity of the nationalities are prohibited.

The higher organs of state should fully safeguard the right of organs of self-government of all autonomous regions, autonomous *chou* and autonomous counties to exercise autonomy, and should assist the various national minorities in their political, economic and cultural development.[43]

The party constitution of 1956 contains a mere two paragraphs, compared to the seven articles noted above. Still, the tone is positive and progressive, particularly in view of the radical tendencies which appeared shortly after with the Great Leap Movement:

Our country is a multi-national state. For historical reasons, the development of many national minorities has been hindered. The Communist Party of China must make special efforts to raise the status of the national minorities, help them to attain self-government, endeavour to train cadres from among them, accelerate their economic and cultural advance, bring about complete equality between all nationalities and strengthen the unity and fraternal relations among them. Social reforms among them must be carried out by the respective nationalities themselves in accordance with their own wishes, and by taking steps which conform with their special characteristics. The Party opposes all tendencies to great-national chauvinism and local nationalism, both of which hamper the unity of nationalities. Special attention must be paid to the prevention and correction of tendencies of Great-Hanism on the part of Party members and government workers of Han nationality.[44]

The results of two decades of demographic dilution, carried out through mass invasion of Chinese into Mongol areas together with the purge of Mongol leaders during the Cultural Revolution, may be seen in the Party Constitution of 1969. Only a passing reference in this document notes that:

The whole Party must hold high the great red banner of Marxism-Leninism-Mao Tse-tung Thought and lead the hundreds of millions of the people of all the nationalities of our country in carrying out the three great revolutionary movements of class struggle, the struggle for production, and scientific experiment...[45]

An alternative explanation of such a downgrading of the nationalities may be that the last statement quoted is "in deference to radical insistence that nationality problems be regarded as no more than class problems."[46] While the Mongol minority has not been erased, it is quite certain that the power structure and decision-making process will never return to the situation in the 1950's when there was some promise of autonomy. The economic, military and administrative structure of Inner Mongolia is now solidly in the hands of the Chinese. While it may be put down as propaganda, there may be more truth than fiction in the allegations by the Russians and the MPR that there is a plot afoot to exterminate the minorities. The Soviet Union has been particularly vocal on this point in Mongolian-language broadcasts beamed into China.[47]

It should be very clear that any attempt to regularize or perpetuate the stratified social system involving the Mongols in China is on the part of the Chinese and not the Mongols. In the writer's experience in over 20 years of contact with Mongols, even Sinicized Mongols, who have lived their entire life among the Chinese and have enjoyed high status, still deplore the persistent Chinese negation of Mongolian rights to self-rule and self-determination. The distress of the Mongols is less the problem of discrimination and maltreatment vis-à-vis the Chinese than the fact of their impotency as a minority in their own native areas and the realization that their political, cultural and economic fate has largely been taken into the hands of the Chinese.

The perpetuation of the present situation depends largely on a number of imponderables. They include:

(1) the nature and intensity of the Sino-Soviet confrontation, much of which takes place in the strategically located Mongol lands, now heavily invested by the People's Liberation Army.

(2) the vicissitudes of Chinese politics and the question whether the dogmatic "Red" Maoist forces prevail, as they did during the Great Leap Forward and the Cultural Revolution, or whether more moderate pragmatists hold power.

(3) whoever is in power, the key factor of policy, upon which depend Communist directives for Mongolia and the way they are applied.

(4) the response of the Mongols themselves. This factor cannot be overlooked, but should not be over-rated. While analysts feel that the radicals have not achieved the Cultural Revolution goal of obliterating nationality differences, still the Mongols have lost military and political power to the Chinese and may be losing cohesiveness as a nationality.

(5) the enormous number of Chinese who have been pushed into Mongolia. This may be the greatest factor in the long run.

(6) whether the Chinese stay and how they relate to the Mongols. There are recent reports of attempts to coerce Mongol women to marry Chinese men, but they cannot yet be verified.

The level of national consciousness of the Mongols is deeper and has a broader base than ever before and there will continue to be resistance to assimilation. Nevertheless, all things being equal, the close proximity of the Chinese with their higher level of status as well as economic and political power should inevitably promote the process of amalgamation or assimilation.

REFERENCES

Cheney, George A., *The Pre-Revolutionary Culture of Outer Mongolia*, Mongolia Society *Occasional Papers*, Vol. V (1968).

Lattimore, Owen, *Nomads and Commissars*, New York: Oxford University Press, 1962.

Bawden, Charles R., *The Modern History of Mongolia*, New York; Praeger, 1968.

Yasuo Misshima and Tomia Goto, *A Japanese View of Outer Mongolia*, New York, 1942.

Rupen, Robert A., *Mongols of the Twentieth Century*, Bloomington, 1964.

Diao, Richard K., "The National Minorities of China and their Relations with the Chinese Communist Regime," in *Southeast Asian Tribes, Minorities and Nations*, Vol. I (ed. by Peter Kunstadter), Princeton: Princeton University Press, 1967, pp. 169–204.

Heaton, William R., "Local Nationalism and the Cultural Revolution in Inner Mongolia," *Mongolia Society Bulletin* (Fall, 1971).

—, "Chinese Communist Administration and Local Nationalism in Inner Mongolia," *Mongolia Society Bulletin* (Spring, 1972).

Hinton, Harold C., "Colonization as an Instrument of Chinese Communist Policy," *Far Eastern Economic Review*, XXIII:17 (October 24, 1957).

—, "The National Minorities in China," Far Eastern Economic Review, XIX:11 (Sept. 15, 1955), pp. 321–5, and XIX: 12 (Sept. 22, 1955), pp. 367–72.

Hudson, G. F., "The National Minorities of China," *St. Antony's Papers* (No. 7) (1960), pp. 51–61.

Hyer, Paul V., "The Re-evaluation of Chinggis Khan: Its Role in the Sino-Soviet Dispute," *Asian Survey*, Vol. VI, No. 12 (December 1966), pp. 696–705.

—, "Ulanfu and Inner Mongolian Autonomy Under the Chinese People's Republic," *Mongolia Society Bulletin*, Vol. II (1969), pp. 24–62.

MacFarquhar, Roderick, "The Minorities," *The New Leader*, Vol. XLII, No. 23 (June 8, 1959), pp. 17–21.

Palat, Augustin, "Some Aspects of the Social Structure of Minority Nationalities in China," *Cina* VIII (1964), pp. 46–8.

Schwarz, Henry G., "Migration to North-West China and Inner Mongolia, 1949–59," *The China Quarterly*, No. 16 (October–December, 1963), pp. 62–74.

NOTES

1. The greater part of the historical and background information set forth in the first part of this essay is available in the general sources on Mongolia. There is no particularly unique interpretation here.

   For more specialized information and interpretations, the author draws upon hundreds of interviews conducted with native Mongols during the past two decades, too numerous to name but including a wide range of former officials of the Mongolian Government (Kalgan) and Mongolian specialists in the Republic of China. Useful information was gained while serving as a Visiting Professor at the Graduate Institute of China Border Area Studies, Chengchi University, 1971–72. The author is grateful for the support in field work of the Joint Committee on Contemporary China (Social Science Research Council and American Council of Learned Societies). Special thanks is due to his closest research colleague, Professor Sechin Jagchid, a most knowledgeable native Mongolian scholar and a man who is a constant source of information.

2. State Statistical Bureau, Peking, "Communique of the Results of the Census and Registration of China's Population," New China News Agency (NCNA), November 1, 1954; translated in *Current Background*, No. 301, November, 1954.

3. *China News Survey* No. 263 (March, 1969) p. 2; also Paul Hyer and William Heaton, "The Cultural Revolution in Inner Mongolia," *China Quarterly* No. 36 (October–December, 1968), p. 117. Yuan, Chia-lo, "Chung-kung K'ung-chih-hsia te wei Nai-Meng-ku tsu-chih ch'u" (The Inner Mongolian Autonomous Region Under Chinese Communist Control), *Fei-ch'ing yen-chiu (Studies in Chinese Communism)*, Vol. I, No. 11, (November, 1967), p. 96.

4. Sechin Jagchid and Paul Hyer, unpublished manuscript; Maurice Freedman, "The Family in China, Past and Present," in Albert Feuerwerker, ed., *Modern China,* (Englewood Cliffs; Prentice-Hall, 1964); Edwin K. Andrus, "Chinese Sedentary Family and Central Asian Nomadic Family – a Comparison," unpublished manuscript.

5. Robert Rupen, "Mongolia in the Sino-Soviet Dispute," *China Quarterly* (London), November-December 1963, pp. 75–85.

6. Robert Miller, *The Socio-political and Economic Aspects of the Monastery in Inner Mongolia.* (Seattle, 1955).

7. *Wen-ko feng-yun* (Cultural Revolution Wind and Cloud) (Canton), No. 2 (February, 1968), in *Survey China Mainland Press*, No. 4151, p. 3.

8. Edwin O. Reischauer and John K. Fairbank, *East Asia, The Great Tradition* (Boston, 1960), pp. 290–294.

9. Owen Lattimore, *Nomads and Commissars* (New York, 1962), pp. 31–32.

10. Personal interview with Takeo Shishido, former Japanese official in Inner Mongolia.

11. Personal interview with Gombojab Hangin.

12. Nei-Meng-ku jen-min ch'u-pan-she (Inner Mongolian Peoples Publishers) Nei-Meng-ku tzu-chih ch'u ch'eng-li shih-chou-nien chi-nien wen-chi (Collected articles commemorating the 10th anniversary of the establishing of the Inner Mongolia Autonomous Region), Huhehot, 1957. See also Kuo-fang pu, ch'ing-pao chu (Intelligence Research Bureau, Defense Department) *Wei Nei-Meng-ku tzu-chih ch'u kai'k'uang* (Survey of the Inner Mongolian Autonomous Region), Taipei, 1961.

13. *Jen-min jih-pao*, 26 December 1956.

14. Zenrin Kyokai, *Soto Moko no gensei* (Tokyo, 1935), p. 62.

15. Hyer/Heaton, "The Cultural Revolution in Inner Mongolia," *China Quarterly* No. 36 (October and December, 1968), pp. 115–128; See also June Dreyer, "Inner Mongolia: The purge of Ulanfu," *Current Scene*, Vol. VI, No. 20 (November 15, 1968). The best detailed account is William R. Heaton, The Politics of Minority Nationalism in Communist China: A Case Study of Inner Mongolia, (unpublished dissertation, University of California, Berkeley, 1972).

16. For a differing interpretation see June Dreyer, "Traditional Minority Elites," as cited in note 18, p. 448.

17. Mao's statement quoted in *Min-tsu t'uan-chieh* (Nationalities Solidarity), No. 8–9 (1961), p. 10.

18. For the best treatment on this subject see June Dreyer, "Traditional Minorities' Elites and the CPR Elite engaged in Minority Nationalities Work," in Robert A. Scalapino, ed., *Elites in the People's Republic of China* (Seattle, 1972), pp. 416–450.

19. A short history of the Yenan Nationalities Institute may be found in Tsung Ch'un, "Cradle of Minority Nationalities Cadres," *Min-tsu t'uan-chieh* (Nationalities Solidarity), No. 7 (1961), pp. 15–19.

20. See Amrit Lahl, "Sinification of Ethnic Minorities in China," *Current Scene*, Vol. VIII, No. 4 (February 15, 1970), p. 9.

21. Dreyer, p. 435; See also Ulanfu, "Success In Nationalities Work and Questions

of Policy," a report to the third session of the First National People's Congress, June 20, 1956, in *Current Background*, No. 402 (July 10, 1956), pp. 17–22.

22. Yeh Hsiang-chih, "Some Questions in the Work of Fostering Minority Nationality Cadres," *Jen-min jih-pao*, Feb. 6, 1957.

23. Nan-fang jih-pao (Southern Daily), May 5, 1957. See also Dreyer, p. 437.

24. Roderick MacFarquhar, *The Hundred Flowers Campaign and the Chinese Intellectuals* (New York, 1960), pp. 255–256.

25. Cheng Hung, "Nationalities Institutes are Political Schools for Fostering Autonomy," *Kuang-ming jih-pao*, Dec. 17, 1958.

26. *Min-tsu t'uan-chieh*, No. 3 (1962) pp. 2–5.

27. Dreyer, *loc. cit.*, pp. 442–43.

28. "The Political Situation in the Provinces –Developments in Inner Mongolia Autonomous Region," *China News Summary* (hereafter cited as CNS) No. 264, April 3, 1969, p. 3. This source is published by the British Regional Information Office, Hong Kong.

29. *CNS* No. 264, p. 3.

30. Radio Huhehot, Aug. 25, 1966.

31. Yuan chia-lo, pp. 93–94.

32. "Decisions of the CCP Central Committee Concerning the Handling of the Inner Mongolian Question," republished in *Documents of the Great Proletarian Cultural Revolution, 1967* (Hong Kong: Union Research Institute, 1968), pp. 415–19.

33. "Comrade T'eng Hai-ch'ing's Speech at Study Class on Jan. 4," in *Wen-ko t'ung-hsun* (Cultural Revolution Bulletin), No. 12, Feb. 1968; SCMP No. 4163, pp. 2–5.

34. Heaton, The Politics of Minority Nationalism, p. 341.

35. Radio Huhehot, June 5, 1967, in Foreign Broadcast Information Services (FBIS), June 8, 1967.

36. Radio Huhehot, June 8, 1967, in FBIS, June 14, 1967.

37. Radio Huhehot, Sept. 22, 1967; "Inner Mongolia in Turmoil," *China News Analysis* No. 687, Dec. 1, 1967, p. 4.

38. *China News Summary*, No. 264, p. 2.

39. Radio Huhehot, June 15, 1967 in FBIS, June 20, 1967; also Heaton, p. 354.

40. Tass dispatch, June 6, 1969; also "Inner Mongolia and the Cultural Revolution" *China Topics* YB 552 July 8, 1960, p. 9.

41. *China News Summary*, No. 264, p. 7 (March 27, 1969).

42. G. Rakhimov, "Troubadours of Chauvinism: Maoists Oppress Small Nationalities," *Komsomolskaya Pravda*, May 20, 1969; *Current Digest of the Soviet Press*, Vol. XXI, No. 21.

43. Amrit Lal, "Sinification of Ethnic Minorities in China," *Current Scene*, Vol. VIII, No. 4, p. 23.

44. Amrit Lal, p. 23.

45. *Ibid.*

46. Dreyer, p. 448.

47. The Russian view is stated by T. Rakhimov, "The Great Power Policy of Mao Tse-tung on the Nationalities Question," *Kommunist*, No. 7 (1967). Also note Radio Peace and Progress (Moscow), September 18, 1968.

Discrimination in Great Britain

WINIFRED CRUM EWING

WINIFRED CRUM EWING, a native of Great Britain, is a professional journalist and has also published a novel and numerous short stories. She has travelled extensively, both before and after World War II, acquiring familiarity with political and social problems in many parts of the world. For several years Mrs. Crum Ewing was Administrator of the Women's Press Club, and she is the British member of the Board of the World Association of Women Journalists and Writers (AMMPE).

Mrs. Crum Ewing was for a number of years associated with the Conservative Central Office, first as an organizer of local and youth groups and later as the party's film and television producer. After leaving the Conservative Central Office in 1959, she formed her own production company, which specializes in documentary films. Among those she has written and directed are a television obituary of Winston Churchill and films on conservation and national defence.

Discrimination in Great Britain

WINIFRED CRUM EWING

For historical reasons the British have long been concerned with the elimination of all forms of arbitrary discrimination. The name, the UNITED KINGDOM, is itself testimony to the fact that different and separate races are one people before the law. Through seven centuries the Scots, Irish and Welsh have retained their positive differences from one another and from the English. The English are a blend of Briton, Anglo-Saxon, Dane, Roman, Norman, plus other European groups who have settled from time to time, and become assimilated.

Since the end of World War II the ethnic groups within the United Kingdom have rapidly increased in number and size. In many ways this has been a surprising development. It was not anticipated that Asians and Africans would choose to leave their newly independent countries in such numbers in order to live and work in the land of the ex-Colonial Power. The West Indians have always been adventurous travellers. When the USA restricted imigration from the Caribbean in 1952, it was natural that the West Indians, who would normally have gone to the States, came to Britain.

British subjects throughout the Empire and Commonwealth had always enjoyed free and unrestricted entry into the United Kingdom. There was a lot of coming and going in both directions. Neither numbers nor race created a problem in Britain. The sons, and a few daughters, of wealthy Africans and Asians frequently came to England to be educated in the best schools and universities, where they were welcomed naturally without prejudice or discrimination. Working class Africans and Asians were mostly found round the sea ports. The numbers were not large enough to have much impact on the local populations, but many facets of race discrimination were common practice. Since the middle of the twentieth century the challenge of race relations and race discrimination has become a major issue. All political parties have earnestly and sincerely searched for solutions to racial tensions and the achievement of social justice. Many voluntary associations are working diligently in the same field.

Discrimination on grounds of race, however, cannot be tackled in isolation. Discrimination on grounds of language, religion, and above all, sex are interwoven. The search for social justice as between one citizen and another is the eternal stuff of politics. Any analysis at any time can never be more than partially correct, as the basic assumptions are always changing. Each generation challenges the concepts of their elders. This has always been so, but modern means of communication have greatly increased the scale and impact of new ideas and expectations.

The British, because of their long association with many races in all continents, feel a special responsibility to eliminate arbitrary discrimination of all kinds. Yet such a simple thought as that raises a number of questions. Is assimilation the ideal, implying much inter-marriage and the elimination of racial differences and cultures? Or is assimilation itself a form of discrimination, imposing the culture and values of the majority on the minorities? Is the pluralist solution the better answer, enshrining the rights of the ethnic groups to preserve all that makes them separate, while living and working within the wider community? If so, how is sex equality to be achieved with some girls and women subject to drastic restriction on religious and customary grounds?

A study of discrimination in Britain today must raise as many questions as it answers. Genuine progress can only be made through the good sense, good will and accumulated experience of ordinary British families. Before the dimensions of the current situation and the details of action taken and projected are examined, a brief summary of the historical origins of British attitudes to inter-group relations may be appropriate.

There is no reason to believe that the British have a greater share of human virtues than the members of any other nation or ethnic group, but the history of the British Empire has given them the kind of experience which leads to the common acceptance of certain principles. To claim that the principles were always applied would be ludicrous. The British have their quota of the cruel and the ruthless as well as the idealistic and kind, but broadly the merchant adventurers, followed by the missionaries, educators, doctors and administrators found that the following rules worked:

1. Help people to like you by liking them and by treating them fairly.

2. Respect their customs and beliefs.

3. Help them to achieve peace and the rule of law.

4. Respect minorities and ensure that all enjoy equality before the law.

These principles add up to a belief in the Plural Society, and also a humble recognition that patience and wisdom are needed to reconcile the conflicting aspirations of diverse groups.

Rather more than other Colonial powers Britain chose the Plural solution for the Colonial territories. In each case, when Independence was negotiated, a constitution enshrining the rights of minorities was agreed, only to be abandoned in almost every territory for some other system within months or years of Independence.

This experience has not shaken the faith of the British that civilization itself depends on evolving institutions which do not inhibit the human right to the basic freedoms of thought, speech, and association, albeit only achievable within a framework of law. Harold Macmillan, speaking as British Prime Minister in the South African Houses of Parliament on 3rd February, 1960, said:

Our judgement of right and wrong and of justice is rooted... in Christianity and in the rule of law as the basis of a free society.

This experience of our own explains why it has been our aim, in the countries for which we have borne responsibility, not only to raise the material standards of life, but to create a society which respects the rights of individuals – a society in which men are given the opportunity to grow to their full stature, and that must in our view include the opportunity of an increasing share in political power and responsibility; a society finally in which individual merit, and individual merit alone, is the criterion for a man's advancement, whether political or economic.

Finally, in countries inhabited by several different races, it has been our aim to find means by which the community can become *more* of a community and fellowship can be fostered between its various parts.

Few would dispute the wisdom of these words. Nearly all would recognise that what was wise for those overseas territories for which Britain had had responsibility was equally applicable to the United Kingdom. The range of opinion as to *means* is wide, but the *end*, in broad terms, is agreed.

I. Race Discrimination

A. Ethnic Groups Identifiable in the United Kingdom

English, Scots, Welsh and Irish

No records are kept to indicate the distribution of these groups, either on their native soil or elsewhere. A story which typifies the free movement arose at a time when Welsh Nationalists were protesting about good Welsh water being piped to Liverpool. The answer came that there were many more Welsh throats to drink the water in Liverpool than there were in the Welsh mountains from whence the water came. The Irish from Ulster and Eire mingle and move freely without any kind of check or discrimination.

The correct definition of Britain is The United Kingdom of Great Britain and Northern Ireland. This paper will cover England, Scotland and Wales. Northern Ireland, where the problems of discrimination are peculiar to that small territory, will be the subject of a later chapter.

Jews

The Jewish Community estimate the number of British Jews as something over 400,000, but as the official census form includes no question which would ask Jews to identify themselves, either by religion or by birth, this total can only be an intelligent guess. There are no exact records of mixed marriages.

In 1974 a group of 350 Jewish students joined in a study to find the answer to the question, "Who and what is a British Jew?" Their report, *Jewish Students: A Question of Identity*, was published by the Social Issues Committee of the Union of Liberal and Progressive Jews. The Educational Correspondent of "*The Guardian*" in reviewing the report on 10th September, 1974, wrote:

The picture which emerges of the typical Anglo-Jewish student is of someone with a certain cynicism "inextricably mixed with basically hopefull attitudes towards the future of the Jewish people in general and the Anglo-Jewish community in particular."
Central to the ambivalence of many of the students towards all things

Anglo-Jewish is what they said was the inadequacy of Jewish education, where it existed at all.

A widespread complaint was that their parents were unable to explain the ritual so that it could have meaning beyond mere tradition. As one student put it: "Our parents don't know the reason why things are done differently by Jews and really aren't terribly interested, so it's not surprising that they haven't taught us anything beyond automatic and, on the whole, meaningless acts."

The one strongly unifying aspect of Jewish experience and identification among the students was their empathy with Israel: The great majority expressed active interest in Israel and optimism for its future.

For some Israel was regarded as an insurance policy, but a minority wished to consider it as another society in the world, with good and bad aspects like other societies.

About half the students thought that inter-marriage could work. All said that the most important criterion in the choice of a partner was concensus of opinion on the essential things of life, but the girls much more than the boys tended to think that this almost necessarily limited their choice of a partner to other Jews.

West Indians, Africans and Asians

The colour of the skin and distinctive bone structure inevitably label these groups at a glance as coming from regions other than Europe. To the British public they are black, brown or yellow, whereas to themselves the groups within each colour are different and separate. The difference between a West African and an East African, and sometimes between nationalities within a single State such as Hausa and Ibo in Nigeria, can be quite as great as between a Scot and an Italian. National, and frequently tribal groups have their own clubs and societies.

Gypsies

According to the Ministry of Housing and Local Government estimates, the Gypsy population doubles every 25 years. Currently it is estimated that the numbers travelling on the roads or in caravan sites are:

England and Wales	20,000
Ireland (Eire and Northern Ireland)	10,000
Scotland	5,000
Living permanently or temporarily in houses	25,000

Because of increasing numbers the problems of the nomadic life become greater. Gypsies were harassed and made to move on from unauthorized camping sites. They felt this to be a discrimination against them. In 1968 Parliament passed the Caravan Sites Act. For the purpose of the Act "Gypsies" were defined as being "persons of nomadic habit of life, whatever their race or origin." This definition covers both Romanies and Irish Tinkers.

The Caravan Sites Act was produced by a Joint Working Party of Local Authority Associations working with the Gypsy Council. It requires all county, county boroughs and London borough councils to provide adequate accommodation for gypsies residing in or resorting to their area, though in heavily built-up areas the Act accepts that very little space will be available. The aim is to provide about 300 sites, but this will take time. About a third of the target has been achieved. Both the Gypsy Council and the Government believe that the provision of permanent sites, with proper facilities, and a Warden on the grounds, may solve some of the problems that Gypsies face.

But modern society has other challenges. To retain their distinctive way of life and also ensure that their children get adequate education to earn a living, and thus be treated as equal citizens without discrimination, may be the most serious challenge. Children are quick to notice differences in life style. If they feel surprise at the differences this may be taken as criticism and rejection, and so a discriminatory attitude of mind can be developed by the very young. It can also be fed by the prejudices of older generations. Traditionally, dealing in scrap metal has been an important source of Gypsy income. Inevitably sorting scrap is an untidy business and not pleasing to the eye. This nuisance may be controlled in properly organised camp sites, but it will take time to eliminate prejudice.

It is fair to say that great efforts are being made to improve the lot and status of Gypsies, to ensure that they get the benefits of the social services, health, housing and education, while retaining their freedom to be different and separate.

B. Size and Location of Immigrant Groups

The flow of Commonwealth citizens admitted to the UK with the right to settle is shown in the following table, published by H.M. Stationery Office:

Year	Employment Voucher Holders	Dependents	Others for settlement	Totals	UK Passport Holders plus Dependents
1965	12,880	41,214	2,968	57,062	6,149
1966	5,461	42,026	2,978	50,465	6,846
1967	4,978	52,813	3,586	61,377	13,600
1968	4,691	43,879	4,499	53,069	18,843
1969	4,010	29.454	3,093	36,557	6,249
1970	3,983	22,941	2,962	29,886	6,839
1971	3,341	20,282	3,032	26,655	11.564
1972	1,718	20,927	3,688	26,333	37,283

The column "Others for Settlement" includes those coming for marriage, those of independent means, as well as doctors and dentists (after June 1971).

The figures for the UK Passport Holders do *not* include the Asians expelled from Uganda, as they were a special case. The figures do include White people from the old Commonwealth countries of Canada, Australia and New Zealand. When the figures for the Old and the New Commonwealth countries are separated, the totals arriving each year are:

Year	New Commonwealth	Old Commonwealth
1965	53,650	3,412
1966	46,602	3,863
1967	57,648	3,729
1968	50,160	2,909
1969	33,942	2,615
1970	26,562	3,324
1971	23,611	3,044
1972	23,587	2,746

These figures show very clearly the shape of the problem confronting successive Governments. The provision of social services, housing, health, education, welfare is a Government responsibility. In a crowded island like Britain, too large an annual increase in population would inevitably overstrain and so destroy the services. Immigration had to be controlled. How to achieve control of *numbers* without adopting a policy of *racial* discrimination was the question which caused deep concern. Successive Acts of Parliament have sought for the right answer or series of answers – not, of course, to the total satisfaction of either the immigrant or indigenous population.

Much has been done to improve community relations and lessen racial tensions. If the immigrants were to settle fairly evenly throughout the UK, community relations would obviously be easier, but it is only natural that new immigrants should want to settle near friends and relations, where there are facilities for the practice of their religion, and social clubs for the enjoyment of their culture and language.

The problem of finding somewhere to live near employment opportunities also has a powerful influence on grouping immigrant families in given areas. Frequently immigrant landlords buy property in city areas ripe for redevelopment, and let accommodation to their fellow countrymen in overcrowded conditions.

The 1971 Census enumerated the population in Britain by birth place. This shows the major location of ethnic groups:[1]

Area	India/ Pakistan	West Indies	New Commonwealth: Total
Greater London	137,000	167,000	477,000
Outer Metropolitan	48,000	20,000	106,000
West Midlands Conurbation	68,000	40,000	121,000
West Yorkshire ,,	38,000	10,000	55,000
South East Lancs ,,	25,000	10,000	47,000
Other Areas	147,000	56,000	351,000
Total	462,000	303,000	1,157,000

The following table shows these figures as a percentage of the total of the enumerated population in each area:

Area	India/ Pakistan	West Indies	New Commonwealth: Total
Greater London	1.9	2.3	6.4
Outer Metropolitan	0.9	0.4	2.0
West Midlands Conurbation	2.9	1.7	5.1
West Yorkshire ,,	2.2	0.6	3.2
S.E. Lancashire ,,	1.0	0.4	2.0
Other areas	0.4	0.2	1.0
Total	0.9	0.6	2.1

In considering the size and location of immigrant groups, and the consequent impact on community race relations, all the figures quoted in these tables are too general and apply to too large an area to give a true picture. Ethnic groups tend to congregate within a small part of any given area. It may be that in a dozen adjoining streets the immigrant population may be 60% to 80% of the total. Even if it is only 25% to 30% the conflict of cultures and tastes in close proximity, twenty-four hours a day, must inevitably impose some strains. Take one small example, the smell of cooking. To a man coming home from work, the smell of the good meal his wife has prepared for him may be most appetising, whereas to the neighbour, who dislikes that form of food, it may be repellant. As in all human relationships, prejudice and friction are born out of the accumulation of small irritations. It is not a question of who is right or wrong, but of incompatibility.

What can the law do to improve the human situation, reduce tensions and increase compatibility?

C. Acts of Parliament Relating to Immigration and Race Relations[2]

1905

The Conservative Government took power to control the admission to Britain of citizens of foreign countries.

1914

Aliens Restriction Act was passed by the Liberal Government. This Act was renewed annually from 1920 to 1972.
Commonwealth citizens were NOT aliens. This was confirmed by the *British Nationality Act of 1948.*

1948

A British National was a person born within His Majesty's Dominions, or born abroad of British parents. Passed by the Labour Government.

1962

The Conservative Government passed the first *Commonwealth Immigrants Act* which laid down that a Commonwealth citizen must first obtain a work voucher (or be a dependent of a voucher holder). Until this Act all Commonwealth citizens had been free to enter, work and settle in Britain.

1965

A Labour Government White Paper restricted the number of vouchers issued to 8,500, and limited the dependents who could settle in U.K. to wives, children under 16 and dependent parents over 65.

The *1965 Race Relations Act* prohibited discrimination on racial grounds in places of public resort: hotels, places where food and drink were served, theatres, cinemas, dance halls, sports grounds etc. or on public transport services. The Act also set up the *Race Relations Board* to secure compliance with the Act. The Board was empowered to set up Local Conciliation Committees.

1968

The *Race Relations Act of 1968* made fresh provisions with respect to discrimination on racial grounds, enlarging the scope of the 1965 Act. It also set up the *Community Relations Commission* to encourage the establishment of harmonious community relations, and to advise the Secretary of State on matters referred to the Commission by him.

1968 the Labour Government passed the second *Commonwealth Immigrants Act* to extend immigration control to Commonwealth citizens who were United Kingdom passport holders, but who had not close personal or ancestral connections with the United Kingdom.

1971

The Conservative Government's *Immigration Act* radically changed the position of most Commonwealth citizens, bringing them more into line with aliens. The Act imposes a single system of control and divides people into "Patrial" and "Non-patrial". Patrials are those with the *right of abode* in the UK, free from controls. They include people who have direct connections with the UK, that is UK passport holders who were born, adopted, naturalized or registered in the UK, or whose parents or grandparents were; UK passport holders who have lived in Britain for five years and are accepted for permanent residence; and Commonwealth citizens who have a parent born in UK. A non-patrial Commonwealth citizen coming to Britain for work no longer has the automatic right to settle, but will need a work permit issued for a specific job in a specific place for a fixed initial period (normally 12 months). Like an alien, he must obtain permission from the Department of Employment to change jobs, and he will have to register at the employment exchange, not with the police. After one year he may apply for extension of his stay. After four years he may apply to register as a UK citizen and be free of conditions. His application will be granted if, among other things, he is of "good character" and has a sufficient knowledge of English. Unlike aliens, non-patrial Commonwealth citizens enjoy the right to vote.

Non-patrial workers who do not need work permits include Irish citizens, ministers of religion, doctors and dentists. The arrangements for alien girls to come to Britain as *au pair* girls to learn English are extended to Commonwealth girls.

The Act gives everyone subject to controls the right of Appeal, including appeals against being refused entry or a change in conditions of stay, and against a deportation order. (These rights are similar to those in the Labour Government's *Immigration Appeals Act of 1969*). The Act gives the Home Secretary, without the Appeals Tribunal, the right to decide cases of entry of someone who is "not conducive to the public good".

The Act gives the Government power to deport the wife and children (under 18) of a person who is to be deported (they have the right of appeal) and removes the exemption of Commonwealth citizens from deportation after five years residence in the United Kingdom, unless they were already resident when the Act came into force. The Act also empowers the Government to give non-patrials assistance to go home if they wish to do so and it is in their interest.

1974

In April 1974 the Labour Government, through the Home Office, offered an amnesty to all immigrants who had entered Britain illegally on or after 9th March 1968 and before 1st January 1973.

D. Historical Background to Acts of Parliament on Immigration and Race Relations

The ideal of the British Empire, and later of the British Commonwealth, was that any British subject, of whatever race or colour, had the right to come to the United Kingdom, without restraint or control. This was accepted by all political parties. The 1905 and 1914 Acts referred only to aliens. It was not until after World War II that circumstances arose to challenge the permanent acceptance of the ideal.

As Hilary Arnott, Information Officer of the Institute of Race Relations, wrote in the introduction to "Facts and Figures 1972",

During the Second World War troops from all over the Empire and Commonwealth were stationed in Britain. For example, 7,000 West Indians enlisted in the R.A.F. The first real sign of coloured settlement from Commonwealth countries was in June 1948 when 492 West Indians arrived from Kingston, Jamaica, in the ship, Empire Windrush. But it was not until 1954 that many immigrants came from the West Indies to Britain. This was partly because two years before, the USA had restricted immigration from Jamaica, leaving Britain as the only major industrial country where West Indians could come and find work. But it should also be remembered that those who came to Britain were brought up in a British Colony with British culture, learning British history at school, and regarding England as the "mother" country. The Government did not control this immigration, and the only direct arrangements to recruit labour were made by such organisations as London Transport and the catering industry, but the numbers of West Indians coming to work did relate to Britain's need for more labour. This situation only changed with the threat of immigration controls.[3]

By 1961 it was not the West Indians who caused the minds of many people of all political parties to consider the necessity of controlling immigration. It was the rapidly growing influx of immigrants from the Indian sub-continent, an area with a staggering annual population increase. Those in Government, whether national or local, whose duty it is to provide for the housing, education, health and social services of the people, realised that an unlimited increase in the number of the people to be served would wreck all the services. This would be as disastrous for the immigrants already in Britain as for the indigenous British. Naturally, with the talk of control, there was a beat-the-ban rush, strengthening the conviction of those who felt it would be irresponsible to take no action. The figures quoted in *Colour, Citizenship and British Society* by N. Deakin, B. Cohen and J. McNeal show the net inward migration from India, Pakistan and the West Indies (number of people entering UK minus the number of people leaving) from 1955 to 1968:

	Source of Immigration			
Net Inward Migration	India	Pakistan	Jamaica	Rest of Caribbean
1955–60	33,070	17,120	96,180	65,270
1961–30 June 1962	42,800	50,170	62,450	35,640
1 July 1962–December 1968	124,260	78,670	32,700	31,310

Asians and West Indians were not the only immigrants wanting to come to Britain. Under the Acts of Parliament granting independence to Tanzania in 1961, Uganda in 1962 and Kenya in 1963 anyone who, before independence, was a citizen of the UK and Colonies would retain that citizenship provided

(a) he did not on the appointed day become a citizen of the new Commonwealth country under its own law, or

(b) he, his father or his grandfather had not been born in that country.

Those who did not obtain local citizenship remained citizens of the UK and Colonies, and therefore were entitled to UK passports. It was not necessary to give specific undertakings to East African Asians, because it was well understood that under the law of the UK at that time they were not subject to immigration control.

Up to 1967 the flow of immigrants was not significant, but in 1968 the flow of UK passport holder immigrants increased, so that the Labour Government found it necessary to extend control to them in their Immigration Act.

During the ten years from 1961, when the first Immigration Bill was under discussion, to 1971 when the last Immigration Bill was going through the House of Commons, the motive of legislation has not been race discrimination, but the logic of arithmatic. During the debate on the Second Reading of the Bill on 16th November, 1961, the Conservative MP Nigel Fisher said,

I oppose the whole underlying idea of this Bill. Not only it brings to an end the long tradition of free entry for all Commonwealth citizens, but I believe it is quite inimical to our whole concept of a multi-racial Commonwealth.

John Hare, Minister of Labour, said,

The Government and others who are prepared to support the Bill do so as an unpleasant duty in the face of facts which to them are unanswerable... We should not forget our obligations towards the immigrants themselves. I do not think it would be right for us to sit back and allow Commonwealth citizens to come into this country in vastly increased numbers unless we are satisfied that reasonable living conditions are available for them.

In "A Better Tomorrow", the Conservative Party manifesto for the 1970 general election, it was stated,

We will establish a single system of control over all immigration from overseas. The Home Secretary of the day will have complete control, subject to the machinery of appeal, over the entry of individuals into Britain. We believe it is right to allow an existing Commonwealth Immigrant who is already here to bring his wife and young children to join him in this country, but future work permits will not carry the right of permanent settlement for the holder or his dependents. Such permits as are issued will be limited to a specific area for a fixed period, normally 12 months.

In the election the Conservative Party was returned to power and the Immigration Act of 1971 followed closely the pledge that had been given in the manifesto. The Act came into force on 1st January, 1973, the same day that Britain became a member of the European Economic Community. This co-incidence focused attention on the complexity of law making. The purpose of the Immigration Act was to control the number of individuals with the right to settle in Britain, without discrimination on grounds of race, but also without depriving patrials of their unrestricted right to return to their own country.

The purpose of the Treaty of Rome in respect of the free movement of people was to allow the citizens of the European Community to settle anywhere within the boundaries of the Community. The nationality laws of each member country, therefore, had to define clearly who were the citizens of the Community, without discrimination on grounds of colour. In such a situation, involving new concepts, it was inevitable that some people would feel that the new laws discriminated against them and their interests. This question of attitudes and belief will be examined in the next chapter.

Probably the most serious criticism of the Immigration Act, and one which certainly troubled the conscience of many Britons, was its retrospective effect, conferring on the Home Secretary the right to deport illegal immigrants who had entered the country before the Act became law. The Law Report in *The Times*, 12th June 1973, explains the legal position as clarified in the House of Lords on the previous day in the case of the Director of Public Prosecution v. Azam and others:

The House of Lords rejected appeals by three illegal immigrants against detention orders under which they may be removed from Britain. They entered clandestinely before 1st January, 1973 when the Act (1971) came into force. They contended that the Act was not retrospective and did not apply to immigrants already settled in Britain *before* it took effect. However Lord Wilberforce said that Parliament could make a law retrospective if it wished and "there was no doubt that it had done so here". The first question which arose was whether the applicants were *illegal immigrants*. Lord Wilberforce ruled that if the appellants were otherwise within the description of *illegal entrants* the Act must apply to them notwithstanding that they had entered illegally before it came into force, and that it had not been argued in the House that the Act had no retroactive force. The second question arose, however, from the fact that even if they were *illegal entrants* under Section 1 (2) of the Act, a person was treated as having indefinite leave to remain if, at the date when the Act came into force, he was *settled* in the UK. Lord Wilberforce said that broadly what Parliament did was to provide that those lawfully here were to be treated as having indefinite leave to remain. Others, if they did not get permission to remain – and the Act provided for their doing so – were liable to be removed. The Act defined the word *settled* so as to exclude those who were in breach of the Immigration Laws – thus conferring on the Home Secretary a new power to deport at any time a person who had come here in defiance of the Immigration Laws then in operation. It follows that *no one will be penalised for doing something which was legal when it was done, and only subsequently made illegal.*

The Home Secretary, Robert Carr, said in the House of Commons on 12th June that the House of Lords decision had confirmed what the Government believed the law to be, and what they had intended it to be. Shortly after the Labour Government took office following the general election of February 28th, 1974, they decided not to operate the Act in respect of illegal entrants before 1st January 1973. The legal situation, however, remained complex and confusing. On 23rd June, 1974, a Staff Reporter on the *Observer* commented:

Many illegal immigrants to Britain who came out of hiding after the Home Office offered an amnesty in April have now learned that they are not covered by the amnesty's terms – and they could be deported at any time.

As a result, the Joint Council for the Welfare of Immigrants is today urging anyone who plans to give himself up, in the belief that the amnesty will protect him, to consult the Council's London HQ before he does so...

So far only 667 immigrants – about 1% of the illegal population – have surrendered. Of those cases the Home Office has examined 88 and has found that only 21 of them have the right to stay here under the terms of the amnesty. In other words, three-quarters of them have moved out of hiding and into trouble...

An official (of the Home Office) added: "The amnesty does not apply to stowaways, seamen, or over-stayers, or to illegal immigrants who entered the country after 1st January 1973".

On 24th June in the *Daily Telegraph* John Grigsby added the information that the Secretary of the Joint Council, Mrs. Mary Dines, thought that there could be 50,000 overstayers, half from the new Commonwealth and Pakistan, and most of the others from Australia, New Zealand, America and Europe.

Two other events in 1972 had direct bearing on the situation of

Asians in or coming to Britain. The end of the Pakistan/Bangladesh war and President Amin's expulsion of Asians from Uganda. The war resulted in Pakistan leaving the Commonwealth and Bangladesh becoming an Independent Republic within the Commonwealth. The Pakistan Bill, debated in the House of Commons on 22nd May 1972, had as its purpose to amend UK law to take account of the withdrawal of Pakistan from the Commonwealth on 30th January, 1972. The effect on Pakistanis in Britain was to turn them into aliens. Transitional arrangements were written into the Bill to ensure that qualified Pakistani citizens were able to apply for citizenship of UK and Colonies. Those who were ordinarily resident in UK before 30th January, 1972 would be able to apply for British nationality when they had lived in Britain for at least five years. Generally aliens are unable to vote in elections or to enjoy the other civic privileges accorded to Commonwealth citizens, such as service on local councils and employment in the Civil Service, the police and armed forces, but for six months after the Royal Assent was given to the Act, Pakistanis continued to enjoy these privileges and, additionally, were able to vote in local and national elections held before 16th February, 1974.

It has been estimated that there are between 100,000 and 130,000 immigrants from former Pakistan who have not registered as citizens of UK and Colonies. This comprises some 60,000 to 80,000 from present Pakistan and between 40,000 to 50,000 from Bangladesh, not including children born in the UK.

When the Asian community was expelled from Uganda the sincerity of British faith in the ideal of the multi-racial Commonwealth was put to the test. Having just passed an Act of Parliament which recognised how essential it was to ensure that immigration should not be the cause of overcrowding, the country was confronted with the urgent request to open the door to something over a quarter of a million people immediately. This cruel and blatant example of racial discrimination on the part of Uganda evoked an unequivocal response. From all political parties and from all social classes came support for the Government's decision to accept responsibility for these unfortunate homeless people. Ad hoc committees were set up. Voluntary organizations called on their members. Dustmen and duchesses alike got down on their knees to scrub floors to make clean and ready reception centres. Other countries, particularly other Commonwealth countries, rallied round to offer jobs and homes. The British do not accept arbitrary discrimination on grounds of race and they made their rejection plain. On the whole the resettlement plans worked well in spite of some short-comings. Without the help of other countries the problems would have been far greater. The experience revealed in stark terms the human suffering and injustice which can be inflicted by ruthless racial policies.

Speaking at a public meeting on 2nd February, 1973, the Home Secretary said:

I do not believe that anyone at home or abroad, looking at the problem in a realistic way, could reasonably expect this or any other country to absorb another such influx. Apart from any other consideration, to do so would cause grave damage to community relations in Britain... While we would continue to honour our obligations in an orderly way, and would look with sympathy at any individual compassionate cause, it was just not possible for us to take on responsibility for accepting another expulsion of a complete community... It is particularly important that children of immigrants born and educated here, should enjoy full equality of opportunity in jobs, in homes, and, in the real meaning of the phrase, *full citizenship* with other young people.

Eighteen days later when the rules governing immigration were debated in the House of Commons the Home Secretary said:

The new rules now, as we see it, complete an overall policy for the future immigration into this country. As we see it, they are based on three principles.

The first principle is the recognition that Britain is a crowded island with a labour force which, for the moment at least, appears ample for her needs, and, therefore, that there must be restriction of all permanent immigration to... the inescapable minimum, and we must have the establishment of effective controls to achieve this.

The second principle is to recognise, within this overall need, a continuing responsibility to those in various parts of the world who remain entitled to UK citizenship. Hence the decision to reserve for these people as large a proportion as possible of the limited numbers we are able to admit year by year, and to do so in a controlled and orderly way.

The third principle is to recognise that it is both right and natural to give easy access to Britain to all those with close and recent family ties with this country, provided that they create no significant pressure for permanent settlement.

Those three principles hang together to create an immigration policy which, within our overall need to limit severely permanent settlement in this country, is reasonably humane, and reasonably and responsibly treats individuals, but also recognises our obligations to other countries.

On the whole the resettlement plans worked to the mutual advantage of immigrant and host, but not without some painful adjustments by both. On 7th February 1975, Peter Evans reported in *The Times* that:

In the next three years most of the United Kingdom passport holders in East Africa should have been allowed to enter Britain. That is expected to be the effect of an announcement in the Commons yesterday by Mr. Jenkins, Home Secretary, of an increase in special vouchers available to United Kingdom passport holders for entry to Britain.

It means that a long-standing obligation by successive governments is now in sight of fulfilment.

The number of special vouchers allocated annually is to be raised from 3,500 to 5,000. The official reckoning is that there are three dependents to each head of family. That would mean a possible entry of 20,000 United Kingdom passport holders and dependents each year.

Although a proportion of special vouchers are available for United Kingdom passport holders in India, most come from East Africa.

The latest official estimate of numbers of United Kingdom passport holders in Africa with no other citizenship is: Kenya, 23,000; Tanzania, 13,500; Malawi, 13,000; Zambia, 7,000; Zaire, 250.

Mr. Jenkins made clear in his reply yesterday that one reason why he was increasing the number of special vouchers was to avoid any possible repetition of the sort of crisis provoked in Uganda.

That fear has been haunting successive governments, and political announcements in East Africa have been watched like a seismograph to see if any increased pressure is being put on Asians to leave.

Great efforts have been made by the Government (of whatever Party) to formulate just laws and to instruct the Civil Service to operate them with common sense. No one is satisfied that in every respect the laws to date have got the balance completely right. Pressure is building up for further legislation to give greater power to the Race Relations Board to go into factories and offices to investigate discriminatory practices, without waiting for an aggrieved individual to lodge a complaint with them, himself, or through his representative.

On August 1st, 1974, the Commons Select Committee on Race Relations and Immigration (all-party) published their Interim Report. They said that before extra powers of compulsory investigation were granted there ought to be more evidence that firms could not be persuaded could not be persuaded voluntarily into taking action to promote equality of opportunity. The Report added:

The record of the TUC is similar to that of the CBI, in that both organisations have declared their opposition to racial discrimination but have taken wholly inadequate steps to ensure that their members work effectively to eradicate it.

A. J. Travers, reporting in the *Daily Telegraph*, wrote on 2nd August:

Mr. William Deedes, Conservative M. P. for Ashford and chairman of the committee, said at a Press conference: "Our main recommendation is that the CBI, the TUC and the Government ought to apply themselves a little more energetically to fulfilling the purpose of the 1968 Race Relations Act.

"Let us try voluntary methods rather more energetically before we give the Race Relations Board strong powers that they would undoubtedly like to have.

"We have not, however, set our face against extra powers for the board – what we say is that the case has not been made out at present for this. But if nothing has happened in a year from now, there will be no answer to the board's request for stronger powers to enable them to do their job."

E. Attitudes and Prejudices Affecting Race Discrimination

In examining laws and their purposes there are well documented facts available. Attitudes and prejudices are a matter of opinion and degree. Another complication in attempting to quantify or evaluate them is that the subconscious attitude or prejudice can be as relevant as the conscious.

Attitudes are derived from personal experience and observation. Prejudices are derived from concepts accepted as the common experience of others. Put together they take on the appearance of *fact*. In other words, what people *believe* to be true becomes *truth* for them. Two very powerful human emotions combine to drive people to seek for this kind of truth, fear and the hunger for power. Both emotions have their virtues and their vices, their constructive and destructive aspects. Fear alerts the mind to approaching danger. Without this state of alertness the way of human progress would never be plotted. The desire for power is essential if any effective action is to be taken. Where fear does not inspire the quest for improvement it becomes the source of entrenched discrimination. Where the desire for power becomes the overriding passion truth and progress disappear, and all the evil side of fear is exploited.

No study of attitudes about race in Britain could omit reference to Enoch Powell. Two facts about him must be taken into account. His mind is unusually dominated by the force of logic, and he is a great classical scholar. To obtain an insight into the effect of classical studies on his character and means of expression one should read his poetry. As a member of the Conservative Party he represented the Parliamentary constituency of Wolverhampton South West, part of the West Midlands conurbation. It is, as the records show, an attractive area for immigrants. When Mr. Powell met his constituents, White, Black or Brown, listening accurately and sympathetically to their daily problems, his logical mind immediately began to work on the significance of the changing trends in the size of ethnic groups. His vivid imagination and human sympathy painted for him a picture of the trends in human problems as well as in statistics. At first he spoke his warnings in moderate terms but it seemed to him that no one was taking any serious notice. Then he turned to the language of imagery familiar to him from the classics. With this he hit the headlines. His dire prophecies of rivers of blood, etc., alerted fear on both sides. Many Whites were fearful lest their home country should be swamped by Asians and Africans, who would take over power and reduce the standard of living to that of the overcrowded areas of the continents which they had left. While Africans, Asians and West Indians feared that they would become the victims of White fears and hatreds, that they might be expelled from Britain or made to suffer every kind of indignity and discrimination.

While this cauldrom of fear was on the boil there were many who charged that Enoch Powell was using the race issue as his own path to power and to take over the leadership of the Conservative Party. Others pointed out that the men who were really using the race issue as their path to power were the extreme Left groups, exploiting the fears of immigrants to disturb community relations and undermine confidence in the police and the establishment. Whether or not there was any truth in these accusations is not important, since there is not a scrap of evidence to suggest that such tactics were likely to meet with success.

It is true that thousands of people look on Enoch Powell as a great leader who has alerted millions to the social dangers that were developing. They believe that by his courage and his oratory he has enabled both major political parties to introduce legislation to avert the impending tragedy. Many of his admirers are far from satisfied that as yet sufficient action has been taken. These admirers are drawn from all classes and from most political parties or from non-partisans. A very similar cross section, this time including the ethnic groups, look on Enoch Powell as the arch-priest of racialism. They believe his oratory has inflamed racial prejudice and fear, and made the human situation much more difficult than it need have been. There is ample evidence available to support either view of him. The one charge that is some times made against him which is certainly not true is that he is personally a racist, disliking or hating peoples of other races. He has nothing in common with Hitler, and he has never been known to preach the superiority of the *Herrenvolk*.

Since the General Election of February 1974, when he gave up his constituency, and publicly declared that he no longer supported the Conservative Party, the charge that he aspires to the leadership of the Party must also be discounted.

Whether or not Enoch Powell should be applauded for the good that he has done or reviled for the bad, it is a fact of history that he has had some influence on public opinion in a decade when the British have had to re-think their cherished belief that free entry to Britain was the right of every Commonwealth citizen.

In the decade 1963–1973 the United Kingdom absorbed three quarters of a million Commonwealth citizens as immigrants. The Race Relations Board, set up in 1965, and the Community Relations Commission, set up in 1968, have brought many people of many races to work together for the common good. In 1973 The Race Relations Board published a progress report under the title of *A Project On Race Relations*. This puts the problems of today in perspective by tracing the movement of peoples through the centuries, concluding that

On the whole, people only leave their homes and uproot themselves because they are dissatisfied with the situation into which they have been born. They want a better life for themselves. Above all, they want a better future for their children. Most immigrants, wherever they come from, have at least these inclinations in common. Another thing they share is the initiative to get up and go, the will actually to do something about an unsatisfactory state of affairs. Immigrants from the Commonwealth present few exceptions to these general observations. But to them must be added another common factor. Most immigrants from the Commonwealth are coloured.

The report analyses the human impulses which distinguish between the formation of personal opinion, which is a human right, and the development of unsocial behaviour, which is infringement of the rights of others. The purpose of legislation, to quote Roy Jenkins as Home Secretary in 1966, was

...not a flattening process of assimilation but equal opportunity accompanied by cultural diversity in an atmosphere of mutual tolerance.

The duty of the Board is to investigate complaints of prejudice and

unfair treatment, and when unfair treatment is established to decide
whether the cause was due to race prejudice or to some other reason.
One of the most serious difficulties encountered by the Board is not
that people complain about unfair treatment that has nothing to do
with colour, but that people are extremely reluctant to complain at
all. The report analyses the cause of this.

Recognising that in a democratic society any law to work must be
based on a large measure of public consent, they conclude that,

The Race Relations Act of 1968, despite the controversy which still sometimes surrounds it, is based on enough public support to enable it to do the
job for which it was designed.

This conclusion, however, should not be taken as a sign of compla-
cency, nor that public support is as yet sufficient to achieve full success.
The 1974 annual report of the Race Relations Board claims that the
Coloured people are not convinced of the public's determination to
root out discrimination, and it makes a powerful demand for stronger
Government action. It warns that:

It is vital to remove this uncertainty form the young 'second generation' of
coloured people, now starting work and setting up homes of their own. The
treatment they receive and the degree of confidence they can reasonably
have in being accepted on equal terms with whites will, to a large extent,
determine the course of race relations in the future.

The report finds that the crude and obvious forms of race discrimination are much less common, but particularly in the area of jobs and
housing there is still much to be done. It asks for power to start its own
investigations where it suspects discrimination, instead of having to
wait for a complaint to be made. During 1973 the Board registered 792
complaints and 93 investigations as compared with in 1972, 845 complaints and 68 investigations.

The report claims that Coloured people do not complain because
they feel the Act gives them little redress, and that complaints may expose them to victimisation.

F. Ethnic Groups and Employment

Sir Geoffrey Wilson, Chairman of the Race Relations Board, told a
Press Conference in London on 19th June, 1974, that progress towards

ending race discrimination in employment would remain far too slow until the Board was armed with stronger legislation. As stated in the annual report, the Board had been investigating two top firms. In one case the Board was told that there were no Coloured supervisors, and in the other the numbers were very small in relation to the number of Coloured workers.

The report says that managements are often reluctant to take positive steps towards equal opportunity because they fear hostile reaction, but in many cases their fears are based on untested assumptions about the attitudes of the unions and the workers, a situation which would be greatly eased if the unions were to show more determination to ensure that all workers should be treated equally.

The same week as the publication of the annual report, an independent research group, PEP (Political and Economic Planning) published *Racial Disadvantage in Employment*,[4] by David J. Smith. This provides extensive evidence to support the need of greater power to be given to the Board to initiate investigations. It accuses both management and unions of unwarranted complacency. With the Labour Government in power, and Roy Jenkins again Home Secretary, this report may provide the evidence for stronger legislation. While there is general agreement between all Parties that equal opportunity for promotion is the right and proper aim, there is some strong difference of opinion as to whether more state intervention is the best way to achieve it.

As reported by Peter Evans, the Home Affairs Correspondent of *The Times*, the PEP publication is based on a survey of nearly 300 plants, in districts where a large number of immigrants are settled. Most workers from the minority groups (74%) are concentrated in only 28% of the plants. More than a fifth of the plants surveyed employed no one from the racial minority groups.

David Wilson, the Labour Correspondent on *The Observer*, reviewed the current situation on 11th August, 1974:

After years of prevarication, the trade union movement is at last taking positive steps to eliminate racial discrimination from its own ranks. This follows an increasing number of strikes in which black workers have been set against white.

The TUC is now ready to recommend that unions should put clauses on equal opportunity into their collective agreements, thereby committing members and managements to protect the rights of minorities.

In the Midlands, the Association of Scientific, Technical and Managerial

Staffs has reported the Walsall Labour Club to the Race Relations Board for operating a colour bar. This is the first time the labour movement has turned on its own kind, to see racial justice done.

The TUC's old attitude was one of laissez-faire. It held the cosy philosophy that black and white workers were united at heart by the struggle against capital and that this common bond transcended the differences of colour or creed. It treated the Race Relations Act with coolness, in the belief that special attention would merely emphasise special problems, and it believed voluntary persuasion could overcome any difference of interest between blacks and whites.

But it has been traumatised by the shop floor unrest of recent months.

The first flashpoint was the strike of 400 Indians at Mansfield Hosiery Mills in Loughborough, in November 1972. They walked out because white workers would not let them into lucrative jobs as knitters. The local chairman of the Hosiery Workers Union threatened a counter-strike if the whites were "flushed out" of the knitting jobs.

At the same time, another TUC union, the Yorkshire Power Loom Overlookers, was being examined by the Race Relations Board over claims that it kept Pakistanis out of apprenticeships. Then at Smethwick, white foremen walked out in support of a colleague who was transferred after antagonising Sikh workers in a foundry.

At Harwood Cash, Mansfield, Notts, the Race Relations Board was called in again when Pakistanis complained about their employment contracts, which were for 60 hours a week while whites were hired for 42 hours.

A year ago, 60 West Indians walked out from a Standard Telephone and Cables factory in North London, when white electricians refused to train one Jamaican as a skilled machine setter.

In each of these cases the coloured workers bitterly criticised their unions for inadequate support. At the recent strike of Asians at Imperial Typewriters in Leicester, a grievance over bonuses rapidly changed into a fight against the Transport and General Workers' Union.

The Asians make up 60 per cent of the labour force, but only two of the 17 shop stewards are coloured: the convenor... is regarded as totally unsympathetic to their cause.

The strikers eventually appealed above the heads of local TGWU officers to union headquarters. Mr. R. Surani wrote that new elections of shop stewards was the first demand: "We make these points to point out how orphaned workers can be without the backing of their union...

At Harwood Cash the Board found nothing wrong in having one shift manned entirely by Pakistanis on 60-hour contracts while the white were on 42 hours, because the Pakistanis voluntarily accepted these terms. Imperial Typewriters was given a clean bill of health, because the complaints were of industrial, not racial, origins.

The Board's job is made harder because the nature of discrimination is changing. The overt forms – such as when the TGWU in Bradford agreed to limit the number of immigrants on the buses to 20 percent or a Coventry hotelier charged coloureds double for their beer – are giving way to passive discrimination, just as insidious but harder to counteract. "This discrimination is so hard to nail, it is the racialism that you feel but cannot overtly see," the strikers at Imperial Typewriters said.

Passive discrimination in effect means that equal opportunity is denied in promotion, recruitment or training often through neglect, not wilfulness. Thousands of coloured workers have qualifications greater then their actual jobs require; thousands feel they do not get deserved promotion or training for the better-paid jobs in factories.

Statistics of the Race Relations Board illustrate the point. Although complaints about discrimination in employment have dropped from 693 in 1968 to 364 last year, the proportion relating to promotion and training within these totals has doubled – from 61 in 1971 to 115 last year.

In other words, white workers have accepted coloureds in their places of employment but not in the better jobs.

"I'm afraid the unions' record is not clean," says Mr. Don Groves, a Birmingham officer of ASTMS who leads a Midlands campaign against discrimination. "In fact, it's a bloody awful record. Often, company management may be prejudiced but they don't discriminate – it's the shop floor which discriminates, and employers fall into line with what their white workers want."

The response of the union movement has been slow. But it is changing.

The vital moment came this summer, when the TUC was preparing evidence to the Select Committee. By a historical quirk the subject is covered by the TUC's international committee – as if race relations stops at Calais – instead of the committee which covers trade union organisation.

In May a draft was presented which reiterated the old TUC argument that intervention merely emphasised racial differences, when the need was to patch them over.

But the committee, under the guidance of Mr. Jack Jones, the TGWU leader, rejected it. It insisted on more positive conclusions, and in July it agreed to back the Race Relations Board's plea for greater investigative power and to instruct unions to bargain for equal opportunity clauses.

This is hardly likely to be enough. The Select Committee was highly critical of the "lack of urgency with which the TUC has approached the subject," and of the fact that the TUC had no separate race relations department...

Equally, there is little evidence that local officials and shop stewards have learnt how to absorb coloured workers into the mainstream of the Labour movement.

The institutional pattern for race relations has been mapped out by the select committee. It is likely that the Race Relations Act will be amended next year, to give the Board subpoena powers and to abolish industrial machinery.

Apart from the type of industrial case referred to in David Wilson's article, and many other individual cases taken up and resolved by the Race Relations Board, there are other aspects of discrimination relating to employment. The Sikhs, on religious grounds, require the right to wear the turban in place of the uniform cap of a bus driver, or the safety helmet worn by motorcyclists. Countless hours of discussion and negotiation have been spent on this issue, from the House of Commons to the shop floor. In Leeds, where two Sikhs were bus drivers, a Union official took the view that it was not a *racial* issue, but one simply concerned with uniform.

Michael Parkin wrote in *The Guardian* on 28th October, 1974:

"Some of the lads," he said, "are arguing that it would be religious discrimination to let them wear the turbans because it would be discriminating against everyone else."

The union is now to discuss the uniform problem with the management. No one seems to have thought of applying the soothing ointment that is said to have worked in an earlier turban dispute. The chairman of the transport committee decided that any busman – Briton, West Indian, Pakistani, Sikh, anyone – should be entitled to wear a turban, if he so desired.

A member of Equity complained about the lack of parts written for Coloured actors, and the consequent lack of opportunities for employment in television. William Raynor explained the extra hazards Black actors and actresses must face, in a profession which is notorious for unemployment, in his article "Caucasian Vicious Circle" published in *The Guardian* 23rd August, 1974. When the physical appearance of an actor is relevant in casting for leading roles the opportunity to acquire experience in leading parts depends largely on the number of playrights producing stories requiring a Black, Brown or Yellow actor or actress.

On 11th February, 1975 the Runnymede Trust and the Workers Education Association published a report, *Trade Unions and Immigrant Workers*. The Home Affairs Correspondent of *The Times* commented:

The report says there is growing trade union militancy involving many black workers, who have become accustomed to the stresses of industrial life and have reconciled themselves to the fact that they are here to stay.

They are becoming much less tolerant of racial discrimination and are beginning to think of their future in Britain and to look for allies.

In many areas, the report says, there is unified effort by workers of all colours and nationalities. But elsewhere, demands by black workers for greater participation in union affairs are a direct challenge to some of the relatively undemocratic union structures.

Trade union membership is proportionately lower among black and immigrant workers, although in well organized factories and industries there is little difference.

Black and immigrant workers are under-represented at every level of trade union leadership. The report argues that some unions, by effectively discriminating against newcomers or the unskilled, make it more difficult for black and immigrant workers to play a full part.

G. Ethnic Groups and the Social Services

Housing

Because the first need of every immigrant is somewhere to live, the 1968 Race Relations Act dealt with housing in great detail. With certain exceptions the Act makes it unlawful to discriminate against a person on grounds of colour, race, or ethnic or national origins in the following circumstances:[5]

Selling

By refusing or deliberately omitting to sell a house, business premises, or land, on the same terms as are available to other applicants.
For example, it would be unlawful to deny to an Indian applicant the opportunity to buy the lease, or freehold, of any property simply on the grounds that he was an Indian.

Renting

By refusing or omitting to rent a property, or rooms, or lodgings on the same terms and conditions as are available to other applicants.
For example, it would be unlawful for anyone to refuse to rent a room to a Pakistani applicant on the grounds only that he was a Pakistani, or on the grounds of his colour. If the grounds of refusal were that children were not acceptable this would not be unlawful provided that this applied to *all* applicants and not only to a Pakistani applicant.

Occupation

By treating occupiers less favourably than other occupiers in similar circumstances.
For example, it would be unlawful for a landlord to refuse to a West Indian occupier access to facilities (such as use of a common bathroom) which were given to other occupiers on the premises having similar letting agreements, simply because he was a West Indian. It would not be unlawful to deny an occupant such facilities if it could be shown that these were denied to all occupants having similar letting agreements. Equally, it would be unlawful for a landlord to harass a tenant in order to get him to leave just because he wanted only West Indian people in his house.

In both cases there might be other legal remedies in addition to those provided under the Act. The Board would not be concerned with these.

Waiting lists

In the handling of waiting lists for accomodation, in the awarding of priorities or in the selection of tenants.
For example, it would be unlawful for a local authority to impose a one year residential qualification for persons born in the United Kingdom and a five year qualification for persons born elsewhere, as a pre-requisite to acceptance on the waiting list or an offer of accommodation.

Estate Agents, Accommodation Bureaux, Services, etc.

By refusing the same services and facilities in connection with sales and letting as are available to other persons.
For example, it would be unlawful for an accommodation bureau to refuse an applicant particulars of premises available, or to refuse to negotiate the rental for a room to let on grounds of the applicant's colour, race, or ethnic or national origins. It would be equally unlawful for an Estate Agent to refuse to negotiate the sale of a house on the same grounds.

Exceptions in Housing

However, none of the above forms of discrimination would be unlawful if, at the time of the disposal of residential accommodation, the following conditions apply:
The *landlord*, or a member of his family, reside and intend to continue residing on the premises *and*
the landlord or a member of his family share some part of the accommodation (other than storage or access) with other occupiers *and*
the premises are "small premises".
"Small premises" are defined in the Act as premises which normally contain accommodation for not more than two households in addition to the landlord's. For example, a house divided into three separate parts with the landlord living in one part and sharing bathroom or kitchen with one or both of his tenants. Or (where this description does not apply) there is not normally accommodation for more than 12 people in addition to the landlord's household. For example, a small boarding house, with sleeping accommodation for ten people, in which the landlord has his own rooms but shares a bathroom with his guests would be excepted. On the 26th November 1970 the number of people in addition to the landlord's household will be reduced to 6 under this exception.
In these exceptions the *landlord* means the person who has the power to provide or dispose of the accommodation.
It is not unlawful for a person to discriminate in the sale of a property owned and wholly occupied by him if he does not use the services of an estate agent for the purposes of the sale, or use advertisements or notices.

Advertisements and Notices

It is unlawful to publish or display any discriminatory advertisements or notices even if the subject of the advertisement is not unlawful. For example, an advertisement or notice announcing "Room to let, no children, no coloureds" would be unlawful even though the accommodation concerned might come within the exemptions mentioned above.

Incitement, etc.

It is unlawful deliberately to aid, induce or incite another person to discriminate unlawfully. For example, a person or persons encouraging a vendor to discriminate in the sale of his house could be acting unlawfully under this section.

Liability

The housing provisions apply to local authorities as well as to private landlords and agents acting on their behalf. An employer is liable for discriminatory acts done by his agent or an employee in the course of his employment, unless the employer can prove that he took such steps as were reasonably practicable to prevent his employee from doing such acts.

Complaints

Complaints of discrimination must be made, normally within two months of the act complained of, to the Race Relations Board or one of its regional conciliation committees. The complaint will then be investigated. If, after investigation, the Board or committee form an opinion that there has been unlawful discrimination they must notify the parties of their opinion and try to get a voluntary settlement of the differences between the parties, and where appropriate, a satisfactory written assurance against further discrimination.
It is only when a settlement cannot be reached, or the respondent refuses to give a satisfactory assurance, that any question of court proceedings arise.

Proceedings

These are civil and not criminal proceedings. The Board (and only the Board) can take proceedings for enforcement in one of the county courts (in Scotland, a sheriff court) designated to deal with them. When considering cases under the Act the Judge will be assisted by two assessors with special knowledge and experience of problems connected with race relations. The Board can also seek damages for actual loss and for "loss of opportunity" (that is compensation for any benefit which a person might have expected to receive but for the act of discrimination).

It was hoped that with these provisions in the Act immigrants would be protected from discrimination in respect of housing. Community

Relations Councils were expected to keep in touch with immigrants and ensure that they knew what their rights were under the Act. "Keeping in touch" is not easy, and requires many hours of work.

Immigrants may arrive with a fear of authority and a general conviction that it is safer to rely on the advice of compatriots than to put questions to official bodies (even when some of their compatriots are members of these bodies). The advice they receive may be influenced by the suffering experienced by illegal immigrants, who first have been fleeced by unscrupulous international operators providing false papers and miserable transport facilities, then fleeced again, largely by their own compatriots, who rent them overcrowded and inadequate accommodation at a greedy price. On top of this there is the constant fear of blackmail – pay up or word can be passed to the police to pick up.

The Asian peasant, unfamiliar with the language, totally without experience of urban living, understandably has difficulty in assessing whom to trust and what services are at his disposal.

Whether fluent in English or not, Muslim immigrants from any territory bring with them beliefs requiring some discrimination, and which impel them to try to live in segregated communities. To respect their human rights while meeting their needs for housing and access to the Social Services imposes immense strains on the whole community. Where human rights conflict, someone has to arbitrate. It is important to understand the dilemma implicit in recognising the difficulty of reconciling different values within one community.

In his study *The Indians of Uganda* H. S. Morris writes:

For Muslims society exists so that men can live correctly, and have a proper relationship with God.[6]

In their book, *Colour and Citizenship, A Report on British Race Relations* E. J. B. Rose and Associates follow through the implications of this faith:

This Canon law defines and governs rules of descent and kinship, inheritance and succession, marriage and divorce, as well as social, economic and political relations amongst believers. Thus, for the Muslim religion, law and social organisation form an inseparable whole, governing not only religious practice and morality, but social relationships, diet and hygene, and those areas of conduct which in Western society are regulated by secular law and civil authority.

Pakistani emigration became highly organised on a system of sponsorship and patronage, which made it selective and confined it to comparatively few villages within a few areas. We have seen the economic success of pioneer Pakistani immigrants, frequently merchant navy deserters. They induced their kinsmen to jump ship and join them. These settlers later sponsored other kinsmen by arranging for their recruitment as seamen. When recruits exceeded demand the Pakistan Government began to restrict the issue of passports, travel agents emerged to fulfil the new and profitable role of supplying forged documents... The organisation of this traffic depended to a large extent to there being sponsors in England who could provide shelter and an introduction to a job, and in many cases, money for the fares.

The effect of the 1962 Commonwealth Immigrants Act was to perpetuate the selective process of immigration, as kinsmen and fellow villagers stood an even better chance of being sponsored under the voucher system than those who had no kinship link with Britain.

A very close reciprocal relationship is set up between sponsor and client. Some form of sponsorship by friends and relations in Britain on a basis of village-kin ties is a necessary insurance for the immigrant against possible hardships during his early days in Britain. This, as much as his desire to be with his kinsmen and friends, explains the emergence of village-kin groups in Britain, which has become a pattern of settlement...

The Pakistan sponsor in Britain becomes the patron of all those he has sponsored, no matter how widely dispersed they might be.

In Pakistan men and adolescent boys spend their leisure hours in the company of male members of the *baradari* at the *baithak* (a kind of club or resting place). Here a family's male guests will often sleep with their host or other male members of the family. Among Pakistani immigrants in Britain the sponsor and his "family", or house group, are the substitutes for the traditional functions of social control exercised at home by the *baithak* and the *baradari*. Within the all male dormitory houses the Pakistani is reminded of his primary obligation, which is to discharge his debt to his sponsor or to his family, who have probably mortgaged their land to pay for his passport and passage, and then to remit sufficient funds to Pakistan to sponsor other kin, and to improve the family's fortunes.

The Pakistani sacrifices material comforts to pursue his economic objectives, and the dormitory house with its low rent is an important means to this end...

Many Pakistanis remit 50% of their earnings. It was estimated in 1963 that remittances amounted to as much as £26 million, or more than the inland revenue of East Pakistan. The dormitory house also serves as a kind of insurance policy or provident society, for rent is excused to a kinsman who falls out of employment. It is, therefore, not surprising that a plan by Bradford City Council to provide hostels for single Pakistanis, with separate cooking facilities in each room (at double the rent of the dormitory house) failed to evoke any response.

Joint living in these dormitory houses with their group activities and total lack of privacy, ensures that the immigrant conforms to the norms of his community in respect of diet, thrift, recreation, language, avoidance of contact outside the sub-group, and modest hospitality to visiting kinsmen. Apart from visiting friends, leisure hours are spent in the house, or watching Indian or Pakistani films. This self-sealing process is carried over into working hours, especially if the immigrant is recruited into an ethnic gang, sometimes on night shift, where contact with employers is maintained through an English-speaking foreman.

These closely knit and self-segregated communities are served by their own shops and other services provided by fellow countrymen.[7]

No one can be surprised that social workers trying to achieve good community relations have difficulty in making contact and advising Asian immigrants what services are available for them. The problem with West Indians and Africans is different. They are mostly Christians, and have not, therefore, the same inducement to segregate themselves. Nevertheless, there is a need for better communication between the immigrants and the Local Authorities whose duty it is to operate the welfare services, including the provision of homes. The House of Commons Select Committee recommended that the Statutory Community Relations Commission should play a more active part in the task of informing immigrants. Many voluntary bodies try to bridge this gap in communications. The motivation for this activity is as mixed as are the individuals who participate in it. Action stems from religion, all Churches; politics, all parties; social service, spontaneous and organised. The result, perhaps, adds as much to enlightenment as to confusion.

Two starkly contrasting examples illustrate the kind of effort which is being made to enable immigrant families to find suitable homes. One seeks to undermine any confidence the immigrants might have in the Local Authorities, and the other explores means to improve and stimulate justifiable confidence.

First, the Black Panthers' line of action. In the broadsheet they publish every fortnight, *Freedom News – north and east London community voice,* they claim:

Throughout the years, from our own involvement and experience, we have learnt to organise in many different areas of black people's life: in housing, education, the work situation, children's creative workshop, youth collective, police courts and general legal advice, immigration and against the new aliens Act.

The line they take on housing in their issue of 16th June, 1973, is to turn the spot light on the suffering of Black families in over-crowded accommodation. (Every voluntary and statutory body working in this field would agree that over-crowding is a social evil which must be tackled, whatever the colour or ethnic group of the families concerned.)

They quote a Mr. A as saying:

"I've been living in this house ever since I came to this country eight years ago, and it is the worst that I have ever seen it. It's worse than when Mr. Best the private landlord had it. You know a month before Mr. Best sold to the Council, a health inspector came here. He said that the landlord had to put in another bathroom and toilet. A little later on when we asked him what was going to be done he said 'nothing, the Council owns the place so there is nothing I can do'."

The editorial comment on this and similar interviews adds up to advice not to co-operate with Local Authority:

It is clear from the experience of the brothers and sisters in Wray Crescent, that the Council does not want to take responsibility for housing people if they can help it. Their policy in acquiring houses is to make the landlord force out all the tenants, so that they won't become the council's responsibility. For every tenant remaining in the house, the Council's policy is to reduce the price offered to the landlord...

Brothers and sisters, we have paid more than enough of our hard earned money in rent for damp houses. We are entitled to a good place to live, in an area of our choice, at a rent we can afford. But we wont get it by hoping and waiting. The Black Panther Movement is calling on all brothers and sisters... to organise and fight.

The contrasting approach is clearly explained in the well-documented report of *The Fair Housing Experiment. Community Relations Councils and the Housing of Minority Groups*. This report is a study of three experimental projects sponsored by the Community Relations Commission and financed by the Calouste Gulbenkian Foundation. The Community Relations Commission asked Political and Economic Planning to examine the progress of the three local Councils for Community Relations – Manchester, Nottingham and Sheffield – in helping immigrants with their housing problems and to report on the efficacy of continuing the Fair Housing Experiment in its present form in these three towns and to extending the approach to other Councils for Community Relations. The study was carried out between autumn 1971 and summer 1972. The report was written by Jane Perry, a member of the Political and Economic Planning research staff.

Mark Bonham Carter, Chairman of the Community Relations Commission, writes in his foreword to the report:

This report is important because of the lessons we can learn from it. It is important also because experiments of this kind require an independent audit. Here, we have an operation launched by the National Committee for Commonwealth Immigrants, financed by a foundation and assessed by an independent research institute – PEP – which has no connection with either of the two bodies.[8]

Some of the lessons to which he draws special attention concern the need to influence the local council if any influence on policy is to be achieved. On some occasions the Fair Housing Groups found themselves in profound disagreement with the municipal council, as, for instance:

Fair Housing Groups and Community Relations Councils have had only limited success in persuading their respective local authorities to keep records of immigrants housed and on the waiting list. This, despite the agreement of both the Cullingworth Committee and the Parliamentary Select Committee on Race Relations and Immigration that only by keeping such records is it possible to monitor the success or otherwise of policies of fair housing in the public sector.[9]

The councils had very strong reasons for refusing to keep records. Housing Departments had an equal responsibility for helping *all* families in need of accommodation. If they gave their applicant families a *race* label this would in itself *imply* a race discrimination, either in favour of, and thereby against, one group as compared with another. Looking at the dimensions of the housing problem as defined by Mark Bonham Carter, the council's dilemma is understandable:

A high proportion of the "immigrant" population is living in the deprived areas of our inner city centres where they share with the indigenous population all the consequences of this situation: poor, decayed and overcrowded housing, social services strained to the limit, rundown schools with a high turnover of teachers, inadequate facilities for leisure and so on. These conditions have a direct effect on the employment opportunities of all who live in, or are brought up in, such areas. If you are fortunate enough to have a good job, you will try to move out of the inner city as soon as possible. But the chance of getting a good job is smaller if you live in such circumstances. Housing and employment, therefore, interact on each other and it is fruitless to argue which of the two is more important. Both have to be tackled.
But there can be no doubt as to the importance of housing.[10]

The PEP assessment of the work of the Fair Housing Groups covers their co-operation with other voluntary bodies, sharing similar aims to identify needs and persuade others, whose responsibility it is, to meet those needs. In this connection they extended their study to cover other cities where Local Authorities had set up housing advice centres, and to which many organisations were contributing.

Shelter Housing Aid Centre (SHAC)

SHAC is very different from the other groups we have looked at and only part of its work is strictly relevant in the context of FHGs as at present understood in this country. It is a regional service for the whole of London concerned with the particular problems of that city, and it deals essentially only with families: single people and elderly couples are referred elsewhere. The families it serves are those who are not catered for by statutory services...

Basically SHAC offers two types of service. It helps those who merely need information or basic advice, the name of an estate agent, the address of a local authority department, how to go about obtaining a mortgage, for example. These are people who with a little help are capable of solving their own problems and do not need an extended interview. But the bulk of its resources are devoted to cases in desperate housing need which require extensive counselling and practical aid at every step. There were 5,080 such cases in 1970... Most cases were then passed to one of three departments, each staffed by a team of professional workers: House Purchase, London Referral (dealing with rented accommodation in London), and Out of London (for those who wish to leave the capital), In 1970, an acceptable solution was found for 1,357 cases; but for 2,866 families there was no solution, either because contact was lost or the accommodation offered was unacceptable. Each week there are about 20 applications for emergency assistance, more than half the families being outside the scope of statutory service and some solution, however temporary, is found for such people before they leave the building.

The London Referral Department has the heaviest case load; during 1970, 2,373 families were referred to it, many of the cases originally coming from statutory bodies and other voluntary agencies. SHAC has managed to establish good relations with building societies, which recognise that a SHAC recommendation means that a borrower's credentials have been thoroughly investigated. One building society indeed, reserves a percentage of its loan money for SHAC clients and is more generous on borderline cases than it would normally be. SHAC is also able to underwrite bank loans if necessary.

The role of fair housing groups

It seems clear from the above that the advisory service provided by FHGs is in many respects very similar to that offered by a number of other bodies. No doubt the allocation of a specific sum to the subject ensures that it will get individual attention and will be regarded as of some importance, but the sums so far allocated have hardly been large enough to make much impact. At present no FHG could, however, hope to match the resources of a good local authority housing service; for example, Lambeth spends £40,000 a year on its centre and can offer a full range of advice on general and particular points as well as direct access to the housing department.[11]

Another aspect of the report to which Mark Bonham Carter draws special attention is the difference between Fair Housing Groups in UK and USA:

The work done by the Fair Housing Groups which are the subject of this study bears very little relation to that done by Fair Housing Groups in the United States. In the States, Fair Housing Groups work largely in the pris vate sector and are primarily concerned with helping middle-income blacks to move into middle-income housing. Public housing in the United States constitutes a tiny proportion of the total housing stock. In the United Kingdom it amounts to about 30 per cent of the total housing stock and in Nottingham to 45 per cent. Local authority housing is therefore a crucial factor in our situation and it was on this area that the three Fair Housing Groups concentrated and here that they achieved the greatest degree of success.[12]

On every page of this broad and candid study of immigrants' housing problems in the UK there is evidence to suggest that ultimate responsibility for improving inter-group relations must depend on the actions of the Local Authority. It should be remembered that every voter in the community has an equal chance to elect the members of that Authority.
The report concludes by emphasizing the need in the future to

...concentrate on the problems of the non-English speaking immigrant in terms both of face-to-face help and advice and of literature, leaflets, posters and so on, while encouraging the English-speaking immigrant to avail himself of existing services direct.[13]

Educational Services

The existing services, of course, cover the whole range of welfare, health and education. As all are administered within the area covered by the Local Authority, where the immigrant is housed has a direct effect on all the other services. This is particularly obvious in the case of education. Provided that immigrant families are not concentrated in a few districts in large cities, the children attending the schools within each district will be a fair mixture of indigenous White, second generation immigrant and new immigrants.

It is important to make the distinction between those children for whom English is their native tongue both at home and at school, and the new immigrants, mostly from Asia, who have first to acquire fluency in English before they can be placed in the class appropriate for their learning ability. The organisation of the state comprehensive school aims to take into account the learning capacity of the pupils, and to place them in a stream suitable for their ability. Lack of knowledge of the native language is not a measure of stupidity or mental limitation. If the fact that immigrant children require special teaching causes their school mates and neighbours to regard them as mentally retarded, the experience is both humiliating and frustrating for parents and children. On the other hand, if the new immigrants with language difficulty do not get special teaching, then the whole class may be held back to accommodate their pace of learning. This in turn generates resentment and stiffens prejudice against the new immigrants. Reconciling conflicting needs within a community demands all the human virtues of patience, tolerance and understanding.

The second generation of Muslim immigrants come up against a different clash of human rights when their daughters take their normal place within the school environment. The family in UK terms is something very different from the Muslim family, with different sanctions and disciplines as far as girls are concerned. The tendency in British schools is to encourage and extend co-education. Even in single sex schools it is customary to combine a number of activities so that boys and girls learn together. This is a practical form of education to equip children to take their place in a society endeavouring to eliminate sex discrimination. This places the Muslim teenage girl in a very distressing situation. She is born a British citizen, entitled to rights equal to those of other citizens. To take up her rights in full, she needs the education which will equip her to do so. In *Colour and Citizenship* the description of the Pakestani family reveals the hard choice she has to make:

The Pakistani family consists of man, wife, unmarried sons and daughters, married *sons* and wives and children. In this extended family authority is vested in the father, and passes on his death to his eldest son. From the age of puberty all girls are secluded from men who are not related... Women may work together in groups... Their lives are spent with other women... When she marries she enters her husband's family as a subordinate and an outsider; she subordinates her will not only to her husband and his father, but also to her husband's mother...

In Pakistan the birth of a son is greeted with great rejoicing – a girl unnoticed. He will work in the fields, when he marries his wife will bring a dowry, and he will support his parents in their old age. A girl is considered a debt to the family; she has to be protected in adolescence, and much money is needed for her marriage and her dowry. This pre-eminence of the male is reflected not only in his authority within the family, but in early separation of boys from girls within the home... Women never eat with their husbands, but only after they have served their meals. Men spend their leisure with other men, so are mostly absent from home during waking hours. This close relationship with their male kin is carried through to emigration in England.[14]

Whose human rights should prevail? Those of the father, impelled by his faith and traditional culture to protect his girl child from the evils of a society in which he has chosen to live, and to have her born a citizen? Or the little girl's rights to take her place in that society, with equal opportunity with her school friends of any race to equip herself for an interesting and fruitful career?

The Community Relations Commission produced a policy paper in 1974: *Educational Needs of Children from Minority Groups*. While the sincerity of the Commission's aspirations for the immigrant children is clear in every paragraph, the cost of the recommendations would be very high. Recruiting of additional qualified staff would also be something of a problem. Such experienced teachers are not unemployed, and immigrant children are not the only ones in need of the best teaching skills. At a time when inflation makes additional public expenditure counter-productive, consider the implications of the following:

a) The importance of language development as a pre-requisite of basic skill acquisition suggesting that action should include language training as part of the initial training of teachers.... To implement this, the following measures are suggested:

i) authorities with large numbers of minority group pupils ought to have on their staff several advisers knowledgeable about and experienced in the educational problems of minority group children;

ii) schools with heavy concentrations of minority group pupils should have specialist teachers of English as a second language and teachers trained to teach dialect-speaking children;

iii) all teachers in such schools should attend in-service training courses on both the general issues involved and on language teaching skills so that the whole staff of a school is equipped to deal with the pupils in it;

iv) a permanent cadre of full-time workers, partly advisory, partly developmental, should be mobilised in authorities with large numbers of settlers to assist schools with their problems and develop and disseminate new ideas, approaches and materials;

v) authorities should examine the role which centres might play for pupils experiencing special difficulty who could attend on a temporary, full-time or on a part-time basis...

b) So far, research and development projects on the needs of immigrant pupils have been limited in scale and funding. Scope I and the recently published Senior Scope, produced by the Schools Council, both present extremely useful and carefully graded teaching programmes, but they deal with the most straightforward part of teaching English as a second language – the first year in a special language class.... Because of this we suggest:

i) funds should be made available for research and development projects in language teaching, to produce suitable materials and to assess the effectiveness of various techniques and the most fruitful points of language intervention;

ii) although there is a need for research into language competence and skills required for the learning situation at all ages, there is a lack of expertise in the teaching of basic skills to secondary age children, and these ought, as a matter of urgency, to be developed;

iii) the importance of mobilising the experience and energies of teachers in schools must be realised: a scheme based upon teachers' centres or special centres to bring together the varied experience of practising teachers and College of Education and University staff is desirable... to meet the needs of minority pupils;

iv) there is a need for teachers to be provided with yardsticks with which they can assess the competence in English of both their native and dialect speakers....

v) one of the difficulties of some children from minority backgrounds is that they are having to acquire a new vocabulary and new language concepts simultaneously at a time when their mother tongue is ill-developed; there is a need for the development of expertise and approaches to assist the language education of children of this type....

c) that the educational performance of pupils in the school is to no small extent the result of experiences, stimulation and support received at home has gained widespread acceptance. Unfortunately its educational implications are only beginning to be worked out. Practical problems facing school staff of involving parents of immigrants and the difficulty of communicating with non-English-speaking parents are real. This suggests two types of recommendations:

i) the involvement in the educational system at professional and lay levels of increasing numbers of people from the minority communities suggests itself as a means of opening channels of communication, as ways of indirectly improving both the language development and performance of pupils, and as a method of creating situations for both white and black pupils in which the authority figure is black. This could mean positive attempts by authorities to recruit teachers from minority groups, as well as social workers, child guidance staff, aides and auxiliaries. In addition the involvement of members of minority groups, on school management committees, local educational authorities, etc is urged;

ii) the need for a more sustained and systematic service linking home and school. The Halsey Report on Educational Priority Areas recommends a home visitors' scheme and this suggestion has obvious potential for minority groups not merely as a means of linking home and school but as an instrument for encouraging the involvement of parents in their children's education and re-directing parental involvement on more effective lines....

d) One of the major findings has been the importance of length of education in determining how well individuals from minority group backgrounds perform; this suggests that priority should be given to the need for pre-school facilities for underprivileged groups generally and for minority groups in particular. However, the need is not merely for more pre-school facilities but also for innovations of two sorts:

i) the importance of building into the pre-school experience (and therefore into the training of the school staff) formal attempts to assist and encourage language development. This means the development and introduction of materials, language kits and programmes relevant to the needs of a pre-school pupil and especially to youngsters from non-English-speaking backgrounds...

ii) ...It is essential that any expansion of pre-school facilities should emphasise the need for lay adult involvement....

How far these points can be met through a general programme for disadvantaged pupils needs careful consideration, but two things must not be lost sight of: first, the special problems and needs of black pupils that are not extensions of the problems of the white disadvantaged, and second, the special needs of schools and areas with heavy concentrations of settlers. What must not be ignored is the extent to which the comparative performances of certain minority groups is a cause for educational concern and the need to mobilise the skills, research, development and action workers to meet their needs. It is in this context that the recommendations of the Select Committee on Race Relations and Immigration of a special fund and an immigrant advisory unit in the DES ought to be given high priority.[15]

Welfare Services

While the function of sociologists may legitimately require them to think out the ideal administration, it is the function of the elected representatives of the people to allocate the people's resources, both manpower and money. Whatever aspect of the immigrants' life in Britain is examined in depth, one clear conclusion arises above all others: the ideal of providing equality of opportunity for all British citizens, of all ethnic groups, can only be achieved if the number of new immigrants is strictly controlled. This is not a policy *for*, it is the only practical policy *against* racial discrimination. Social justice cannot be increased if the social services are over-burdened by weight of numbers. It is not difficult to imagine the feelings of a White woman and her husband if she cannot get a bed in a local hospital to have her baby, because all the beds are taken by immigrant mothers, whose need, of course, is just the same as hers. At a moment of need, and an emotional one at that, to be frustrated is to seek for a scapegoat. At such a time is prejudice re-born.

Commonwealth immigrants are not predominantly on the receiving end of social services, as one might assume from studying the problems of some of them. They are, in fact, very important contributors to the British economy and social services. Doctors, dentists, nurses and social workers of all races have earned the gratitude and respect of the British public for their kindness and skill. The opportunities which the teaching hospitals offer in training and experience works to the mutual advantage of the sending and receiving country. Some doctors, dentists and nurses settle in Britain but many return to raise the standard of health services in their own countries.

During the '60s the Economic Commission of the European Union of Women studied the situation of social workers and nurses in the eleven member countries of Western Europe. Apart from some significant variation in pay and conditions, the outstanding conclusion was that Britain did not suffer from the shortage of these workers in the same way as other countries, because so much talent from the Commonwealth was attracted to Britain for in-service training.

The immigrants in the labour force are, of course, paying taxes on their earnings; they have contributed more than they have received to pension funds. They have also filled some very important gaps in the labour force, particularly in the service industries. This has brought them into daily contact with the public. The friendliness, courtesy and efficiency of the majority has earned them public respect, one of the essential bases of a multiracial society.

II. Sex Discrimination

Perhaps one of the most important studies in depth, seeking to discover the reason why sex prejudice persists, was sponsored by the Leverhulme Trust and undertaken by Political and Economic Planning in collaboration with the Tavistock Institute. The book, *Women in Top Jobs*: *Four Studies in Achievement*,[16] was written by Patricia Walters, A. J. Allen, Isobel Allen and Michael Fogarty. Instead of covering again the well worn ground, it took case histories of women in two large companies, women as directors, women in the BBC and women in the administrative class of the Civil Service. By following the various careers the reader is able to come to some conclusion as to which decision at which point either supported or inhibited the woman in her efforts to acquire the skills necessary to hold down a top job. One of the main reasons why able women get so far up the ladder but do not reach the top is that they fail to recognise the moment when they must move from a satisfying appointment, often of a specialist nature, in order to get wider experience. It was generally found that once a woman achieves the top level sex prejudice disappears. It is easy to work with first class minds because they are not afraid of or jealous of ability in others.

A. Historical Background

The movement for Women's Rights is more than a century old. One suspects that it has been a recurring theme throughout recorded history. The modern movement grew out of the industrial revolution and the development of democracy in the Nineteenth Century. As the franchise was extended, it became logical that women should also have the right to vote in Parliamentary elections.

The Suffragette movement before 1914 cut right across party lines. The leadership was essentially middle class and upper class, though a few admirable working-class women did actively participate. The Liberals were in power, therefore, as the Government they were the main target. The 1914–18 war dramatically changed the climate of opinion. Women were given the opportunity to show their natural ability. For the first time on a large scale achievement rather than protest captured attention. At the first general election after the war women over thirty were given the vote.

The social and industrial troubles of the 1920's set the clock back for women in many ways, though the Conservative Government of 1924–29 lowered the voting age of women to the same as men, twenty-one. The suffragettes had won their case, but that was just the beginning of the battle for women's rights. Innumerable laws on the Statute Book discriminated against women one way and another.

The suffering caused by unemployment generated the idea that such jobs as there were should be left to men, who had the responsibility of supporting their families. Many people thought that a woman who sought a career of her own was depriving some man from having a job. The real causes of unemployment, of course had nothing to do with women working.

The trade agreements signed in Ottowa in 1932 turned the tide of unemployment, and by the end of the '30s many improved opportunities for women were becoming available. World War II again demonstrated women's capabilities and made nonsense of many entrenched attitudes and prejudices. Yet again, when the war was over, many of the old attitudes crept back. One illustration was the attitude of the London Press Club, strictly male. During the blitz on London when women were working in Fleet Street, ably doing the men's jobs while they were in the forces, a club was a matter of some importance. Journalists' hours of work are different from those of most other workers. Naturally places of refreshment closed early to let people get home before the blackout and probable air raids. The men's Press Club still refused to open their doors to women, and so the Women's Press Club of London was born. It was not until 1972 that the two Press Clubs of London became one, even though for many years the Women's Press Club had admitted men journalists to membership.

The real progress in the battle for women's rights, however, has been fought in Parliament. Step by step laws have been amended and new laws have been passed. As in the days of the suffragettes, the pressure has been about equal on all political parties.

B. Influence of Voluntary Organizations

Many and various are the Women's Voluntary Organizations looking after the interests and expressing the views of women in all walks of life. Some are political but most are by their constitution non-political. Politicians of all parties choose to consult them in order to understand the women's point of view. With 52% of the electorate female, the women's vote is a matter of great importance. (The suffragettes did not labour in vain!)

Many of the Voluntary Organisations are members of the Status of Women Committee so they can mount collective pressure for the reforms they deem desirable. It was founded in 1935 on a strictly non-

party basis. Its aims are to achieve equality of status and equality of
rights for women in all fields – legal, social, marital, political, technical
and in employment. The Committee works for the removal of any kind
of discrimination on grounds of sex, whether this should operate in
favour of women or against them.

The many strands of British life represented on the Status of Women
Committee are illustrated by the names of its member organizations:

Amalgamated Union of Engineering Workers; Association of Head-
mistresses; Association of Assistant Mistresses; British Association of
Women Executives; Commonwealth Countries League; Fawcett Society; Institute of Qualified Private Secretaries; Josephine Butler So-
ciety; Married Women's Association; League of Jewish Women; Na-
tional Council of Women; National Union of Townswomen's Guilds;
Open Door Council; Women in Media; Pilot Club International; St.
Joan's Alliance; Scottish Council of Women Citizens Association; Six
Point Group; Society of Women Writers and Journalists; Society
for the Ministry of Women in the Church; Suffragette Fellowship;
Women's Farm and Garden Association; Union of Women Teachers;
Women's International League for Peace and Freedom; Women's
Liberal Association; Women's Travel Club of Great Britain; Zonta
Club of London.

Looking at their Manifesto published in the '6os it is interesting to note
how much the declared policy of both the Labour and Conservative
Parties is in line with their aims.

To all citizens concerned with human rights:

MANIFESTO
of the
STATUS OF WOMEN COMMITTEE

"THE PEOPLES OF THE UNITED NATIONS have, in the Charter, re-affirmed
their faith in fundamental human rights, in the dignity and worth of the
human person and in the equal rights of men and women."

"DISCRIMINATION against women is incompatible with human dignity and
with the welfare of the family and of society, prevents their participating, on
equal terms with men, in the political, social, economic and cultural life of

their countries and is an obstacle to the full development of the potentialities of women in the service of their countries and of humanity."

"The full and complete development of a country, the welfare of the world and the cause of peace require the maximum participation of women as well as men in all fields."

The Declaration on the Elimination of Discrimination against Women adopted by the General Assembly of the U.N. in 1967.

Despite this affirmation and the Universal Declaration of Human Rights of 1948, and despite the progress made over many years towards equality, there still exists considerable discrimination against women as regards political, economic, cultural and social rights.

The Status of Women Committee, representing a number of women's organisations, exists for the special purpose of securing the removal of the discrimination against women that exists in Britain today.

Equal Educational Facilities

These should include access to every kind of educational institution, the same choice of curricula, irrespective of sex, with teaching staff, premises and equipment of the same quality and range, regardless of whether the institution is co-educational or not.

Equal Provision for Training

This should include equal access to apprenticeship and equal opportunity for vocational training and day release. Retraining, both part-time and full-time, and for older workers is needed for both men and women.

Equal Opportunity in Employment

(i) A woman, married, or unmarried, should have the same right as a man to contract for any employment, to receive promotion in it and to work under the same conditions. This involves an equal right to enter any vocation, profession, occupation or industry, the removal of all restrictions as to hours, conditions, overtime, nightwork and age of retirement, which at present are imposed on women only.

(ii) Safeguards against bad conditions and dangerous processes should apply to all workers, men as well as women.

(iii) Incapacity for work on account of maternity should be dealt with under the same regulations as apply to absence owing to illness, injury or disability.

Equal Pay for Work of Equal Value

This should cover all forms of pay including allowances, bonuses, and such other financial benefits as pensions and gratuities derived from employment. The Equal Pay Act provides for equal pay for the same or broadly similar work by 1975. This will not eliminate the exploitation of women as cheap labour in so-called "women's work".

Equality in Social Security

All differences based on sex or marital status in Social Security, Insurance, and similar schemes should be abolished. Conditions of retirement should be the same for men and women. Married women should be insured in their own right and entitled to the full rate of benefits.

Equality in Taxation

The incomes of husband and wife should be individually assessed and taxed. There should be equal treatment with regard to allowances, assessment, payment and reliefs.

Equal Moral Standards

Section 1(i) of the Street Offences Act 1959 penalises for solicitation only the "common prostitute" while ignoring the menace of solicitation by men (e.g. kerb-crawlers). The section should be repealed and replaced by a clause applicable to all citizens, whether men or women, who cause annoyance or nuisance by loitering or soliciting in the streets and public places.

Equal Status Within Marriage

(i) Both husband and wife should be entitled to share the family home, goods, and income, and on death or in the event of a breakdown of a marriage, there should be an equitable distribution of such assets as have been acquired and accumulated as a result of the work of the two spouses.

(ii) The domicile of a married woman, which determines the law applicable to nullity proceedings, divorce or separation, and to the disposal of her property on death should not automatically follow that of her husband: she should be entitled to acquire a domicile of choice in the same way as a man or single woman.

(iii) The mother of every legitimate minor child should at all times have joint guardianship and custody with the father and have equal authority, rights and responsibilities.

The Status of Women Committee demands full equality for women in social, political, economic and cultural life, and asks your support for the reforms set out in this Manifesto.

This Manifesto gives a good indication as to the wide range of laws which need amendment in addition to any Act specifically outlawing sex discrimination. The build-up of pressure has a very powerful effect on law making, even if each step in itself is not spectacular. Many people believe that "Women's Lib"-demonstrations and protest are the influences which give rise to action. There is no question but that the Women's Liberation movement has played a useful part, catching the attention of the media, making news, disturbing the air, shaking the complacent, making those thousands of women working constructively aware that they are justified in their dedication to achieve equality. The Women's Liberation tactics, however, in some ways discredit the cause and alienate support. Individuals make exaggerated claims, not asking for equality, but for discrimination in favour of women. They appear to know little of the virtue and happiness to be found in loving and giving, and instead generate a philosophy based on "because *I* want, *I* must have." They are painfully lacking in a sense of humour, a lack which really does make family life intolerable. Some also use women's liberation as a step towards their goal of changing the capitalist system to one of "true socialism". From their writings and speeches it is clear that they equate progress with changing the order of society, apparently under the illusion that women in Eastern Europe have achieved equality. Anyone who reads Walter Dushnyck's paper in volume two of this collection will realise the extent to which this idea is based on a myth.[17]

Those who study facts rather than mythology are the real pioneers. Their balanced presentation is what moves the minds of men to think again. A good example of this kind of work was the *Report of an Inquiry into Women's Careers* organised by the British Federation of Business and Professional Women.[18]

This 80 page report is a model of how to present carefully researched evidence in a form highly readable for the non-expert. The ludicrous stereotype picture of a women as an emotional creature, given to excessive use of detail, incapable of assessing the importance of one detail as compared with another, should be destroyed by this report. Unfortunately, of course, those who enjoy believing in the stereotype are unlikely to read it.

The general situation faced by the business or professional women in Great Britain is suggested by the Introduction to the report, which is worth quoting *in extenso*:

INTRODUCTION

Adaptability of Women

Are women able to adapt their training and be flexible in their careers? This was a question often raised on public platforms wherever there were discussions about automation, new technical developments in industry, and rapidly changing methods in commerce and management. The Executive Committee of the British Federation of Business and Professional Women discussed this idea, and decided to set up a small working party to explore the possibility of a survey of business and professional women to test the idea.

The result of these deliberations led on to the decision to organise a limited inquiry into women's careers. It was hoped to explore the circumstances where careers were interrupted or switched and for what reasons. It was also hoped to find out what use women made of their education and training.

The Survey

From the very first it was realised that the British Federation had not the resources to mount a professional survey that would be statistically valid. The "survey" was to be more in the nature of an inquiry from a limited number of members of women's organisations.... The inquiry could only claim to be exploratory, something in the nature of a pilot study.

Further Research

...This inquiry emphasises how a woman's career will always be tempered by her fundamental functions as a wife, mother, and her role as her parent's daughter. This inquiry shows the necessity for some comparative exploration and research into the male employment situation. For example, how far does a man's functions as husband and father contribute to the interruption and diversion in *his* employment situation? What particular stresses and strains does he encounter?

Womens Education and Training

In this inquiry many women with difficulty "went back to school" when circumstances arose which made their support of themselves and families an economic necessity. Some of these women were ill-equipped to become breadwinners, and it was only change, determination and considerable personal effort which enabled them to succeed. This inquiry also indicates the need to know more about the relationship of girls' education, and training after marriage.

If this inquiry lacks statistical validity it has a qualitative value. Many simple statements of fact only hint at the human situation which lies beneath many hundreds of women's lives. The impression which shines through is the resilience of women; their ability to cope with crippling accidents of life, and their astonishing capacity for coming out on top in so many cases.

Countless married women stressed the importance to them of some employment which provides a discipline, an opportunity to use their skills, to be stretched as persons and to be kept in touch with developments in their field of interest. To neglect such needs is to diminish personal happiness, ignore a resevoir of ability, and waste intensive capital investment in female education.

C. Influence of the Trade Unions

In theory trade unions have always been in favour of women's rights, equal pay and equal status. Practice is very different from the theory. In a number of unions the rate for the job, irrespective of who is doing it, is firmly established. That sounds like equal pay, but it so happens that those grades carrying the highest pay are regarded as men's jobs, while some of the lower paid ones, though not barred to men, happen to be regarded as women's jobs.

One trade union with a good record in respect of women's work, the Association of Cinematograph and Television Technicians, decided to take very positive action to investigate discrimination. In April 1975 they produced a report some 80,000 words long, which examined thoroughly the position of women in the film, television and laboratory industry. As an introduction to the report, Alan Sapper, the General Secretary of the Union, wrote:

At its Annual Conference in 1973, the Association of Cinematograph, Television and allied Technicians passed a resolution calling on the union to undertake a major investigation into the position of women employed in the film and television industries. The decision resulted from growing pressure from women in the union for a detailed analysis of the inquality and discrimination which they felt existed in the communications and entertainment industries. The analysis was considered indispensible if the union was to formulate and pursue collective bargaining demands which were relevant to the rights and position of women.

After two years of sustained research and discussion, the union's Committee on Equality and the specially appointed researcher Sarah Benton, have completed their work. I believe that the Report adds immeasurably to our knowledge of the concrete position of women in our industry. It explains in meticulous detail the patterns of discrimination which exist and its roots in the class structure and ideology of our society and in the way in which those are reflected in the operation and assumption of the employment structures in film and television.

ACTT achieved equal pay for women in the late 1930s. That principle has been incorporated in every agreement that the ACTT has signed since that time. What this report does is to indicate with dramatic force, that, in isolation, equal pay is merely the first stage in an effective trade union offensive whose objective is to gain equality of opportunity for women. In this instance, the achievement of equality is dependent upon acknowledging the role of woman as a childbearer and in acknowledging it to forge policies and agreements which guarantee that she will not be penalised or discriminated against as a consequence of it.

There is however another and broader dimension to this report which transcends the frontiers of the communications and entertainment industries. The Report is a major, and thus far unique, contribution to the debate on the position of women which in International Women's Year is properly mobilising the labour movement behind hard edged policies to achieve women's rights as citizens and as workers. I believe that the work which we have undertaken will assist all trade union members.[19]

Although the film and television industry is not exactly like any other, the evidence presented in the opening part of the report justifies the kind of pressure that has been building up for legislation against sex discrimination:

1. The proportion of women in the ACTT has been dropping in the last 20 years. From 18 per cent of the members in 1953-4, through a peak of 19 per cent in 1963-4, to the present 14.8 per cent of membership. Today, the union has 18,458 members, of which 2,739 are women. Whereas the number of men in the union has increased by 5,730 (57 per cent) in the last ten years, the number of women has risen by only 437 (19 per cent). This is largely explained by the Laboratory Branch, where female membership has dropped by almost one third in the last ten years. In film production the proportion of women members dropped by over 10 per cent whilst the proportion of men rose by 2 per cent. The increases that there have been in the number of women in the union is almost entirely due to television, but it has not been enough to offset the drop in the laboratories and film production. Twenty years ago women worked in a fairly wide variety of grades. Now they work in almost complete sexual "ghettoes".

2. There are two main reasons why the gap between the numbers of men and of women is increasing.

(i) The opportunities for women to work in many areas of the industry where they do not work now – particularly the skilled grades in the laboratories, and post-production grades in the film studios, came during and shortly after the Second World War. As men returned from the forces, many women were eased out. They have now been almost entirely replaced by men. The division between men's and women's jobs has become more and more rigid in the last twenty years, so that many young members assume it has always been, and will always be, like this.

(ii) The move away from permanent employment and studio-based film production to an almost entirely freelance film production Branch has hit women especially hard. Many of the departments they once worked in in film studios, such as art and post-production departments, no longer exist. And for women, with their greater domestic commitments, it was far easier to take a permanent job in a studio than to cope with the location filming or the irregular hours that are so often demanded in casualised film production.

THE SEXUAL DIVISION OF LABOUR

3. The industry is largely divided between a wide range of "men's" jobs and a very narrow range of "women's" jobs. For many women, opportunities for work in the industry open-and-close with secretarial jobs. In film and television production, out of over 150 grades covered by ACTT agreements, 60 per cent of the women are concentrated into just three of those grades: production secretary, continuity girl and ITV Production Assistant. In film processing, the proportion of women concentrated into the secretarial and clerical grades drops to nearer 50 per cent. In half of the existing ACTT grades there is not a single woman, and the proportion of women in a further 12 per cent of the grades is less than 1 in 10.

4. In television and film production there is a tiny handful of women (less than ten per cent) amongst the several thousand technicians. In the top production grades, the ratio of women to men is less than 1:10; although work on high-budget and/or "important" films and programmes is an almost exclusively male preserve. In film processing, women who are not in clerical or secretarial grades are concentrated in ten of over 120 grades in the ACTT agreements. These ten grades include none of the most highly skilled technical jobs.

5. This rigid division of labour has far-reaching effects. It is not just that women have so much less "freedom of choice" than men – it also means women are denied the opportunity to earn as much as the "average" man.

7. The top grades in the jobs primarily done by women are usually considerably lower, in terms of earnings and authority than the top grades in jobs done by men. Thus, a principal PA in ITV earns less than a fourth year sound technician, and the top rate for continuity is usually about the same as the rate for follow focus cameraman (one of the starting grades in camera). On the whole, the grades in which women work are concentrated in the lower half of the scales of grades and pay, even where women are working in the top grades of their type of work.

The Nature of Discrimination

8. "We don't discriminate here" was the first response from employers to questions about job segregation in their companies. And shop stewards often confirmed this statement. By this they mean that the most blatant form of prejudice – telling a woman that she cannot have a particular job because she is a woman, does not exist in their company. In fact, this sort of open discrimination is fairly widespread throughout the industry, but it is by no means the most important cause of women's inequality. The same causes exist throughout the field of employment, though of course they vary in importance according to the type of industry. The most significant ones are:

(i) Blatant discrimination against women applying for particular jobs in any firm, or for any jobs in particular firms.

(ii) The undervaluing of jobs primarily done by women in terms of pay and respect.

(iii) Educational and social "conditioning" and the lack of training facilities, which deny women the opportunity, both before and after they enter the industry, to work in a wide range of grades.

(iv) The job structure of the industry, which makes most women's jobs dead-end jobs, and makes movement across and up from the type of work they normally do, extremely difficult.

(v) Job insecurity and work relationships which make men see women (and other minority groups) as a threat not only to the existence of their jobs, but also to the work relationship they have with other men doing the job.

(vi) The denial of women's "right to work" by both State and employers which exclude many women from various rights and conditions of employment because it is assumed their main responsibility lies in the home.

(vii) The lack of trade union activity against discrimination because women have less time to attend union meetings, and many of their needs and concerns are not regarded as suitable union concerns.

(viii) The economic and social structure and the inadequacy of legislation which force women into certain economic roles which the law does not alter.

Many of these points are clearly connected and themselves create an enormous number of discriminatory attitudes and ideas which are not listed above. They reinforce each other in such a way that progress to equality of rights for women and men involves not only a change of "male chauvinist" ideas, but also far more wide-reaching changes in both the film and television industry itself, and in the structure of the society in which the industry exists.[20]

Most of the conclusions of the report with recommendations for action concerned the internal operation of a non-discrimination policy. The final recommendations are in line with attitudes shared by women in many other unions:

3. Union Policy in the TUC and Labour Party

The Union should take a lead in all struggles to create equality for women, and in particular to put resolutions to the Annual TUC for concrete changes that will help break down the existing patterns of discrimination. This is particularly important as the union has decided to withdraw from the Women's TUC.

The union should fight for the statutory enforcement of maternity and paternity leave provisions for all parents; a massive increase in Local Authority child care facilities; the extension of protective legislation (especially on hours and safety at work) for women and men; the amendment of the Equal Pay Act to eliminate the many loopholes; the rapid introduction of legislation on Equal Opportunities, to cover discrimination in all areas including pensions, with enforcement machinery that is simple to use, and which effectively penalises those who break the law.[21]

The reason why some trade unions now will not send delegates to the Women's Trade Union Congress is that they disapprove of the segregation of the sexes in trade union matters. When the writer was a delegate some years ago she was horrified to hear it stated from the platform that the Conference would not discuss TUC *policy* issues in case the women should differ from the men, and so undermine the solidarity of the TUC. On the General Council of the TUC, which is the governing body, just two places are reserved for women.

The policy of the TUC to prevent women working in factories at night, and restricting overtime, was derived from Nineteenth Century conditions. Today it is highly discriminatory. Modern technology often requires round the clock working. If qualified women are not permitted to take their turn on the night shift they are automatically excluded from promotion, and are deprived of the chance to earn the extra pay. Conditions for working on the night shift are legitimate concern of trade unions, and a number of men as well as a number of women may wish to forego advantages of cash and promotion for personal or domestic reasons, and it is reasonable that they should be enabled to do so. What is not reasonable is to deprive women of the choice. A childless couple, with the man working on night shift, could well find it highly desirable that both should work the same hours so that they could enjoy their leisure together.

In spite of the lip service the trade union movement pays to the rights of women, its record of action is far from impressive. Of the 129 trade-union sponsored MP's on the Labour side of the House, only 5 are women. Of the 16 MP's sponsored by the Co-operative Movement only 1 is a woman. Yet many of the members of the Co-operative Society are women. It was reported to the 1972 TUC Congress that women's membership in unions affiliated to the TUC was 2,417,711 at the end of 1971 and that also 23 of the 129 organisations within the TUC had more women members than men. Despite this there was only one woman General Secretary of a trade union; she was Mrs M. Fenwick of the Jute, Flax and Kindred Textiles Operatives Union, which had only 3,380 members.

D. *Women's National Commission*

The Women's National Commission is an advisory committee to H.M. Government attached to the Cabinet Office. After the United Nations published their Declaration on the Elimination of Discrimination against Women the Secretary General wrote to the Government of each member state asking if they had a Status of Women Commission and if not why not. The Prime Minister of the day, Mr. Wilson, referred the letter to the Women's Consultative Council which said that much of the legislation preventing discrimination was already passed or was about to come before Parliament but that there was still need to ensure that these Acts were implemented. They felt therefore that there was need for a Commission which would keep watch on affairs and which would keep the Government informed of its views on matters of public concern. The Commission was accordingly set up and consists of 48 members drawn from 10 trade unions, 10 professional organisations, 4 political parties, ten religious organisations and 14 voluntary organisations. The Government determines which organisations shall be represented on the Commission but each organisation chooses its own representatives and determines how long she shall stay on the Commission. Members elect their own executive committee. The Government appoints one of the Co-Chairmen, who is a member of the Cabinet, the Commission elects the other Co-Chairman from the members of the Executive.

Two important points to notice are that the Commission does not speak for the women of this country. It speaks for itself but does have the advantage of a wide range of experience and knowledge among its members. The second point is that there is no Men's National Commission to ensure that the informed opinion of men is given its due weight in the deliberations of Government. The reasons are obvious. Only 26 of the 630 members of Parliament are women. Out of the 52,000 members of Government appointed councils, committees, advisory bodies and so on only 500 are women. The first and continuing duty of the Commission therefore is to do everything it can to ensure that when any committee or board or council is appointed by any government department, the names of women with the necessary knowledge and experience are considered on the same terms as those of men. It will take time to ensure proper representation but already there is good co-operation between the Commission and the various government departments at national level and local level. The Commission works through ad hoc committees when dealing with matters of public concern.

One popular device is to pass resolutions at the annual conference of national associations or federations and to send these with a covering letter to the appropriate Secretary of State. Some 7000 are sent each year. They produce little effect beyond a non-committal and faintly encouraging reply expressing the agreement of the Secretary of State and outlining perhaps something of what might be done to meet the demand. It is therefore the practice to follow up at three-monthly intervals requesting specific information about progress on the lines suggested or ask the minister to receive a deputation to discuss the question. If the resolution deals with something at local government level the preferred approach is to ask local officials or councillors to come and speak about the particular problem and find out whether they are well ahead in dealing with the problem or need further information to deal with it effectively.

The Women's National Commission and its constituent organisations make it a point to generate systematic input into the policy making process. They make use of the following opportunities:

(a) The Government invites public co-operation when it sets up a Committee to deal with a particular problem, e.g. The One-Parent Family. When such a committee is set up there is an opportunity for organisations to prepare and send in written evidence, and if the Committee appreciates the value of that written evidence they will invite members of the organisation to meet them and discuss their report in detail. Such Government Committees generally produce recommendations leading to legislation.

(b) The Government also produces for public discussion Green Papers or White Papers which set forth ideas about possible major changes in Government policy, for example "Proposals for a Tax Credit System" which seeks to combine income tax with social security and to help the lower paid with a form of negative income tax which adds to their pay packets.

(c) A third opportunity for discussion at an early stage of controversial matters comes through the publication by the Law Commissioners of Working Papers. In the one dealing with the Assessment of Damages in Personal Injuries Litigation it was possible to put forward a strong plea for a realistic appraisal of the economic value of the work of a mother in the home when assessing damage for injury which prevents her carrying out these duties. It was also pointed out that if a mother had to give up a paid job in order to come home and nurse an injured member of her family, the damages for that persons' injury should take into account the mother's loss of earnings and of promotion prospects.

(d) When Bills are before Parliament organisations may make representations to the Committee in Charge of a particular Bill about amendments they would like made to it. At the same time individual members of the organisation can approach their own M.Ps about amendments.[22]

E. International Action

Women in all countries who have given some thought to the subject of women's rights have always recognised that it is a universal and not a national cause. Many voluntary organizations both large and small have established international links. Laws and customs will not be changed until attitudes have been shifted from rigid acceptance of the status quo to an awareness that a different set of values could be more fruitful.

When the United Nations gave the cause a good push forward they found organizations ready to take advantage of the impetus. The Treaty of Rome required the original six countries of the E.E.C. to establish equal pay for equal work. Four years before the men got down to signing the Treaty, European women politicians had recognised the need for regular contact and study of all social and political issues from a European point of view. On the initiative of an Italian and an Austrian the European Union of Women was formed, with Britain one of the founder members, in 1953.[17]

In May 1973 the National Council of Women of Great Britain held a seminar *Women of Europe – Partners in Progress*, where the operations of the Women's National Commission were explained. This was followed up by the Status of Women Committee, to study the relationship between sex equality and social progress. A year later the Status of Women's Committee organised a conference in London on *Sex Discrimination in the European Community – Where does Britain Stand?* The leading speakers came from the E.E.C. Commission, Netherlands, Belgium, France, Italy, Denmark, Republic of Ireland, Germany and Britain. The main subjects discussed were Status of Women, conditions of employment, equal opportunity, social security (widows, single women, pensions, family allowances, maternity grants), mortgages and loans, property law and trade unions.

There are two world organizations of women journalists. The World Association of Women and Home Page Journalists (AIJPF) based in Brussels, and the World Association of Women Journalists and Writers, (AMMPE) based in Mexico City. They complement rather than duplicate each other. British women writers and journalists participate actively in both.

The National Council of Women is very active within the International Council of Women.

This very brief and far from comprehensive outline of international action and co-operation in the campaign for women's rights is included as an integral part of the British scene.

F. Equality and the Law

Amendment to the law is the acid test of progress. The patient work of thousands of people using the mechanisms of Parliamentary democracy help M.P.s to assess the need for change. Everyone recognises that the law alone cannot change attitudes, yet when an Act of Parliament is placed on the Statue Book, attitudes do change.

Barbara Castle's Equal Pay Act at the end of the Labour Government's term of office in 1970 was supported at the time by the Conservatives and Liberals. In 1973 when Maurice Macmillan was the Conservative Secretary of State for Employment he wrote individually to 400,000 industrial and commercial organizations to launch a campaign to remind them of the need to make progress towards equal pay, and prepare for its full implementation by the end of 1975. Helpful advice on achieving this, and an explanation of the Act, were sent out in a booklet, *Equal Pay – What Are You Doing About It?* with Mr. Macmillan's letter.

After the Equal Pay Act was passed there was widespread concern that pay was only the first step, equal opportunity was even more important.

In 1971 Mrs Joyce Butler, M.P. Labour, supported by six other Labour M.Ps and two Conservative M.Ps. presented an Anti-Discrimination Bill, but for several reasons it was not successful. In 1972 Lady Seear, a Liberal, introduced the Anti-Discrimination Bill in the House of Lords. A month later Lord Jellicoe on behalf of the Conservative Government proposed a motion to refer Lady Seear's Bill to a Select Committee, who decided that because of the evidence they had received, legislation on the problem was justified. On 14th May, 1973 Lord Colville announced in the House of Lords that the Government would be putting forward a paper on sex discrimination in employment, training and education. In September 1973 the Conservative Government Consultative Document, *Equal Opportunities for Men and Women* was published. Before a Bill could be presented the general election of 1974 had returned the Labour Party to power. The new Government published their White Paper, *Equality for Women*, on which the Bill was based, in September 1974.

Comparison between the Conservative Consultative Document and the Labour White Paper and Bill is interesting because it shows the response of both Parties to the evidence which has been presented to them, what they have in common, and where there is a distinct difference in emphasis.

Conservative

The scope of the legislation proposed in the document was limited to education, training and employment and the machinery for enforcement was that already existing in the industrial tribunals.

(1) *Education* – the document felt that discrimination on the ground of sex did exist and that Her Majesty's Inspectors should undertake a study of the extent to which curricular differences and customs contribute to unequal opportunities." Single sex education would be continued.

(2) *Employment* – the Government proposed to make discrimination

in employment against women unlawful in all firms employing more than 25 people with this figure being reduced and eventually eliminated. Exemptions were only justifiable on the ground of sex being "a genuine occupational qualification." Protective legislation was to be repealed except in the cases of women working underground and the restriction on men becoming midwives.

(3) *Enforcement Procedure* – It was considered that an individual should refer any complaint to an industrial tribunal similar to the Equal Pay Act procedure. This would be a convenient and speedy method of redress with conciliators playing an important part in the settlement of many cases. This would be the job of the Department of Employment Conciliation Officers who have a similar function in cases of unfair dismissal. Statutory procedures would be suspended if the Parties decided to try and reach a voluntary settlement. At the Industrial Tribunal it would be up to the complainant to prove his or her case.

(4) *The Equal Opportunities Commission* – Its role was limited to conducting inquiries, publishing reports and educating and persuading public opinion. The Conservative Government's view was that the Equal Opportunities Commission would be "more able to secure co-operation, and thus to be a more influential body, if it were not also encumbered with enforcement responsibilities." The Commission could not act on behalf of a complainant, but would inform him or her of the procedure of obtaining redress.

Labour

The Labour White Paper and the new Bill differs from the Conservative Consultative Document as regards its scope, its enforcement powers and the role it gives to the Equal Opportunities Commission.

1. *Scope.* The White Paper extended the anti-discrimination laws to cover the provision of housing, accommodation, goods, services and facilities.

The Bill has further widened the scope of legislation by:

(a) only permitting exceptions for a firm of five or fewer employees.

(b) introducing the concept of *"unintentional discrimination"*.

(c) permitting employers to take positive action in the form of single-sex training to encourage women and men to make use of the op-

portunities offered by the Bill in employment situations previously
dominated by the other sex.

Education. In education the Bill makes discrimination unlawful in
admissions to and provision of facilities in educational establishments.
Facilities for education must be provided without sex discrimination.
Concerning single-sex education Roy Jenkins said:
"It is not the Government's intention to end single sex schools but our
proposals will probably accelerate the general trend towards co-education." ("News of the World", 28th July 1974).

2. *Enforcement.* The greatest difference between the Labour and Conservative approach to this legislation is over the enforcement powers.
Whereas the Conservative Consultative Document concentrated on the
Equal Opportunities Commission being a research and conciliation
body the Labour Government propose to extend the functions of the
Equal Opportunities Commission to cover some elements of enforcement.

In *employment*, complaints will be dealt with by the industrial tribunals. In *housing*, accommodation, education, the provision of goods,
services and facilities, complaints will be dealt with by specially designated county courts, and by sheriff courts in Scotland. However the
Commission has powers to apply to the courts for an injunction on its
own initiative. Only the Commission is entitled to bring complaints
before the courts concerning discriminatory advertisements or in cases
where instructions have been given or pressure exerted to discriminate.

The enforcement powers of the Commission also include power on
completing an investigation, to serve a non-discrimination notice, requiring the recipient to cease the practice. Non-compliance with the
terms of the notice will render the recipient liable to legal proceedings.

The Bill has also introduced two other *new* provisions:
– help for aggrieved persons to obtain information regarding their
complaints.
– quicker access for the Equal Opportunities Commission to obtain
an injunction against persistent discrimination.

In the Committee stage some of the drafting may be clarified. As it
stands there is room for confusion. "Unintentional discrimination"
could give rise to diverse interpretations. One wonders what imagina-
tive arguments would be put forward by the top tier of the governing
body of the City of London Corporation, the Court of Aldermen, who

recently exercised their right under a Fourteenth Century ruling to refuse to admit a democratically elected member to their Court. She happened to be the first woman ever to be so elected. In two elections her ward voted her in, and twice the Aldermen stressed that she was not turned away because she was a women, but they gave no other reason. As she had served on the Court of Common Council, the lower tier of the governing body, with great distinction and with no discrimination problems, the natural inference was sex discrimination. Would the Aldermen now argue that it was *unintentional*? The result of this sad story is that the Court of Aldermen lost an able and charming colleague in Mrs Edwina Coven, the electorate lost the representative they wanted, the Court lost respect.

Apart from the major legislation pending on the Anti-Discrimination Bill many other measures have been passed since 1970 which have considerably improved the position of women. These include better pensions in many categories. Family Income Supplement to help low wage earners with children, increasing the income of many women – widows and deserted wives – bringing up children on their own. Women in England and Wales, deserted by their husbands, can get their maintenance money through the post instead of having to go to court for it. In guardianship of children, mothers have now been given equal rights with fathers.

III. Social and Political Discrimination

A. Discrimination and the Social Classes

To consider discrimination on grounds of *class* requires very careful analysis of what is meant by the term *social class*.

Those engaged in surveys and opinion polls often divide the population into five classes, A. B. C. D. E. For some purposes they compress the five to three, A-B. C. D-E. An arbitary line is drawn to establish the correct size of each class as a percentage of the population. It is left to the judgment of the interviewer to decide into which class to place the interviewee.

How should upper class (A or A-B), middle class (C or B-C or B-C-D) and lower class (D-E or E) be defined? Is education, property or earnings the determining factor?

At the beginning of this century lower class could be used to describe those who had no more education after 14, who earned their living in low-paid, unskilled or semi-skilled jobs, and who never had more than enough money to pay for the basic necessities of food, clothing and a rented house for their families. The situation of the country poor was different in many respects from the urban poor, but for this purpose both could be classified as lower class, though Britain never had a peasant population. It is important to note that *lower class* did not imply inferior human beings, but those barred by material circumstances from developing their full human potential.

The middle class were those who earned enough to have some margin above the basic necessities. To do so required education to qualify. The reward was having some choice as to how the family income above basic necessities should be spent. Some domestic help in the home was often the first choice because the drudgery of running the home before the days of vacuum cleaners, washing machines and electric fires was so great that it was often too much for the physical strength of the mother, who might be producing a new baby every year. The premature aging of lower-class women was witness to the suffering endured by those who had no help.

As the middle-class income increased, so time and money became available to buy more services and enjoy art, music, literature, travel and sports. The boom in the piano trade during World War I when munition workers suddenly found they had the money to buy music in the home, even though many of them still had to learn how to play the instrument, was a touching testimony to the hunger for the cultural side of life.

The upper class could be described as those who had inherited wealth, titles and land, or those who were highly successful in commerce or the professions. As acknowledgement for the services they had given, one way or another, many had been honoured with a title. Because of the strength of the hereditary factor, wealth was of secondary importance. Some members of the upper class were seriously impoverished.

Karl Marx and his followers, looking at the class structure in Britain at the height of the Industrial Revolution, saw the potential power base in the proletariat, or lower classes. Because they had to work to live, they were named the *working* class, and in political terminology the *class struggle* was born. Marx's famous prophesy that under capitalism the rich would get richer and the poor poorer revealed his lack of appreciation as to what machines and technology were going to do to the whole process of the production of wealth.

The leaders of the working class, like Marx himself, have always been predominantly middle class. Leading Marxists today have a terrible guilt complex about their middle-class or even aristocratic background. The TUC extended the definition of working class to "workers by hand or brain". The white-collar unions are rapidly increasing their membership to the point of overtaking the manual workers.

In terms of the 1970s, therefore, how should *class* be defined? Education, access to cultural pursuits, form of speech, dialect or accent, earnings, property, family name – all are possible criteria, collectively they add up to something, separately they have little relevance.

Education is available to all. Some make good use of their opportunities to learn, others do not. This applies to children in all social classes.

Access to cultural pursuits is available to all through radio and television. Artistic gifts have never been the prerogative of any class. In the past much in-born talent in working-class children was never discovered or developed, because they had no means to hear great music, to see ballet or great plays. Modern technology of communication has changed all that.

The working-class accent has almost disappeared, though fortunately regional accents and intonations have survived to add colour and humour to the language. Wider education, radio and television has equipped all children with the chance to bring their vocabulary and pronunciation up to the standard.

Earnings most certainly are not the hallmark of class. Many middle class families would rejoice and deem themselves rich if they could enjoy the pay of *some* manual workers. Such affluence is by no means available to all manual workers.

Property increasingly becomes less of a distinguishing mark as more and more families buy their own homes, cars and boats.

Family name and the social standing of parents and grandparents have some relevance in some connections, but none whatever in others. Family ties operate in every class, but few outsiders care who is related to whom. When Sir Alec Douglas Home became Prime Minister, having relinquished his title of Earl so that he could sit in the House of Commons, the class-conscious sought to devalue him by referring to him as the "fourteenth Earl". Their implication was that he had only been chosen by the Conservative Party as leader because of his family name through fourteen generations. With a smile he neatly disposed of this fallacious argument by referring to the "fourteenth Mr. Wilson".

The Conservative Party next chose Edward Heath as leader, a grammar-school boy, inheriting neither wealth nor social status, and ten years later they chose Margaret Thatcher, the daughter of a small shop-keeper in a provincial town. All three were chosen for their personal merit, by people who knew them well, and who had had the most opportunity to assess their quality as individuals.

It is very difficult to find criteria by which social class can be defined, making it doubly difficult to present evidence of discrimination on grounds of class. There are snobs and class-conscious people, but they are to be found in every class. That which they have in common is a desire to impress other people, not by their own merit, but by the merit of the people they know. Joining the right club in their eyes, therefore, becomes a passport to success. How many individuals who have applied for membership of a club and have been rejected by the membership committee is not recorded. Even if some figures were available there would be no reliable evidence for the reason of rejection. The basic purpose of most clubs is to allow people with some interests in common to meet where they can *relax*, and where they will *not* be pressured and used as a step ladder by other members for their own purposes. Clubs are an expression of the basic freedom of assembly.

Clubs as instruments of discrimination have drawn much public attention because of the case brought by the Race Relations Board under the 1968 Race Relations Act against the Preston Dockers' Labour Club for refusing to admit a coloured man. He was an Associate member, one of about a million Associate members of the 4,000 Working Men's Clubs in the country. The Race Relations Board won their case in the first Court and again in the Court of Appeal, but when it finally came before the House of Lords the verdict was reversed in favour of the Working Men's Club. The reasoning was strictly legal, turning on the correct interpretation of the words in the Act. Naturally passions were aroused because of the wide sympathy felt for the coloured man in question. Many were shocked that the Law Lords should side with the Preston Dockers to humiliate a good citizen, whose only disadvantage was the colour of his skin. Lord Hailsham brought out the depth of this misconception in his letter to *The Times* on 25th October, 1974:

I write to protest against the assumption that the Appellate Committee of the House of Lords is free to make its decision on policy grounds. In the recent case it was bound by the words of the Race Relations Act which it was its duty to interpret and by the rules of construction in accordance with the words of an Act of Parliament have in law to be interpreted.

Neither the House of Lords in its appellate jurisdiction nor any other Court of Law in England is free to decide cases on purely policy grounds, and the contrary assumption is, I believe, largely based on the quite different terms of reference of the American judiciary...

Our Courts are there to apply the Law as made by Parliament, not to improve it.

The crux of the case was that Clubs do not "serve the public", they are owned privately by the members, and were thereby excluded from the Act. Whether they should be is another matter. Parliament is free to amend the Act if the Members think fit. In a publication concerned with human rights and the Law there is much food for thought in this case. The precise wording of any anti-discrimination measure is of vital importance to the freedom of every citizen. Any such Act must limit in some way freedom of speech and of assembly. It may well be expedient to pass such a law in given circumstances, but it must be recognised what is being done.

This case also invalidates the claim that Clubs are used for discrimination on a class basis. On the other hand if the word *discrimination* is considered in its dictionary sense it means nothing derogatory. It is the act or quality of distinguishing; acuteness, discernment, judgment. It is fair to say that all clubs, youth clubs, airmen's clubs, old age pensioners clubs, sports clubs, all distinguish, exercise judgment and decide who is to be admitted to membership. Because some London Clubs in a convenient location for members have very costly property to keep up, and have to charge a subscription to cover the cost, it is hardly logical to reckon this as Class discrimination.

In *Colour and Citizenship*, chapter 3, "England, Whose England?", quotes T. E. Marshall's definition of citizenship in his book, *Citizen and Social Class*. The three components of a citizen's rights are:
Civil rights: rights under the law to personal liberty, freedom of speech, association, religious toleration and freedom from censorship.
Political rights: rights to participate in the political processes.
Social rights: access to social benefits and resources, education, economic security and the range of welfare state services.

Marshall argues that the basic structure of civil rights, had been completed by 1832, and extended to all recognised minorities – except women – by the middle of the Nineteenth Century. Taking his definitions, there is no discernable discrimination on grounds of social class in Britain. His statement does imply, however, that those who have money are free to spend it on those things they value, whether these be private education for their children, private health service or insurance, or even a club subscription. As people with money pay their taxes to provide the state services, it would be a positive discrimination against them if freedom to spend their own money were to be inhibited.

B. Discrimination and Nationalism

The Welsh Nationalists and the Scottish Nationalists are both claiming that they have the right to control the resources of their own countries, and to take decision as to how those resources should be used. By some it is deemed to be an example of discrimination that these rights are not already theirs. The Labour, Liberal and Conservative Parties pay some attention to their claim that more decisions should be taken in Scotland and Wales, and they have all put forward plans for Devolution. These plans far from satisfy the ardent Nationalists. The problem is different in kind from separatist movements in Africa or Asia, but there are underlying principles in common. Where there is a profound difference of opinion, and a totally different interpretation of economics, someone has to take decision as to whose judgment should prevail. The discovery of oil under the sea around British shores has given a new twist to the economic arguments. Many millions of words will be spoken and written before institutions will be devised to reconcile the conflicting factions.

C. Impact on Politics, Law and Order

Progress towards greater social justice and respect for human rights can only be made by politicians, acting on their assessment of (a) what is good for the people, or (b) what the people want. This responsibility is shared by the professional communicators, who help to mould the attitudes expressed in inter-ethnic relations.

The two conflicting political trends apparent in every continent are also present in Britain:

1. To draw the nation states together into larger units, so that solutions can be found to problems that would be insoluble without international co-operation.

2. To assert the right of small ethnic groups to achieve independence and autonomy over their own affairs.

These trends are apparent in the movements for independence in

Scotland and Wales at the same time as the country is adjusting to full membership in the European Economic Community. Politicians have to take judgment as to whether it is right to bow to the wishes of a minority, or whether it is their duty to prevent fragmentation in what they deem to be the interest of the whole nation in the long run. These volumes reveal that the latter political judgment serves as a rationale for the Russification of the USSR described in Walter Dushnyck's chapter, as well as the national unification policies discussed by Mohan Das in his chapter, "Discrimination in India," and by Justus M. van der Kroef with reference to Indonesian rule over Irian Jaya (West New Guinea).

The British are also deeply involved in this conflict in Africa. "One man, one vote" in Rhodesia appeals to many concerned with human rights in principle. Reality requires an understanding of the facilities which may be available to the leaders of each collective group to ascertain that they are correctly representing the aspirations of those they claim to represent. If unbiased information on complex political issues, involving both the long term and the short term, is difficult to communicate in Britain, it has to be infinitely more difficult where language and tribal custom and vast distances are a part of the problem.

The experience of the British throughout their Imperial history that peace and the rule of law are the first essentials for any community to achieve social justice and economic progress causes them to view with deep concern the new element which technology has brought into the political arena – the power of the terrorist. This could be summarized as the activities in a variety of causes of the "my way, or else – brigade". Unless their convictions are made to predominate, they will use death and destruction as a weapon to prevent any other way succeeding.

All too often this attitude is excused as "idealism", on the assumption that violence is clearly the quickest road to success. Here again, great responsibility rests on the communicators. Violence makes news, but if social justice and human rights are to be reconciled, the idealism in the rule of law has to be better understood.

The attitude of the British remain what it has always been, in the words of Harold Macmillan,

To create a society which respects the rights of individuals – a society in which men are given the opportunity to grow to their full stature... To find means by which the community can become *more* of a community, and fellowship can be fostered between its various parts.

Notes

1. Census data drawn from publications of H.M. Stationery Office, London.
2. Acts of Parliament and Reports of Commons Select Committees are published by H.M. Stationery Office.
   Halsbury's Statutes of England, Public General Acts and Measures, volumes produced annually and identified by the year, London: Butterworth & Co., Ltd.
3. *Facts and Figures 1972,* British Council of Churches' Community and Race Relations Unit and the Institute of Race Relations, London, 1972, p. 3.
4. David J. Smith, *Racial Disadvantage in Employment,* jointly financed by the Gulbenkian Foundation and the Home Office, London: Political and Economic Planning, 1974.
5. *The Race Relations Act 1968, Housing,* explanatory leaflet published by the Race Relations Board.
6. H. S. Morris, *The Indians in Uganda,* London: Weidenfeld and Nicholson, 1968, p. 64.
7. E. J. B. Rose and Associates: Nicholas Deakin, Mark Abrams, Valerie Jackson, Maurice Peston, A. H. Vanags, Brian Cohen, Julia Gaitskell, Paul Ward, *Colour and Citizenship, A Report on British Race Relations,* London: Oxford University Press, 1969, pp. 440/3.
8. Mark Bonham Carter, foreword to *The Fair Housing Experiment,* by Jane Perry, London: Political and Economic Planning, 1973.
9. *Ibid.*
10. *Ibid.*
11. *The Fair Housing Experiment,* p. 36.
12. Mark Bonham Carter, *op. cit.*
13. *The Fair Housing Experiment,* p. 42.
14. *Colour and Citizenship, op. cit.,* p. 61.
15. Community Relations Commission, *Education Needs of Children from Minority Groups,* London, 1974, pp. 9–12.
16. Patricia Walters, A. J. Allen, Isobel Allen, Michael Fogarty, *Women in Top Jobs, Four Studies in Achievement,* sponsored by Leverhulme Trust, undertaken by Political and Economic Planning in collaboration with the Tavistock Institute, London: Allen and Unwin, 1971.
17. Walter Dushnyck, "Discrimination and the Abuse of Power in the USSR", Vol. II of this collection, IX, The Status of Soviet Women.
18. British Federation of Business and Professional Women, *Report of an Inquiry into Women's Careers.* Affiliated organizations: The Individual Members' Group, The Adwomen, National Gas Federation, Women's Advertising Club of London, Women's Engineering Society: Inquiry Analysis and Report, B. Okely. Unpublished, property of the British Federation.
19. Association of Cinematograph and Television Technicians, *Patterns of Discrimination Against Women in the Film and Television Industries,* London, 1975.
20. *Ibid.,* pp. 3–4.
21. *Ibid.,* p. 23.
22. Dr. Edith Young, Co-Chairman with Margaret Thatcher, M.P. of the Women's National Commission, notes of address to the National Council of Women Seminar, London, May, 1973.

About the Editors

WILLEM A. VEENHOVEN, a native of the Netherlands, studied history at the universities of Amsterdam and Leiden, obtaining a Ph.D. from the former and a Doctor in History degree from the latter university. He was for many years professor of history in secondary schools, from which position he recently retired. He is a member of the Royal Institute of Linguistics and Anthropology in Leiden; the Historical Society in Utrecht; and founder and chairman of the Foundation for the Study of Plural Societies (Stichting Plurale Samenlevingen), the Hague, as well as editor-in-chief of its quarterly journal *Plural Societies*.

Dr. Veenhoven's publications include: *The Battle for Deshima*, an investigation of American, British and Russian attempts against the Dutch trade monopoly in Japan, 1800–1817 (Bloemendaal, 1950); *From Batavia to Djakarta*, a set of articles about the Dutch-Indonesian conflict from 1945 to 1950 (The Hague, 1969–70); and numerous articles on de-colonization and European minorities.

WINIFRED CRUM EWING, a native of Great Britain, is a professional journalist and has also published a novel and numerous short stories. She has travelled extensively, both before and after World War II, acquiring familiarity with political and social problems in many parts of the world. For several years Mrs. Crum Ewing was Administrator of the Women's Press Club, and she is the British member of the Board of the World Association of Women Journalists and Writers (AMMPE).

Mrs. Crum Ewing was for a number of years associated with the Conservative Central Office, first as an organizer of local and youth groups and later as the party's film and television producer. After leaving the Conservative Central Office in 1959, she formed her own production company, which specializes in documentary films. Among those she has written and directed are a television obituary of Winston Churchill and films on conservation and national defence.

DR. CLEMENS AMELUNXEN, a native of Munster in Westphalia, studied law and theology in Germany, Belgium, and the United States, where he held a Fulbright Fellowship at Duke University. After obtaining his Doctor of Laws degree in 1953, he entered the German judicial service and served as a judge in various district and circuit courts. Since 1962 he has been a judge of the High Court of Appeal (Criminal Division) of the Rhineland, and he is now a member of the Presidency of that Court. His other public-service activities include the chairmanship of the Youth Welfare Council in Leverkusen, serving as counsel to the Road Safety Committee at Dusseldorf, and membership on the Judicial Examinations Commission.

Dr. Amelunxen has published 13 books on various aspects of public law, among which the following may be mentioned (titles translated from German): *Criminality in Women* (1959); *Suicide* (1962); *The Child and Criminality* (1963); *Political Criminals* (1964); *The Small States of Europe* (1965); *The Victim of Crime* (1970); *Industrial Protection and Industrial Crime* (1973); *Society and Its Rights* (1974); and *The Individual in Modern Criminal Law* (1975). Dr. Amelunxen's decorations include the Commander Cross, Order of Merit, Principality of Liechtenstein, and the Pontifical Medal awarded by the Holy See.

KURT GLASER, a native of the United States, studied at Harvard University, obtaining the B.A. and M.A. degrees in government and the Ph.D. in economics and government. His early career was spent as an Organization and Management officer in the U.S. Federal Government. After service in the U.S. Military Government in Germany, he shifted to research and teaching in international relations. He conducted a research project for the Governmental Affairs Institute and served on the overseas faculty of the University of Maryland. Since 1959 he has been with Southern Illinois University at Edwardsville, where he is Professor of Government. He has also held a Fulbright Lectureship at the University of Kiel.

Professor Glaser's interest in the problems of plural societies began with a study of Central European problems and those of expellees and refugees while in Germany. His book, *Czecho-Slovakia: A Critical History*, which deals at some length with nationality problems of that state, was published in Caldwell (Idaho) in 1961 and in Frankfurt/Main in 1964. He is co-editor (with David S. Collier) of five books of articles on Eastern European problems sponsored by the Foundation for Foreign Affairs (Chicago) and (with John Barratt, Simon Brand and David S.

Collier) of *Accelerated Development in Southern Africa* (London: Macmillan 1974), to which he contributed an article on "Development Problems of Multi-Ethnic Societies". Professor Glaser is an Associate Editor of *Modern Age* (Chicago) and has contributed to that journal a number of articles and reviews dealing with American foreign policy.

STEFAN T. POSSONY was born in Vienna, Austria, and completed his Ph.D. at the University of Vienna in 1935. An outspoken opponent of National Socialism, Dr. Possony left Austria for France, from where he proceeded to the United States and joined the Institute for Advanced Study in Princeton, New Jersey. During World War II he served as a psychological warfare officer in the U.S. Office of Naval Intelligence. From 1946 to 1961 he was a Professor of International Politics at Georgetown University in Washington, D.C., and in 1961 he joined the Hoover Institution on War, Revolution and Peace at Stanford University, where he is now Senior Fellow. He has served on the editorial boards of eleven publications and is the Strategic Affairs Editor of the American Security Council. His books include *A Century of Conflict*; *Lenin, the Compulsive Revolutionary*; *Aggression and Self-Defense*; *The Geography of Intellect* (with Nathaniel Weyl); and *International Relations* (with Robert Strausz-Hupé). His titles written in German include *Strategie des Friedens* (Strategy of Peace) and *Zur Bewältigung der Kriegsschuldfrage* (Mastering the War-Guilt Question). His most recent book, *Waking Up the Giant*, a comprehensive analysis of the problems faced by current American foreign policy, was published by Arlington House in 1974.

JAN PRINS, a native of the Netherlands, studied Indonesian law at the University of Leiden, and served in the East Indies civil service from 1928 to 1949. He was Professor of Non-Western Sociology at the University of Utrecht from 1949 to 1972. His publications include: *Adat and the Doctrines of Islamic Duties in Indonesia* (1948, 1950, and 1952); *Christian Influences on Adat Law* (Utrecht, 1949); and *The Coloured Between the Devil and the Deep: South African Four-Stream Policy* (Assen, 1957). He has also contributed to the "Recueils de la Societé Jean Bodin," Brussels.

NIC. J. RHOODIE, a native of South Africa, is Head of the Department of Sociology of the University of Pretoria. A graduate of the same university (M.A. and Ph. D. cum laude), he specializes in race and minority relations. He is the Chief Editor of the *South African Journal of Sociology*, and an Associate Editor of *New Nation* (South Africa); and the United States based *International Review of Sociology* and *International Journal of Sociology of the Family*. He has visited both the United Kingdom and the United States as guest of their respective governments.

Professor Rhoodie has written extensively on intergroup relations in general and ethnic and racial problems in particular. His main publications include: *Apartheid and Racial Partnership in Southern Africa*, Pretoria: Academica, 1969; "The Social System of Communism", in Cronjé, G. (Ed.), *Communism. Theory and Practice* (Pretoria: Van Schaik, 1969); *Social Demography*, Academica, Pretoria, 1973 (Dr. C. F. Swart co-author); (Editor) *South African Dialogue: Contrasts in South African Thinking on Basic Race Issues* (Johannesburg: McGraw-Hill, 1972); and "The Coloured Policy of South Africa," *African Affairs*, London, Vol. 72, January 1973. Professor Rhoodie is co-author (editor) of *Homelands: The Role of the (Black Development) Corporations*, (Johannesburg: Chris van Rensburg Publications, 1974); and *Stepping Into the Future. Education for South Africa's Black, Asian and Coloured Peoples*, (Johannesburg: Erudita Publications, 1975).

JIRO SUZUKI, a native of Japan, studied at Waseda University and has held a variety of positions in both sociology and anthropology. In 1952 he became Associate Professor and in 1964 Professor of Social Anthropology at Tokyo Metropolitan University. While on leave from this position he worked as a Research Associate in Anthropology at the University of Illinois from 1957 to 1959, specializing on problems of Blacks and American Indians. He has also conducted field research on minorities in Malaysia, India, Mainland China and the Philippines.

Professor Suzuki is the author of eight books, including *Black Americans* (Tokyo, 1957); *Race and Prejudice* (Tokyo, 1969); *The Buraku Problem in Tokyo* (Tokyo, 1971); and *White, Black and Yellow* (Tokyo, 1973). All of the foregoing titles are in Japanese, as are a number of articles Professor Suzuki has written about the status of various minorities in Japan. His articles available in English are "Some Observations on Malaya," in J. C. Darvala, ed., *Tensions of Economic Development in South-East Asia* (Bombay: Allied Publishers, 1961,); "*Burakumin*: Japan's 'Untouchables'," in Schneps and Cox, eds., *The Japanese Image* (Tokyo and Philadelphia: Orient/West, Inc., 1965); and (with Hugh H. Smythe) "Human Relations: A Report on the *Burakumin*," Journal of Human Relations: (Central State College, Ohio), Vol. 10, No. 2 (1965).